The Metastatic Cell

The Metastatic Cell

Behaviour and biochemistry

CLIVE W. EVANS

University of Auckland, New Zealand

CHAPMAN AND HALL

LONDON • NEW YORK • TOKYO • MELBOURNE • MADRAS

UK	Chapman and Hall, 2–6 Boundary Row, London SE1 8HN
USA	Chapman and Hall, 29 West 35th Street, New York NY10001
JAPAN	Champan and Hall Japan, Thomson Publishing Japan, Hirakawacho Nemoto Building, 7F, 1–7–11 Hirakawa-cho, Chiyoda-ku, Tokyo 102
AUSTRALIA	Chapman and Hall Australia, Thomas Nelson Australia, 480 La Trobe Street, PO Box 4725, Melbourne 3000
INDIA	Chapman and Hall India, R. Seshadri, 32 Second Main Road, CIT East, Madras 600 035

First edition 1991

Type set in 10½/12pt Sabon
Disc conversion by Columns Typesetters of Reading
Printed in Great Britain by Page Bros (Norwich) Ltd

ISBN 0 412 30300 0

British Library Cataloguing in Publication Data

Evans, Clive W.
The Metastatic Cell
1. Man. Cancer. Metastasis
I. Title
616.994

ISBN 0–412–30300–0

Library of Congress Cataloging-in-Publication Data

Available

In memory of my father,

Lyall William Harrison Evans,

who died of cancer

Contents

Acknowledgements

A book of this nature would simply not be possible without its source material. Wherever possible I have endeavoured to refer to primary sources, but largely for reasons of space I have not always been able to do so. To the authors I have overlooked by quoting review papers, I can only apologize, but it needs to be said that their research is hardly belittled by recognizing its established status.

To the thousands of authors I have read, for their results and ideas which form the basis of this book, I offer my greatest thanks. Of course, I have not always been in agreement with published work, but then controversy is a major part of scientific progress and there is no reason why it should not play its role in this book as elsewhere.

Numerous colleagues have assisted me in bringing this book into fruition, particularly during its final stages. These have included Professor Ray Ralph, Professor Stan Bullivant and Dr Warren Judd in Auckland, Dr Corrado D'Arrigo, Dr David Sinclair and Dr Gordon McPhate in St Andrews, and Professor Adam Curtis in Glasgow. I am especially indebted to Dr Clare Ward (Auckland) for her sterling efforts in reviewing the entire manuscript, and to Mr Simon Evans (Auckland) for his assistance with compiling the reference list. It goes without saying that whatever gremlins have crept into the text originate solely from my word processor.

I should also like to pay a tribute to my research students over the years for the contributions they have made both directly and indirectly to the contents of this book.

My own research mentioned in the text would not have been possible without the financial support of a number of bodies, including my host institutions (the Universities of Auckland, St Andrews and Glasgow), the Medical Research Council (UK), the National Foundation for Cancer Research, and the Association for International Cancer Research. For their generous support of my research endeavours, I remain exceedingly grateful.

Finally, completion of this book has not been without a considerable burden on my family. To Barbara and Simon, for their endurance, patience and forbearance, I offer my most sincere thanks.

Preface

This book provides a basis for approaching the problem of cancer through examination of the biochemistry and behaviour of malignant cells. It stems from my experiences with medical students who, despite a solid grounding in general pathology and now also in molecular and cellular biology, nevertheless remain relatively ignorant of a large body of research in fundamental aspects of cancer biology. Its purpose is to redress this imbalance, providing a framework for critical analysis. It should also prove comprehensive enough to offer a basis for science students contemplating a career in cancer research. It must be emphasized, however, that this topic is exceedingly vast and this book reflects only a few selected areas, particularly those which fall within the sphere of my own research interests.

I do not apologize for the fact that this book offers no answers to the problem of cancer. If it serves only to make the reader ask more questions, then in many ways it will have achieved its purpose. It contains a wealth of experimental results from which a plethora of hypotheses relating to cancer may be proposed. In dealing with this information it is important to keep in perspective that cancer is not a single disease, and that there is unlikely to be any universal answer to the cancer problem. For this reason, sweeping generalizations based on single model systems are to be avoided at all costs. As the reader will find out, scientists who have causes to champion and an insatiable desire for funding do not think twice before launching into the most improbable of generalizations. I have endeavoured to keep these in perspective, but readers contemplating a career in cancer research should nevertheless appreciate the fundamental importance of hypotheses to scientific enquiry.

1 The tumour cell phenotype

1.1 WHAT IS A TUMOUR?

Strictly speaking, the term **tumour** (Latin *tumor*) refers to a swelling, and many of us will recall learning that Celsus recognized *tumor*, along with *rubor* (redness), *calor* (heat) and *dolor* (pain), as the cardinal signs of acute (non-persistent) inflammation. However, it is now common to use the term tumour as a synonym for **neoplasm**, meaning a new growth. This new growth results from an inheritable change in a cell (or cells) which allows them to escape from many of the normal homeostatic mechanisms which control proliferation. It seems likely that this change can occur in any cell capable of dividing. When it has taken place, the cell is said to be **transformed**, although this term was originally used when the cell change occurred in an *in vitro* experimental system. There are a number of provisos in using the term 'transformed' to describe tumour cells under both *in vitro* and *in vivo* conditions, the most important being that *in vitro* transformed cells do not necessarily form tumours *in vivo*. Cells transformed experimentally by certain viruses, for example, may not grow when injected into syngeneic hosts. The most likely reason for the failure of such cells to form tumours is that they are rejected by the host as a consequence of its defence systems recognizing viral determinants on the cell surface. Experimentally, we may suppress the defence systems of animals in order to allow such cells to grow *in vivo*. This satisfies the main criterion for neoplastic change, since in practical terms confirmation of *in vitro* transformation is best obtained by witnessing the *in vivo* growth of a tumour from the corresponding cells.

It is of some importance at this early stage to differentiate between cell proliferation of normal tissue (even when extensive as seen in wound healing, for example) and cell proliferation associated with transformation. In the former case, proliferation is under some form of control and ceases when a regulated end point (e.g healing) is reached; transformed cells, however, usually continue to proliferate to the detriment of the host. We shall see later that even under *in vitro* conditions, normal cells respond to proliferation signals in a manner which is usually distinct from that of transformed cells, although many specific aspects of their responses share common molecular pathways.

The escape of tumours from many of the normal regulatory control mechanisms of the body also clearly distinguishes them from other tissue swellings such as may be seen, for example, in a lymph node draining an inflammed site. Despite the origins of the term 'tumour', it should be noted that not every tumour will necessarily result in a swelling. Leukaemias, for example, may extend diffusely in the bone marrow and circulate in the blood without producing an overt swelling. Very occasionally tumours may be seen to regress, which might suggest that in certain cases some tumours may not have fully lost, or may be able to regain, sensitivity to the normal proliferation regulatory mechanisms.

Finally, it is worth noting that what is clinically called a tumour is more than just a mass of transformed cells. First, as will be decribed in due course, a tumour is heterogeneic in that over time its cells are likely to change. Some tumours may do this at different rates or along different paths from others, resulting in a phenotypically mixed bag of cells. Second, as described by Woodruff (1980), a tumour may be viewed as a highly complex ecosystem which contains not only transformed cells but normal host cells as well. These are likely to include macrophages, lymphocytes and fibroblasts to varying degrees, and the tumour will also contain blood vessels and other stromal elements. There is usually also a good deal of necrotic material, particularly in large tumours. It may be expected that there will be interplay between these various components and that the ultimate pattern of tumour growth will depend as much on this interplay as on other contributions from the host such as the supply of nutrients and removal of waste products.

1.2 GENERAL ASPECTS OF CELL GROWTH

Before discussing aspects of tumour cell growth further it is pertinent to review some general aspects of cell division.

1.2.1 The cell cycle

The cell division process is best described as a cycle from which cells may enter or leave via a 'resting' phase (G_0). As illustrated in Fig. 1.1, cells enter the cycle in a 'gap' known as G_1 prior to their commencing DNA synthesis in S phase. This phase is followed by another 'gap' (G_2) which leads into the mitotic or M phase. The 'gaps' are thought to represent preparatory phases prior to DNA synthesis and mitosis: protein and histone synthesis occur in G_1, for example, and these are essential for progression through the cell cycle. The daughter cells arising from mitotic division enter G_1 from which they may continue to cycle or they may enter the resting phase in which they could take on specific functions (**differentiate**). Some normal cells, such as the polymorphonuclear neutrophil leukocytes (PMNs), are thought not to leave G_0 and therefore they are not capable of division. Tumour cells may also enter G_0 and this may be the basis of dormancy in which tumour cells seem to be in a non-proliferative state.

The regulated movement of cells to and from and within the cycle controls cell division. Control is thought to be mediated by various types of intra- and intercellular signals acting at certain positions in the cell cycle called control points. Detailed studies of the responses of confluent, serum-starved BALB/c 3T3 cells to fresh serum or growth factors has lead to the demonstration of several such points (reviewed in Denhardt, Edwards and Parfett, 1986). The addition of certain growth factors such as platelet-derived growth factor (PDGF) acts on cells in G_0 to induce a state referred to as **competence** from which the cells can **progress** through cell division following the addition of other factors (Pledger *et al.*, 1978). About halfway through G_1 there is a control point referred to as the V point beyond which cells will not progress unless they have been treated with appropriate factors such as somatomedin C, also known as insulin-like growth factor 1 (IGF–1). Later in G_1, about 2 h before the commencement of S phase, there is a restriction point (R) where cells are thought to commit themselves to DNA synthesis. Another control point at which cells may be arrested, the W point, exists immediately before S phase. As far as transformed cells are concerned, some are thought to lack the requirement for a competence-inducing signal and some may not display an obvious restriction point. It should be noted that the details of cell cycle events worked out for one particular cell type in response to particular growth factors need not fully describe the processes involved in other cells, and that the functions of control points in tumour growth are far from being understood fully. Experimental studies, particularly from research on the development of chemically induced squamous carcinomas in the hamster cheek pouch, suggest that transformation is associated with an increase in the rate of cell proliferation, but as yet there is little reliable data from human studies.

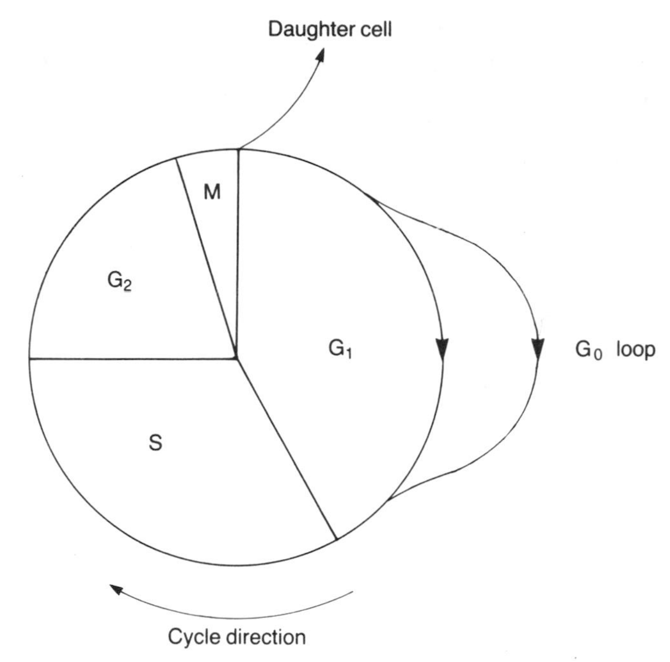

Figure 1.1 The cell cycle. Each cell can be considered to enter the cycle at G_1 after the mitotic division of its parent. Under appropriate conditions, DNA synthesis commences and the cell proceeds into M phase via G_2. Some cells may enter a 'resting' phase (G_0) usually leading to differentiation, and from which they may or may not re-enter the cycle. A typical mammalian cell in culture has a cycle (generation) time of around 18 h, with about 8 h being spent in G_1, 6 h in S, 4 h in G_2 and less than one hour in M phase.

1.2.2 Growth control

Different organs and cells may respond to growth controls in different ways. Generally speaking, however, we may hypothesize a steady state under normal circumstances in which cells are either stabilized, having reached an appropriate population size, or they turn over to match the number of cells lost for one reason or another. This steady state may be perturbed under various circumstances. The removal of one kidney, for example, may lead to compensatory growth in the other; the application of exogenous hormones may lead to target organ growth as seen in the adrenal gland following treatment with adrenocorticotrophic hormone (ACTH); and wounding of organs such as the skin may lead to reparative growth. These growth responses are nevertheless still under some form of control, and it is only following transformation that this appears to break down.

Studies of normal renewing cell populations, such as that of the epidermis and bone marrow, have led to the formulation of a **stem cell model** of tissue maintenance. There are many refined versions of this model, but all of them are based on a hierarchical level of organization (Fig. 1.2). Generally speaking, the loss of normal differentiated tissue cells is maintained by a steady state balance in which new cells are supplied from a small pool of stem cells. Although their progeny may enter the differentiating pathway, stem cells are the only cell type capable of self renewal. As shown in Fig. 1.2, stem cells may also shift in and out of a non-proliferating state (G_0). The stem cell progeny which enter the differentiation pathway (**transitional cells**) have a finite proliferative ability, but the divisions they do undergo result in considerable expansion of the original clones. The transitional cells can assume varied differentiation characterisitics which are determined for each tissue. Progression along the differentiating pathway results in a population of **end cells**, which are fully differentiated but non-dividing. The stem cell concept has been applied to neoplastic tissues and this is discussed further in section 1.6.3.

What mechanisms can be proposed to control the steady state of cell growth? It must be admitted at the outset that there is no single, widely accepted biological model of growth control. It is possible to envisage a steady state of growth under the influence of factors which stimulate and/or inhibit cell division and a considerable amount of research is proceeding in this direction. What controls these possible factors, however, is even more of an enigma. A summary of various growth factors implicated in the control of cell division is provided in section 1.7.1. At this stage it is worth pointing out that there are three main pathways by which factors might influence cell division:

(a) *The endocrine pathway*

This is the pathway reflected in the activities of the typical hormones, which are secreted by one cell type and are carried in the blood to influence the functioning of another cell type some distance away.

(b) *The paracrine pathway*

Factors produced by one cell type diffuse locally to influence another cell type nearby.

(c) *The autocrine pathway*

This pathway is characterized by the synthesis of a factor for which the

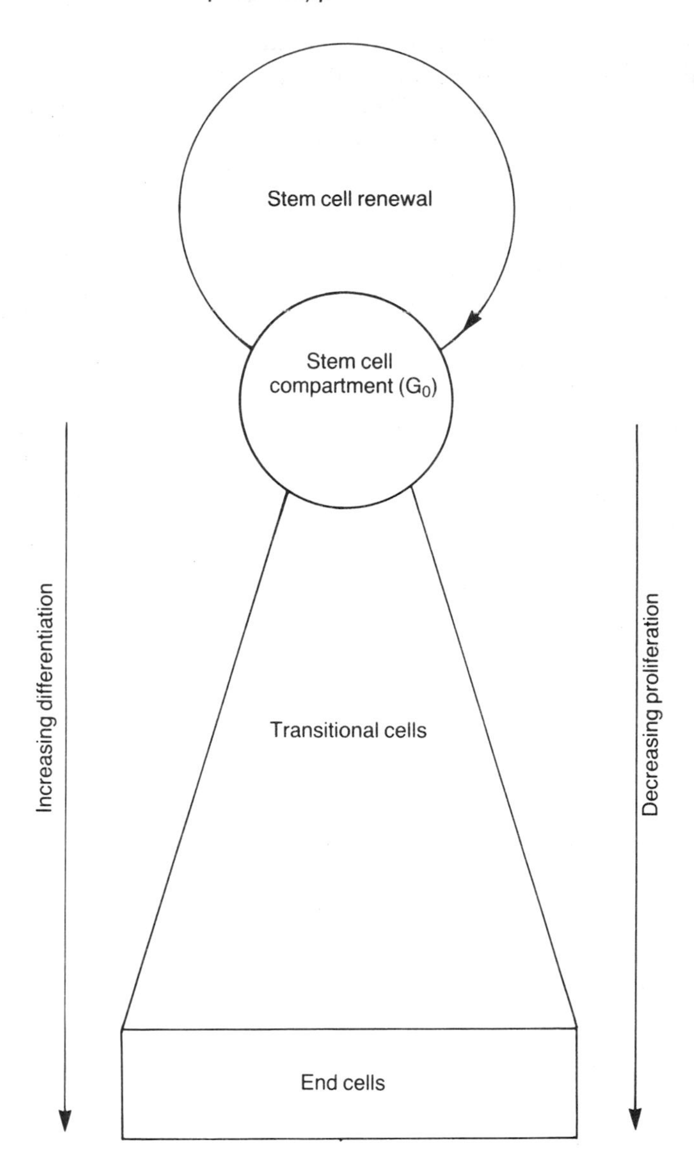

Figure 1.2 The stem cell model of cell renewal and differentiation. In this model, stem cells are the only cell type capable of self renewal. Under appropriate conditions, stem cell progeny enter the differentiation pathway passing via transitional cells into a terminal end cell stage. Clonal expansion of the original stem cell progeny occurs during differentiation, but the cells become progressively restricted in their division potential (after Buick and Pollack, 1984).

producing cell also has receptors (Sporn and Todaro, 1980). Although autocrine secretion is a normal physiological process (occurring in embryonic development and wound healing, for example), its unregulated activity has been reported in a number of tumours.

1.3 HOW ARE TUMOURS CLASSIFIED?

Tumours are classified with reference to a number of criteria including their behaviour, their appearances, and their origins. Clinical experience indicates that there are two fundamental types of tumours, **benign** and **malignant**, which behave in different ways. Benign tumours remain localized and do not spread to different parts of the body. Malignant tumours, on the other hand, usually invade and destroy host tissue and may spread throughout the body. The term **cancer** is used to mean a malignant tumour. A summary of the major differences between benign and malignant cells is provided in Table 1.1.

1.3.1 Behaviour

There are two major behavioural traits of tumours which assist in their classification, namely their growth and their invasive and metastatic behaviour.

(a) *Growth*

Unlike the growth of normal tissue, the growth of a tumour seems not to be under control by the normal regulatory systems of the body. This

Table 1.1 Major differences between benign and malignant cells*

Feature	*Benign*	*Malignant*
Cytoplasm	slight basophilia	marked basophilia
Mitotic figures	few and normal	many and abnormal
Nucleus	predominantly normal	pleomorphic
Nucleoli	little altered	often swollen
Tissue structure	usually normal	dysplastic/anaplastic
Functions	usually normal	lost or deranged
Capsule	usually intact	often lacking
Metastasis	never	often
Local invasion	rare	common
Fatalities	rare	often

* Note that benign lesions never metastasize, but exceptions exist for every other feature listed.

is not to say that the growth of a tumour is entirely autonomous, since it is still dependent on the host for the supply of nutrients and the removal of waste products. Although uncontrolled growth is taken as the hallmark of a tumour, tumours vary in their individual rate of growth, an observation which suggests that this behavioural trait may not be particularly reliable as a basis for classification. Indeed, in classifying tumours it should be borne in mind that there are few hard and fast rules and that there are many exceptions to the generalized features used in classification. Thus, although many (but certainly not all) malignant tumours have a higher mitotic rate than benign tumours (often as judged by the presence of more numerous mitotic cells), some tumours of either type may not be growing actively at all. In addition to having generally a relatively high mitotic rate, malignant cells often have nuclei which are more variable and irregular (**pleomorphic**) when compared with the more typical nuclei of benign cells. Furthermore, the nucleoli of malignant cells may be relatively larger and more prominent than those in benign cells, and their cytoplasm may be more markedly basophilic indicating high levels of cytoplasmic mRNA which is typical of active cells. It should also be noted that benign tumours seldom kill the host unless they grow to press on some vital structure such as a major artery or an airway. According to impeccable sources (*The Guiness Book of Records* quoted in *The Times*!), the largest benign tumour ever removed from a human (an ovarian cyst, which usually contains a high fluid content) weighed a remarkable 328lb (about 149kg).

(b) *Invasion and metastasis*

Cells of benign tumours are often enclosed in a fibrous capsule and they usually remain at their site of origin. Malignant tumours, on the other hand, rarely have a complete capsule and they frequently invade the local tissue. Although invasiveness is a characteristic feature of malignant cells it is not peculiar to them. Normal cells in the embryo can be invasive as seen, for example, in the invasion of the endometrium by trophoblasts. In adults, leukocytes such as the polymorphonuclear neutrophils (PMNs) display invasive behaviour when they exudate from the blood towards inflammatory foci. Superimposed on their locally invasive character, malignant tumours often spread to remote sites after gaining access to the lymphatic or blood systems or to body cavities. The spread of malignant tumours culminating in the establishment of one or more secondary tumours at a remote site is known as **metastasis**. Although trauma may occasionally result in the implantation of a benign tumour cell at a distant site this is extremely rare and does not result from a behavioural property inherent within the cell itself. Consequently, the ability to metastasize probably represents the most reliable criterion for distinguishing malignant from benign tumours. The metastatic behaviour of malignant cells is discussed in more detail in the following chapters.

1.3.2 Appearances

(a) *Differentiation*

Both the gross and the microscopical appearances of tumour cells are used in their classification. The term **differentiation** refers to the degree of resemblance of the tumour to the normal tissue from which it was derived. **Dysplastic** tumours show partial resemblance to the parent tissue, whereas **anaplastic** tumours show almost no resemblance. Generally speaking, the more highly differentiated a tumour is, the better is the prognosis. In pathological terms, tumour differentiation is usually assessed 'naked-eye' or by microscopical observation of histological sections. Whereas benign tumours usually resemble their tissues of origin in both naked-eye and microscopical appearance, malignant tumours are often anaplastic and their origins may not be immediately obvious by either means. Changes in the state of cell differentiation may be associated with changes in function, and it is often found that whereas benign tumours may retain the functions of the parent tissues, these functions may be occasionally lost or abnormal in malignant tumours.

Although the introduction of more sophisticated techniques (such as those based on antibodies) has helped in determining the origins of cells in some poorly differentiated tumours, their advent has not significantly altered the ground rules of classification. Monoclonal antibodies (McAbs) are available commercially against specific intermediate filaments such as the cytokeratins which are useful in identifying carcinomas. Other McAbs are available against desmin for identifying muscle cell sarcomas, against vimentin for identifying other sarcomas, against glial fibrillary acidic protein for identifying gliomas, and against neurofilaments for identifying certain tumours of the nervous system (see section 1.7.15).

(b) *Cytology and tumorigenicity*

Cytological changes occurring during the spontaneous transformation of a number of rodent cell lines were recorded by Barker and Sanford

(1970). Most cell lines which became transformed showed a progression of changes including an increase in cytoplasmic basophilia, an increase in the number and size of nucleoli, an increase in the nuclear:cytoplasmic ratio, retraction of the cytoplasm, and the formation of clusters and cords of cells. A number of cultured rat liver epithelial cell lines were examined by Montesano *et al.* (1977) in order to test for any correlation between specific cytological characteristics and tumorigenicity. Of the fifteen characteristics they evaluated, increased cytoplasmic basophilia and increased nuclear:cytoplasmic ratio correlated best with tumorigenicity, although other criteria including variation in cell size and shape, nucleolar variation and reduced cytoplasmic spreading were also thought to be useful in characterization. Many of these criteria are similar to those found useful in characterizing the tumorigenic state of tissues *in vivo*.

1.3.3 Origins

Tumours originate from different tissues (**histogenetic** differences), from different organs, and from different cells and these are taken into account in their classification. It is worth noting seven important types of malignant tumours which will be repeatedly referred to in the following pages:

(a) *The carcinomas*

These are malignant epithelial tumours and they represent by far the most common type. **Carcinoma** *in situ* refers to an epithelium (particularly of the cervix or skin) which shows many malignant characteristics but does not invade the underlying tissue. Strictly speaking, carcinoma *in situ* is a premalignant change since invasion is lacking. Not all premalignant lesions of this type necessarily progress to full malignancy and the term carcinoma is thus considered by some authors to be a misnomer for this disease.

(b) *The adenocarcinomas*

Malignant tumours of glandular epithelial tissue.

(c) *Malignant melanoma*

A malignant tumour of pigmented epithelial cells (melanocytes).

(d) *The sarcomas*

Malignant tumours of connective tissue.

(e) *The leukaemias*

These are malignant tumours of precursor cells within the haemopoietic system.

(f) *The lymphomas*

Tumours of lymphoid cells.

(g) *Teratocarcinomas*

These malignant teratomas are tumours of multipotential embryonic tissue cells (embryonal carcinoma or EC cells).

1.4 GRADING AND STAGING

The clinical diagnosis and course of treatment of malignancy is aided considerably by classifying the tumour with reference to defined grades and stages. Grading (low, intermediate or high) is an estimate of malignancy based on histological criteria including mitotic activity, the degree of pleomorphism, and the extent of differentiation. Staging records the amount of disease progression and is based on clinical criteria. One staging scheme, the TMN system, takes into account the status of the primary tumour (T), its metastases (M), and the regional lymph nodes (N). The reader is referred to Rubin (1973, 1974) and to Beahrs and Meyers (1983) for further information on clinical staging, the details of which need not concern us here.

1.5 WHAT CAUSES TUMOURS?

Transformation may be triggered in a number of ways, including exposure to chemicals, certain viruses, and radiation. In most, if not all cases, the basis of transformation is probably a **mutation** (a change in the primary structure of DNA) but it is also likely to be influenced by **epigenetic** events (shifts in gene expression). It must be admitted, however, that the extent to which each of these forms of cellular change is involved in transformation is open to question. A role for mutational events in transformation is supported by the observations that most carcinogens are also mutagens and *vice versa* (Miller and Miller, 1971), and that many tumours show chromosomal abnormalities (Harnden, 1976). Furthermore, many **oncogenes** (cancer causing genes) are found at or near sites of chromosomal abnormalities. On the other hand, some involvement of epigenetic events is suggested by changes in the expression of surface antigens, such as the oncofetal antigens discussed below. Whatever its basis, transformation must involve a heritable change since the daughters of the originally transformed cell retain the transformed nature, except under extraordinary circumstances (which may offer the possibility of a novel form of therapy as discussed in section 6.3).

Interestingly, humans appear to be characterized by a relatively stable phenotype and malignant change often appears to occur after a relatively long latent period. As a general observation, animal (particularly rodent) cells are relatively easy to transform *in vitro*, whereas human cells are much more difficult (reviewed in DiPaolo, 1983). The underlying reason for this difference in transforming ability is not known: it seems not to be a simple matter of the more efficient DNA repair capabilities of human cells, since all repair-deficient human cells do not generally transform much more readily than normal cells. It may be that the selection and inbreeding of experimental animals has led to an increased susceptibility to cancer, but further study is obviously required before any firm answer is reached.

1.5.1 Tumour origins

Most tumours (but perhaps not all) are thought to be **clonal** in origin: that is, each is thought to be derived originally from a single, transformed cell. There are four lines of evidence in support of this conclusion:

(a) The presence of the same marker chromosome on all cells of a single tumour.
(b) The cells of a plasma cell tumour all produce the same immunoglobulin.
(c) The cells of certain tumours possess antigens in common.
(d) The expression of sex-linked markers. **Isoenzymes** (or isozymes) are enzymes which exist in two or more forms due to the presence of multiple gene loci or multiple alleles. The isoenzyme glucose-6-phosphate dehydrogenase (G-6-PD) exists in two forms (A and B) which are the products of a gene on the X chromosome. Women who are heterozygous at the G-6-PD locus (about 25% of black women are so) will produce cells with either the A form or the B form of the isoenzyme (but not both). The reason for this, as expounded in the Lyon hypothesis, is that one of the two X

chromosomes is inactivated randomly in each nucleated cell. In G-6-PD heterozygous women, monclonal tumours will contain either G-6-PDA or G-6-PDB, whereas only polyclonal tumours are likely to express both isoenzymes in the same neoplasm. Of course some polyclonal tumours may be derived from two or more cells with the same isoenzyme, but statistical corrections may be applied to satisfy this particular problem.

A number of human tumours have been classified as being of clonal origin through isoenzyme analysis (Fialkow, 1974; Friedman and Fialkow, 1976). A somewhat similar situation exists in the mouse with respect to the X-linked isoenzyme phosphoglycerate kinase (PGK-1). The A form of this isoenzyme was discovered in a feral mouse, whereas most inbred strains have what is referred to as the B form. F_1 hybrids between such mice will be heterozygous, and again 50% of their cells will be PGK-A and 50% PGK-B.

More recently, Vogelstein and colleagues (Vogelstein *et al.*, 1985; Fearon, Hamilton and Vogelstein, 1987) have analysed the clonal origins of human tumours using both X-linked and autosomal **restriction fragment length polymorphisms** (RFLPs). RFLPs are pieces of DNA generated by the action of specific bacterial endonucleases (**restriction enzymes**) which cut DNA precisely following the recognition of particular nucleotide sequences. Now a normal cell inherits half its chromosomes from each parent and thus it usually has two copies of a given piece of DNA. If a mutation was to occur at a restriction site on one chromosome (either maternal or paternal in origin), then subsequent restriction enzyme treatment would yield DNA fragments of different lengths. After electrophoresis and Southern blotting, heterozygous cells (containing the mutation) could be distinguished from homozygous cells, thus providing the potential for an analysis of the clonal origins of tumours. During **Southern blotting**, DNA fragments are transferred to nitrocellulose after electrophoretic separation. They are then denatured and hybridized with radioactively labelled specific complementary DNA sequences. Fragments which hybridize with the radioactive probe are subsequently recognized by autoradiography.

Rather ingeniously, if somewhat laboriously, Fearon, Hamilton and Vogelstein (1987) extracted the DNA for RLFP generation from cryostat sections of tumours. They estimated the amount of normal tissue within the tumour mass microscopically using stained frozen sections, and physically removed this material before DNA extraction from alternate unstained sections. Their results led them to suggest that studies based on G-6-PD purporting to show polyclonal origins for some colorectal carcinomas may have been due to the presence of normal tissue present in an otherwise monoclonal tumour.

It is possible, of course, that tumours may be polyclonal in origin but in time one clone begins to dominate, and this particular problem cannot be resolved at present. Heim and colleagues (1988) have suggested that in some circumstances multiple clones may emerge intitially, only to converge subsequently so that a single clone predominates. Ultimately, this clone may diverge as the tumour continues to progress in its development. In the specific case of HTLV-1 induced acute T cell leukaemia, viral infection initially induces a polyclonal expansion of T cells. The tumour results, however, from a single cell which undergoes a subsequent karyotypic change, probably involving the T cell receptor gene. Like their primaries, most metastases also seem to be monoclonal in origin (section 2.14.13).

1.5.2 Initiation and promotion

It has been proposed that the process of transformation leading to the development of a tumour (**carcinogenesis**) is a multi-stage phenomenon. Most of the evidence in support of this point of view stems from studies associated with chemical carcinogenesis, although some viral oncogenes appear to cause tumours in a single step. Chemically-induced transformation is thought to begin with some form of **initiation** that subsequently undergoes **promotion** to reach the overt stage. Initiation is relatively rapid and irreversible, whereas promotion may be reversible and usually occurs over a long time interval known as the latency period. Working within this framework, it is conceivable that initiation involves mutational change while promotion involves epigenetic change, but as yet there is no convincing evidence for such a straightforward viewpoint. In fact, the tumorigenic process may progress to full malignancy and as we shall see later there is some evidence to suggest gene involvement in relatively late steps in this process. We should note here the key difference between epigenetics and genetics: epigenetics refers to all the processes which are required or influence the implementation of genetic information.

The classical experimental system for demonstrating initiation and promotion is based on the mouse skin. In this model system, application of a low dose of initiator (usually a polycyclic aromatic hydrocarbon) leads to tumour growth only after repetitive application of a promoter (such as croton seed oil). The time between application of initiator and promoter can be extended up to many months indicating the irreversible nature of initiation. The time between repetitive applications of the promoter, however, cannot be similarly delayed suggesting that the cellular changes associated with promotion may be reversible.

1.5.3 Progression

During their life history, tumours are commonly observed to change in both their behaviour and in their histological appearance. For example, they may begin to grow more rapidly, they may become more invasive, and they may appear more anaplastic. Various clinical and experimental observations suggest that tumours may also convert to an ascitic form of growth, they may develop drug resistance, show an increased tendency towards aneuploidy, or gain the ability to grow in allogeneic hosts. Furthermore, although many tumours may be of clonal origin in that they are suspected to have arisen from a single transformed cell, they are usually phenotypically diverse. The term **progression** has been used to decribe the process whereby the cells of tumours show increasing phenotypic diversity (reviewed in Foulds, 1975). Originally, progression was defined as 'the acquisition of permanent, irreversible qualitative changes of one or more characteristics in a neoplasm'. Contrary to this original concept of progression, however, many of the phenotypic changes which individual tumour cells undergo are not permanent in nature and nor are they necessarily qualitative. It is also generally thought that the changes associated with progression tend towards inducing autonomy from the host, and that the acquisition of metastatic potential may be the final step in tumour progression. It should be noted, however, that progression need not always lead to the evolution of more malignant phenotypes. Indeed, tumour regression may occur although (as suggested from clinical observations) it is unlikely to be a common event. Despite these modifications, the concept of progression has proven invaluable in providing a theoretical framework to study many aspects of the development of tumours. It seems likely that progression arises from the selection of tumour cells which have adapted to environmental influences, and thus we may envisage progression as a cellular form of evolution. Nowell (1976) has suggested that the basis of diversification during progression may relate to the genetic instability of tumour cells, and indeed as tumours progress they show increasing chromosomal aberrations. Progression may also reflect epigenetic events to some degree, however, since genetic instability coupled with selection is unlikely to proceed at a rapid enough rate to explain the phenotypic divergence displayed by most tumour types (Nicolson, 1984 a,b; 1987a). Even the fastest spontaneous gene mutation rates in metastatic mouse cells are considerably smaller than the rates of variant formation of biochemical and biological properties of tumour cells. Thus Cifone and Fidler (1981) have reported gene mutation rates of about 7×10^{-5} mutations per cell per generation, but the rate of appearance of albumin variants in rat hepatoma cells is in the vicinity of 10^{-2} variants per cell per generation (Peterson, 1983).

Several lines of evidence indicate that progression in some tumours might proceed in a reproducible way. The serial transplantation of C3H mouse mammary tumours, for example, can lead to the scheduled changing of phenotypic characteristics (Vaage, 1980). This suggests that in some cases progression might involve a programmed selection of changes, but to what extent this is a function of the *in vitro* assay system is not clear. The phenomenon of phenotypic instability and the relevance of the heterogeneity resulting from both genetic and phenotypic instability are discussed further in section 1.7.26.

In clinical terms, benign tumours seldom progress to become malignant, which could be construed to suggest that the two types of tumour have different origins. In the mouse skin model, however, initiation with a chemical carcinogen followed by the repeated application of a promoting agent leads first to benign tumours (papillomas), some of which may subsequently become malignant. These results suggest that benign and malignant tumours may have similar origins and that there may be progression from the former to the latter which is not always evident in clinical cases. Indeed, it is conceivable that all malignant tumours proceed through a benign stage. For some tumours the benign period could be remarkably short, while for others the progression might take place at an early stage such as that represented by a single or just a few cells. Either way, such changes could be easily overlooked, and supporting evidence is obviously hard to accrue for this hypothesis. The age-dependent incidence of tumours in humans suggests that five or six steps may be involved in progression to full malignancy, and the time involved is clearly a key factor in why many benign tumours may not be seen to progess at the same rate as tumours in experimental animals.

1.5.4 The carcinogenic pathways

It is possible to identify three major pathways which can lead to the development of tumours:

(a) *Chemical carcinogenesis*

Chemical carcinogens may be either complete or incomplete in nature: complete carcinogens are capable of forming tumours on their own (they thus have initiating and promoting properties) whereas incomplete carcinogens require promoting agents subsequent to the initiating step. Not all initiated cells ultimately form tumours even in the presence of tumour promoters, which presumably reflects subtle but important differences between cells and/or their microenvironments. Chemical carcinogens are markedly diverse structurally, but they all share the common property of being able to give rise to electron deficient forms which can covalently bind with electron rich sites on molecules such as DNA. Binding to DNA leads to the formation of adducts which perturb DNA structure. This altered DNA is usually repaired and corrected, but if it replicates first then the alteration may become permanently established. It is not clear how the promoting activity of complete carcinogens operates, but current hypotheses suggest that promoting activity involves stimulation of cell division in such a way that growth becomes essentially autonomous. Some carcinogenic chemicals may activate oncogenes (see below), but precisely how the two processes may be related is not yet clear.

Many chemical carcinogens are perhaps best described as procarcinogens in that they have to be enzymatically converted into active carcinogens within the body. Polycyclic hydrocarbons such as benzo(*a*)pyrene, for example, are activated metabolically by cytochrome P450 mixed function oxidases associated with intracellular membranes. If cells lack the appropriate enzymes for activation of the procarcinogen then effective transformation will not result.

(b) *Radiation carcinogenesis*

Like chemical carcinogenesis, radiation carcinogenesis is thought to be a multistage process. The absorption of radiation induces the formation of reactive species such as free radicals (molecular fragments with an unpaired electron in their outer shell), which are probably the main source of biomolecular damage. Much of this damage is repaired and appears to have little significant effect on the cell, but some is lethal while other damage (usually involving DNA lesions) may lead to transformation. Radiation may also activate oncogenes, and some radiation induced leukaemias are associated with retroviral gene expression.

(c) *Viral carcinogenesis*

That both RNA and DNA viruses are involved in cancer first became clear from studies on viral infection in animals and cells in culture, but evidence now exists to implicate viruses in human cancer. Most viruses with RNA genomes are not associated with cancer since they do not need to invade the nucleus to replicate. One group of RNA viruses (the **retroviruses**), however, has acquired (or possibly retained) the ability to replicate via a DNA intermediate stage and these viruses are associated with the development of cancer (reviewed in Varmus, 1988). Functionally, retroviruses belong to the group of mobile genetic elements known as **transposons**. Morphologically, the retroviruses may be classified into four types (A–D); major interest focuses on the C-type viruses which are implicated in various tumours including carcinomas, sarcomas, lymphomas and leukaemias.

The external coat or envelope of a retrovirus consists of a lipid bilayer (derived from the host cell plasma membrane) which incorporates various molecular entities including glycoproteins. The envelope overlays a matrix layer which itself encompasses the viral core of various proteins and two single strands of RNA (making retroviruses diploid). As illustrated in Fig. 1.3, the retroviral genome contains a number of sequences which are involved in the regulation of viral transcription (promoter and enhancer regions), in the formation of the proteins of the internal viral core (*gag*), in the production of enzymes including reverse transcriptase (*pol*) and in the production of viral envelope glycoproteins (*env*). Soon after infection of a cell the RNA genome of a retrovirus is 'reverse copied' into double-stranded DNA by the reverse transcriptase (an RNA-dependent DNA polymerase). This double-stranded DNA copy of the viral RNA genome is referred to as the **provirus**, and it may be either free or integrated into the host cell DNA. The DNA of the provirus is actually longer than the RNA of the retrovirus from which it was derived because of duplication of regulatory elements at both the 3' and 5' ends to form long terminal repeat (LTR) structures. The 3' and 5' terminal regions of the normal retroviral genome actually include a sequence of nucleotides repeated at both ends known as the R segment, and another sequence which is unique for each terminal region (U_3 and U_5). During reverse transcription, a copy of the U sequence from the opposite end is added to each terminal region to form the LTR. The presence of an LTR at the 3' end of the provirus enables it to act as a regulator for adjacent cellular genes (**downstream promotion**).

Historically, several milestones can be recognized in the research which led to the identification of retroviral oncogenes that are active in human cancers:

(a) The work of Rous (1911) which identified filterable agents that could cause cancer in chickens and which ultimately led to the characterization of RSV (the Rous sarcoma virus). Although

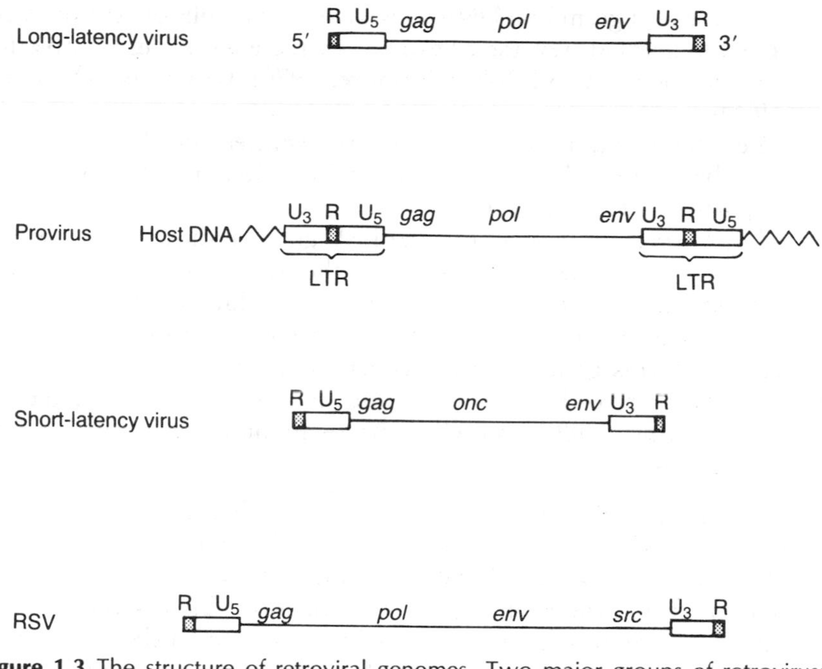

Figure 1.3 The structure of retroviral genomes. Two major groups of retroviruses known as the short- and long-latency forms have been identified. Long-latency forms have a full complement of replicative genes represented here by *gag* (encoding viral core structures), *pol* (encoding reverse transcriptase) and *env* (encoding viral envelope proteins). Short-latency forms lack all or part of their replicative genes, and these have been replaced by *onc* genes. Note that the deleted sequences for short-latency viruses differ for each virus, as does the site of insertion of the *onc* gene. The Rous sarcoma virus (RSV) is the only retrovirus known to contain an *onc* gene in addition to its normal complement of replicative genes. R denotes a repeated sequence at each end of the genome, and U refers to unique sequences involved in regulation. The long terminal repeat (LTR) structure regulates transcription of the integrated provirus. Since this sequence is found at both the start and the end of the viral sequence, 'downstream' cellular genes may also come under its promoting influence. For further details refer to the text.

viruses had already been discovered by the time Rous performed his experiments, he was reluctant to use the term to describe his filterable agents. It has been said that this was a consequence of peer pressure, but whatever the underlying reason, recognition of his work finally came with the award of a Nobel prize.

(b) The studies of Bittner (1942) which led to the recognition of a substance (the Bittner factor), transmitted through the mother's milk, which influenced the development of mammary tumours in mice. This factor is now known to be a retrovirus, the mouse mammary tumour virus (MMTV), and it has since been shown to be present in germ line DNA as well as in the milk of certain mice.

(c) The discovery of reverse transcriptase, the enzyme responsible for converting RNA to DNA (Baltimore, 1970; Temin and Mizutani, 1970).

(d) The identification of proviruses transmitted in the germ line (Huebner and Todaro, 1969), which led to the oncogene hypothesis of cancer formation.

(e) The observation that part of the RSV genome (now known as the *src* oncogene) was necessary for the transformation of RSV-infected cells (Duesberg and Vogt, 1970; Martin, 1970).

(f) The discovery by Stehelin *et al.* (1976) of proto-oncogenes, cellular forms of retroviral oncogenes (see below).

(g) The discovery of human retroviruses involved in T cell leukaemia (Poiesz *et al.*, 1980) and AIDS (Barre-Sinoussi *et al.*, 1983; Gallo *et al.*, 1984).

The cancers induced by retroviruses may take either a short time (days or weeks) or a long time (months or years) to develop. It turns out that most short-latency (or acutely transforming) retroviruses have actually lost part of their genome involved in replication. This is replaced by new genomic material, however, which is responsible for their relatively rapid transforming ability. As a result of this transforming effect, the new genomic material introduced into each virus is referred to as an **oncogene** or as an *onc* gene. More specifically, the gene is known as a viral oncogene or v-*onc* gene, since similar *onc* sequences are found in normal cellular DNA (Stehelin *et al.*, 1976). In fact, it is believed that the short-latency retroviruses evolved after the chance incorporation of what we now refer to as c-*onc* sequences from ancestral host cells. Evidence in support of this stems from the fact that retroviruses do not contain introns (intervening sequences which, in normal DNA, are excised when forming mRNA). Since the positions of both introns and exons in c-*onc* genes are consistent, it is highly unlikely that they could have been derived from the viral gene. Because there is no evidence that normal c-*onc* genes cause cancer when expressed at normal levels, some authors prefer to refer to c-*onc* genes as **proto-oncogenes**. Note that the Rous sarcoma virus is the only known retrovirus which contains a full set of replicative genes and an oncogene (Fig 1.3), and thus that most acutely transforming retroviruses are replication deficient requiring a reverse transcriptase from a non-deficient (e.g long-latency) helper virus. On the other hand, since these retroviruses contain both an oncogene and a promoting region which can switch the gene on or off, then their oncogenic behaviour is independent of the site of insertion in host DNA. This probably accounts for their relatively rapid transforming effects.

How might long-latency retroviruses, which lack an *onc* gene, transform cells? The general process is referred to as **insertional mutagenesis**, namely that proviral insertion or integration into host DNA overrides normal control processes and this may ultimately lead to tumour development. One possible molecular mechanism, as illustrated by avian leukaemia virus (ALV), involves downstream promotion. Occasionally, ALV may integrate into host DNA close enough upstream from the c-*myc* gene (the cellular homologue of the avian myelocytomatosis virus oncogene) to induce increased *myc* transcription. Increased production of the *myc* product could affect cell proliferation and perhaps predispose the cell to other tumorigenic events.

In summary, there are two common types of oncogenic retroviruses: one carries host-derived oncogenes and causes tumours relatively quickly, whereas the other lacks a viral oncogene and requires a long latency period before causing a tumour. A list of some of the known oncogenes, of which there are now over 50, is presented in Table 1.2.

Table 1.2 A partial list of retroviral oncogenes*

Oncogene	*Major product*	*Activity*	*Virus*
abl	$p120^{abl}$	TK	Abelson murine leukaemia
erbB	$gp65^{erb}B$	tEGFR	avian erythroblastosis
fes	$p85^{fes}$	TK	Snyder-Theilen feline sarcoma
fms	$gp120^{fms}$	M-CSFR	Susan McDonough feline sarcoma
fos	$pp62^{fos}$	transcription	FB3 murine sarcoma
jun	$p39^{jun}$	transcription	avian sarcoma–17
mos	$p37^{mos}$	S/TK	Moloney murine sarcoma
myb	$p45^{myb}$	?	avian myeloblastosis
myc	$p110^{myc}$	transcription	myelocytomastosis–29
ras	$p21^{ras}$	G protein	Harvey and Kirsten murine sarcomas
ros	$p68^{ros}$	TK	UR2 avian sarcoma
sis	$p28^{sis}$	PDGF	simian sarcoma
src	$pp60^{src}$	TK	Rous sarcoma

* Notes:

1. In shorthand terms such as $p28^{sis}$, the p refers to its protein nature, 28 refers to its relative molecular mass (M_r) in kilodaltons (kD), and *sis* refers to the particuar *onc* gene or oncogene associated with its production.
2. The molecular masses quoted in this table and generally throughout the text are largely determined by electrophoresis under reducing conditions. The reader should be aware that different techniques can yield widely disparate molecular mass values.
3. Other abbreviations: tEGFR: truncated epidermal growth factor receptor; gp: glycoprotein; M–CSFR: macrophage colony stimulating factor receptor; p: protein; pp: phosphoprotein; S/TK: serine/threonine kinase; TK: tyrosine kinase; PDGF: platelet-derived growth factor.

We can ask ourselves a number of important questions with respect to the role of oncogenes in cancer, aspects of which are discussed in Hunter (1984) and Bishop (1985):

(1) If normal cells have c-*onc* genes, why are they not transformed? There are a number of possible answers to this question, but generally speaking they are covered by two hypotheses, which are by no means mutually exclusive:

(a) The dosage hypothesis This assumes that transformation is associated with changes in the level of oncogene expression. These changes may result from chromosomal aberrations such as the translocation seen in Burkitt's lymphoma, or the gene amplification seen in some examples of neuroblastoma (section 1.7.2).

(i) Ectopic expression. It is possible that certain c-*onc* genes are not expressed in normal cells, and that their ectopic expression could lead to transformation. The product of the c-*sis* gene, for example, has not been found in normal cells, although it may exist in megakaryocytes (the cells which secrete PDGF). It is possible that the expression of the c-*sis* product could be associated with the establishment of fibrosarcomas and gliomas, two tumours which contain detectable levels of c-*sis* RNA (see section 1.7.1).

(ii) Overexpression. Alternatively, the c-*onc* genes may be expressed poorly in normal cells and their overexpression could lead to cancer. There is evidence suggesting that cell lines can be transformed with c-Ha-*ras*, for example, providing there is considerable overexpression of the normal p21ras product. Thus the introduction into NIH 3T3 cells of a c-Ha-*ras* gene driven by a retroviral LTR can lead to the production of cells capable of causing tumours in nude mice (Chang *et al.*, 1982). As will become clear below (section 1.5.7) there are many restrictions on how we should interpret these results using cell lines and nude mice for testing tumorigenicity. Indeed, c-*onc* gene expression is not always related to tumorigenicity: some normal cells show high levels of oncogene expression without any signs of transformation. Amnion cells, for example, have an abundance of $p62^{c-fos}$ (Muller, Verma and Adamson, 1983), and post-mitotic neurons have high levels of $pp60^{c-src}$ (Sorge, Levy and Maness, 1984).

(iii) Unscheduled expression. Although under normal circumstances c-*onc* genes do not induce tumours, it is possible that the tumorigenic state could be associated with their unscheduled expression. The c-*myc* gene, for example, is cell-cycle regulated and its continuous expression may be associated with tumorigenicity, as seen in Burkitt's lymphoma. When Daudi Burkitt lymphoma cells were treated with interferon (either IFN-α or IFN-β), c-*myc* expression was reduced and the cells accumulated in the G_0/G_1 phases of the cell cycle (Einat, Resnitzky and Kimchi, 1985).

(b) The mutational hypothesis This hypothesis suggests that c-*onc* genes are activated by a mutational event, and that these mutated genes lead to cancer. In 1981, it was reported that the tumorigenic phenotype could be induced in non-transformed cultured cells by transfection of the 'normal' cells with DNA from a human bladder cell line. Since the tumour cell line had not been knowingly infected with an oncogenic virus, what had induced the tumorigenic change? It turned out that the gene responsible for the transformation was the cellular *ras* gene, with an important difference: it had been 'activated' by a single nucleotide change which ed in one amino acid substitution. Activation of c-*onc* genes by mutation has now been detected in a wide variety of tumours.

(2) Are all cells transformed equally by retroviral oncogenes? It seems that some cells are more susceptible to transformation by particular oncogenes than others, but the processes involved in this susceptibility are not known. Interestingly, v-*myc* is the only known viral oncogene which can induce tumours in cells from epithelial, mesenchymal and haematopoietic tissues.

(3) How do retroviral oncogenes exert their effects? The products of the retroviral oncogenes can be divided functionally into two groups, those that act within the nucleus and those that act in the cytoplasm, although the functional locations of some are still unknown (Weinberg, 1985). It should be noted that the location of a particular oncogene in the nucleus or cytoplasm does not necessarily restrict its activities to one or the other:

(a) The nuclear acting products These include the proteins of v-*fos*, v-*myc* and v-*jun*. Tumorigenicity is associated with the loss of normal proliferative controls and thus it seems an absolute requirement that the transformation process should affect the regulation of cell division. Recent research with the *fos* proto-oncogene has shown that its product activates gene transcription in association with that of the *jun* proto-oncogene. One developing idea is that the *fos* and *jun* products bind together via leucine-rich domains (the 'leucine zipper') and this association allows them to bind to a wide range of sites on DNA thereby influencing transcription (Landschulz *et al.*, 1988). The products of c-*fos* and c-*jun* have recently been shown to be components of transcription factor activator protein-1 (AP-1), formerly thought to be a single polypeptide. The cellular levels of AP-1 increase after cell stimulation with a variety of agents including serum, and it is thought that its components (such as the products of c-*fos* and c-*jun*) function in the induction of competence (section 1.2.1). Exactly how this is brought about is not clear. Recent work focussing on *fos* and *jun* suggests that they may act as sensors for incoming short-term signals, converting them to long-lasting responses by regulating specific target gene expression (Curran and Morgan, 1987). It is not without interest that transcription of c-*fos* is influenced by both the phosphatidylinositol- and cAMP-based intracellular signalling pathways (section 1.7.22).

(b) The cytoplasmic acting products These proteins appear to subserve various functions including:

(i) Tyrosine specific protein kinase activity as shown by the products of v-*abl* and v-*src*, for example. A kinase is an enzyme that phosphorylates a protein by transferring the terminal phosphate group of ATP to amino acids with a free OH group, namely serine, threonine or tyrosine. Most proteins are phosphorylated on serine (90%), some on threonine (10%), and thus very few (about 0.1%) on tyrosine. Although several cellular proteins can be phosphorylated on tyrosine, no specific phoshorylated substrate has been detected in transformed cells, thus leaving the significance of tyrosine kinase activity in some doubt. All of the tyrosine protein kinases share a similar domain of about 30 kD which is responsible for their enzymatic activity, and they are usually membrane-associated in some way. The product of v-*src*, for example, is myristylated at its *N*-terminus, and this is probably involved in its attachment to the plasma membrane and certain cytoplasmic membranes such as that of the Golgi apparatus. How the product of v-*src* might bring about transformation is unclear: it may have increased specific activity over the product of c-*src*, or it may act on different substrates. There is clear evidence nevertheless that reductions in tyrosine kinase activity alter the transforming ability of v-*src*.

(ii) Serine and threonine specific protein kinase activity (e.g.v-*mos*). Very little is known about how this activity might be implicated in transformation.

(iii) Guanine nucleotide binding as shown by the v-*ras* family. As discussed elsewhere (section 1.7.22), the product of v-*ras* binds GTP, thereby mimicking the activity of G proteins which transmit signals across the plasma membrane.

(iv) Growth factor activity as illustrated by the product of v-*sis* whose chains are homologous with the B subunit of PDGF (section 1.7.1). The v-*sis* product may act in an autocrine manner binding to either internal or external receptors. The problem as to why PDGF itself does not transform cells is discussed in section 1.7.1.

(v) Growth factor receptor activity as shown by the product of v-*erb*B, which is a truncated version of the receptor for epidermal growth factor (EGF) that has constitutive tyrosine kinase activity. Since the product of v-*erb*B cannot bind its ligand, it probably is not down regulated from the surface of the cell. On the other hand, it seems that most of the product remains in the membranes of the Golgi apparatus in an immature mannose-rich form, but the relevance of this in transformation is uncertain.

The overall conclusion is that although in some cases we may know a lot in detail about the products and activities of various oncogenes, we are in fact not able to pin-point exactly how these induce transformation. There is little doubt that normal cell behaviour and division are under the control of complex, interacting regulatory events, and subtle changes which shift the balance of control might have knock-on effects which result in transformation. Several other points are worth noting: viral oncogenes are not the major cause of human cancer; other cancer-causing genes probably remain to be detected; and finally, it has been estimated that around 1000 previously inactive genes may be transcribed when cells are transformed by v-*src* (Groudine and Weintraub, 1980).

Recent studies have implicated oncogenes in tumour initiation and progression. Quintanilla and colleagues (1986), for example, have used the mouse skin model to show that most tumours initiated with DMBA (dimethyl benzanthracene) have mutations around codon 61 of the mouse Ha-*ras* gene, although a minority have mutations in position 12. The principal mutation (an adenine to thymine transversion) appeared to be heterozygous in most premalignant papillomas, but was homozygous or amplified in some carcinomas. It occurred very soon after initiation, and was not influenced by the nature of the promoter. Interestingly, the v-Ha-*ras* gene can act as an initiator in two stage skin carcinogenesis, thus confirming the importance of Ha-*ras* gene mutation in initiation. In summary, the observations are in keeping with the idea that initiation in this model system involves Ha-*ras* mutation, and that progression is associated with further chromosomal changes involving this locus.

Of interest here are the studies of Gilbert and Harris (1988) who have suggested that a complex genomic destabilization follows oncogene transfection, and that this allows the progressive emergence of new phenotypes. Data from their experiments with activated c-Ha-*ras* indicate that transfected cells have a low incidence of tumour take, and that those tumours which do grow *in vivo* require a long latent period. Subsequent cytogenetic analysis showed that tumour production resulted from the overgrowth of variant cells whose chromosome constitution was markedly different from that of the original cells. Other experiments by these authors indicated that once tumorigenic variants had been selected, the continued production of $p21^{ras}$ was no longer necessary for maintenance of the transformed state. Thus the crucial aspect of transformation by an activated oncogene seems to lie in the spectrum of genetic changes which follow its expression, and once these have taken place continued oncogene expression is no longer necessary. How oncogenes might affect other aspects of tumorigenicity (such as malignant cell spread) will be discussed later (section 6.1).

Viruses with DNA genomes are also implicated in the development of cancer. After the pioneering studies of Rous (1911), research focused on the DNA viruses because a mutation in polyoma virus was found which affected its ability to transform cells *in vitro* (Fried, 1965). The use of DNA viruses, however, had a number of technical problems which tended to slow progress down: the oncogenes of DNA viruses are tied up with the expression and replication of viral DNA and are thus not easy to work with, and when the viruses replicate they kill their host cells thereby terminating experimental observations. As illustrated in Table 1.3, there are six main families of DNA tumour viruses, most but not all of which have been shown to contain specific oncogenes.

Compared to the retroviruses the genome of the DNA viruses is

Table 1.3 DNA viruses and their known oncogenes

Virus	*Oncogene*
Hepadnaviruses	
Hepatitis B	unknown
Papillomaviruses	E5, E6, E7
Papovaviruses	
Polyoma	Large T, middle T, small t
SV40	Large T, small t
Adenoviruses	E1a, E1b
Herpesviruses	
Herpes simplex 1 and 2	unknown
Epstein-Barr virus	EB nuclear antigen II
Cytomegalovirus	unknown
Poxviruses	vaccinia growth factor

quite complex, but it may be divided functionally into early and late regions associated more or less with genome replication and structural components respectively. Occasional reference will be made later in this book to some of the genes from the early regions of certain papovaviruses and the adenoviruses. The early regions of polyoma and SV40 viruses encode proteins known as T antigens (tumour antigens originally detected using immune sera from tumour bearing animals). It should be noted here that the polyoma virus has three T antigens, referred to as large T, middle T and small t, whereas SV40 has only two, large T and small t. The equivalent regions in adenoviruses represent the left hand 8–12% of the viral genome, and this is subdivided into two further regions (referred to as E1a and E1b) which code for two separate sets of proteins (reviewed in Bishop, 1985). As will become clear later (section 7.1), some of these early genes result in cell immortalization (polyoma large T; adenovirus E1a), whereas full transformation may require cooperation with other early genes (polyoma middle T and adenovirus E1b respectively), or other genes including retroviral oncogenes and the retinoblastoma gene.

Except in experimental situations, it has proven difficult to establish a causal link between DNA virus infection and cancer. Indeed, most of the evidence relating to human cancers has a predominant epidemiological basis. Examples of clinical relevance in which associations between DNA viruses and cancer are likely include the Epstein-Barr virus (EBV) which is a major risk factor in nasopharyngeal cancer and in Burkitt's lymphoma, and hepatitis B which is associated with a risk of developing liver cancer. In these examples, viral infection is much more widespread than the related tumour incidence, and this illustrates an important point in the development of cancer: namely, that it is a complex process likely to involve a number of interacting factors. With respect to Burkitt's lymphoma, for example, epidemiological data indicate geographical clustering and suggest endemic malaria as a possible cofactor. As will be discussed elsewhere (section 1.7.2), Burkitt's lymphoma is also associated with a chromosomal translocation which brings c-*myc* under the control of immunoglobulin genes and leads to its elevated expression. Although still not completely clear, it seems that EBV infection alters the growth properties of B lymphocytes, and that this may be more common in individuals who may have perturbed immune systems (such as those living in areas where malaria is endemic). The affected cells are then thought to be more susceptible to other changes, some of which could include the chromosomal translocations which lead to malignancy. Epidemiological data suggest that the ingestion of certain Cantonese-style salted fish may be related to the distinct regional distribution of nasopharyngeal cancer in China, although there is also evidence of a correlation between the use of particular medicinal plants containing naturally occurring phorbol esters and this form of cancer. Recently, normal primary human skin epithelial cells were found to be able to grow in soft agar (i.e. anchorage independently) after treatment with EBV and phorbol esters (Tomei *et al.*, 1987). Anchorage-independent growth (see section 1.7.24) is often taken as a criterion for transformation, although some normal cells can grow in soft agar. These results may be interpreted in terms of the initiation/progression concept of carcinogeneis, with the phorbol esters serving as promoters following EBV-associated initiation. Interestingly, the cells which grew anchorage-independently in this study showed no signs of viral replication within them, nor did they express EBV receptors. The goal now in this work must surely be to show the full tumorigenic nature of these *in vitro* transformed cells.

The strongest correlation of an association between a DNA virus and human cancer is that shown in a prospective study by Beasley, Lin and Hwang (1981) for the hepatitis B virus and cancer of the liver. Indeed, the corelation has been deemed strong enough to warrant large-scale prevention trials of at-risk individuals. Acording to Beasley and colleagues (1981), the risk of a hepatitis B carrier developing liver cancer is 100× greater than that of non-carriers, whereas the risk difference for smokers and non-smokers developing lung cancer is around 20-fold. The role of hepatitis B in the development of liver cancer is not completely clear, but it has been postulated that infection induces chronic, immune-related liver damage and that the repeated cycles of liver regeneration and further damage which result increase the possibility of cellular transformation occurring.

With respect to the human adenoviruses and papovaviruses, not one has been shown definitively to be related to tumours in man. Of the papillomaviruses, recent evidence suggests that benign warts may progress to various forms of malignancy. Genital warts in women (in which HSV appears to be a risk factor), for example, may progress to carcinoma of the cervix. These observations are supported experimentally, since it has been shown in the mouse skin model that chemically induced papillomas can progress to carcinomas. There is at least one known example of a human tumour (a benign plantar wart) which has been attributed to a poxvirus.

Interestingly, infection with DNA tumour viruses does not always lead to transformation of cells either from susceptible species or from the original viral host. Thus cells infected *in vitro* with certain DNA viruses may display numerous features of the transformed phenotype and may even form tumours in immunoincompetent hosts, but not in normal, syngeneic ones. It has been tacitly assumed that this is due to the immunogenic nature of the virus-specific T antigens which are involved early in infection. Studies by Lewis and Cook (1985), however, suggest that T antigen immunogenicity of DNA virus transformed hamster cells plays little or no role in determining tumorigenesis in syngeneic animals. Instead, they found that tumorigenicity correlated with resistance to immunologically non-specific lysis by NK cells and non-activated macrophages. The significance of the immune system in tumour development is discussed further in section 1.8.

1.5.5 Carcinogen testing

It is important to be able to test for potential carcinogens in a rapid, reliable and cost-effective manner. To this end, a number of short term tests, such as the **Ames test**, have been designed for assaying potential carcinogenicity. The basis of the Ames test is a mutant strain of *Salmonella typhimurium* which cannot synthesize the amino acid histidine needed for protein synthesis. In order to grow, this bacterial strain requires exogenous histidine, but exposure to a mutagen may restore its ability to synthesize the amino acid (by a simple base-pair mutation) thus allowing it to grow in normal medium. Although a reverse mutation of this sort (his^- to his^+) is likely to be relatively rare, tens of millions of bacteria are used and a single revertant can be detected because it will eventually grow to a visible colony on a plate. The basic bacterial mutant has now been improved upon in order to increase its sensitivity to mutagens, and the test is usually performed in the presence of the supernatant (S9 mix) obtained from a rat liver homogenate centrifuged at 9000 g. S9 mix is thought to contain many of the enzymes which are responsible for metabolizing procarcinogens into carcinogens. Although the Ames test is actually based on mutagenic potential it has proven to be reasonably reliable in identifying carcinogens. Nevertheless, it does yield both false positive and false negative results. Cairns (1981) has recently proposed that point mutations may in fact not represent the most common underlying cause of human cancer. His argument is based on the observation that patients with xeroderma pigmentosum (who have defects in their DNA repair mechanisms) may show a high incidence of malignant melanoma (attributable to mutations induced by sunlight) yet they do not show a significant increase in other common cancers, such as cancer of the lung, colon and breast. Furthermore, patients with Bloom's syndrome (who show a high frequency of pronounced chromosomal aberrations) do show an increase in the occurrence of the common cancers. Based on these observations, and various other lines of evidence, Cairns (1981) concluded major chromosomal changes (section 1.7.2) rather than point mutations are important in the origin of human cancers. If this proves to be the case, the Ames test (which is based on a point mutation) may only be of limited significance in determining human carcinogens.

1.5.6 DNA transfection

Most of the research activity associated with applying molecular biological techniques to tumour biology has involved the isolation and transfer of oncogenes. The process of DNA transfection (namely the insertion of a new gene(s) into host cell DNA whilst assuring its expression) is amazingly simple when viewed with hindsight, but a review of the literature over the last 25 years would show the phenomenal amount of ground work involved. That the transfer of DNA from chemically transformed cell lines could result in the transformation of recipient cells from an untransformed cell line was first shown by Shih *et al.* (1979) and confirmed in different systems (Krontiris and Cooper, 1981; Shih *et al.*, 1981). Soon after, Shih and Weinberg (1982) identified a single gene in the DNA of a human bladder tumour cell line that could transform recipient cultured cells. This gene was subsequently shown to be an activated form of a cellular oncogene that was homologous to v-Ha-*ras*, the transforming gene of the Harvey sarcoma virus (Der, Krontiris and Cooper, 1982; Parada *et al.*, 1982).

Although there are many ways in which genes can be transferred into cells (including microinjection), here we shall consider the technique often referred to as **cotransfection** which has proven popular with a large number of research groups. Briefly, the technique involves the use of plasmids containing a gene (such as the *neo* gene) for resistance to an antibiotic (here a neomycin analogue such as G418, which is toxic to cells through its effects on ribosomes). The antibiotic resistance gene in experimental plasmids is usually ligated to another gene sequence which assumes transcriptional control. Now plasmids exist as independent genetic elements or 'minichromosomes' in bacteria, and some of them have the ability to jump in and out of the main bacterial chromosomes. In the presence of calcium phosphate, plasmids (and indeed other pieces of DNA) can also insert themselves into eukaryotic chromosomes and it is this ability which provides the basic mechanism for transfection. How then do we transfer a chosen gene (perhaps an oncogene) into a eukaryotic cell? At first sight, the best approach might seem to be to ligate the oncogene to the antibiotic resistant gene in the plasmid. In the long run, this approach might turn out to offer more control over the transfection process, but it is not necessary since DNA sequences added to cells along with the plasmid and calcium phosphate are also taken up into the host's chromosomes (hence **cotransfection**). How DNA insertion is actually achieved during transfection remains uncertain, but it is assumed that the DNA is phagocytosed and subsequently stably integrated into the host DNA. Now some degree of selectivity is necessary since we cannot be sure that all the cells treated with the oncogene have been transfected. It is here that the value of the antibiotic resistance gene is seen, since this acts as a selectable marker. Normal cells will not grow in a medium containing G418, but if a plasmid containing a neomycin resistance gene is functionally integrated into the cells then they will grow in this medium and the appropriate cells can be selected for further study. Since in our experimental situation these cells are also likely to have incorporated an oncogene, then this too will be expressed in the transfected cell and we may now proceed to study its effects.

There are several questions that we need to ask about this experimental procedure. Are there any adverse effects of the experimental manipulations which are necessary during transfection? How do the transfected genes insert into the host DNA? Is it entirely random? Are multiple copies inserted? Does insertion affect the expression of normal host genes? Most of the early work on transfection did not attempt to answer these questions, and it may be that the correct interpretation of the effects of transfection will require an intimate knowledge of what happens to the DNA after plasmid insertion.

Some of the questions raised above, however, have been tackled in more recent studies, and new developments in technique are beginning to offer more control over the transfection process. It is worth noting that at least one report has suggested that the process of calcium phosphate mediated transfection itself (i.e. transfection with the neomycin resistant plasmid pSV2neo without an oncogene) can induce heritable changes in tumorigenic and metastatic behaviour (Kerbel *et al.*, 1987). Some of the pSV2neo plasmid transfected cells grew in nude mice but failed to grow in immunocompetent mice, possibly as a result of the expression of histocompatibility antigens which were not detectable on the original cells. In another study, transfection of cells with oncogene-containing plasmids was found to result in secondary chromosomal changes that could lead to transformation in the absence of stable integration of the oncogene (Lau *et al.*, 1985). This 'hit and run' form of tumorigenesis clearly implies a need for an appropriate analysis of the genome after transfection before valid conclusions can be drawn.

Finally, we should note that the transfection technique can be highly variable on a day-to-day basis and furthermore that the outcome of transfection is influenced by the nature of the recipient cell. We might consider the ideal recipient cell to be one recently isolated from fresh tissue, one which could be called 'normal' in every respect other than it has been isolated from the body. Unfortunately, primary cells with these characteristics are not readily available and most studies have utilized the NIH 3T3 cell line which has proven to be relatively easy to transfect. These cells are already immortalized, and success in their transfection might be related to their 'partially transformed' nature.

1.6 HOW DO TUMOUR CELLS DIFFER FROM NORMAL CELLS?

Although tumour cells clearly differ from normal cells in their responsiveness to the homeostatic processes which regulate cell division, a considerable amount of detailed research has identified many other differences between the two cell types. These differences are largely thought to arise from shifts in the tumour cell of the normal pattern of gene expression which, in some cases, are associated with recognized changes in the genetic material.

1.6.1. *In vitro* and *in vivo* comparisons

As a prelude to further discussion, it should be noted that tumour cells can be compared with normal cells under *in vivo* and *in vitro* conditions. Under *in vitro* conditions, the normal tissue relationships of cells are lost and they are obliged to grow remote from the body under very artificial circumstances. Of course *in vitro* growth could affect both normal cells and tumour cells, and one of the biggest problems with studies of this sort is that the original population may change during the course of culture. Many tumour cells are particularly sensitive to inadvertent selection pressures which operate *in vitro*, but they may also change significantly during serial passage *in vivo*. For these reasons a case may be made for the use of spontaneous tumours (Alexander,1977; Hewitt, 1978) and freshly isolated normal tissues in comparison studies. Although there are undoubted reservations in the use of cells maintained under *in vitro* conditions, the practicalities of the situation have demanded their use in experimental systems. The artificial nature of *in vitro* growth should nevertheless caution us on any conclusions we might derive from observations made under such conditions.

For practical reasons then, many of the features thought to be characteristic of tumour cells have been determined from the detailed study of cells maintained *in vitro*. In such studies the tumour cells are often not derived from primary tumours but from established cell lines following their transformation by viruses, chemicals or radiation. The parental cell lines in such experimental systems are taken to represent 'normal' cells, but they often share little in common with cells in normal tissues or in primary cultures. As noted above, normal tissues and organs maintain fixed sizes and relationships in the adult and these are usually lost when cells are grown *in vitro*. Cells prepared directly from tissues usually have a limited lifespan and they become senescent after a number of generations characteristic of the species of origin (Hayflick and Moorhead, 1961). For reasons which are not yet fully understood but are presumed to have a genetic basis, it is possible to grow cells under conditions which enable them to become established i.e. the cells avoid senescence thereby becoming immortal (Todaro and Green, 1963). Cells (usually of embryonic origin) which are derived under such conditions are able to be transferred from one culture vessel to another (passaged) indefinitely and they are the source of the established cell lines such as the mouse 3T3, often used in comparison studies with transformed cells. These cells are similar to normal tissue cells maintained in primary culture in many ways: they grow to a relatively low density and they require a relatively high concentration of serum for optimum growth. However, although thought of as 'normal', established cell lines do not become senescent in culture, they are often not diploid, and in certain growth conditions (high culture density, for example) they may show a relatively high frequency of spontaneous transformation. Cells from established lines usually do not form tumours on injection into syngeneic hosts, although they may do under certain circumstances such as injection in association with substrate (Boone, 1975). Such cells may be best described as being 'partially transformed' although the molecular connotations of this term are unclear. It should be noted here that the baby hamster kidney (BHK) cell line, often used as a source of 'normal' cells for comparison studies with virally transformed cells, is in itself tumorigenic and hardly ideal for this purpose. Furthermore, recent evidence to be discussed later suggests that even the widely used NIH 3T3 cell line may be tumorigenic in experimental animals.

Interestingly, some recent research has shown that the senescence which overcomes most normal embryonic cells when attempts are made to grow them *in vitro* may be due to negative factors present in the serum. Loo *et al.* (1987) succeeded in growing mouse embryonic cells in serum-free medium containing EGF (50 μg ml^{-1}): the cells grew to high density and formed multilayers, but karyotype analysis showed no detectable chromosomal abnormalities. When serum was added to embryonic cells growing under serum-free conditions growth was inhibited and the cells accumulated in G_1. This suggests that established cell lines develop mechanisms for overcoming the negative effects of culture in serum-containing medium.

In vitro, tumour cells show a number of features which are thought to be characteristic of the transformed state. The fact that many of these characteristics relate more to *in vitro* transformed cells rather than to tumour cells which arose *in vivo* has important consequences in our understanding of tumorigenicity since they may yield a misleading picture. Furthermore, these properties do not necessarily relate to malignant cells (at the expense of benign cells) nor do they relate to the degree of metastatic capability. Indeed, Ossowski and Reich (1980) found that the metastatic potential of Hep-3 cells correlated with low or reduced expression of several features thought to characterize the transformed cell.

Although there clearly are reservations on how we interpret these so-called characteristic features of transformed cells, at this stage four deserve particular mention:

(a) Transformed cells have escaped senescence i.e. they are immortal (Todaro, Wolman and Green, 1963). This characteristic is shared, however, with non-transformed, established cell lines, which are often termed 'normal'.
(b) Such cells often grow in suspension (Macpherson and Montagnier, 1964).
(c) They usually grow to high density (Temin and Rubin, 1958).
(d) They usually have reduced growth requirements for serum (Smith *et al.*, 1971).

The two latter points are probably interrelated, particularly if the inability to form multilayers is due to a lack of nutrients or appropriate growth factors as mentioned above.

1.6.2 Tumour markers

Interest in studying biochemical differences between normal cells and tumour cells has focused on the possibility of using specific molecules as markers for malignancy. These biochemical markers may be located cytoplasmically or on the cell surface, or they may be released into the body fluids. Tumour markers have considerable potential in various aspects of the detection and treatment of neoplastic disease. Cell surface markers, as illustrated by the surface immunoglobulin of B cell tumours, may be useful as target molecules for immunotherapy. Surface markers may also be useful in locating both primary and secondary tumours within the body, and with highly anaplastic tumours they may also be instructive in indicating their origins.

Providing the detection of markers is both sensitive and relatively specific, they may provide insight on:

(a) *The extent of tumour burden*

Detectable levels of the tumour marker should reflect the size of the tumour present.

(b) *Prognosis*

In some cases the amount or mere presence of a particular marker may provide an indication of outcome. The presence of a particular marker, for example, may correlate with tumour aggressiveness.

(c) *The effectiveness and possible choice of a course of treatment*

The appearance, disappearance or quantitative change in detectable levels of a marker may indicate the outcome of a particular course of treatment.

(d) *Recurrence*

The reappearance over time of a marker lost after seemingly effective treatment may provide an indication of recurrence of the tumour.

Unfortunately, few of the characteristic features of tumour cells listed in Table 1.4 actually function as markers in this sense, particularly with respect to the carcinomas which make up the majority of human tumours. Mention should be made here of the possible use of either oncogenes or their products as potential tumour markers. There is some evidence, although at this stage not very extensive, that oncogene expression may correlate with tumour progresion. Since oncogene expression yields a specific product, its detection by McAbs may aid in identifying both the presence and extent of neoplastic disease. It should be mentioned here that the majority of studies of tumour markers have been concerned with positive qualitative changes, although negative changes (both quantitative and qualitative in nature) may be equally or more important. The idea that malignancy may be associated with the lack of key differentiative features is discussed in a review paper by Buckley (1988).

1.6.3 Tumour stem cells

There is at least one other point concerning the comparison between normal cells and tumour cells which warrants some elaboration, and that is the actual origin of the tumour cell itself. Largely as a result of studies with normal renewing cell populations (such as the bone marrow, intestinal mucosa and epidermis) and with their associated tumours, the idea has been put forward that many neoplasms may arise from tissue stem cells (Mackillop *et al.*, 1983; Buick and Pollak, 1984; Greaves, 1986). Generally speaking, the stem cell concept of tissue homeostasis (see section 1.2.2) implies that there are only two major categories of cells which could give rise to a tumour: the stem cells and their differentiated (or partially differentiated) progeny. Since differentiated cells are believed to have limited proliferative potential, then some form of dedifferentiation would seem to be necessary if this category is to give rise to tumours. Stem cells, on the other hand, are undifferentiated cells and they also have the capacity for self renewal, making them ideal candidates for the transforming process. There are several lines of evidence in support of the validity of a stem cell model for tumour growth (reviewed in Selby, Buick and Tannock, 1983; Buick and Pollack, 1984):

(a) Most human tumours arise in tissues which are believed to normally proliferate by cell renewal (stem cell) mechanisms.
(b) Cells from human tumours are heterogeneous with respect to their proliferative capacity and their differentiation. Indeed, human tumour cells may be fractionated into different subpopulations with defined properties such as self renewal and the expression of differentiated features.
(c) Radiation therapy suggests that only a small fraction of cells (the stem cells?) need be destroyed in order to abolish tumour growth.

(d) Clonal markers (such as isoenzymes) in certain haematopietic malignancies are present in a wide variey of differentiated cell types, and this is consistent with their origin from multipotential stem cells.

(e) In clonogenic assays, only a small fraction (about 0.001% to 1.0%) of the total cells from a tumour can form colonies. Colony-forming cells are thought to have considerable proliferative potential and since this is generally inversely related to the state of differentiation, it is thought that such cells may be equivalent to stem cells. It should be noted, however, that there is no definitive proof that this is so.

If the validity of the stem cell origin of tumours is established, then the appropriate normal cell for comparison with a tumour cell is not a typical tissue cell, but a relatively rare stem cell. This is also not without considerable consequence on how we should treat tumours. Consider a human tumour of 10^{10} cells: destruction of all its cells (say by cytotoxic drug therapy) is likely to put an intolerable burden on the patient. A better chance of patient survival might be possible, however, if the same end result could be achieved by killing only a relatively small number of stem cells. Therapy which is designed to reduce tumour bulk may be directed towards the transitional and end cell compartments rather than towards the stem cells; by failing to eradicate the stem cells, such treatment will not be curative in nature. However, it is entirely possible that a transforming event may render a differentiated cell immortal, and thus the actual source of a tumour need not be a pre-existing stem cell *per se*. It would not be unlikely for a compromise position to exist *in vivo*: thus some tumours may derive from stem cells, while others may result from transformation of cells in the transitional or even fully differentiated compartments (providing they undergo mitotic division). In the latter situations, it would nevertheless seem likely that the transformed source cell would take on most of the characteristics of a stem cell, particularly its capacity for self renewal.

1.7 CHARACTERISTIC FEATURES OF TUMOUR CELLS

A summary of some of the features possibly altered in tumour cells is provided in Table 1.4. It should be made clear that these features are not absolute since many may be shared by both normal cells and tumour cells. In fact, a large proportion of the differences between tumour cells and normal cells are quantitative rather than qualitative in nature. Many of the so-called 'characteristic' features of tumour cells,

Table 1.4 Altered features reported for tumour cells*

1. Loss of normal growth conrol
2. Growth to high density *in vitro* i.e. a relatively high nuclear overlap index (>0.5 *versus* <0.2 for normal *versus* transformed cells)
3. Reduced heterotypic contact inhibition *in vitro*
4. Decreased cell-surface fibronectin
5. Expression of TATAs, TAAs and TSAs, including oncofetal antigens and viral determinants
6. General alterations in cell surface components (proteins, glycoproteins, proteoglycans, sialic acid)
7. Altered glycosylation
8. Increased plasma membrane shedding
9. Increased sensitivity to lectin agglutination
10. Increased receptor mobility
11. Altered permeability and transport of nutrients
12. Decreased mitochondrial content
13. Unregulated autocrine secretion of growth factors
14. Decreased requirements for serum and exogenous growth factors
15. Altered cell junctions
16. Altered cell communication
17. Increased deformability
18. Cytoskeletal alterations
19. Changes in cell-surface projections.
20. Altered cell motility (including chemotactic behaviour)
21. Changes in enzyme content and secretion
22. Altered production and secretion of hormones
23. Changes in hormone dependency
24. Increased histological heterogeneity (e.g. cytoplasmic basophilia, nuclear pleomorphism)
25. Altered adhesiveness
26. Decreased anchorage dependence
27. Altered ploidy
28. Increased chromosomal aberrations
29. Changes in intracellular signalling activity
30. Presence of activated oncogene products
31. Excessive production of proto-oncogene products

* Many of these changes reflect alterations in cell metabolism/behaviour without any readily obvious direction of change (i.e. as to whether the alteration is increased or decreased). Even where a general trend may be apparent, there exist many specific exceptions so that it is difficult to define any universal tumour cell characteristic.

for example, are cell-cycle related, reflecting their relatively high division rate or their proportion of cycling cells rather than the tumorigenic state *per se*. Some characteristic features are also relevant only for certain experimental systems, and indeed many are probably of dubious relevance to human tumours. When studying the characteristics of tumour cells derived directly from a tumour mass it is not always possible to separate the tumour cells from other host cells such as macrophages, which are abundant in some tumours, and this may also present a misleading picture. Tumours are also heterogeneic, so that not all cells within the same tumour mass will necessarily display the same features (Hart and Fidler, 1981). Furthermore, malignancy is not a static phenomenon: it changes as the tumour cell progresses and thus it would be surprising if it was possible to define a universal phenotype.

Since, by definition, tumour cells should form tumours in the appropriate host, it might seem logical to accept this feature as the *sine qua non* of the tumour cell. As mentioned previously, however, tumour cells are clearly affected by the immune defence system of the host and thus it is not always certain that a tumour cell will grow when injected into a syngeneic recipient. Many cells transformed *in vitro* are not necessarily tumorigenic in normal, syngeneic recipients since they often express virally coded (or other) antigens on their surfaces which render them highly immunogenic.

The major outcome of the reservations discussed above is that there may be little universality of application of the possible characteristics of tumour cells listed in Table 1.4. Nevertheless, we need to know what a tumour cell is in order to be able to study it, and in this respect many of the listed characteristics are helpful, even if of somewhat limited application.

1.7.1 Growth factors and the control of cell division

Tumour cells may be thought of as originally normal cells which by the process of transformation have escaped the normal homeostatic systems which control cell division. The exact nature of these regulatory systems is uncertain (see sections 1.2 and 1.7.22) and this topic remains a very active and somewhat speculative research area. It is clear, however, that cells respond to extracellular signals and that messages of this sort may stimulate cell division (reviewed in Rozengurt, 1983; Gospodarowicz, 1985; Goustin *et al.*, 1986; Ross *et al.*, 1986). Bioactive molecules which influence cell division may be peptides or polypeptides or they may be steroidal in nature. They belong to either of two general categories, the hormones and the growth factors, which differ in the responses they elicit and in their modes of delivery. The hormones include insulin, the sex hormones, and the corticoids, whereas the growth factors include the insulin-like growth factors IGF-I (somatomedin C) and IGF-II (somatomedin A), transferrin, fibroblast growth factors (FGFs), platelet derived growth factor (PDGF), epidermal growth factor (EGF), the transforming growth factors (TGF_α and TGF_β), and nerve growth factor (NGF).

Numerous haematopoietic growth factors such as erythropoietin, the colony-stimulating factors and the interleukins are also included in the latter category. These represent just a few amongst a plethora of factors, many of which are poorly characterized.

It seems that the growth of many cell types requires multiple growth factors, although on transformation this requirement may not be necessary. As will become clear below, the decreased serum requirement of many transformed cell types *in vitro* probably relates to their ability to divide in the presence of reduced amounts of exogenous growth factors. Of the reasonably well defined growth factors, four (PDGF, EGF, the TGFs and the FGFs) are of considerable relevance to tumour growth and warrant some consideration. Collectively, the roles played by these four growth factors can be used to illustrate many important points relating to the control of cell division. The intracellular signals involved in mediating the effects of growth factors are discussed more fully later in this chapter (section 1.7.22).

(a) *Platelet-derived growth factor (PDGF)*

Serum is a source of many of the components required for cell division, and cells will stop growing in serum-depleted medium unless other essential components have been added. Despite the need for serum components which stimulate cell division, however, most tumour cells are able to grow in levels of serum which do not support the growth of their normal counterparts (Holley, 1974). There are several reasons why this may be the case. As discussed in more detail below, tumour cells may, for example, endogenously produce the same or similar components found in serum which serve as growth factors, or they may have evolved means of short-circuiting the need for such factors. PDGF, despite its origins, is a potent mitogen found in serum (reviewed in Cochran, 1985; Rozengurt, 1986). Most serum PDGF is synthesized in megakaryocytes, stored in platelet α granules, and released during blood clotting which explains why it is found more in serum (15–20 ng ml^{-1}) than in platelet-poor plasma (1–2 ng ml^{-1}). Indeed, plasma does not support the growth of many cell types whereas serum does, even when prepared from the same pool of blood. Nevertheless, plasma is required in order to show optimal stimulation of DNA synthesis in PDGF-treated cells (Pledger *et al.*, 1981). It is thought that PDGF treatment renders cells competent, that is, sensitive to factors in plasma (such as somatomedin C) which enable them to undergo progression from G_0 or G_1 into S phase. Cells which have not been rendered competent cannot undergo progression and therefore do not proceed through the cell cycle.

The presence of PDGF in platelets probably ensures the delivery of growth factors to wounded sites where rapid tissue repair is essential, but it should be noted that several normal cell types including endothelial cells, smooth muscle cells and macrophages also produce PDGF. Apart from its growth factor effects, PDGF is chemotactic for fibroblasts, smooth muscle cells, monocytes and PMNs, and it also has potent vasoconstrictor activity. Furthermore, secretion of PDGF by endothelial cells is stimulated by components of the coagulation cascade including thrombin and Factor X. Taken together, these various responses add up to a powerful repair mechanism which would be brought into action following blood vessel damage. Most transformed mesenchymal cells can both produce and respond to PDGF, while some non-responsive malignant cells such as the T24 bladder carcinoma and the hepatoma cell line Hep G-2 can nevertheless secrete it. PDGF-responsive tumour cells would clearly benefit from the growth factor effects of PDGF, but non-responsive tumours may also benefit, albeit indirectly. Some tumours, such as scirrhous carcinoma of breast, are associated with a desmoplastic reaction characterized by the extensive deposition of connective tissue. The secretion of PDGF by both responsive and non-responsive malignant cells may contribute to this reaction through the mitogenic and chemotactic effects of PDGF on host fibroblasts.

PDGF is a glycoprotein existing in at least two forms (PDGF-I and PDGF-II), each of which are dimers (A and B chains) with approximate molecular masses of 28 30 kD and 30 32 kD respectively. It binds to a cell surface receptor of about 175 kD with primarily tyrosine specific, protein kinase activity (Deuel *et al.*, 1981; Nishimura, Huang and Deuel, 1982; Glenn, Bowen-Pope and Ross, 1982). Although a number of receptors for growth factors, as well as various oncogene products, have tyrosine specific protein kinase activity, the actual role of this activity in triggering cell division is not clear (Hunter, 1986). PDGF receptors have been found on most mesenchymal cells including fibroblasts, smooth muscle cells and glial cells, but not on epithelial cells (except trophoblastic cells), lymphocytes or endothelial cells (even though these secrete PDGF).

The B chain of the PDGF dimer bears a close resemblance to the transforming protein $p28^{sis}$ of simian sarcoma virus (SSV) and it has been suggested that this similarity underlies the abnormal growth characteristics of cells transformed by SSV (Deuel *et al.*, 1983; Doolittle *et al.*, 1983; Waterfield *et al.*, 1983). In fact, SSV-transformed cells actually secrete a B–B homodimer rather than the A–B hetereodimer produced by normal cells. It seems that the $p28^{sis}$ product is rapidly processed into a number of fragments, with most mitogenic activity residing in a $p20^{sis}$ form. If a cell which produces an active *sis* product also has a receptor for it, then the protein may exert its activities on the producing cell in an autocrine manner. The human osteosarcoma cell line U-2 OS, for example, has PDGF receptors and produces PDGF, the function of which can be inhibited by polyclonal anti-PDGF antibody. Since SSV-transformed NP1 marmoset fibroblastic cells do not actually secrete $p28^{sis}$, it has been suggested that autocrine stimulation in some cases may be mediated through the intracellular recognition of $p28^{sis}$, perhaps within the endoplasmic reticulum. It should be noted here that autocrine stimulation by growth factors is not an attribute which is exclusive to transformed cells: smooth muscle cells from young rats, for example, respond to PDGF in an autocrine manner (reviewed in Sporn and Roberts, 1985).

The treatment of responsive cells with PDGF leads to the induction of mRNA from the c-*fos* and c-*myc* genes (Cochran *et al.*, 1984; Kelly *et al.*, 1983). Since both of these proto-oncogenes code for nuclear proteins it has been suggested that their products may be involved in the transduction of the mitogenic signal within the nucleus (Persson and Leder, 1984). It should be noted here that c-*fos* and c-*myc* are only two of literally scores of genes which are likely to be activated on PDGF stimulation of quiescent cells, and thus we are far from fully understanding the molecular processes involved. Indeed, there is some evidence that stimulation with PDGF can lead to the expression of growth inhibitors such as IFN-β_2 (thought to be identical with interleukin 6), which supports the existence of controlling feedback loops for cell growth. Further events which follow PDGF binding to responsive cells (particularly the turnover of phosphatidylinositol, the release of arachidonic acid, and prostaglandin formation) are discussed in section 1.7.22.

(b) *Epidermal growth factor (EGF)*

This is a single chain polypeptide (M_r 6045 D) which has been isolated from mouse submaxillary glands (Cohen, 1959, 1962) and, as urogastrone, from human urine (Gregory, 1975). It is a potent mitogen for a variety of cultured cells of ectodermal and mesodermal origin under anchorage-dependent conditions, but not for cells of haematopoietic origin (Carpenter and Cohen, 1979). The failure of EGF alone to allow anchorage-independent growth suggests that its activity is restricted to growth rather than to transformation. There is some evidence that EGF can have inhibitory effects on certain cell types such as squamous carcinoma cells, and as will become clear below, EGF bears a close relationship to TGF_{α}.

The mitogenic ativity of EGF is potentiated by the hormone insulin, and it acts on some cells in synergism with PDGF. As with many other growth factors, EGF binds to its cell-surface receptor (an integral membrane protein of M_r 170 kD) which then aggregates into patches and is internalized by an energy-dependent process (endocytosis) before undergoing lysosomal degradation (Haigler, McKanna and Cohen, 1979). The clearance and degradation ('down regulation') of bound EGF may represent a mechanism limiting the response to the hormone, which is mediated via a transmembrane signalling system. The heart of this system is the EGF receptor, which is a tyrosine-specific protein kinase (Buhrow, Cohen and Stavros, 1982) capable of self phosphorylation. As illustrated by Downward *et al.* (1984) a fragment of the receptor for EGF (which includes the tyrosine kinase domain but lacks the external EGF binding domain) shows striking homology with the protein encoded by the v-*erb*B oncogene of the avian erythroblastosis virus (AEV). Apparently, this truncated version of the EGF receptor does not require EGF binding for activation. Thus transformation by AEV may be brought about by the production of a protein which functions much like the EGF receptor, but induces proliferation in the absence of binding of the ligand i.e. it is constitutively active.

Velu *et al.* (1987) have recently succeeded in inserting the EGF receptor gene into a retrovirus vector (in which the viral oncogene was deleted). NIH 3T3 cells were then transfected with the viral DNA and injected into nude mice. Transfected cells overexpressing EGF receptors (4×10^5 per cell) were capable of forming tumours under these conditions indicating that increased numbers of normal EGF receptors can contribute to the establishment of the transformed phenotype.

(c) *Transforming growth factors (TGFs)*

The transforming growth factors are polypeptides found originally in conditioned medium from cells transformed by retroviruses, such as the murine sarcoma virus (MSV). TGFs enable non-transformed cells to express at least some of the phenotype characteristic of transformed cells, particularly their ability to show anchorage-independent growth and to reach high density in culture (DeLarco and Todaro, 1978). Sarcoma conditioned medium actually contains two TGFs which are very different in their composition, receptor binding and functional activities. TGF_α (M_r 5600 D) competes with EGF for receptor binding, whereas TGF_β (about 25 kD) seems likely to bind to a different receptor. In fact, it now turns out that TGF_α shows significant homology to EGF (Marquardt *et al.*, 1983, 1984), and that the ability of sarcoma conditioned medium to stimulate anchorage independent growth resides mainly with TGF_β. TGF_α is produced by the human placenta and by rodent embryos, but in adult tissue it is only produced by certain transformed cells. It has been suggested that TGF_α may be an embryonic form of EGF. Since most tumours produce TGF_α rather than EGF, and if TGF_α is an embryonic form, then we may have another example of an oncodevelopmental antigen serving as a marker for the transformed phenotype (section 1.7.5.1). In at least one study, TGF activity was found in the urine of untreated cancer patients but not in the urine of non-cancer patients, suggesting that it may serve a diagnostic role in neoplastic disease (Twardzik *et al.*, 1982).

There are three variants of TGF_β resulting from different combinations of two subunits, and at least three different receptors (65 kD, 85 kD and a 280 kD complex). TGF_β is of particular interest as a growth factor since not only can it allow the anchorage-independent growth of normal cells, but it can both stimulate and inhibit cell division depending on various conditions and the cell type under study. Generally speaking, TGF_β inhibits the division of most cell types with the exception of some fibroblasts and osteoblasts. This yin-yang effect of TGF_β complements that shown by EGF and by other growth factors, indicating that a dual role for such molecules may be fundamental to their nature. In terms of controlling cell division, it may be that the expression of a particular sequence or pattern of growth factors could be more important than the simple expression of a single such factor.

As with many growth factors, the exact mode of action of TGF_β is uncertain. The induction of division by TGF_β requires almost twice as long as does that induced by EGF or PDGF, and in fact it is associated with the appearance of PDGF in the medium. It is suggested from these results that TGF_β induces the production of PDGF and that this latter molecule actually acts as the mitogen. Thus TGF_β appears able to affect cell growth by altering gene expression. The inhibitory activity of TGF_β was first shown by Holley and his colleagues (Holley *et al.*, 1980, 1983; Tucker *et al.*, 1984). The proposal has been put forward that negative autocrine effects mediated by TGF_β may influence division as the cells approach confluency, but the role that TGF_β might play in density control of cell division is not yet clear (section 1.7.3). Although many transformed cells are sensitive to the inhibitory effects of TGF_β some normal cells actually lose their responsiveness on transformation. Normal retinal cells, for example, are sensitive to TGF_β whereas retinoblastoma cells are not, but whether this is due to the actual loss of appropriate receptors or to the presence of abnormal ones is not clear.

What should be obvious from the foregoing is that the two TGFs are such diverse molecules that they should not be classified together. In fact, the term 'transforming growth factor' is rather a misnomer, given the bifunctional role of $TGF_β$, the fact that the ability to influence cell growth in agar is not necessarily predictive of transformation, and that this characteristic is not the sole preserve of the TGFs in any case.

(d) *Fibroblast growth factors (FGFs)*

The fibroblast growth factors are represented by a family of at least five related molecules. Original observations by Trowell, Chir and Willmer (1940) showed that crude brain homogenates contained mitogenic activity for fibroblasts, and later work by Armelin (1973) identified similar activity in the pituitary. The active principles were partially purified by Gospodarowicz, Mescher and Birdwell (1978), and subsequent research identified that different but related molecules were responsible for the activities of the different organs (Thomas *et al.*, 1980; Lemmon *et al.*, 1982). The two principal FGFs are identified as acidic (brain) or basic (pituitary) on their isoelectric points. They are both 16–18 kD monomers with mitogenic activity for mesodermal and neuroectodermal cells. It would seem that several supposedly different growth factors are probably FGFs. Endothelial cell growth factor, for example, is probably equivalent to a-FGF. The fact that FGFs stimulate angiogenic activity renders them of considerable significance in tumour development (section 1.7.20).

Production of b-FGF, which is abundant in some tissues, does not lead necessarily to excessive cell division. Like EGF and $TGF_β$, b-FGF has both stimulatory and inhibitory effects on cell division stimulating endothelial cells, for example, but inhibiting certain sarcoma cells (Schweigerer *et al.*, 1987). In culture, NIH 3T3 cells appear not to secrete b-FGF into the medium, a strategy possibly adopted to avoid autocrine stimulation. Rogelj *et al.* (1988) have recently reported transfecting NIH 3T3 cells with a plasmid containing b-FGF cDNA linked to a sequence which codes for an immunoglobulin secretory signal peptide. Transfection led to expression of a transformed phenotype, with the cells forming tumours in syngeneic mice. Interestingly, b-FGF was still not apparent in the medium, leading to the hypothesis that it might bind its receptor at an intracellular location.

Another member of the FGF family (*int-2*) has recently been identified following insertional mutagenesis of cells with MMTV. This oncogene is induced when the virus integrates adjacent to it on the host cell genome, since it is not carried by the virus itself. Another member is the oncogene *hst* which was first identified in a human stomach tumour and has since been shown to be identical with KS3, an oncogene of Kaposi's sarcoma which is a tumour often found in patients with the acquired immune deficiency syndrome (AIDS). The significance of the relationship between these oncogenes and the original FGFs is uncertain, but their expression may lead to angiogenic induction therby facilitating tumour growth and spread.

Now if we accept that extracellular signals communicate with the cell through some form of receptor, then we can postulate a number of ways by which the normal regulatory mechanisms could be altered to promote cell division:

(a) There could be an increase in production or availability of positive controlling factors which could be produced by the tumour itself (autocrine stimulation) or by a secondary cell (paracrine stimulation). In many instances of paracrine stimulation, growth factor production is probably by normal cells which, like macrophages, may be integral within the stroma of the tumour. Clearly, the removal of such normal cells as often happens during the *in vitro* culture of malignant tissues could have profound effects on the behaviour of the tumour cells, and will tend to increase the artificiality of the *in vitro* system.

(b) The expression of receptors for such positive factors on the tumour cells may increase either in terms of their number, their accessibility, affinity, or in terms of their activity.

(c) Alternatively, there may be a decrease in the level of production or availability of a specific **chalone** allowing the tumour to escape from negative control. Chalones are thought to play a role in growth control by inhibiting cell division, mainly through arrest in G_1 or G_2 (Bullough and Laurence, 1960; Laurence, 1979). They are thought to be produced by differentiated cells but to act on cycling cells through negative feedback inhibition. Thus it is envisaged that chalone levels reflect the number of differentiated cells in a tissue, with the presence of more tissue cells leading to greater suppression of division. Although the chalone theory has many attractive features, current research is directed more towards the study of positive control mechanisms. Nevertheless, the demonstration that a transforming growth factor (TGF_{β}) can inhibit cell division (including the division of cells which secrete it) has brought research on chalones back into focus. Indeed, there is mounting evidence that in normal cells negative controlling factors dominate over positive ones and that perturbation of this relationship leads to transformation. Strictly speaking, these controlling factors may not turn out to be true chalones, which are classically characterized on the possession of four attributes:

1. Total cell specificity
2. Species non-specificity
3. Reversible effects
4. Non-cytotoxic action

(d) Receptors for the postulated chalone may decrease in number or change their affinity, accessibility or activity.

(e) There may be some alteration within the tumour cell so that negative signals fail to be acted upon and thus the cell will only respond to positive signals. This possibility and its variants will be described in more detail later when intracellular signals are discussed more fully (section 1.7.22), but the lack of an inhibitory response to TGF_{β} shown by some tumours may be an appropriate example.

(f) Some combination of these possibilities may be involved.

Evidence from the studies with virally transformed cells has offered support for a number of these possibilities. Thus transformation with SSV leads to the production of a molecule with similarities to a growth factor, while transformation with AEV leads to the expression of a molecule which is similar to that part of a growth factor receptor thought to be involved in the generation of the intracellular signal. In the latter case, the signal for cell division could be generated autonomously in the absence of EGF binding. Interestingly, very high numbers of EGF receptors occur in squamous carcinomas of the head and neck, as well as in the A431 squamous carcinoma cell line. Monoclonal antibodies (McAbs) directed against high-affinity EGF receptors on A431 cells can inhibit their division *in vitro* and also their growth *in vivo* in nude mice (Masui *et al.*, 1984). In a similar approach, McAbs directed against bombesin (another growth factor) have been tested for their effects on a cell lines derived from small cell carcinoma of lung. Bombesin is a 14-amino-acid peptide first found in the skin of the frog *Bombina bombina*. In mammals, bombesin-like activity has been attributed to a 27-amino-acid-long molecule known as gastrin-releasing peptide (GRP) which is partially homologous with amphibian bombesin. At least some human small cell carcinoma of lung cell lines secrete bombesin-like material and have receptors for it, indicating a possible autocrine role. McAbs directed against bombesin block the binding of bombesin-like material to receptors on cells from these lines, decrease their clonal growth *in vitro*, and reduce tumour growth in nude mice (Cuttitta *et al.*, 1985). Although these studies illustrate the importance of growth factors in tumorigenesis, the antibody based approach used is unlikely to be clinically significant unless the McAbs can be directed exclusively towards tumour cells.

One other protein thought to be involved in growth control deserves mention here, and that is p53 which has been found in SV40 transformed cells amongst others (reviewed in Rotter and Wolf, 1985). Cells transformed with SV40 express two tumour antigens (known as large T and small t) which are proteins encoded by the virus. The large T antigen is essentially the viral oncogene product and it immunoprecipitates in conjunction with p53. Although p53 is also found in non-transformed cells, it is usually in small amounts and has a shorter half life than when it is found in transformed cells. Furthermore, its distribution appears to be different in normal and transformed cells, being found mainly in the cytoplasm of normal cells and the nuclei of transformed cells (Rotter, Abutbul and Ben Ze'ev, 1983). The nuclear location of p53 is not dependent on large T, since it is still located in the nuclei of cells which have been transformed by viruses which do not code for the large T antigen. Embryonic cells appear to produce more p53 than adult cells suggesting that differentiation may be accompanied with a reduction in the level of its expression. It also has been suggested that p53 is involved in immortalization as well as in other aspects of transformation. Thus p53 is able to immortalize rat chondrocytes which normally become senescent *in vitro* (Jenkins, Rudge and Currie, 1984) and its overexpression in the immortalized Rat-1 cell line is associated with the induction of tumorigenicity as assayed in nude mice (Eliyahu, Michalovitz and Oren, 1985). It should be noted here that Rat-1 cells themselves can form tumours in nude mice, albeit rarely and usually with a long latency, and thus p53 may only enhance latent tumorigenic potential. Parada *et al.* (1984) were able to show that p53 could complement activated c-Ha-*ras* in successfully transforming rat embryo fibroblasts, but the significance of this observation is open to question since it has now been shown that the activated oncogene alone can transform primary cells. Rotter, Abutbul and Wolf (1983) have described an apparently immortalized cell line (L12) transformed by Abelson murine leukaemia virus which lacks detectable p53. When these cells were injected into syngeneic mice, tumours grew at first but they were subsequently rejected. Prior treatment with the tumour promoter 12-O-tetradecanoyl-phorbol-13-acetate (TPA) resulted in the establishment of a variant cell line which expressed p53 and formed aggressive tumours. Other immortalized cells such as the human promyelocytic cell line HL-60 also fail to produce p53 leaving some uncertainty as to its precise role. Nevertheless, some function for p53 in DNA synthesis seems likely given the observation that anti-p53 antibodies inhibit DNA synthesis when injected directly into cell nuclei at or around the time of mitogenic stimulation (Mercer *et al.*, 1982). Despite these intriguing correlations, it must be admitted that precisely how proteins such as p53 affect division and transformation is far from clear. It would seem that in many cases the final story will be quite complex, involving cooperative activity with other cell products. Inteestingly, as research into the function of p53 continues information is coming to light which suggests that this molecule may be the product of an anti-oncogene rather than that of an oncogene. Apparently one early study implicating an immortlizing role for p53 unsuspectingly utilized mutant p53 genes, and thus it was p53 malfunctioning which allowed *ras* transfected cells to express the transformed phenotype (see above). In other words, p53 may act under normal circumstances to suppress tumourigenicity, although the situation is not yet fully resolved.

The growth responses of cells are not only influenced by soluble extracelluar signals, and indeed there is considerable evidence that populations of cells may also respond to signals delivered through intercellular contact. This form of communication may involve membrane-bound signalling molecules which interact when cells come into contact, or it may involve cytoplasmic messengers which pass between cells through channels such as those provided by gap junctions. Plasma membranes from 3T3 cells inhibit DNA synthesis in non-transformed variants but not in their SV40-transformed counterparts (Whittenberger and Glaser, 1977). The proliferation of endothelial cells is inhibited by plasma membranes with some evidence for specificity in terms of cell type and species (Teitel, 1986). The possibility that the added membranes might compete for growth factors thereby reducing cell division was tested experimentally by adding excess endothelial cell growth factor (ECGF) which was without effect. Other factors or phenomena (including cytotoxicity) may be involved, however, and it would be premature to conclude that membrane–membrane interactions limit proliferation when non-transformed cells reach confluence. The role of membrane–membrane interactions in controlling *in vitro* cell density is discussed later (section 1.7.3).

1.7.2 Chromosome abnormalities

The number of chromosomes in malignant cells often deviates from the diploid state and there may be frequent abnormalities in chromosome structure and shape, some of which are consistently associated with particular types of tumours (summarized in Mitelman, 1983). These abnormalities reflect genetic instability, and it is pertinent to ask its source: does it result from an inherited trait or is it an acquired phenomenon? Many patients with cancer show no signs of generally increased chromosomal breakage, suggesting an acquired basis for the phenomenon of instability. Patients with genetic disorders such as ataxia telangiectasia, however, inherit a tendency for increased chromosomal breakage and they are known to be at high risk for developing neoplastic disease. Such evidence points towards a constitutional basis for genetic instability, and thus we must conclude that both acquired and inherited phenomena probably influence genetic instabilty.

Considerable research is currently being undertaken in the general area of chromosomal abnormalities and cancer, but it will suffice to refer to a few selected examples in order to illustrate the direction and extent of progress. Chromosome aberrations result from point mutations, gene rearrangements or from chromosome duplications. The first transforming gene isolated from human tumour cells was a point mutated c-Ha-*ras* oncogene whose product ($p21^{ras}$) had a single amino acid substitution at either position 12 or 61 (reviewed in Duesberg, 1985). A similar activated c-Ha-*ras* gene was later found in clinical tumour samples, but not in normal tissues from the same patients, suggesting that the point changes involved arose by somatic mutation (Fujita *et al.*, 1985). It now turns out that point mutations at various positions including 12, 13, 59, 61 or 63 of p21ras can all activate the transforming potential of c-Ha-*ras*, with the proviso that the substitution of some amino acids (such as proline) may not result in transformation. The role of the *ras* gene product in transformation and metastasis is discussed further in section 1.7.22.

Structural changes leading to gene rearrangements involve chromosomal breakpoints which show an association with identifiable fragile sites on the chromosome (reviewed in Le Beau, 1986). There are two main classes of fragile sites: the heritable fragile sites which are rare and segregate in a simple Mendelian fashion, and the common fragile sites which are frequent and may well be caused by environmental factors such as chemicals, radiation and viruses. The association between fragile sites of either type and cancer-specific chromosomal breakpoints has led to the suggestion that the presence of fragile sites may predispose an individual towards the development of cancer. If this is so, it might be expected that genes involved in cancer may be at or near the fragile sites. Many breakpoints (and therefore presumably many fragile sites) have been shown to occur at or near c-*onc* genes but other genes (including those for IFN-α and IFN-β) are also present near some fragile sites associated with particular tumours. At present, the precise functions of these latter genes in cancer development are not clear, although there is some evidence that their products might regulate oncogene expression.

The most common structural rearrangement is a translocation of DNA between chromosomes, although inversions, deletions, insertions and duplications also occur. Duplications occur in a number of forms. Some may involve the loss of one arm of a chromosome with duplication of the other leading to the formation of an isochromosome. Others may involve the duplication of DNA segments leading to gene amplification, and there is evidence that this occurs at a higher frequency in tumour cells relative to normal cells (Sager *et al.*, 1985). Amplified DNA is usually (but not always) associated with two main chromosomal abnormalities which may be detected cytochemically in chromosome spreads (Schwab, 1985):

(a) *Double minutes (DMs)*

These are small, spherical acentric chromosomes.

(b) *Homogeneously staining regions (HSRs)*

Located on single chromosomes, as their name suggests they do not show typical chromosome banding patterns after Giemsa staining. It should be noted that DNA amplification is not synonomous with elevated gene expression, although the latter usually follows as a consequence of the former. Indeed, c-*onc* gene amplification is often associated with their increased expression. The amplification and increased expression of c-*myc*, for example, is seen in the human acute promyelocytic leukemia cell line HL60, and N-*myc* may be amplified several hundred times in neuroblastoma (reviewed in Alitalo, 1984). Generally speaking, however, the frequency of amplification of oncogenes is not high, being found in less than 10% of human tumours so far examined. Interestingly, greater than 90% of all amplifications studied so far involve *myc* genes, and amplifications tend to be associated more with tumours in advanced rather than early clinical stages. Duplications of whole chromosomes rather than chromosomal fragments are also found in certain tumours. Duplication of chromosome 8 (referred to as trisomy 8), for example, has been found in association with some forms of acute (AML) and chronic (CML) myelocytic leukaemia: noteworthy here is that chromosome 8 in the human contains c-*myc*.

Translocations (t) occur between two chromosomes at breakpoints on the short (p) or long (q) arms of each chromosome and this can be referred to in shorthand as shown, for example, in t(9;22)(q34;q11). In its decoded form this refers to a reciprocal translocation between chromosomes 9 and 22 with breaks on the respective long arms at bands 34 and 11. Bands are identified by histological staining of spread chromosomes with Giemsa (G bands) or with quinacrine derivatives (Q bands). The translocation t(9;22)(q34;q11) is found in most cases of CML. This haematopoietic tumour is characterized by a chronic phase lasting about 3–5 years, in which there is expansion of committed myeloid progenitor cells leading to the accumulation of granulocytes in the blood. Most patients then develop a fatal 'blast crisis' in which there is accumulation of immature myeloid and lymphoid blast cells. Karyotypic analysis of patients with CML has identified an unusual chromosome, the Philadelphia chromosome (Ph^1), which is a short chromosome 22 resulting from the translocation.

Translocation from chromosome 9 to 22 involves an *onc* gene (c-*abl*) which is the cellular homologue of the Abelson murine leukaemia virus (Ab-MLV). In fact, c-*abl* from chromosome 9 is fused to the breakpoint cluster region (bcr) on chromosome 22 to generate a *bcr-abl* hybrid gene whose product (p210) has an N-terminal region derived from *bcr* and a C-terminal region from c-*abl*. The function of *bcr* is unknown, and although p210 has elevated tyrosine kinase activity (compared to the normal c-*abl* gene product) its actual role in CML is also uncertain. Interestingly, the reciprocal translocation (chromosome 22 to 9) involves c-*sis* (the cellular homologue of v-*sis*, the *onc* gene of SSV), but its transcription appears not to be affected by the translocation. As CML progresses, additional karyotypic abnormalities become apparent. These include a second Ph^1, trisomy 8, and the development of a 17q isochromosome, but their functional significance remains uncertain.

Burkitt's lymphoma is a lymphoid cell tumour in which there is a high association with a translocation between chromosomes 8 and 14 (80% of cases) or 8 and 2 (5%) or 8 and 22 (15%). In t(8;14)(q24;q32) patients, the cellular *onc* gene c-*myc* on band 24 of the long arm of chromosome 8 is translocated to the immunoglobulin (Ig) heavy chain locus on chromosome 14. In the other two variant translocations, c-*myc* remains on chromosome 8 and the Ig light chain loci (kappa on chromosome 2 and lambda on chromosome 22) are translocated to its vicinity. In these translocations the regulatory sequence of c-*myc* is cut and control shifts to the new chromosome. Since immunoglobulins are produced extensively in lymphoid cells of B lineage, it has been suggested that this will result in enhanced transcription of c-*myc* because of its fortuitous position (Rabbitts *et al*., 1983; Robertson, 1983). In some Burkitt lymphomas the c-*myc* oncogene is altered structurally, while in others it is apparently normal.

In most cases, however, there is a change in c-*myc* regulation leading to elevated constitutive expression. Since the product of c-*myc* is thought to be involved in cell proliferation, the translocation probably drives the unrestricted proliferation of the affected B lymphocytes, although exactly how is not clear.

Further studies on the genetic basis of cancer have made use of the technique of cell hybridization in which normal cells can be fused with malignant cells by agents (fusogens) such as inactivated Sendai virus or polyethylene glycol (PEG). Cell hybrids at first contain the chromosomal complement of both cells, but with time chromosomes are lost. The retention or loss of particular chromosomes can be correlated with certain cell characteristics such as malignancy, allowing such behaviour to be mapped by chromosomal analysis (reviewed in Harris, 1988). Cell hybridization between normal and malignant cells yields non-malignant hybrids as long as they retain the complete chromosome complement of both parent cells. However, malignancy is expressed in cell hybrids if some normal chromosomes are lost and these results together therefore suggest that malignancy in some cases behaves as a recessive character. Chromosomal analysis has suggested that loss of chromosome 4 is implicated in malignant reversion in some mouse hybrid cells (Evans *et al.*, 1982) and loss of chromosome 1 in some human hybrids (Benedict *et al.*, 1984). This is an intriguing observation since extensive homology exists between these two particular chromosomes, but it must be noted that the loss of other chromosomes has been implicated in malignant reversion of other hybrid crosses. The expression of a normal phenotype in hybrids of D98AH2 human uterine carcinoma cells and normal human fibroblasts seems to be under the control of chromosome 11 (Stanbridge *et al.*, 1981). Furthermore, a single copy of a normal fibroblast chromosome 11 has been shown to suppress tumorigenicity in cultured HeLa cells following microcell fusion (Saxon, Srivatsan and Stanbridge, 1986). There is some evidence here for a gene doseage effect in that hybrids with two copies of this chromosome show more stable suppression of the malignant phenotype than those with one. It is worth raising the problem of monoclonal antibody producing hybridomas here. These cells result from the fusion of a normal lymphocyte with a myeloma cell, but they are clearly malignant in nature. In keeping with the general hypothesis of Harris (1988), it is assumed that hybridomas have undergone a degree of chromosome loss in order to allow the expression of malignant characteristics. As yet, however, definitive chromosome analysis of these cells has not been reported.

Recessively acting genes are also thought to be involved in Wilms' tumour and in retinoblastoma, two tumours which are associated with the deletions del(11)(p13) and del(13)(q14) in chromosomes 11 and 13 respectively (Cavanee *et al.*, 1983). Retinoblastoma is a highly malignant paediatric tumour which metastasizes by direct extension along the optic nerve into the central nervous system, although it also may give rise to secondaries in the liver. It occurs in both non-heritable and inheritable forms, with inheritance in the latter occurring in an autosomal dominant manner (reviewed in Nyhan, 1987). The disease develops unilaterally in non-heritable cases, whereas about 80% of children with the heritable form develop tumours in both eyes. Some time ago Knudson (1971) put forward the suggestion that *two* changes in the genome were required for expression of dominantly inherited cancers such as retinoblastoma. His general proposal was that the heritable and non-heritable forms of these tumours resulted from mutations in the same gene, and that the mutations acted recessively at the cellular level, with both copies of the gene having to be lost for cancer to develop. In non-heritable cases of retinoblastoma, both the required mutations are thought to occur somatically in a single retinal cell. Since this is an unlikely event it probably explains why retinoblastoma is relatively rare. In heritable cases, retinal cells are predisposed towards the malignancy because they carry a specific mutation, namely del(13)q(14), which affects one of the two copies of the gene involved. Retinoblastoma would develop in such cases after a somatic mutation in the homologous allele.

Current theory proposes a regulatory gene (the *RB1* gene) on the long arm of chromosome 13 whose product $p105^{RB}$ normally prevents the development of the tumour. The *RB1* gene is thus an example of a tumour suppressor gene. Normal individuals carry the wild-type regulatory gene on both members of the chromosome 13 pair. In the first step towards malignancy, the regulatory gene on one of the chromosome pair is believed to be lost or altered as a consequence of the long arm deletion. Such an individual would be heterozygous, with the normal allele dominant. The second genetic event may involve any of several alterations, including deletion, mutation or inactivation, but the end result is the functional loss of the remaining normal allele which allows the malignancy to be expressed.

Essentially similar events are believed to occur in Wilms' tumour, a paediatric nephroblastoma which also occurs in heritable and non-heritable forms. The first step involved in the development of this tumour is a deletion of the short arm of chromosome 11, and the second step involves alteration of the remaining wild type chromosome 11. Recently, Weissman *et al.* (1987) have succeeded in introducing a single, normal human chromosome 11 into a Wilms' tumour cell line. The hybrid cells generated had similar morphology to the parental

Wilms' tumour cell line, they produced plasminogen activator in similar amounts, and they had a similar distribution of cell surface fibronectin. There was, however, a decrease in the ability of the hybrids to grow in soft agar and (unlike the parental line) they were completely unable to form tumours in nude mice. These results appeared to be specific for chromosome 11, since introduction of the X chromosome or chromosome 13 was without effect on the *in vitro* and *in vivo* behaviour of the Wilms' tumour cell line. This study clearly supports other observations that chromosome 11 controls the expression of malignancy in Wilms' tumour, and it also indicates the poor correlation between many *in vitro* transformed cell characteristics and tumorigenicity *per se*, as discussed elsewhere in this chapter.

Interestingly, the hybrid cells used in the study of Wilms' tumour by Weissman *et al.* (1987) did not show any obvious correlation with the expression of a number of cellular oncogenes including *ras, myc, src* and *erb*B, which leaves some doubt as to the precise role of their products in the regulation of tumorigenicity. Furthermore, when malignant cells containing known oncogenes are fused with diploid fibroblasts, the expression of malignancy in the hybrid does not correlate with oncogene activity (Geiser *et al.*, 1986). Yet cell transformation has been demonstrated experimentally by transfection techniques in which DNA from tumour cells is transferred to non-transformed cells (Cooper, 1982), and in some cases transformation has correlated with the presence of specific oncogenes (reviewed in Bishop, 1987). These transfection studies suggest that neoplastic transformation with oncogenes occurs in an essentially dominant fashion, in contrast to the recessively acting effects noted in the studies with hybrid cells referred to above. As pointed out by Harris (1988) and Skuse and Rowley (1989), however, the terms 'dominant' and 'recessive' are not used here in the classical Mendelian sense.

A compromise hypothesis would suggest that carcinogenesis might involve recessive genetic lesions which lead to the inactivation or removal of a regulatory gene which then allows the expression of a dominantly acting oncogene. Based on their study of the clonal nature of human colorectal tumours, Fearon *et al.* (1987) proposed that the initial event in neoplastic growth arises as a benign lesion in a single colonic cell, possibly arising as a consequence of DNA hypomethylation (section 6.1). Most human colon carcinomas progress through three recognizable stages: a benign, tubular adenomatous (polyp) stage, a villous stage (which contains patches of carcinoma tissue), and an invasive stage. During benign adenomatous growth a mutation of the c-Ki-*ras* gene may take place, and final progression from adenoma to carcinoma may result from the loss of tumour suppressor genes on chromosome 17p or occasionally on other chromosomes. Interestingly, since not all experimentally transfected cells can form tumours *in vivo*, it has been speculated that those which do might represent a highly selected subpopulation conceivably lacking effective supressor genes. Although this scenario brings together many of the features of transformed cells described above and elsewhere, it is obviously a rather speculative model in need of more detailed study. Other studies with familial adenomatous polyposis (FAP) have located the gene responsible to chromosome 5q (Bodmer *et al.*, 1987). Like the *RB1* gene, it is thought that the gene involved in FAP is a suppressor gene (reviewed in Astrin and Costanzi, 1989). Available evidence suggests that the loss of a single FAP allele leads to polyp formation, whereas the loss of both alleles is associated with the development of colon cancer (in about 40% of cases). Interestingly, there is some evidence that the loss of the gene associated with FAP correlates with deregulation of c-*myc* expression.

Children with the heritable form of either retinoblastoma or Wilms' tumour are at risk for the development of other tumours such as osteosarcoma and hepatoblastoma respectively. It would thus seem that the gene loci involved in these paediatric malignancies may also be associated with other cancers.

1.7.3 *In vitro* density

When an apparently normal tissue is dissociated and the constituent cells (e.g. fibroblasts) are cultured *in vitro* (usually in the presence of 10% serum), the most striking feature is that the cells grow to confluence (i.e. they form a monolayer on the surface of the culture vessel). Not all normal tissue cells form monolayers of this sort as is shown, for example, by cells from the bone marrow which form cultures several layers deep (Dexter and Testa, 1976) and by BHK cells which can form layers 5–10 cells thick (Stoker, 1967). Of course, most tissue cells such as fibroblasts do not form monolayers *in vivo* so this observation is by and large an *in vitro* phenomenon. However, the ability of cells (particularly those of fibroblastic origin) to pile up on the vessel surface is characteristic of cultures prepared from tumours or from normal cells which have been transformed experimentally.

It is still controversial as to exactly what mechanisms underlie this *in vitro* difference in growth behaviour between normal cells and tumour cells and it may be that several phenomena are involved. Since cells which pile up in culture reach a higher density, the control of cell division in normal cells may reflect some form of density-dependent control of the division process (Stoker and Rubin, 1967). It would thus be necesssary to suppose that this mechanism of control is defective in some way in transformed cells. Density-dependent inhibition of division should not be confused with contact inhibition (section 1.7.4) which is a behavioural response elicited when certain cells come into contact. BHK cells, for example, display homotypic contact inhibition of movement even though they can pile up to form multilayers *in vitro*. The term **topoinhibition** has been used to refer to the suppression of cell growth at confluence, and it is taken to be synonomous with density-dependent inhibition of division. Interestingly enough, most cells decrease in size as they grow to increasing density, but whether this is a cause or an effect of topoinhibition has not been examined. Some possible control mechanisms of topoinhibition include:

(a) *Soluble signals*

Cells may secrete messenger molecules which restrict the density of cell growth. This could be achieved by a balance of positive and negative signals and it has been postulated that chalones may act in this manner (see section 1.7.1). In this situation, transformed cells would be envisaged to have escaped from the effects of the negative signals or have become more sensitive to the effects of positive signals.

Recently, Lipkin and colleagues have prepared serum-free supernatants from a confluent hamster melanoma cell variant which does not overgrow *in vitro*. Some component(s) of this supernatant can induce serum-, density- and anchorage-dependent growth in cells of the parental RPMI 1846 hamster melanoma line (Lipkin and Knecht, 1974; Lipkin, Rosenberg and Klaus-Kovtun, 1986). These results imply the involvement of some sort of soluble signal, but despite many years of study virtually nothing is known in detail about the active principles involved. Interestingly, the vitamin A analogue retinoic acid has been shown to restore density- and anchorage-dependent growth to transformed mouse L929 cells (Dion, Blalock and Gifford, 1977, 1978) which at least provides a parallel for the results achieved with melanoma supernatants.

(b) *Contact dependency*

Since normal cells *in vitro* do not usually grow beyond contact, the controlling mechanism may be mediated just as well via direct membrane–membrane interaction rather than through the effects of soluble signals. Early evidence in support of this possibility came from studies of ‘wound healing’ in confluent cultures of 3T3 cells. When a narrow strip of a confluent monolayer of 3T3 cells is removed with a razor blade, new growth begins in those cells which migrate from the edge of the confluent sheet into the denuded area (Todaro, Lazar and Green, 1965). For an individual cell, the initiation of division can be correlated with loss of contact with its neighbours (Dulbecco and Stoker, 1970). These results suggest that cell contact induced the cessation of growth, but as will become clear in due course other studies have suggested that more than mere cell contact is involved in establishing confluence.

The mouse 3T3 cell line has been used in many studies relating to the effects of density on cell growth. This line was originally selected by repeated passage at low cell number of mouse embryo cells, and its saturation density is dependent upon serum concentration (Todaro and Green, 1963). The growth of 3T3 cells, however, can also be inhibited by 3T3 cell membrane preparations and these result in the accumulation of cells in G_1 (Whittenberger and Glaser, 1977). This result suggests that cell contact may somehow regulate cell division and, by stopping cells from dividing when they come into contact, effectively induce density-dependent inhibition of growth. The active component(s) may be solubilized from 3T3 cell membranes by extraction with the nonionic detergent octylglucoside (Whittenberger *et al.*, 1978). The extracted component has been found to be cross-reactive in its activity with cells from the IMR91 normal human fibroblast line (Lieberman, Raben and Glaser, 1981), but like 3T3 cell membranes it does not inhibit the rate of division of SV40-transformed 3T3 cells. However, because the growth of 3T3 cells can be reinitiated by treatment with growth factors (Vogel, Ross and Raines, 1980), it would seem that phenomena other than contact-mediated ones also contribute towards limiting the density of cell growth of this particular cell line. Indeed, the inhibition of 3T3 cell growth by 3T3 cell membrane extracts is counteracted by the addition of excess serum (Whittenberger *et al.*, 1978). More recently, Hsu and Wang (1986) have isolated a 13 kD polypeptide from densely cultured Swiss 3T3 cells which inhibits division by 50% at as little as 3 ng ml^{-1}. Its effects are non-cytotoxic and reversible, but its relationship with other postulated controlling factors is uncertain.

Schmialek *et al.* (1977) followed the kinetics of normal and spontaneously transformed hamster embryo cell division after cell contact. These authors reported that on contact the cells were induced to divide once more, after which about half the culture died to achieve a particular cell density. These intriguing results need to be explored in more detail, but they are indicative of cell contact influencing division.

Although endothelial cells respond to growth factors during their growth phase, their saturation density is independent of serum concentration and growth cannot be reinitiated by the addition of growth factors (Haudenschild *et al.*, 1976). The cell-surface proteins of endothelial cells change in composition as the cells attain confluence (Vlodavsky, Johnson and Gospodarowicz, 1979) and a urea extract of such cells yields material which inhibits endothelial but not smooth muscle cell division in a reversible manner (Heimark and Schwartz, 1985). Since the extract contains a number of proteins it is uncertain as to where the activity resides, but evidence of this sort strongly implies a role for cell membrane proteins in the density-dependent regulation of cell growth.

Interestingly, there is some evidence that the post-transcriptional regulation of c-*myc* expression may be modulated, at least in part, by cell–cell contact (Dean *et al.*, 1986). Quiescent 3T3 cells obtained at confluence or at subconfluence by serum restriction respond to the addition of fresh serum by the induction of c-*myc* transcription. The level of transcription then falls, but the steady-state accumulation over 18 h of *myc* RNA in the subconfluent cells is much greater than that in the confluent cells. The activation of c-myc is thought to be involved in competence, but exactly how cell contact influences its expression is not yet known.

Monolayers of confluent fibroblasts fixed with glutaraldehyde inhibit the growth of freshly seeded cells (Culp *et al.*, 1978). This inhibition is blocked after treatment with agents such as heparitinase, possibly suggesting some involvement of heparan sulphate (HS) in the control of cell division and the establishment and maintenance of confluent monolayers *in vitro*. Heparan sulphate from normal rat liver inhibits the growth of confluent hepatoma cells (Kawakami and Terayama, 1981) and the growth of smooth muscle cells is inhibited by HS prepared from confluent (but not sub-confluent) cells of the same type (Fritze, Reilly and Rosenberg, 1985). Although HS might influence cell division through its effects on cell adhesion (see later), it may also be involved more directly since HS inhibits DNA polymerase activity *in vitro* (Furukawa and Bhavanandan, 1983). Whether membrane-located HS can influence this activity or not, and indeed whether the interaction between HS and DNA polymerase has any physiological relevance at all remains uncertain. Because of its chemical nature, HS can be expected to interact with many proteins and the possibility that its effects may be due to non-specific binding of molecules such as growth factors (thereby limiting their access to the cell) remains to be studied in detail.

Hakomori (1970) and others have reported that some glycosphingolipids (section 1.7.5.2) may increase in length when non-transformed cells such as the BHK line come into contact. This 'contact extension' of glycosphingolipids is apparently not seen when transformed cells come into contact. The phenomenon is thought to result from increased activity of enzymes associated with oligosaccharide synthesis (glycosyltransferases) and/or a decrease in enzyme activity (such as sialidase) associated with their breakdown. However, some tumour cells show a similar response to normal cells in their glycosphingolipid changes and the relationship between this response and cell density in culture is far from clear. Furthermore, there is some evidence that BHK cells may be able to form tumours in appropriate hosts.

(c) *Adhesion and cell shape*

Anchorage dependency (i.e the inability to grow in suspension) is a characteristic feature of many normal cells. This suggests that cell division normally requires an appropriate substrate, and it may be that a number of cytokinetic events are based on the generation of intracellular tensions which could benefit from the anchoring of a cell to its substrate (Maroudas, 1973). In a rather ingenious experiment, Curtis and Seehar (1978) grew chick embryo fibroblasts as 'sails' on nylon mesh, which they then repeatedly stretched by coupling it to an oscillating piezoelectrode. The mesh area remained the same on distortion, thereby avoiding changes in cell density, but both the mitotic frequency and the proportion of cells in S phase were found to increase on stretching. These experiments suggest that there are mechanical effects on division, but their precise contributions remain to be determined.

It is possible that the surfaces of some cells are not appropriate substrates for the adherence and spreading of other cells, and indeed the apical surfaces of epithelial cells are considered to represent a very poor substrate for other cells (Vasiliev and Gelfand, 1978). When epithelial cells reach confluence, it is highly unlikely that they will continue to divide if they cannot use the upper cell surface as a substrate. If other normal cells are seeded onto an already confluent monolayer of epithelial cells they too should fail to divide because the nature of the epithelial surface effectively generates non-anchorage conditions. Interestingly, Dulbecco and Elkington (1973) found that the division of epithelial cells was limited by the surface area of the culture vessels whereas this was not the case for fibroblasts, which were limited by the supply of medium components (see below).

Carter (1965) suggested that multilayering may be inversely related to substrate adhesiveness. Transformed cells may show a considerable degree of overlapping of each other (perhaps leading to multilayering and therefore to greater density) because they are often poorly adherent to the substrate. Normal cells are thought to be more tightly adherent to the substrate, particularly around their edges, and are thus less likely to overlap which could serve to limit their density (Harris, 1973; Bell, 1977). It must be admitted, however, that these interpretations are rather speculative since ovelapping depends not only on the adhesiveness of the substrate, but also on that of the cell surface.

The effects of adhesion and cell shape on division were examined in other experiments by Folkman and Moscona (1978). These authors altered the adhesiveness of tissue culture plastic by treatment with different concentrations of poly (2-hydroxyethyl methacrylate) i.e. poly(HEMA). They found that this affected 3T3 cell shape and that the less adhesive (i.e. taller or more spheroidal) the cell the less DNA synthesis took place. In their experiments, sparsely seeded cells could be growth-inhibited by culturing on relatively high levels of poly(HEMA). The height of these cells was similar to that seen in normal confluent cultures where division had ceased, and where the cells were relatively tall because overcrowding did not allow them to spread properly. Folkman and Moscona (1978) argued that cell height affected cell division, and that this could be controlled either by altering the substrate or by allowing the cells to reach confluence. The growth of SV40-transformed 3T3 cells (SV-3T3) was not found to be significantly affected by poly(HEMA), from which it was suggested that transformation somehow uncoupled the implied relationship between cell shape (height) and growth. It seems likely, however, that something else other than cell shape is involved in the establishment and maintenance of confluence, and many would argue that this is a consequence of the inadequate supply of nutrients and growth factors.

(d) *Nutrients and growth factors*

It has been argued by a number of authors that cells which can get by with lower nutrient requirements should reach higher densities (Todaro *et al.*, 1965; Stoker and Rubin, 1967; Dulbecco and Elkington, 1973). Autocrine stimulation, which has been shown to occur in some transformed cells, may allow them to grow to high densities in the absence of exogenous growth factors, although clearly there will still be a limitation to growth set by the level of available nutrients and the build up of toxic metabolites.

Stoker (1973) has argued on hydrodynamic grounds for the existence of a diffusion boundary layer at the cell surface. The boundary layer arises because fluid convective movement approaches zero near a solid surface. Thus the transfer of molecules to and from a cell surface must occur predominantly by diffusion, which is a relatively slow process, and key molecules may be used up by the cell before they can be replaced locally. In other words, the presence of a boundary layer would be expected to impair the access of growth factors and nutrients to confluent cells thus serving to limit cell division. Alternatively, the boundary layer could prevent the escape of inhibitory molecules secreted by the cells which could again result in the inhibition of division. In practice these two possibilities cannot be separated, but suppose we assume that cell division is limited by the restricted access of a growth factor to the cell surface. Three parameters control the nature and effects of the boundary layer. These are the diffusion coefficient of the growth factor involved, its rate of uptake by the cell, and the flow rate of the culture medium over the cell surface. For similar cultures the effects of the first two parameters are invariable, but the flow rate over the culture can easily be varied. Stoker (1973) used a small pump to produce a jet of medium over a culture of 3T3 cells and he found that this increased division locally. This is in accordance with the presence under static conditions of a more effective boundary layer limiting the supply of the postulated growth factor. Dunn and Ireland (1984) provided further evidence that the growth of 3T3 cells is a diffusion limited process by using a rotating disc to establish a laminar flow over the surface of a wounded monolayer. These authors found that cell division was greatest in the wounded edges where the direction of flow was off a denuded region rather than a confluent region, presumably because the proposed growth factor was not taken up in the denuded zone. Whittenberger and Glaser (1978), on the other hand, have argued against a role for diffusion in limiting cell density. They reasoned that if Stoker's hypothesis was correct, then increased fluid viscosity should decrease cell saturation density. However, they were unable to show any effects of as much as a 25-fold increase in viscosity on both saturation density or on the rate of growth of 3T3 cells. Unfortunately, we must admit a certain amount of ignorance at what occurs in cell cultures at the molecular level, and this does not really allow us to reach any firm conclusions in this area.

The studies of Loo *et al.* (1987) in which embryonic mouse cells were grown as multilayers in serum-free medium supplemented with EGF support the idea that a lack of growth factors may contribute to the failure of established cell lines to form multilayers *in vitro*. Not all cells, however, need necessarily be sensitive to growth factors and nutrient levels to the same extent. Indeed, Dulbecco and Elkington (1973) found that whereas 3T3 cells were limited in growth by medium components, SV-3T3 cells were affected little by either

Table 1.5 Factors influencing cell density *in vitro*

1. The extent of general nutrient supply and the requirements of the cells involved
2. The rate of production and build up of metabolic waste
3. The requirements of the cells for specific growth factors, which may be met endogenously or exogenously
4. The requirements of certain cells for particular substrates such as collagen, coupled with the availability of such substrates
5. The extent of anchorage dependency and independency of the cells involved
6. Responsiveness to contact phenomena such as membrane–membrane signalling
7. Responsiveness to soluble mediators spreading through the medium by diffusion, or more directly between cells via gap junctions
8. The phenotypic adaptability of the cells involved

nutrient levels or substrate availability. BSC-1 epithelial cells, on the other hand, were limited primarily by the avalability of substrate. These latter results might depend on the origin of the epithelial cells, since some epithelia exist *in vivo* as monolayers or sheets (e.g. squamous epithelia) whereas others are stratified (e.g. transitional epithelia). The possibility that topoinhibition of epithelial cells *in vitro* might be related to their *in vivo* topography, however, remains to be tested rigorously.

In yet another study, Rizzino and colleagues (1988) have shown that the binding of four growth factors to five different cell lines decreases as cell density is increased. Since this relationship was shown for both normal and transformed cells, however, its relevance to density differences between normal and transformed cells is not clear.

In light of the discussion above, it is difficult to generalize about the factors which might control cell density *in vitro*. It would not be surprising if the answer to what controls cell density turned out to be a compromise between the various possibilities which have been discussed. Table 1.5 summarizes some of the more important factors which might influence cell density *in vitro*.

1.7.4 Contact inhibition

Over the years, a general degree of confusion has accumulated with respect to cell contact-related phenomena *in vitro*. We have seen that many normal cells tend to form monolayers in culture, and that this is related in some way to the inhibition of their division. Although Abercrombie and Heaysman (1954) coined the term **contact inhibition** in relation to monolayer formation, in its original sense the term referred to the changes in locomotory behaviour seen when cells come into contact: it was not intended to explain inhibition of cell division. Indeed, the locomotory behaviour of cells when they come into contact can be examined independently of the effects of contact and/or cell density on division. The term 'contact inhibition' will be used here to refer to a set of behavioural changes seen when certain cells come into contact with each other (Abercrombie and Heaysman, 1953, 1954; Abercrombie, 1965). Either one or both of these cells may change their direction of locomotion, or one may crawl over or under the other, or one or both may cease movement, or some combination of these possibilities may result. If a cell crawls over or under another then it is said to have failed to have shown contact inhibition of locomotion. Note that the terms 'overlap' and 'underlap' can be used interchangeably unless a frame of reference (one particular cell type) is provided. Within a monolayer of cells, a considerable amount of movement still exists (Martz, 1973; Garrod and Steinberg, 1975) and it is therefore important to recognize that the term 'contact inhibition' refers to a **pattern of behaviour** when cells come into contact, a part of which may (but need not) include the actual inhibition (cessation) of movement by one or other or both of the cells in contact.

The term **contact inhibition of division** was introduced by Rubin (1961) on the grounds that there might be a common basis between contact inhibition *per se* and the inhibition of division seen in many confluent cell monolayers. There is no evidence in support of this possibility, however, and the term contact inhibition of division will not be used further.

Some authors refer to two main types of contact inhibition (Abercrombie, 1970b; Heaysman, 1978). In many examples, however, it is not yet clear which (if either) type of contact inhibition dominates:

(a) *Contact inhibition Type 1*

This type of contact inhibition involves paralysis of locomotory activity in the contact area (the 'contact paralysis' of Gustafson and Wolpert, 1967) which naturally leads to the failure of cells to overlap. Cells which display Type 1 contact inhibition cease local membrane ruffling, appear to form localized cell–cell adhesions, and then contract at the leading edge to partly separate the cells. Meanwhile a new leading edge may form elsewhere in one or other or both of the cells and, if free space is available, a change of direction of movement may occur leading to cell separation.

(b) *Contact inhibition Type 2*

This too is expressed as the failure of cells to overlap, but it is thought to result from differences in cell–cell and cell–substrate adhesion. One possible basis for this type of contact inhibition is that the upper cell surface of cells in culture is thought to be poorly adhesive (DiPasquale and Bell, 1974) and thus overlapping cannot occur. This is not true for all cell types, however, since pigmented retinal epithelial (PRE) cells can adhere and spread on choriod fibroblasts while the latter can adhere but not spread on the former (Parkinson and Edwards, 1978). Indeed, the formulation of a distinct type of contact inhibition based exclusively on the poor adhesiveness of the dorsal cell surface is rather dubious.

Numerous theories have been proposed to explain contact inhibition of locomotion (Harris, 1974 a and b), but it seems likely that several phenomena are involved and these are likely to include mechanical obstruction, adhesive differences between cell-surfaces and the substrate, and contact-induced cytoskeletal changes.

Many early reports of contact inhibition were actually based on the measurement of nuclear overlap, with the assumption being made that cells which display poor overlap are contact inhibited. The overlap index relates the number of observed nuclear overlaps to the number of overlaps which would be expected if the nuclei were randomly distributed. It is thus calculated from the area of the nucleus and the number of nuclei in a given area of substrate and it is expressed as a ratio of maximum value 1, which implies no inhibition of overlap (Abercrombie and Heaysman, 1954). Many fibroblasts show nuclear overlap indices of less than 0.2, whereas many transformed cells may show indices of greater than 0.5. Predictably, the actual values vary with the cell type, culture density and nature of the substrate. Turbitt and Curtis (1974) used the overlap index to study interactions within cultures of either BHK C13 cells or their polyoma transformed counterparts. Stoker (1964) had reported earlier that the transformed variant displayed more overlap than its normal counterpart, whereas Curtis (1967) had shown these two cell types to have similar overlap indices. According to the later study of Turbitt and Curtis (1974), however, the index changes with cell culture density and age and unless these conditions are kept constant between experiments, it is possible to get conflicting results when the two cell types are compared. By and large, at comparable densities the transformed cells always displayed greater overlap than the control cells. However, as indicated above, contact inhibition is really a dynamic behavioural response and it is somewhat misleading to use the overlap index to describe this phenomenon. The overlap index provides a measure of monolayer formation. Of course, if cells cannot crawl over or under each other then a monolayer will result, but unless the locomotory behaviour of the cells is actually observed it is impossible to conclude that contact inhibition was involved in monolayer formation.

It is important to note here that the phenomenon of contact inhibition may be examined in two situations involving either homotypic or heterotypic interactions. There is every indication that the cells behave differently in the two situations, and indeed Abercrombie (1960, 1967) noted that homotypic contact inhibition by tumour cells (unlike its heterotypic counterpart) is not consistently high.

1.7.4.1 *Homotypic contact inhibition*

Bell (1978) made a thorough study of the homotypic contact interactions of 3T3 cells and of their polyoma transformed counterparts (Py3T3). He found that 3T3 cells which came into contact adhered to each other, often showed a local cessation of locomotor and ruffling activities, and tended to change their direction of locomotion. When he examined the homotypic contact interactions of Py3T3 cells, Bell (1978) found that these too could display contact inhibition as described for the non-transformed 3T3 cells. The transformed Py3T3 cells nevertheless differed from 3T3 cells, first in their morphology (their ruffles are localized at the end of long processes rather than on broad lamellipodia) and secondly in their nuclear overlap ratio (0.422 for Py3T3 *versus* 0.096 for 3T3 cells). From his detailed observations, Bell (1978) suggested that the thin processes of Py3T3 cells are more easily able to underlap neighbouring cells than the broad lammellipodia of 3T3 cells. He also suggested that 3T3 cells have more of their margins in adhesive contact with the substrate than Py3T3 cells which, although it agrees with the general notion that transformed 3T3 cells are less adhesive than their non-transformed counterparts, nevertheless was based on rather speculative morphological criteria. Bell (1978) also found that ruffle-to-ruffle contacts tended to result in the cessation of locomotion for either cell type, whereas ruffle-to-side (i.e ruffle-to-non-ruffle) contacts resulted in more underlapping. Because of their differences in morphology, Py3T3 cells experienced more ruffle-to-side contacts than 3T3 cells. Presumably, these features of Py3T3 cells contribute to the more extensive nuclear overlap ratio observed for these cells but, of course, they tell us nothing of why the cell types are so different.

Although it has been argued that homotypic contact inhibition might contribute to the directed outgrowth of a tumour, its relevance is difficult to put into perspective since normal cells also show homotypic contact inhibition. Finally, a somewhat bizarre situation may occur when a single cell extends two processes which come into contact: in this case contact inhibition of self may be seen (Ebendal and Heath, 1977).

1.7.4.2 *Heterotypic contact inhibition*

In the early studies of Abercrombie and colleagues (Abercrombie and Heaysman, 1954; Abercrombie, Heaysman, and Karthauser, 1957) it was reported that mouse sarcoma cells showed reduced contact inhibition relative to normal cells when the two cell types were cultured together. Experiments typically involved the confrontation *in vitro* of out-growths of cells from two tissue explants placed about 1 mm apart. In order to standardize the assay, one of the explants was always a fragment of embryo chick ventricle from which fibroblasts (CHFs) emigrated, while the other explant was either from normal or tumorigenic tissue. Now over the years, Abercrombie and his colleagues used different methods to assay what happened when outgrowths from the two explants collided (Abercrombie, 1979). These methods were designed to provide information on invasion as well as on contact inhibition. First, these authors looked at the dimensions of the invasion zone, which refers to the area occupied by cells from *both* explants after a standard period of time (24 h after first contact). The invasion zone (characterized by radial outgrowth) was typically in the order of 20 μm when normal mouse muscle fibroblasts collided with standard CHFs, but it was around 130–290 μm for various tumours and CHFs. Does this mean that the tumour cells are unobstructed in their invasion of normal CHFs? In order to answer this question, the researchers determined 'between-to-side' (BTS) ratios. A BTS ratio is determined for each explant by measuring the distance its leading cells have migrated towards the other explant relative to the mean distance the leading cells have migrated from the free sides of the explant. Each experiment yields two BTS ratios, so that the effects of the control CHF cells on the invasion of experimental cells as well as the effects of experimental cells on control cells can be determined. Calculated BTS ratios suggested that some tumours (such as 311 and BAS56 sarcomas and a mouse melanoma) were obstructed in their invasion of the CHF outgrowth whereas others (such as sarcoma 180) were not. In fact, the situation is somewhat more complicated than this. Thus for the above four tumours the BTS ratios were the same as that for their paired CHFs: in other words the invasive behaviour was reciprocal. Nevertheless the cells of each pair of explants differed in the extent of obstruction that occurred, with cells from the S180/CHF combination being free to invade each other with minimal obstruction while cells from the other combinations showed various degrees of obstruction. Cells from two other sarcomas (FS9 and MC1M) showed non-reciprocal behaviour when confronted with CHF cells: that is, cells from these two sarcomas invaded CHFs much more extensively than the CHFs invaded them (see section 4.3).

The original analyses of the invasive patterns described above, however, were based on static observations and we have already seen that contact inhibition refers to a behavioural response in which cell movement is altered after contact. In order to study the more dynamic aspects of contact inhibition, Abercrombie and colleagues resorted to time-lapse cinematography. Detailed studies revealed that cells could underlap or overlap each other to a considerable degree before contact inhibition was observable, thus confirming the importance of direct observation methods. In S180/CHF confrontation experiments, direct visual analysis showed that both the sarcoma cells and the fibroblasts were not contact inhibited, as might have been predicted from the BTS ratios. A slight decrease in the motility of the S180 cells was noticeable, however, and direct observation also showed that the sarcoma cells tended to migrate over the upper surface of the fibroblasts. Observation of collisions between FS9 and MC1M sarcoma cells showed non-reciprocal contact inhibition, again as might be expected from the pattern of invasion. Analysis of films showed that a head on collision between an FS9 cell and a CHF resulted in immediate retraction of the fibroblast while the sarcoma cell was essentially unhindered in its movement. Abercrombie (1979) concluded from these and other studies that there is a relationship between invasive ability and defective (not necessarily absent) contact inhibition. The work of various other groups, however, suggests that this view may be oversimplistic.

Although Abercrombie and his colleagues worked with mouse tumours, others have studied similar phenomena using human tumours. Heterotypic contact behaviour between explants of human malignant melanoma (MM96) with human skin fibroblasts (HSFs) was studied in detail by Stephenson and Stephenson (1978). These authors also studied the contact behaviour of MM96 cells with CHFs. Interestingly, in the MM96/CHF and MM96/HSF combinations both fibroblast types and the melanoma cells were capable of invading, although the melanoma cells did so to a much greater extent. These authors also found that the melanoma cells invaded HSFs better than CHFs, due in part to the lower density of the HSF cells and also to the movement of the MM96 cells over the upper surfaces of the HSFs but not that of the CHFs. Invasion of CHFs by MM96 cells occurred in spaces vacated by the former. These observations on how invasion took place are of crucial importance, since they illustrate how careful the experimenter must be in order to avoid possibly misleading interpretations.

Stephenson and Stephenson (1978) also found that normal CHF cells strongly invaded normal HSF cells, which seemed to retreat leaving space for the forward movement of the CHF cells. Thus normal cells can display invasive behaviour in *in vitro* confrontation systems, and this leads us to question the use of this type of assay as a model system for studying *in vivo* invasion by tumour cells. When these authors analysed cell behaviour on contact, they found non-reciprocal contact inhibition in tumour cell/normal cell pairs, whereas reciprocal contact inhibition took place in the normal cell pairing even though only the CHFs invaded. These results suggest that invasion is not a simple outcome of contact inhibitory behaviour, and it is clear that considerably more research is required in this area before we fully understand all of the implications.

An attempt to study early ultrastructural changes associated with contact inhibition of locomotion was made by Heaysman and Pegrum (1973 a and b). These authors observed collisions between cells and then immediately fixed them for electron microscopy. Of course, it is hard to be sure of the exact time scale involved and the addition of fixatives may induce some changes in the cells, but nevertheless experiments such as these might offer some indication of the changes which occur when cells come into contact. According to Heaysman and Pegrum (1973 a and b), the main ultrastructural features of collision involve a specialization in the contact region of both cell membranes (recognizable as an increase in electron density) which occurs within 20 sec and an alignment of microfilaments in each cell within 30 sec which inserts into the contact zone. Distinct bundles of microfilaments become visible within 2 min of contact first being observed. The impression is given that the cells attempt to form contacts or junctions, but of what type is uncertain. It is conceivable that junctional failure could lead to cell retraction as a consequence of forces generated by the microfilaments, but this fails to explain the other events associated with contact inhibition.

Interactions involving epithelial cells seem to be different from that shown by fibroblasts, presumably reflecting the observation that an epithelium cannot tolerate a free edge. Thus epithelial cells grow *in vitro* as sheets and they prefer not to exist as single, wandering cells. Collisions between epithelial cells lead to paralysis and the formation of adhesions, but there apparently is no retraction phase (Middleton, 1982). Collisions between epithelial cells may also lead to contact-induced spreading in which the area occupied by the two cells in contact is greater than the sum of areas they occupied when single (Middleton, 1977). This behaviour is presumably related to epithelial sheet formation, and it may contribute to tissue stabilization which would seem crucial to the proper functioning of an epithelium. Given the differences in behaviour following contact of either epithelial cells or fibroblasts, we may ask ourselves what happens when these two cell types meet? Parkinson and Edwards (1978) studied this possibility using two cell types from the developing chick eye: pigmented retinal epithelial (PRE) cells and choroid fibroblasts. They found that collisions between these two types resulted in contact inhibition and retraction of the fibroblasts while the behaviour of the PRE cells remained essentially unaffected. This might seem logical, since invasion of epithelial barriers *in vivo* by fibroblasts would lead to the breakdown of normal tissue relationships. On the other hand, epithelial cells do not normally invade fibroblastic tissues. Parkinson and Edwards (1978) suggested that the inability of normal pigmented epithelial cells to invade the choroid *in vivo* may be due to their integrated, sheet-like nature. It should be recalled, however, that *in vivo* a relatively well established basement membrane (Bruch's membrane) exists between the pigmented epithelium of the eye and the choroid, and thus contact inhibition of cell movement may have little relevance in this situation. Furthermore, malignant invasion by epithelial cells can occur in sheet-like protrusions, and thus growth in coherent sheets may not by itself explain the failure of non-contact inhibited epithelial cells to invade.

Interestingly, contact inhibition of live cells can be induced by dead cells, although such behaviour has been detected so far only in a particularly contrived experimental situation (Heaysman and Turin, 1976). The key seems to lie in fixing the dead cell with zinc chloride, although why this particular agent should work and not others is not known. Perhaps zinc chloride maintains or confers some property to the cell surface which is important in signalling.

Finally, several authors beginning with Abercrombie and Heaysman (1953) have reported that mean cell speed is inversely related to the number of contacts ('contact inhibition of speed'). Reductions in speed of up to 50% have been recorded on contact but, apart from the possible slowing down of *in vivo* movements, its relevance is uncertain. Interestingly, Garrod and Steinberg (1975) reported that chick liver cells actually moved faster on contact; in fact, as single cells they did not move at all. Whether most epithelial cells behave differently in this respect to fibroblastic cells remains to be determined.

It should be clear from the foregoing that studies of contact inhibitory behaviour demand continuous observation and recording of actual contact events and that extrapolation from single observations of fixed material may provide a misleading interpretation. Furthermore, the outcome of homotypic cell contact interactions can be markedly different from that of heterotypic interactions and it could be argued that the latter are more relevant to the invading tumour cell. The experiments of Stephenson and Stephenson (1978) described above suggest that tissue stability may not result from the expression of contact inhibitory behaviour, since normal CHF cells can invade normal HSF cells even though both express heterotypic contact inhibitory behaviour. Even in tumour cell/normal cell pairings in which non-reciprocal contact inhibition may be quite frequent, there is evidence that both cell types invade. Of course, contact inhibition need not result in total cessation of movement and the observed outcome may reflect some compromise in this direction.

It must be admitted, however, that the biggest problem of all with the concept of contact inhibition is that we do not understand how (or, in some situations, whether at all) this phenomenon is manifested *in vivo*. Indeed, for practical reasons most studies of contact inhibitory behaviour have involved 2D studies, and it is not sure how these relate to the 3D situation either *in vitro* or *in vivo*. Since the frequency of contact will vary with cell number, then the pattern of invasion will clearly be a function of cell density and this makes it very difficult to relate *in vitro* observations to the *in vivo* situation. Furthermore, it seems that the outcome of cell contact is dependent on the nature of the collision, with head-on collisions leading to more dramatic changes than head-to-side collisions. According to Erickson (1978), the outcome of a head-to-side collision between two BHK cells depends upon the angle of contact. At angles less than 55° the cells deflect to come beside each other, but at greater angles the cells often show retraction and reversal of locomotion. Whether head-to-head and head-to-side collisions occur with similar frequencies in 2D as in 3D is doubtful, although head-to-head collisions may be emphasized if cells are contact-guided along collagen tracts and the like.

Possible evidence for the *in vivo* operation of contact inhibition was provided by Bard and Hay (1975) who filmed movements of corneal fibroblasts *in situ*, but the demonstration of contact inhibitory behaviour *in vivo* does not necessarily indicate a significant role for this process in invasion. If it is possible to extrapolate to the 3D *in vivo* situation, then many authors have suggested that contact inhibition could play a pivotal role in invasion. Thus if cells lose their contact inhibitory behaviour on transformation, we might expect to see the acquisition of the invasive behaviour typical of malignant tumour cells. Such changes would be dependent on the cells being mobile, and indeed Abercrombie and Ambrose (1962) have argued that two changes are necessary for normal cells to become invasive: namely, they must lose their contact inhibited nature and they must become mobilized if they are not already so. However, any attempt to describe tumour cell invasion solely in terms of the loss of contact inhibition is likely to prove overly simplistic.

Most organs are not solid blocks of cells in that they are usually associated with varying amounts of connective tissue. They are also pervaded by nerves and blood vessels which in many instances are known to offer paths of least resistance for invading cells. Thus when invasion takes place *in vivo* there is no guarantee that cell–cell collisions wil be involved. Using a 3D model based on invasion into the embryonic chick wing, Tickle, Crawley and Goodman (1978 a and b) were unable to show any simple relationship between invasive ability and lack of contact inhibition. In fact, these authors reported that both normal (e.g. Nil 8) and transformed (e.g. HSV Nil 8) cells could show invasive behaviour, as could cells which exhibited contact inhibitory behaviour in 2D systems (e.g. BHK). Interestingly, when they examined the invasive behaviour of CHF cells and sarcoma 180 cells, their results paralleled those of Abercrombie *et al.* (1957). Thus CHF cells which display contact inhibtion of movement in culture do not invade the limb, whereas sarcoma 180 cells which are not contact inhibited can do so. Whilst this similarity of results is somewhat satisfying, studies with the other cell types mentioned above point towards this being rather a special situation. It should be mentioned here that the studies by Erickson (1978 a and b) on contact inhibition in BHK cells related to homotypic interactions, whereas we have seen that Abercrombie and his colleagues emphasized the importance of heterotypic interactions. Nevertheless, when tested against PyBHK cells both normal BHK and their virally transformed counterparts displayed contact inhibitory behaviour. The earlier work of Stoker (1967), which he notes was in conflict to the observations of Abercrombie *et al.* (1957), also confirmed that PyBHK cells displayed contact inhibition when in contact with non-transformed BHK cells, although Stoker (1967) found no evidence of homotypic contact inhibition for the transformed cell type.

It would seem that the results of Abercrombie's group relate to a specific set of conditions, and deviation from these can lead to different results. Taken as a whole, we must conclude that there seems to be no unversality in the commonly held belief that lack of contact inhibition is predictive of invasive behaviour. Furthermore, both normal and transformed cells may display or lack contact inhibitory behaviour and thus the expression of this trait does not always allow discrimination between normal and transformed cells.

For further information on contact inhibition (with a slightly different perspective) the reader is referred to an early review by Martz and Steinberg (1972).

1.7.5 Cell surface composition

The plasma membrane of animal cells is classically described as a lipid sea in which globular proteins float like icebergs. Without wishing to dispel such romantic notions, it must be said that the truth of the matter is actually something quite different. The composition of cell membranes varies with cell type and with intracellular location, but generally they are about 20% water, with 60% of their dry weight being protein and 40% being lipid. This does not take into account carbohydrates which may be attached to either the proteins or the lipids in quantities which vary from about 1–10% of the total membrane dry weight.

Membrane lipids are amphipathic in that they have both hydrophilic ('head') and hydrophobic ('tail') regions. The hydophilic and hydrophobic regions are separated by a bridge, which is usually a glycerol moiety or a sphinganine derivative (or homologue). Each head region is attached to the bridge either via a phosphate ester (to form phospholipids, the most common membrane lipid type) or via a glycosidic linkage from a sugar molecule (to form the glycolipids). In the sterols, a hydrophilic group is attached to one end of the hydrophobic sterol backbone. The phospholipids exist in bilayer form, with the 'water hating' hydrophobic tails sequestered within the bilayer interior. The packing of the long hydrophobic tails in the interior of the bilayer is a function of their conformation, which may be essentially straight (actually a planar zig-zag) as in saturated and trans-unsaturated tails (the latter being rare in nature) or bent, as in cis-unsaturated tails. The bend (of about 30°) in cis-unsaturated tails disrupts the more or less orderly packing of the hydrophobic tails and tends to make the membrane more fluid. There are in fact three major features of membranes which govern their fluidity:

(a) *The degree of acyl chain saturation*

This is referred to above.

(b) *The lipid composition*

Of particular importance here is the molar ratio of sphingomyelin to the generally more unsaturated phosphatidylcholine (these are the two most common lipids of nucleated eukaryotic cells).

(c) *The presence of cholesterol*

Cholesterol sits in the bilayer near the head region of the phospholipids and tends to stabilize this zone, possibly as a result of condensed packing.

The fluidity of the membrane at physiological temperatures allows the phospholipids to rotate and to diffuse laterally, and they may even 'flip-flop' between the inner and outer leaflets, although this is at a considerably reduced rate compared to the other two processes. It has been estimated that the half-time for flip-flop is in the order of two weeks, whereas phospholipids could diffuse the length of a cell in a matter of seconds.

One interesting aspect of all cells studied so far is that membrane lipids are not symmetrically distributed between the two leaflets of the bilayer. The choline-containing lipids of the human erythrocyte, for example, tend to segregate to the outer leaflet, while the amino-lipids tend to show the opposite distribution. Virtually all glycolipids are located on the outer leaflet, with their sugar groups facing the extracellular space. The function of this distinct membrane asymmetry is unknown, although it has been speculated that the phospholipid distribution may serve to keep membrane proteins oriented properly within the membrane, while the external location of glycolipids may be related to a signalling role.

Like the membrane lipids, membrane proteins are also amphipathic. Experimentally, membrane proteins are classified into two types, the peripheral membrane proteins (which are extractable by gentle procedures such as treatment with salt) and the integral membrane proteins (which require membrane disruption for extraction). Generally speaking, it is thought that membrane proteins either span the entire membrane (transmembrane proteins) in which case they may be held in the bilayer by hydrophobic interactions with the fatty acid tails of the phospholipids, or they are loosely bound to the extremities of such transmembrane proteins. Alternatively, membrane proteins may be bound covalently to certain lipid moieties such as glycosylated phosphatidylinositol by a process referred to as glypiation (reviewed in Low, 1987). As will become clear later, it is thought that some molecules involved in adhesion such as N-CAM (the neural cell adhesion molecule) and heparan sulphate proteoglycan may be anchored in this way. N-CAM (at least in the chicken) in fact exists in three forms. Two ($N\text{-}CAM_{130}$ and $N\text{-}CAM_{160}$) are typical transmembrane proteins whereas the third ($N\text{-}CAM_{120}$) is the one which is anchored to glycosylated phosphatidylinositol (Hemperley, Edelman and Cunningham, 1986). A fourth possibility for membrane protein anchoring has been postulated to exist in order to account for proteins (such as cytochrome B_5) which only have one hydrophilic region. In this situation, a hydrophobic domain would anchor the protein in the lipid bilayer without extension across the membrane. It should be noted that some transmembrane proteins cross the lipid bilayer more than once, somewhat like a condensed running stitch in needlework. A good example of this type of topology is the liver gap junctional protein (connexin 32) which crosses the membrane four times. Note that in this particular example both the N and C terminal regions are located in the cytoplasm (Zimmer *et al.*, 1987).

Why should different forms of membrane protein anchoring exist? In the first place, transmembrane anchored proteins could allow communication across the membrane and such proteins have been shown to act as ion pumps and receptors. Connection with cytoskeletal components, either directly or indirectly, will not only serve to anchor the proteins in the membrane but will also provide a mechanism by which the cell can relocate its membrane proteins as required. Such a function could be important in establishing and maintaining cell polarity and in cell locomotion. Finally, anchorage to lipid components may facilitate the lateral mobility of membrane proteins since the anchoring region could be located entirely within the outer leaflet of the bilayer. Furthermore, since glycosylated phosphatidylinositol anchored proteins may be selectively released by phospholipase C (PLC), it is possible that this form of anchorage may be relevant to dynamic processes such as cell adhesion, where the adhesive bond must be broken for the cell to move. Interestingly, PLC mediated hydrolysis of glycosyl-phosphatidylinositol linked proteins will generate 1,2 diacylglycerol (DAG) which could diffuse into the cell from its location in the outer lipid leaflet. Since DAG acts as an intracellular messenger activating protein kinase C, then processes associated with PLC-mediated release of membrane proteins may have wide ranging consequences (see sections 1.7.1 and 1.7.22).

The fluidity of the plasma membrane allows its proteins to be cross-linked by bivalent antibodies directed against them. The resulting antigen–antibody complexes form two-dimensional patches within the plane of the membrane. Patched proteins are swept to one area of the cell in an active process termed 'capping' that requires ATP and an intact actin microfilament system. Capped material may then be ingested by the cell in a process referred to as endocytosis. The role of the cytoskeleton in capping is not yet clear, but it has been suggested that actin and myosin interact to pull the patches together, thereby implying a connection (either pre-existing or induced) between plasma membrane proteins and the cytoskeleton.

Table 1.6 Major sugars present in glycoproteins

D-glucose (Glc)	D-galactose (Gal)
D-mannose (Man)	L-fucose (Fuc)
L-arabinose (Ara)	D-xylose (Xyl)
N-acetyl-D-glucosamine (GlcNAc)	*N*-acetyl-D-galactosamine (GalNAc)
sialic acid (*N*-acetylneuraminic acid)	

Virtually all of the carbohydrate components of the plasma membrane, whether on proteins or lipids, are exposed on the outside of the cell where they may be stained and visualized under the electron microscope. Stains such as ruthenium red show the plasma membrane of most cells to be covered by a carbohydrate-rich coat, known as the glycocalyx, which consists of the oligosaccharide side-chains of membrane proteins and glycolipds, as well as of adsorbed glycoproteins and proteoglycans. The question which may now be asked is: How far does the plasma membrane extend? Evidence suggests that many of the adsorbed glycoproteins and proteoglycans may play crucial roles in cell behaviour, and this has led many researchers to favour the concept of 'a greater membrane' which includes the material of the glycocalyx, although it must be admitted that the separation of the glycocalyx from components of the extracellular matrix is not always distinct.

Although there is considerable diversity in the oligosaccharide content of different glycosylated proteins and lipids, there are only nine sugars involved in their construction (Table 1.6). Oligosaccharide chains are attached to proteins in two ways: either as *N*-linked sugars to the amide nitrogen of asparagine via GlcNAc, or as *O*-linked sugars to the hydroxyl oxygen of threonine or serine via GalNAc, to serine via xylose, to hydroxylysine via Gal, or to hydroxyproline via Ara. Most of the glycolipids of animal cells are glycosphingolipids (GSLs), particularly glycosylceramides in which the glycosidic linkage is to a ceramide (an *N*-acylated sphinganine derivative). The two most common glycosylceramides are galactosylceramide (which is rich in myelin) and glucosylceramide (which is found more in non-neuronal plasma membranes), and together these are referred to as the cerebrosides. The gangliosides are GSLs derived from various ceramides by the addition of one or more sialic acid residues. Sialic acid, on both glycolipids and glycoproteins, is responsible for most of the negative charge of the cell surface under physiological conditions.

By and large, it might be supposed that the biochemistry of the cell surface is likely to be similar for both normal and transformed cells since both types must carry out certain basic functions in order to survive. There have been a number of reports, however, of both quantitative and qualitative differences in the cell surface composition of normal and malignant cells, including the expression of oncofetal antigens and tumour-associated transplantation antigens (TATAs). Since the expression of many cell surface components changes during the cell cycle, and because more tumour cells are likely to be dividing at any one time than their normal counterparts, it is perhaps not surprising that quantitative differences have been detected, many of which may have limited applicability in relation to tumorigenicity.

A number of studies have emphasized the heterogeneity of tumours which is likely to be reflected in their surface composition. It also seems clear that metastases can differ antigenically from their primary tumours (Sugarbaker and Cohen, 1972). These differences in surface composition both within a primary and between the primary and its secondaries make it even more difficult to reach any generalized conclusions on the relationship between specific cell surface changes and the acquisition of the tumour state. Much of the early research on cell surface changes associated with transformation and proliferation has been reviewed in papers by Pardee (1975) and Nicolson (1976, 1982a, 1984a) which contain many of the early references.

1.7.5.1. *Proteins and glycoproteins*

Most molecules of interest within this category are glycosylated to some degree. One way of testing the importance of oligosaccharides in tumour biology is to interfere with oligosaccharide synthesis, and an interesting drug in this respect is tunicamycin which blocks GlcNAc addition during *N*-linked oligosaccharide formation. Early studies, however, were more concerned with examining the glycoprotein nature of normal and transformed cells. Glycopeptide fragments obtained by proteolytic digestion of glycoproteins, for example, were found to be longer on transformed cells than on non-transformed cells, and longer glycopeptides were also seen on rapidly growing cells (Buck *et al.*, 1971 a and b). More recent work has confirmed that the membrane glycoproteins of many transformed cells contain carbohydrate residues which are more highly branched or elongated and which contain increased amounts of sialic acid (Smets and Van Beek, 1984). Other studies have focused on changes in particular classes of proteins and glycoproteins, and even particular molecules:

1. *Tumour-associated transplantation antigens (TATAs), tumour-associated antigens (TAAs) and tumour-specific antigens (TSAs).* The relationship between these three categories of tumour antigens is not particularly clear since the terms are used differently by different research groups. Both TATAs and TAAs are cell surface

molecules which elicit rejection responses in autochthonous (original) or syngeneic (inbred, genetically identical) hosts. First attempts to show tumour rejection involved the grafting of tumours between outbred mice, which is actually akin to studying transplantation differences. This point was made abundantly clear by Woglom (1929) but appropriate experiments had to await the development of inbred mice. The presence of TATAs on tumour cells was first identified by Gross (1943, 1945) and later by Foley (1953) and Prehn and Main (1957). Using different experimental approaches, these authors showed rejection of methylcholanthrene induced tumours in inbred mice, thus overcoming earlier problems associated with histocompatibility differences. In an even more definitive experiment, Klein *et al.* (1960) showed that a mouse could be made resistant to its own tumour. This they did by removing the primary tumour (a sarcoma induced with methylcholanthrene) and then injecting the host with irradiated tumour cells before challenging again with viable tumour which was rejected.

The distribution of TATAs and TAAs need not necessarily be restricted to tumour cells and they are often found on embryonic cells suggesting that their true function may be in differentiation. They can be induced by chemical carcinogens, or by the expression of normally silent embryonic genes (such as in the oncofetal antigens described below). Various TATAs and TAAs will be referred to in the following sections.

Strictly speaking, TSAs are antigens found only on tumour cells and this criterion will be used here to distinguish them from other surface antigens. There is some debate as to whether certain antigens are properly described as TSAs or not. In some cases, for example, the antigen may be present on both tumour cells and normal cells, although not in an immunologically accessible form on the latter. In other cases, the antigen may be present on normal cells but at an extremely low level, or it may be present only on rarely occurring normal cells. There is some evidence that the 'new' antigens elicited by many chemical carcinogens are actually present on normal cell surfaces and thus these antigens would be excluded from the specific marker category. An interesting point here is that if each chemically induced tumour does have unique antigens then it might not be practical to use therapeutic techniques based on immunorecognition for the detection and treatment of such cancers because of the enormous number of reagents that would be required.

Unlike many chemically induced antigens, the virally induced antigens are good candidates for TSAs. Any given tumour virus will induce the same antigen on any cell, but the antigens induced by chemical carcinogens will all be different, even for a single chemical carcinogen acting on a single cell type. In some circumstances, chemically induced tumours may express virally related antigens on the cell surface presumably as a result of cryptic viral gene activation. There are essentially two categories of viral induced TSAs: the nuclear (T) and surface (S) antigens induced by oncogenic DNA viruses, and the actual viral proteins specified by RNA viruses.

A relatively strong case has been put forward for mammary tumour glycoprotein (MTGP) as a TSA since it is found in the cytosol and particularly the membranes of many malignant but not benign human breast tumour cells (Leung *et al.*, 1978, 1979). Further studies on the distribution of MTGP are required, however, to exclude it from embryonic cells and other normal adult cells.

The question of tumour heterogeneity is important when considering tumour antigens since (apart from virally coded antigens) it does not follow that all cells of a given tumour will possess the same antigens. Wikstrand, Bigner and Bigner (1983) tested a number of clones of a glioma cell line using a panel of 10 monoclonal antibodies and found six of the clones to differ quite markedly in their immunological reactions. The phenotypic variability of TATAs (defined by the generation of specific cytotoxic T cells) has been studied by Schirrmacher (1980) using two variants (Eb and Esb) of the methylcholanthrene induced murine lymphoma L5178Y. Both Eb and Esb express TATAs which are distinct and non-cross reactive, but studies of clones of these cells indicate that tumour cells may switch from one phenotype to another. This phenotypic switch is rapid, apparently occurring within three cell generations. When twice-cloned Esb cells were injected into a tumour immunized host (thus containing specific cytotoxic T cells), the Esb cells rapidly adapted by losing their TATA. In this way the injected tumour cells were able to escape the immune response of the host and this illustrates dramatically that there are many advantages for a tumour cell in having a rapidly changeable phenotype.

2. *Fetal or embryonic proteins (oncofetal antigens).*
The cells from some malignant tumours are characterized by the expression of proteins which are normally only found in embryonic tissue (Coggin and Anderson, 1974). The reason for this is not clear, but it has been postulated to reflect a dedifferentiation

process in which the tumour cell has reverted to a more primitive cell type. Alternatively, the transforming event may have occurred in a tissue stem cell which, unlike other tissue cells, may not display the full differentiated phenotype. The two best examples of fetal proteins which are expressed on malignant cells are carcinoembryonic antigen (CEA) and alpha-fetoprotein (AFP). CEA is a heterogeneic glycoprotein (c. 200 kD) found on the membranes of cells from the embryonic digestive tract and from various tumours including cancers of the colon, pancreas, stomach and bone. It was first identified when rabbits were immunized with human colon cancer extracts (Gold and Freedman, 1965). After the resulting immune serum was adsorbed against normal human colon, it was still able to react with cancers of the gastrointestinal tract and pancreas, but not with other adult tissues as determined by techniques such as immunodiffusion. Since the anti-serum was also found to react with fetal gut, pancreas and liver, the antigen shared by the tumour cells and by these fetal tissues was termed carcinoembryonic antigen. Small amounts of CEA can actually be found in normal plasma and tissues using more sensitive techniques such as radioimmunoassay (RIA) and the differences between normal, and diseased levels may not always be enough to allow screening for the presence of CEA as a diagnostic aid. The functional role of CEA remains enigmatic although structural studies suggest it belongs to the immunoglobulin supergene family (Paxton *et al.*, 1987), some members of which play significant roles in cell adhesion.

AFP is a normal product of the fetal liver, but it is also found in the plasma of patients with cancer of the liver or testicular teratoma. AFP (70 kD) was first identified in the serum of neonatal mice and in a transplantable mouse hepatoma by Abelev *et al.* (1963), and it was subsequently identified in the serum of patients with primary liver cancers by Tatarinov (1964).

Recent molecular studies have identified developmentally regulated oncogenes which may be important in tumorigenesis. The *int-2* oncogene, for example, is a transforming gene induced by insertional mutagenesis with the mouse mammary tumour virus (MMTV). This gene is expressed in mouse embryos but not in adults unless activated by MMTV. The product of *int-2* shares considerable homology with basic FGF, a known growth factor and angiogenic molecule (section 1.7.1), but how it functions in both embryogenesis and tumorigenesis is not clear. Interestingly, *int-1* (another gene induced by MMTV) is identical to the *Drosophila* segment polarity gene *wingless*, which clearly has a significant role in development but how this relates to mammalian systems is not clear.

3. *Fibronectin.*
Fibronectin exists in two forms, one soluble in the plasma (pFN) and the other cell-surface bound (csFN). The insolubility of csFN may be attributable to the presence of a specific, short repeated sequence which is lacking on pFN. The two main forms occur predominantly as dimers (although csFN does form multimers) with subunits of about 220–250 kD linked together by disulphide bridges. The polypeptide chains of both types of fibronectin contain tightly folded regions or domains joined by more flexible segments less resistant to proteolytic attack. Each domain subserves one or more functions, such as cell binding or collagen binding, which contribute to the ability of fibronectin to act as an adhesive molecule. A common change seen in many transformed cells when compared to normal cells is a decrease in the amount of csFN (Hynes, 1973). Since the expression of csFN varies with the cell cycle, being greatest in G_1 and least in M phase (Hynes and Bye, 1974), a reduction might be expected to be seen in rapidly dividing tumour cell populations. However, there is no simple inverse relationship between the presence of fibronectin and malignancy (Stenman and Vaheri, 1981).
Several reasons for the decreased levels of csFN on tumour cells may be put forward:

(a) There may be a general failure to retain fibronectin on the surfaces of tumour cells through an alteration involving its receptor or the molecule itself, or arising as a consequence of increased membrane turnover and/or shedding. Ehrlich ascites tumour cells, for example, secrete fibronectin but do not retain it on the surface in significant quantities (Zardi *et al.*, 1979).
(b) There may be general decrease in the synthesis of fibronectin or its receptor. Transformed chick embryo fibroblasts, for example, show a decreased rate of fibronectin synthesis (Olden and Yamada, 1977).
(c) There may be an increase in fibronectin degradation, possibly due to the elevated production of proteolytic enzymes following transformation (Hynes, 1974).

Studies made by Taylor-Papadimitriou, Burchell and Hurst (1981) indicate that both normal and malignant epithelial cells synthesize fibronectin, but that malignant cells lose more than 90% of the molecule. Exogenous fibronectin induces transformed cells to flatten and adhere more strongly to the substrate (Yamada, Yamada and Pastan, 1976) and it also is associated with the appearance of well-defined bundles of microfilaments (Ali *et al.*, 1977). The role of fibronectin in the adhesion of tumour cells is discussed in more detail in section 5.4.2.

4. *Surface immunoglobulin.*
   The expression of surface immunoglobulin (sIg) by tumours of B cell origin provides a specific example of the relevance of cell surface markers for tumour cells. Since individual tumours are by and large monoclonal in nature, all the cells of any one tumour of B cell origin should express sIg with identical variable domains (V domains). Highly selected antibodies raised against V domain determinants (anti-idiotype antibodies) and which identify the cells of particular lymphoid tumours can be prepared by monoclonal antibody techniques. Such antibodies (McAbs) can then be used in direct anti-idiotype therapy or as 'magic bullets' to target cytotoxic drugs. In anti-idiotype therapy, anti-idiotype McAbs are delivered to patients who have B-cell-derived tumours that express sIg without appreciable Ig secretion (which would mop up the McAbs). There has been some therapeutic success with this form of treatment although the underlying mechanism is unknown (Miller *et al.*, 1982). The possibilities of antibody therapy are discussed further in section 6.4.

5. *Other normal proteins and glycoproteins.*
   Tumour cells express a range of normal proteins and glycoproteins, other than those above, whose expression may occasionally be altered. In some cases such as certain murine leukaemias, for example, the expression of major histocompatibility complex (MHC) antigens has been noted to decrease on transformation (Boyse and Old, 1969). Changes in the expression of MHC antigens may influence the immune response of the host against the tumour thus allowing it to escape from the immunosurveillance system of the body. However, there may be other ways in which a tumour can subvert the immunosurveillance system such as by shedding TATAs and by the formation of 'blocking factors' with antibody (Alexander, 1974; Hellstrom and Hellstrom, 1974) so that MHC changes are not necessarily universal. A large, highly glycosylated protein known as epiglycanin has been identified in the plasma membranes of TA3-Ha mouse adenocarcinoma cells where it appears to functionally cover the MHC antigens, presumably by some form of steric hindrance, thus allowing the cells to be allotransplantable (Codington *et al.*, 1985).

1.7.5.2 *Glycolipids*

Changes in glycolipid nature and content have been detailed in a number of experimental systems (Hakomori, 1973; Brady and Fishman, 1974; Critchley, 1979; Yogeeswaran, 1983). The majority of reports deal with quantitative or qualitative changes in glycosphingolipids (section 1.7.5).

There are three major types of GSL alteration associated with transformation: reduction or elongation in the oligosaccharide chains, and the appearance of new GSLs which were perhaps only present in trace amounts before transformation. One of the more common changes which has been reported to be associated with transformation is a ganglioside simplification. In early studies Hakomori and Murakami (1968), for example, showed that transformed BHK-21 cells contained less of the ganglioside G_{M3} and more of its precursor (from which G_{M3} is derived by the addition of sialic acid) than did non-transformed cells. Merrit and colleagues, using a model system based on chemically induced hepatoma variants, showed that although total ganglioside levels were increased in all but poorly differentiated tumours, there was a progressive simplification of ganglioside pattern from hyperplastic to malignant tissue (Merritt *et al.*, 1978 a and b). Unfortunately, there are many transformed cells which do not exhibit a simplified ganglioside pattern and generalizations are not valid (Brady and Fishman, 1974; Critchley, 1979). Some of the changes in glycolipid nature may result from changes in glycosyltransferase activity, but again their general applicability is in question. A group of patients with adenocarcinomas was found by Hakomori *et al.* (1967) to accumulate less complex fucose-containing glycolipids, and in a subsequent study this was found to correlate with reduced levels of some glycosyltransferases (Stellner *et al.*, 1973).

1.7.5.3 *Sialic acid and other sugars*

Sialic acid is a common terminal sugar on many cell surface glycoproteins and glycolipids (on which it may also occur internally). There appear to be no consistent changes in sialic acid content in transformed cell lines but *in vivo*, in both animals and humans, there is some evidence that tumours are associated with elevated levels of sialic acid (Yogeeswaran, 1983) and sialyltransferase activity (Bosmann and Hall, 1974; Bramwell, 1985). This possible difference between cell lines and *in vivo* tumours illustrates the need for caution when characterizing tumour cell properties by comparing different sytems since there may be little universality of application. Of course, this apparent difference itself may become less obvious as more *in vitro* and *in vivo* transformed tumours are studied, and indeed recent research is furthering support for altered carbohydrates on many *in vitro* transformed cell lines as well as on *in vivo* tumours.

It should be noted that many studies on sialic acid content do not distinguish between cell surface- and total cell-sialic acid, possibly masking any crucial differences which might be involved in cell surface interactions. Furthermore, differences in sialylation may exist for different classes of membrane components such as the glycoproteins and glycolipids, for example. The extent of cell surface sialylation will also be affected by neuraminidase (sialidase) activity and it is usually difficult to ascribe the final extent of sialylation to specific changes in sialyltransferase or neuraminidase activity.

The CA antigen or epitectin (Bramwell, 1985) is a sialated, mucin-like molecule which runs as two bands (M_r 350 kD and 390 kD) after polyacrylamide gel electrophoresis (PAGE). Epitectin was first identified as a consequence of binding to the monoclonal antibody CA1 which was raised against cell membrane components with affinity for wheat germ agglutinin (WGA). CA1 has been shown to bind to many human malignant cells but not to most normal cells (Ashall, Bramwell and Harris, 1982). Exceptions amongst the malignancies include prostatic carcinomas, testicular teratomas, some sarcomas, and malignant melanoma, while normal tissues which react with the antibody include bladder epithelium, type 2 pneumonocytes of the lung, and apocrine sweat glands (Bramwell, 1985). It is postulated that epitectin may have some sort of protective role, possibly serving an anti-immune function.

1.7.5.4 *Glycosaminoglycans and proteoglycans*

The glycosaminoglycans (gags) are saccharide chains usually composed of dimers of uronic acid and an amino sugar (the exception is keratan sulphate which does not contain uronic acid). The gags are usually covalently linked to a serine residue on a protein core forming a structure referred to as a proteoglycan (see Table 2.1), which may be conveniently classified according to gag type. The exception here is hyaluronic acid which is not covalently bound a protein core; it is also not sulphated (reviewed in Kraemer, 1979; Gallagher, Lyon and Steward, 1986). Proteoglycans may show wide variation in the structure of the protein core and the number and length of gag chains and thus comparison between normal cells and tumour cells is difficult. Heparan sulphate (HS) proteoglycan, like most of the others, is found in the extracellular matrix (ECM) but it also is the most abundant proteoglycan on the cell surface. Cell surface HS may be relatively loosely associated with the plasma membrane, binding to a receptor via the gag side chains or via the protein core, or it may be fully integrated utilizing a hydrophobic sequence on the core structure. Alternatively, HS may bind to glycosylated membrane phosphatidylinositol. HS has been reported to be poorly sulphated in transformed cells (Underhill and Keller, 1975; Winterbourne and Mora, 1981) and this may influence various cellular interactions since many glycoproteins of the ECM such as laminin and fibronectin have binding sites for HS. Proteoglycans, particularly HS and CS (chondroitin sulphate) containing types, have been found amongst the substrate-attached material (SAM) which is left after cells have been stripped off with EGTA (ethylene glycol bis[2-aminoethyl ether] tetra-acetic acid), a divalent cation (Ca^{2+}) chelator (Terry and Culp, 1974). The SAM proteoglycan prepared from Balb/c 3T3 is about 90% of the CS type with the remainder of the HS type, whereas in their SV40 transformed counterparts most of the proteoglycan is of the HS type (Wightman, Weltman and Culp, 1986; Lark and Culp, 1982, 1983). It is not yet certain how applicable this observation will be to other tumour cells, but most of the available information suggests that tissue gag content is extremely variable and there are no consistent differences in gag composition between normal cells and tumour cells. The role of proteoglycans in tumour cell adhesion will be discussed further in section 5.4.6.

1.7.6 Cell surface shedding and antigenic modulation

The shedding of cell surface material may be important in normal membrane turnover. Shedding increases with cell activation and cell growth as seen, for example, after mitogen stimulation of normal cells. In tumour cells, however, shedding appears to be continuous: it has been estimated that about 50% of human melanoma cell surface antigens could be shed within 3 h and about 44% of B16 melanoma antigens over 48 h (Bystryn, 1976). Shedding from cells is usually in the form of particular molecules or as membrane fragments (reviewed in Black, 1980; Taylor and Black, 1986). Studies using B16 cells have shown various antigenic molecules to be shed: Gersten and Marchalonis (1978) identified antigens of 59–69 kD, Nishio *et al.* (1981) found molecules generally of 50–100 kD, whereas Bystryn *et al.* (1974) isolated shed antigens primarily in the 150–200 kD range. The importance of these different molecules remains to be elucidated, and to what extent they reflect different technical procedures is not clear. The shedding of large membrane fragments of $M_r > 10^7$ D has been reported for tumour cells but not for normal human fibroblasts or 3T3 cells. It seems that shed membrane represents selected microdomains, since the fragments released contain more protein, cholesterol and sphingomyelin, but less phospholipid, than does the normal (averaged) plasma membrane of the cells in question.

The shedding of tumour cell surface components may be related to the level of activity of these cells, but it can also serve a protective function by facilitating tumour escape through, for example, the mopping up of anti-tumour antibodies or by frustrating cytotoxic cell activity. Dvorak and his colleagues (1981) have shown that three experimental carcinomas (Lines 1 and 10 guinea pig hepatocarcinomas and the TA3-St mouse breast carcinoma) shed plasma membrane vesicles which contain procoagulant activity. The *in vivo* significance of this activity is unfortunately not fully understood (section 5.7.22).

The term **antigenic modulation** describes the loss of a surface antigen which may result in the avoidance of the host's immunosurveillance system (see section 1.8 below). When Boyse and his colleagues injected TL+ve leukaemia cells into mice immunized against TL antigen they found, somewhat contrary to their expectations, that the cells were not rejected. Closer examination showed that the injected cells had lost their TL antigenicity and that this arose not as a consequence of masking nor as a result of the selection of TL-ve cells, but because the TL antigen was actively lost from the cell surface (Boyse, Old and Stockert, 1965; Boyse, Stockert and Old, 1967).

As with many other properties under discussion here, shedding is not the sole preserve of tumour cells, but it may be expressed at a different level than in normal cells and it may serve a number of functions in tumour cells beyond that thought to be normal. Interestingly, shedding is a property of a number of fetal cells, and it is has been proposed that shedding may represent a common mechanism by which fetal cells and tumour cells may escape immune destruction (Alexander, 1974).

1.7.7 Lectin binding and agglutinability

Lectins are proteins which are commonly derived from plants, although they appear to be ubiquitous in their distribution. One of their characteristic features is their ability to bind specifically to sugars such as those found on glycoproteins. Because they are often polyvalent in nature, lectins may bridge across the surface glycoproteins of different cells leading to cell clumping. The term **agglutination** as distinct from the term **adhesion** is used exclusively to delineate cells which bind together following the addition of exogenous lectin. Most tumour cells generally agglutinate at a lower concentration of lectin than that required to agglutinate normal cells (Aub, Sanford and Cote, 1965; Burger and Goldberg, 1967; Pollack and Burger, 1969; Burger, 1973) although there are many exceptions. For example, no differences in lectin agglutinability have been shown between human fibroblasts and sarcoma cells (Glimelius, Westermark and Ponten, 1974) and between human lymphocytes and chronic lymphocytic leukaemia (CLL) cells (Glimelius, Nilsson and Ponten, 1975). Despite their often reported differences in lectin agglutinability, both normal cells and tumour cells by and large possess similar numbers of lectin surface receptors (Cline and Livingston, 1971; Inbar, Ben-Bassat and Sachs, 1971). However, quantitative studies of lectin binding using RAW117 cell lines have shown a correlation between reduced lectin binding for some lectins (concanavalin A) but not others (wheat germ agglutinin) and increased metastatic potential (Reading *et al.*, 1980a). Similar observations have been made using con-A-resistant variants of the B16 melanoma (Tao and Burger, 1982), but other studies with rat hepatocarcinoma variants have shown a positive correlation between both con A and WGA binding with lung colonization for three variant cell lines (Talmadge *et al.*, 1981; Stanford *et al.*, 1986). Interestingly, the binding of these three hepatocarcinoma variants to endothelial monolayers was inhibited by the presence of con A.

Non-transformed cells may be rendered more lectin agglutinable by brief treatment with proteolytic enzymes (Burger, 1969, 1970). This does not seem to be due to the unmasking of cryptic lectin receptors since protease treatment increases lectin agglutinability without any obvious increase in receptor number (Sela *et al.*, 1971). The enhanced lectin agglutinability of tumour cells seems also not to be due to differences in microvilli density as originally proposed (Willingham and Pastan, 1974, 1975) since there appears to be no positive relationship between microvilli abundance and agglutination (Collard and Temmink, 1975, 1976). Differences in membrane fluidity also seem unlikely to underlie the differences in lectin agglutinability of non-transformed and transformed cells since both electron spin resonance (ESR) and biochemical analysis of lipid composition do not indicate any consistent differences between these two categories of cells (Gaffney, 1975). The lectin agglutinability of tumour cells may be a function of other cell properties, however, including receptor mobility and transmembrane interactions between receptors and cytoskeletal components. Interestingly, lectin agglutinability may be related to cell cycle events since studies on the con A agglutinability of mouse 3T3 and SV3T3 cells (their simian virus 40 transformed counterparts) has suggested that it may be higher in M phase and G_1 than in the other phases (Smets, 1973). Indeed, both the non-transformed and the virally transformed cells were highly agglutinable in M phase.

Recently, Raz and colleagues have identified an endogenous, galactoside-specific lectin on the surfaces of various tumour cells of mouse and human origin (Raz, Meromsky and Lotan, 1986). This lectin, which agglutinates erythrocytes in the presence of a reducing agent, exists on B16 mouse melanoma cells in two forms of 14.5 kD and 34 kD as determined by immunoblot analysis with an appropriate McAb. The authors speculate that tumour cells adhere to themselves or to other cells through the binding of galactosyl residues on one cell to the lectin on the other. An analysis of lectin levels on various normal cells and tumour cells was performed by the same authors using a fluorescence activated cell sorter (FACS). They found that normal, diploid cells such as rat embryo fibroblasts bound less McAb directed against the lectin than did essentially the same cells recently transformed by transfection procedures. Furthermore, the authors also showed that poor lung colonizing variants of two transformed murine cell lines (K-1735 melanoma and UV-2237 fibrosarcoma) expressed less lectin than did their high lung colonizing variants. However, no difference in lectin expression was detectable between the low (B16F1) and high (B16F10) lung colonizing variants of the B16 melanoma, which suggests that there may not be any generalized relationship between the expression of this particular endogenous lectin and lung colonization. Nevertheless, pretreatment of B16F1 melanoma cells with the specific anti-lectin McAb resulted in a reduction in the number of lung colonies which developed following i.v. injection of the treated cells into syngeneic mice. Because the parental B16F1 vaiant was used in this study rather than the B16F10, the actual effects observed involved rather small lung colony numbers (e.g. a reduction from an average 11 colonies per mouse in controls to an average four colonies in experimental animals). A better understanding of the precise role of endogenous lectins in many aspects of tumour cell behaviour will only be gained as more experimental studies are reported.

1.7.8 Receptor mobility

When ferritin labelled con A was applied at 20°C to virally transformed mouse SV3T3 cells, the labelled lectin (bound to its cell surface receptor) was found to be more clustered than that on non-transformed 3T3 cells (Nicolson, 1971). However, in similar experiments in which the cells were fixed first the labelled lectin was randomly dispersed in both cell types (Nicolson, 1973). Results from many similar experiments have lead to the proposal that receptors may be more mobile on tumour cells when compared to that on normal cells. Receptor mobility has been studied further by the use of cell fusion techniques (Edidin and Weiss, 1974). These authors fused pairs of cells in three combinations (transformed–transformed, transformed–normal, normal–normal) and looked at the rates of receptor intermixing in the resulting heterokaryons (hybrids with non-fused nuclei). Receptor mobility was greatest in heterokaryons from transformed–transformed pairs and least in normal–normal heterokaryons.

One of the major problems with studies of receptor mobility is that if the probe used to mark the receptor is not univalent then the receptors will patch and subsequently cap under physiological conditions (Taylor *et al.*, 1971). The capping process is essentially blocked if the cell microfilaments and microtubules are disrupted by a combination of cytochalasin B and colchicine (De Petris, 1974). This result suggests that any differences in receptor mobilty detected under such circumstances may be a function of cytoskeletal involvement rather than of simple receptor diffusion within the membrane. For some receptors at least, mobility is probably a function of interaction between themselves (either directly or via an external molecule such as a lectin) and between themselves and internal molecules such as cytoskeletal components. Other receptors may not be connected either directly or indirectly to the cytoskeletal system and thus they would be free to diffuse in the plane of the membrane. When capping of surface immunoglobulin (sIg) was studied in human lymphocytes and CLL cells, it was found to occur faster and to a greater extent on the normal cells (Cohen and Gilbertsen, 1975). Essentially the opposite was found by Stackpole, Jacobson and Lardis (1974) when they studied patching and capping of TL alloantigens on TL^+ thymocytes and T cell derived RADA1 leukaemia cells. These and many other results again do not allow any general conclusions to be drawn. The interpretation of cytoskeletal involvement in receptor mobility during capping is further compounded by the observations of Yahara and Edelman (1972) which suggest that in some cases the cytoskeleton may serve to anchor or restrict the movement of receptors.

1.7.9 Membrane fluidity

As mentioned previously, there do not appear to be any significant and consistent changes in membrane fluidity associated with cell transformation as determined by electron spin resonance (ESR) techniques or by analysis of lipid composition (Gaffney, 1975; Yau *et al.*, 1976). In ESR, synthetic phospholipids containing a nitroxide group are experimentally introduced into cell membranes. The nitroxide group or 'spin label' contains an unpaired electron which absorbs energy with a characteristic spectrum when the cell is placed in an external magnetic field. As the labelled lipids disperse within the cell membranes the absorption spectrum alters, and in this way an estimate can be made of membrane fluidity.

1.7.10 Surface charge

Generally speaking, there do not appear to be any consistent differences in surface charge between cells obtained from normal tissues or from tumours, although occasional reports have suggested otherwise. According to the studies of Weiss (1979 a and b) surface charge is not related to malignancy or to metastatic potential.

1.7.11 Permeability and transport

Cells which escape the normal growth regulation controls are still dependent upon the supply of critical nutrients. Although the rate of transport of certain essential nutrients is likely to influence the rate of cell proliferation, increased transport is unlikely to be essential for transformation. The transport of some sugars (glucose, mannose, galactose and glucosamine) has been shown nevertheless to be increased in certain tumour cell lines, while the transport of other sugars remains little changed (Hatakana, 1974). Similarly, the transport of some amino acids such as glutamine and arginine (Foster and Pardee, 1969; Isselbacher, 1972) and of certain ions such as phosphate (Cunningham and Pardee, 1969) may be increased in tumour cell lines, but all of these changes are quantitative in nature and many may relate to the relatively high level of proliferation of tumour cells. Two proteins (75 kD and 92 kD) which are related to sugar metabolism have been shown to increase on transformation (Pouyssegur *et al.*, 1979).

It should be borne in mind that some experimental studies relate to uptake (i.e. incorporation into some product) rather than transport *per se* and that these two processes are separable phenomena. The transport events consequent upon triggering of a growth factor receptor (sections 1.7.1 and 1.7.22) appear not to be peculiar to tumour cells although there may be quantifiable differences between normal cells and tumour cells.

1.7.12 Cell junctions and intercellular communication

Like many other of the parameters used to illustrate possible differences between normal cells and tumour cells, differences in the nature of cell junctions vary for the cellular systems under examination. Intercellular junctions in tumour cells may differ in a number of ways from that shown in their equivalent normal cells:

(a) They may be present in lesser or greater numbers.

(b) They may be structurally altered.
(c) They may be defective functionally.
(d) They may be expressed inappropriately for a particular cell type.

In some studies, cell junctional formation has been reported to be diminished in both tumours and tumour cell lines (McNutt and Weinstein, 1967; McNutt, Culp and Black, 1971). In a review of the descriptive literature, Weinstein, Merk and Alroy (1976) concluded that although tumours displayed a wide range of junctions there was some indication that they were often less frequent than in normal tissues. However, there appears to be no consistent pattern and a correlation between junctional deficiency and the tumorigenic state is not always obvious. This is almost certain to reflect the qualitative nature of the information, and more detailed, quantitative studies looking at particular junctions in particular tumours are required.

Although there may be a general lack of consistency, any junctional defects which might exist in cancer cells could be expected to be reflected in three abnormal cellular functions, namely decreased intercellular communication mediated by gap junctions, decreased cell sealing mediated by tight junctions, and decreased intercellular adhesion mediated in part by most junction types, but particularly desmosomes (Weinstein and Pauli, 1987).

(a) *Effects on communication*

Intercellular communication can take place in several ways via soluble signals, either at a long distance (e.g. hormones) or at a short distance when cells are apposed (e.g. gap junctions). Small ions and molecules can move between cells coupled via gap junctions and in this way energy-rich metabolites such as nucleosides and sugar phosphates could move between cells. Other small molecules (typically <1000 D and as yet unidentified) which contribute to homeostasis, proliferation and differentiation could also pass between cells via gap junctions. It is thus possible to visualize gap junctions as linking cells into functional units which share parts of a common cytoplasm. Since tumours involve perturbation of normal patterns of cell proliferation and differentiation, changes in junctional communication might not be unexpected in light of the possibilities described above. McNutt, Hershberg and Weinstein (1971) estimated gap junctional frequencies in human cervical epithelium and concluded that there was a decrease in junctional number from normal squamous epithelium through metaplastic and dysplastic states to carcinoma *in situ* and invasive carcinoma. Both carcinoma *in situ* and invasive carcinoma had markedly low gap junctional frequencies, but these were not significantly different from each other, suggesting that the acquisition of an invasive character is not simply correlated with gap junction depletion. It may be that gap junction changes are more relevant to earlier events in tumour progression, but further studies are needed to support this possibility. Interestingly, tumour promoters such as the phorbol esters have been shown to reduce intercellular communication (Yotti, Chang and Trasko, 1979).

In a recent report from Yamasaki and Katoh (1988) it was shown that chemically transformed Balb/c 3T3 cells form a communicating network *in vitro* which is independent of adjacent normal cells. Perhaps some form of selective communication might be important in tumour cell growth? This is not always the case, however, since transformed C3H 10T1/2 cells fail to display a homotypic communicating network. In fact, they are communication deficient (Boreiko, Abernethy and Stedman, 1987). The studies of another group have shown that the formation of channels between normal cells and transformed cells can lead to restriction in growth of the latter (Mehta *et al.*, 1986). These observations suggest that in some tumours progression might be restricted by the passage of growth inhibitory molecules from normal cells. Indeed, whether an overt tumour develops or not may be a factor not only of transforming events but also of environmental effects mediated by neighbouring cells.

The studies reported above lead us to consider that there may be two aspects to the role of cell communication in tumour development:

1. Heterotypic junctions between normal and transformed cells might allow negative growth signals to pass between the two, thus keeping the latter type in check. Disruption of this junctional type might be proposed as a pre-requisite for tumours to grow.
2. Homotypic junctions between transformed cells, on the other hand, might promote tumour growth by encouraging the sharing of stimulatory signals and essential metabolites. The maintenance of this type of junction would facilitate tumour growth.

The different effects which might result from the formation of homotyic or heterotypic gap junctions adds another dimension to the study of cell communication in tumour development which as yet is little studied.

Azarnia and Loewenstein (1973) have reported a parallel correction of malignant growth and a genetic defect of intercellular communication through cell fusion. Thus when non-coupling tumour cells were fused with coupling normal cells, the hybrids formed were able to form communicating junctions, they grew in a density-dependent manner as do normal cells, and they were unable to form tumours when injected into host animals. More recent studies by Lowenstein and colleagues suggest that gap junctional alterations may follow transformation events which involve changes in protein phosphorylation activities, as has been suggested for many oncogenes. These authors have shown that cells transformed by temperature sensitive mutants of Rous sarcoma virus have their permeability down-regulated at the permissive (transformed) temperature (Wiener and Lowenstein, 1983; Azarnia and Lowenstein, 1984). Down-regulation is thought to be mediated by the src gene product $pp60^{src}$ which has tyrosine phosphorylation activity. According to Chang *et al.* (1985), transfection of NIH 3T3 cells with v-*src* increases protein C kinase activity and this is associated with decreased intercellular communication. More extensive studies are obviously required before we can evaluate fully the role of phosphorylation activity in gap junctional communication.

Further discussion on the nature of the gap junction is provided in section 5.3.1.3, but it seems that although some tumours have deficiencies in gap junctional structure and communication many may not, and such changes are thus unlikely to be characteristic of the tumorigenic state.

(b) *Effects on cell sealing*

Tight junctions act as barriers to the diffusion of ions and molecules across most epithelia and endothelia (see section 5.3.1.1). Pauli and Weinstein (1982) have studied the organization of tight junctions in three cell lines (derived from a chemically induced transitional cell carcinoma of the rat bladder) which differ in their invasive capabilities: line RBTCC-2 is a pre-invasive carcinoma, line RBTCC-5 is a locally invasive carcinoma, and line RBTCC-8 is derived from a metastasis of the parent tumour. Generally speaking, the level of tight junctional organization in these cell lines decreases with increasing invasive ability. Thus in line RBTCC-2 the tight junctions resemble those in normal rat bladder epithelium, whereas in line RBTCC-8 the junctional components (although present in near-normal amounts) are unable to be assembled appropriately to form the intramembrane fibrils which characterize tight junctions as visualized by freeze fracture electron microscopy. The tight junctions of the locally invasive cell line RBTCC-5 are intermediate in organization between these two extremes. It is likely that these structural changes will correlate wih altered barrier functions, but the significance of this in the progression of the tumour is not clear.

(c) *Effects on adhesion*

Since they link cells together, all cell junctions are likely to contribute to cell adhesion to some extent, but there is evidence based on their resistance to disruption (Chambers and Renyi, 1925) that desmosomes make a major contribution to the ultrastructurally recognizable means by which cells are held together (see section 5.3.1.4). Although our attention is drawn towards cell junctions when using the electron microscope, it is important to remember that cells may also adhere through relatively non-specialized regions of their surface. Since tumour cells must escape from the primary lesion during the process of metastasis, it is often thought that they are likely to display decreased homotypic (mutual) adhesiveness (Coman, 1944; McCutcheon, Coman and Moore, 1948). This generalization appears not to hold for all tumours since different research groups have reported markedly different levels of adhesiveness for various tumours. It should be borne in mind, however, that for technical reasons it is not possible to measure directly the actual force of intercellular adhesion and that the results obtained when attempts are made to quantify cell adhesion depend very much on the techniques employed (Elvin and Evans, 1982, 1984) and the state of the cell (Elvin and Evans, 1983). According to Pauli *et al.* (1978), cell junctions are sometimes abundant in clusters of tumour cells that are actively invading and thus junctional loss need not be a pre-requisite for all types of tumour cell invasion. In the chemically induced bladder transitional cell carcinoma model (RBTCC) there appears to be no correlation between desmosome number and invasiveness (Pauli *et al.*, 1978) although a relationship between these two characteristics has been suggested for human bladder carcinoma (Alroy, Pauli and Weinstein, 1981).

The rather limited techniques available for studying the development, structure and function of cell junctions has hindered progression in this research area. Although some tumours may show junctional changes which correlate with transformation or the acquisition of invasive potential, available evidence does not allow for any generalizations to be made. We may ask of what relevance is a detailed knowledge of tumour cell junctions? Of interest here is that they may be used as a guide in classifying certain types of tumours prior to clinical treatment. Discrimination between lymphomas and poorly differentiated carcinomas, for example, may be made on the presence of desmosomes in the latter, although antibodies directed against intermediate filaments may be more advantageous. In their study of over 100 different types of tumours, Weinstein, Merk and Alroy (1976) did see some specific patterns of junctional expression, but there were many exceptions particularly with respect to the state of differentiation and degree of anaplasia of the tumours. It may be that the main relevance of junctional studies will lie in gaining an understanding of the role of cell communication in transformation and progression, but much further research is required before a proper evaluation can be made. It is worth noting at this stage that some junctions are likely to be altered during cell division (although interphase and mitotic cells can be electronically coupled), and that many of the junctional changes reported for tumours may only reflect their mitotic state.

1.7.13 Deformability

Although probably not very important in tumour growth, cell deformability is likely to be important in tumour cell spread. Sato and Suzuki (1976), working with a series of rat ascites cell lines, showed that extrapulmonary tumours were found only in rats injected intravenously with highly deformable cell variants. Pulmonary tumours, on the other hand, were formed whether the injected cells were highly deformable or not. Cells collected from the left ventricle after direct inoculation into the pulmonary artery (i.e. after a single pulmonary passage) were tested for viability by dye exclusion. The less deformable cells showed greater cell death which was to some degree independent of cell size. The authors concluded that deformability promoted capillary passage and contributed to the more widespread metastases of such cells. Of course, normal cells also express a range of deformabilities with some (such as the PMNs) being markedly so. Nevertheless, deformability may be a trait which confers considerable advantages to metastasizing tumour cells, particularly those that spread through the blood.

1.7.14 Cell projections

Allred and Porter (1979) classified cell projections (or cytopodia) into several types:

(a) **Microvilli** (0.1 μm in diameter × 2–5 μm long).
(b) **Filopodia** (0.1–0.2 μm × 5–30 μm) which include attachment and detachment fibres and the larger lobopodia (0.5–3.0 μm wide).
(c) **Ruffles** or **lamellipodia** (sheet-like cytopodia often found at the leading edge of a cell).
(d) **Blebs** (round cytopodia 1–10 μm in diameter).

In general terms, transformed cells often have more cytopodia which are also more variable than those in non-transformed cells. However, both the density and the form of cytopodia vary with the cell cycle, and in this respect the cytopodia of transformed cells resemble those more characteristically seen on mitotic, non-transformed cells. Cytopodia are likely to function in motility and adhesion, but their appearance on rounded cells (either in M phase or after trypsin treatment, for example) suggests that they may also act as membrane stores for use when the rounded cell re-adheres and spreads on the substrate. The abundance of cytopodia on many transformed cells may be correlated with their reduced adhesiveness which leads to a rounder and less flattened cell. When fibronectin is added to transformed cells they become flatter, adhere more strongly to the substrate, and show a reduction in their number of microvilli and ruffles (Yamada, Ohanian and Pastan, 1976).

1.7.15 Cytoskeleton

The cytoskeleton is composed of three main structural components:

(a) **Microtubules** (25 nm in diameter)
(b) **Intermediate filaments** (10 nm)
(c) **Microfilaments** (7 nm)

A fourth category, in which fibres of various dimensions (2–15 nm) appear to form a microtrabecular lattice within the cell (Wolosewick and Porter, 1979; Porter and Tucker, 1981) is probably artifactual. This microtrabecular lattice is seen in cells prepared for electron-microscopy by critical point drying rather than by standard plastic embedding techniques. According to Ris (1985), traces of water and ethanol in the CO_2 used for critical point drying result in surface tension related distortions of other cytoplasmic filaments to yield the lattice-like meshwork.

Some transformed cells are characterized by a reduction in organization of the microfilamentous cytoskeleton as visualized using both immunofluorescent and ultrastructural techniques, but there are exceptions. Goldman *et al.* (1976), for example, did not detect any gross differences between the microfilament bundles of Swiss mouse 3T3 cells and their SV40 transformed counterparts. The transition of human colonic tumours from benign to malignant, on the other hand, has been associated with decreased actin organization (Friedman *et al.*, 1984). The relevance of microfilament changes in transformation is brought into question by the observation that artificial flattening of cells seems to result in a more elaborate cytoskeleton (Ali *et al.*, 1977).

There appears to be a general relationship between cell adherence and spreading and microfilament organization, all of which may be reduced in transformed cells but restored by fibronectin (Ali *et al.*, 1977), but the general consensus is that there are no consistent, qualitative changes in the cytoskeleton of neoplastic cells compared to normal cells. It has been claimed that tumour cells may contain more F actin than their normal counterparts (Low, Chapponier, and Gabbiani, 1981), and that this might contribute to their increased motility and invasiveness. There is in fact no incontestable evidence of any functional relationship between F actin content and invasiveness, and since increased actin polymerization is not a property of all tumour cells it may actually reflect phenomena unrelated to transformation *per se*.

Studies with metastatic variants of tumour cell lines such as the mouse K1735 melanoma system suggest that the less metastatic variants have a more elaborate stress fibre system than do the highly malignant variants (Volk, Geiger and Raz, 1984). As will become clear later, there is some evidence suggesting that stress fibres are more pronounced in stationary cells than in motile ones, but although locomotion may be advantageous to the metastatic cell, it appears not to be an absolute requirement for tumour spread (section 1.7.16).

Attention has already been drawn to the possibility of using intermediate filament type to classify tumours. The functional role of intermediate filaments is not yet properly understood, but their cellular disposition and their relative insolubility suggest a role in distributing intracellular shearing forces. Recent evidence indicates that the fibrous lattice of the nuclear lamina contains a class of intermediate filaments, collectively referred to as the lamins (not to be confused with the basement membrane protein laminin). This nuclear lattice (often referred to as the karyoskeleton) apparently connects with the perinuclear ring of intermediate filaments, which themselves can extend through the cytoplasm to interact with submembranous components (such as erythrocyte ankyrin). These submembranous components in turn connect to transmembrane receptors and may also interact with the microfilament system. This continuity of the cytoskeletal network is no doubt crucial to its proper functioning.

Table 1.7 summarizes the five main intermediate filament classes (based on sequence analysis) and illustrates their relationship to particular cells and tumours. Tumours generally express the same pattern of intermediate filaments as their cells of origin. In this respect, intermediate filaments are clinically useful in the diagnosis of metastases whose origins may otherwise be uncertain. On the other hand, it is clear that apart from one exception there is no evidence to suggest that malignant cells differ from normal cells in their

Table 1.7 Classification of principal intermediate filaments

Class	*Type*	M_r *(kD)*	*Cell types/tumour categories*
I	acidic keratins	40–60	epithelial; carcinomas
II	neutral/basic keratins	50–70	epithelial; carcinomas
III	vimentin	57	mesenchymal; haematopoietic cells; sarcomas; leukaemias; lymphomas
	desmin	53	muscle; myogenic tumours
	glial filament protein	51	astrocytes; gliomas
IV	neurofilaments	57–150	neurons; neural tumours
V	lamins	60–70	nuclear envelope

intermediate filament content. The one exception relates to ascitic carcinoma cells which express vimentin in addition to the keratin normally expresssed by epithelial cells *in vivo*; this difference may reflect nothing more than different growth conditions. Interestingly, B16 melanoma cells grown as spheroids display increased metastatic capacity and decreased vimentin synthesis relative to cells of the same tumour type grown in monolayer culture (Raz and Ben-Ze'ev, 1983), but there is no evidence that the two changes are causally related. A more recent report from Ben-Ze'ev and Raz (1985) has shown that cycloheximide treatment of B16 cells leads to a reversible reduction in synthesis of vimentin, tubulin and actin, although only intermediate filament organization is detectably altered. When injected i.v. into syngeneic mice, cycloheximide treated cells form fewer lung colonies. It is difficult to compare these two studies with respect to the role of vimentin in tumour cell behaviour, however, since cycloheximide (a potent inhibitor of protein synthesis) has multiple cellular effects.

1.7.16 Cell motility

The expansion of a primary tumour and indeed invasion of blood vessels can take place without recourse to cell motility since the cell division process itself leads to translocation. The underlying reason for this is quite obvious: two daughter cells cannot occupy the space previously occupied by the single parent. Eaves (1973) has suggested that mechanical pressure resulting from growth may underlie the invasive process, but this would not seem to be the whole story since rapid growth and invasiveness are not always correlated (Willis, 1973). Local invasion and metastasis would seem to benefit from cell motility (Strauli and Weiss, 1977) and in fact it is difficult (although possible) to conceive of metastasis not involving motility, particularly at the extravasation stage during haemogenic spread. Many normal tissue cells display little movement *in vivo* although the proper functioning of others clearly demands active motility. This is not to say that many normal tissue cells are not capable of movement, only that they may respond to certain signals which, by limiting cell movement, ensure the appropriate topography and functioning of particular organs and tissues. Tumour cells may fail to respond appropriately to such signals, although it is conceivable that non-motile tissue cells gain motility on transformation.

Whatever the process underlying changes in cell motility, available data would seem to suggest that cell motility is advantageous in metastatic spread but not absolutely necessary to the transformed cell. Changes in cell motility have been noted on transformation *in vitro* nevertheless. Bowman and Daniel (1975) found that the motility of SV40-transformed WI-38 human lung fibroblasts (0.8 $\mu m\ h^{-1}$) was less than that for young (4.6 $\mu m\ h^{-1}$) or old (2.6 $\mu m\ h^{-1}$) WI-38 cells, but similar changes have not been seen for other cells on transformation (Goldman *et al.*, 1974). Schor and colleagues have studied the ability of normal and transformed cells to migrate into 3D gels of type I collagen prepared from rat tail tendons. They found that the migration of WI-38 cells was inversely proportional to cell density, whereas the migration of their SV40-transformed counterparts increased in a proportional manner (Schor *et al.*, 1985c) and that a wide range of transformed fibroblastic cell lines showed this latter behaviour (section 4.7).

1.7.17 Enzymes

Weber (1977 a and b) made an attempt to correlate enzyme changes with the extent of malignancy through the use of the minimal-deviation hepatomas developed by Morris (1965). Minimal-deviation hepatomas, which vary in their growth rates and degrees of differentiation, were originally induced in rats by ingestion of *N*-(2-fluorenyl)-phthalamic acid. The idea behind the selection of these tumours was to produce hepatomas which differed only minimally from normal liver. Studies of the biochemical changes in these minimal-deviation hepatomas relative to normal liver would hopefully offer some insight into the process of transformation uncluttered by the myriad of changes presumably associated with progression. Selection of a panel of hepatomas whose character deviated from normal liver over a wide range might then also offer the possibility of comparing biochemical changes associated with transformation with those associated with progression. The general conclusion from the studies of Weber (1977 a and b) was that malignancy correlates with changes in nucleic acid synthesis, increased anabolism and decreased catabolism, which is perhaps not surprising given the nature of malignant cells. Despite the extensive nature of this work, the effects of cell division and transformation *per se* on enzyme changes are still not clear. Although it is often stated that the regulation of enzyme synthesis is abnormal in neoplasms, the universality of this claim remains to be fully substantiated.

The secretion of plasminogen activators (PAs), serine proteases which convert serum plasminogen to plasmin which then degrades fibrin, may be increased in tumour cells relative to that in normal cells (Unkeless *et al.*, 1973; Nagy, Ban and Brdar, 1977), but this is not always seen to be the case (Mott *et al.*, 1974). Interestingly, tumour promoters such as the phorbol esters and agents that can induce mutations by damaging DNA (such as UV light) enhance PA production. In a study of a number of rat liver epithelial cell lines, however, Montesano *et al.* (1977) could find no correlation between PA production and tumorigenicity, since both non-tumorigenic and tumorigenic cell lines displayed a wide range of PA activity. It would thus seem unlikely that elevated PA production is a universal feature of the transformed state. Although generally an extracellular enzyme, tumour-produced PA may be intimately associated with the cell membrane thus initiating local effects. Fibronectin is sensitive to treatment with plasmin (Hynes *et al.*, 1975) but the low level of csFN on transformed cells does not always correlate with plasminogen activator secretion (Pearlstein, 1976).

Many, but not all, malignant tumour cells show raised levels of lysosomal enzymes, such as the collagenases and cathepsin B, which may assist in the digestion of connective tissue during invasion (Dresden, Heilman and Schmidt,1972; Sylven, Snellman and Strauli, 1974; Poole *et al.*, 1978). The production of type IV collagenase, which digests basement membrane collagen, has been correlated with metastatic behaviour as shown by Turpeenniemi-Hujanen *et al.* (1985). These authors formed cell hybrids between normal cells and metastatic cells in which both the metastatic phenotype and type IV collagenase secretion were suppressed. The work of Hicks, Ward and Reynolds (1984), on the other hand, has shown that many invasive tumours do not synthesize metalloproteinases, the group of enzymes to which type IV collagenase belongs.

In transformed cell lines, proteolytic activity may be increased both *in vitro* and *in vivo* (Sylven and Bois-Svenssen, 1965; Unkeless *et al.*, 1973). Using clinical material, however, Strauli and Weiss (1977) found that although certain benign and malignant tumours (including fibroadenomas and breast carcinomas) secreted proteolytic enzymes, there was no relationship between enzyme release and invasiveness. This observation is supported to a degree by the study of Strauch

(1972), who found that benign tumours often showed increased levels of collagenase activity when compared to that in normal tissue. This increase, however, was seldom as significant as that seen with malignant tumours.

Interestingly, exogenous proteases have been shown to initiate cell division in some cultured cell lines thus allowing them to reach higher saturation densities (Burger, 1970), although there is some conflicting evidence on this point, particularly with respect to the effects of proteases on 3T3 cell division (Holley, 1975). Treatment of transformed cells with protease inhibitors can induce changes in phenotype towards that more typical of non-transformed cells (Schnebli and Burger, 1972) but again some research is at variance with this observation (Chou, Black and Roblin, 1974). Whether protease production by tumour cells is related to their loss of growth control or not still remains speculative.

In other studies, oligosaccharide degrading enzymes such as β-galactosidase, α-mannosidase, and neuraminidase were found at higher levels in malignant human breast and colon tissues when compared to that in normal tissues, but to what extent this was due to contaminating host cells is uncertain (Bosmann and Hall, 1974). Studies based on tissue fragments also suffer from the need to take into account endogenous protease inhibitors, many of which are found in the blood and tissue fluids, and this has often not been done. Furthermore, the presence of dead and dying cells in tumour tissues may contribute significantly to the overall enzymatic content. In many studies it is not clear what efforts were made to distinguish between total cellular enzymes, cell surface enzymes, or secreted enzymes, and some of these may be more relevant to tumour cell behaviour than others.

The activities of various glycosyltransferases have been studied in a variety of human and animal tumour systems. The glycosyltransferases catalyze the transfer of a sugar residue from a donor (usually a sugar nucleotide) to an acceptor (a sugar on a growing oligosaccharide arm of a glycoprotein or glycolipid). As mentioned above, sialyltransferase activity may be increased in some human tumours but this is often not the case in experimental tumour cell lines. Likewise, the activities of other glycosyltransferases seem to be variable in different tumour systems. Viral transformation of mouse 3T3 cells reduces the transferase activities associated with sialic acid, galactose and *N*-acetylgalactosamine but not that for mannose, which is actually increased (Patt and Grimes, 1974).

Increased levels of the cytochrome P450 mono-oxygenases may be associated with some human tumours. These enzymes turn inactive carcinogens into their active forms. Aryl hydrocarbon hydroxylase, for example, converts polycyclic aromatic hydrocarbons found in cigarette smoke into carcinogens and it may be elevated in lung cancer patients (Kouri *et al.*, 1982; Ayesh *et al.*, 1984). Such changes, however, are not universally detectable.

Some transformed cells express ectopic enzymes which are not present in their tissue of origin, while a number have been reported to secrete fetal isoenzymes. The expression of particular isoenzyme forms is often developmentally regulated as seen with fructose-1,6-diphosphate aldolase, for example, in which different forms are found in fetal and adult livers. In certain liver cancers, however, the fetal form of this isoenzyme rather than the expected adult form may be expressed (Schapira, Dreyfus and Schapira, 1963). A wide range of fetal isoenzymes have been shown to be expressed in human cancers, but they are not always specific and since some are also found in normal tissues, the changes are often quantitative rather than qualitative in nature. It is thought that perturbed regulation of gene expression may be responsible for the changes seen in those tumours which do express fetal isoenzymes (Weinhouse, 1983). The expression of fetal isoenzymes by such tumours is clearly another example of oncofetal antigen expression (section 1.7.5.1).

Finally, a number of serum enzymes may show differences in activity as seen, for example, with serum acid phosphatase which is often elevated in patients with prostatic carcinoma (Shaw *et al.*, 1982).

1.7.18 Hormones

Elevated levels of human chorionic gonadotrophic hormone (HCG) are found in the plasma of most women with choriocarcinoma. HCG is normally produced significantly only during pregnancy when it is secreted by the syncytiotrophoblasts of the placenta, so that its presence in the plasma of non-pregnant women is usually a good indicator of a tumour of placental origin. Some men with testicular tumours also show elevated levels of HCG.

A number of tumours produce hormones which are inappropriate considering their tissue of origin (**ectopic**). In some instances, perhaps due to aberrant synthesis, such hormones may not be fully functional. Oat cell carcinoma of bronchus, however, has been found in some cases to secrete functionally active adrenocorticotrophic hormone (ACTH) which is normally produced by the pituitary. ACTH controls the synthesis and secretion of the glucocorticoids within the adrenal gland. The clinical effect of functional ACTH production by oat cell carcinoma is manifested as an ectopic form of Cushing's syndrome, which develops in response to excessive secretion of cortisol. The ectopic secretion of hormones such as ACTH in oat cell carcinoma may result from the re-expression of suppressed genes (**derepression**). Alternatively, some hormone secreting tumours may arise in non-endocrine tissues from cells which actually have endocrine functions. These are the so-called **apud** cells which have the common property of **a**mine- **p**recursor **u**ptake and **d**ecarboxylation, and which give rise to tumours known generically as **apudomas**. Obviously, hormones produced from apudomas are not ectopic in the strict sense at all. Since the cellular origin of oat cell carcinoma of bronchus is still under debate, it is not yet clear whether this is truly an ectopic hormone-secreting tumour. It should be noted here that ectopic hormone production is not a universal attribute of tumours, and nor is it exclusive to them since a number of non-neoplastic diseases may also display this particular trait.

Other tumours may be hormone dependent, examples of which include prostatic carcinoma which is androgen dependent, some thyroid carcinomas which are dependent on thyroid stimulating hormone (TSH), and certain breast carcinomas which are sensitive to manipulation by various sex hormones. Certain strains of mice such as RIII, GR, and BR6 frequently display pregnancy-related mammary tumours which are sensitive to levels of prolactin, progesterone and oestrogen (reviewed in Matsuzawa, 1986). When hormone levels are insufficient for growth, many of these tumours may become dormant. Mammary tumours in these mouse strains nevertheless progress towards autonomy, that is, they escape progressively from hormonal control and acquire resistance to endocrine therapy. There is considerable evidence that hormones influence the growth and progression of tumours and some, such as many of the steroidal hormones, may also act as tumour promoters. The hormonal environment can also influence metastasis, as shown experimentally in nude mice where the metastatic ability of MCF-7 human mammary carcinoma cells is apparently enhanced by the ectopic implantation of oestrogen-containing pellets (Liotta, Lee and Morakis, 1980).

1.7.19 Serum glycoproteins

Some serum glycoproteins, particularly the acute phase reactants such as α_1-antitrypsin and transferrin, may be elevated in cancer but they tend to lack specificity (reviewed in Drummond and Silverman, 1982). In many instances elevation of serum proteins may be in response to the presence of the tumour rather than a result of production by the tumour itself and as such they are not representative of the tumour cell phenotype.

The shedding of tumour membrane proteins referred to previously may lead to their detection in the serum, and ganglioside bound serum sialic acid (Kloppel *et al.*, 1977) and serum glycosyltransferases (Weiser and Wilson, 1981) may be elevated in association with the presence of a tumour.

1.7.20 Tumour angiogenesis factors

The induction of new blood vessel growth (**angiogenesis**) is essential if most solid tumours are to grow to a size beyond about 3 mm in diameter, since the transport of nutrients and waste products cannot take place efficiently beyond this distance by diffusion alone. These new blood vessels not only promote the growth of the tumour, but they also facilitate the possibility of tumour spread.

When a tumour is grafted into a slit made in the cornea of an experimental animal (about 1–2 mm from the normal vascular bed), blood capillaries are induced to grow towards the graft at about 0.2 mm per day. Similarly, when a tumour is implanted into the chorio-allantoic membrane of the chick egg (usually on day 9), the growth of new capillaries is induced within 72 h. Using assay procedures such as these, many groups have shown a range of tumours to be angiogenic and the hypothesis has been put forward that tumour cells release a soluble factor(s) which acts on endothelial cells to induce new blood vessel formation (reviewed in Folkman and Klagsbrun, 1987a; Folkman, 1985; Vallee *et al.*, 1985).

Angiogenic activity is acquired or markedly increased during transformation. Tsukamoto and Sugino (1979), for example, reported that transformation of Balb/c 3T3 cells by fragments of bovine adenovirus type 3 DNA correlated with the expression of angiogenic activity. Furthermore, the mammary gland of rodents does not normally have significant angiogenic activity but its acquisition correlates with the development of a malignant mammary tumour. In fact, elevated angiogenic activity may be detectable before any morphological indication of malignancy (Gimbrone and Gullino, 1976; Maiorana and Gullino, 1978). Jensen *et al.* (1983) have shown from their studies with human breast tissue that normal lobules from cancerous breasts are more likely to induce angiogenesis than are lobules from non-cancerous breasts. They suggest that preneoplastic transformation in the breast is diffuse, possibly involving more than 25% of all lobules, and that it precedes morphological alteration. Like normal mammary tissue, benign mammary tumours have little angiogenic activity, but it should be noted that some benign tumours (such as adrenal adenomas) may be associated with high angiogenic levels.

Furthermore, some tumours (particularly low density tumours such as the ascites type, in which the cells are essentially in suspension within a body cavity) may continue to grow in the absence of angiogenesis.

Recent work has lead to the identification of a number of angiogenic factors, but many still remain to be characterized and others probably still await discovery. Four major groups of angiogenic factors have been isolated and characterized from tumours:

(a) *Fibroblast growth factor (FGF)*

Both acidic and basic forms of FGF are angiogenic (section 1.7.1). A cationic polypeptide of about 18 kD which stimulates the migration and proliferation of endothelial cells was isolated from a rat chondrosarcoma by Shing *et al.* (1984, 1985). Since the same molecule stimulates the proliferation of mesenchymal cells, the authors speculated that angiogensis may be associated with fibroblast migration and proliferation. FGF directly promotes endothelial cell proliferation, and it also stimulates angiogenesis when as little as 0.33–0.54 pM is applied to the chick chorio-allantoic membrane.

(b) *Epidermal growth factor (EGF) and transforming growth factor-α (TGF_α)*

In the hamster cheek pouch model, both EGF and TGF_α have been shown to be active angiogenic agents at around 2 nM and 50–170 pM respectively.

(c) *Angiogenin*

This angiogenic protein (M_r 14 400 D) was isolated from the human adenocarcinoma cell line HT-29 by Fett *et al.* (1985). It is active in inducing angiogenesis when as little as 3.5 pmoles are added to the chick chorio-allantoic membrane. The amino acid sequence of angiogenin has been described (Strydom *et al.*, 1985) from which it has been determined that angiogenin is about 35% homologous with human ribonuclease. The exact significance of this relationship, however, remains uncertain and angiogenin appears not to have any enzymatic activity. Angiogenin cDNA has been isolated from a human liver cDNA library (Kurachi *et al.*, 1985), but despite this wealth of detail on the molecule itself little is known of precisely how it acts on endothelial cells to induce angiogenesis.

(d) *Low molecular weight products*

Several reports have suggested that low molecular weight extracts from cancer cells have angiogenic activity. A recent study by Kull *et al.* (1987) has shown such extracts from W256 carcinosarcoma cells to include two peaks of activity. One apppears to be due to nicotinamide, while the other is a molecular complex containing nicotinamide or its derivatives. Nicotinamide is a vitamin which is found in the ubiquitous cofactor nicotinamide adenine dinucleotide (NAD) and its derivatives, so important as electron carriers in the oxidation processes which yield energy for the cell. According to Kull *et al.* (1987), nicotinamide is active in inducing angiogenesis in the rabbit cornea in amounts equivalent to that found in normal tissue, but the mechanism by which it works is unknown.

A fifth angiogenic factor known as **platelet-derived endothelial cell growth factor** (PD-ECGF) has recently been characterized from platelet extracts. This 45 kD protein appears to be the only angiogenic factor isolated so far which has more or less exclusive specificity for endothelial cells (Ishikawa *et al.*, 1989).

Since some of the molecules above are known to be produced by normal cells, it follows that the ability to induce angiogenesis is not an exclusive attribute of the transformed state. Activated macrophages (Polverini *et al.*, 1977) and lymphocytes, particularly lymphokine-secreting T cells (Adelman *et al.*, 1980), are known to produce angiogenic factors. Indeed, there is a possibility that angiogenic effects attributable to transforming growth factor-β (TGF_{β}) might in fact be mediated by products from activated macrophages, since this molecule is an active chemoattractant for these cells. Recent research by Leibovich and his colleagues (1987) has shown that the angiogenic factor released by activated macrophages is identical to tumour necrosis factor-α (TNF_{α}). This raises an interesting paradox. On the one hand TNF_{α} promotes blood vessel growth which should be beneficial to the tumour, while on the other it is capable of destroying the tumour. The overall outcome is presumably a result of a balance between the two, which may depend on local environmental parameters. The suggestion has been made that extravascular TNF_{α} may induce angiogenesis, while intravascular TNF_{α} may stimulate procoagulant activity associated with thrombosis, small vessel destruction and haemorrhage (reviewed in Folkman and Klagsbrun, 1987b). TNF_{α}, like most of the other agents listed above, is active at low concentrations (when around 0.2 pmol is applied to the CAM).

During wound healing and the menstrual cycle, angiogenesis is a perfectly normal physiological response. The main difference between normal and tumour induced angiogenesis seems to be the lack of any limiting control in the tumour situation, although more angiogenic factors need to be characterized to see if normal tissues and tumours produce similar or different factors. Thus although the induction of angiogensis is not peculiar to tumour cells, the possibility remains that some tumours may elaborate particular angiogenic factors not usually associated with normal cell types. The postulated role of angiogenesis in metastasis and the concept of anti-angiogenesis as a technique for limiting tumour growth and controlling metastasis are described in section 6.4.

1.7.21 Polyamines

Polyamine synthesis is essential for progression through the cell cycle. Synthesis is via the diamine putrescine which is produced from the amino acid ornithine under the action of the enzyme ornithine decarboxylase (ODC). Putrescine then yields spermidine and subsequently spermine under the action of appropriate synthesases. It has been suggested that the intracellular spermidine:spermine ratio may be a guide to cellular activity. Low ratios (<1) are found in steady-state tissues, while higher ratios (>2) are found in more active tissues such as growing tumours. Like spermidine, intracellular putrescine may increase during tumour growth, but these changes are probably also cell cycle related and not related to transformation *per se*. For some tumours, such as medullablastomas, polyamine levels (specifically in the cerebrospinal fluid) may be of diagnostic significance.

ODC may be elevated in proliferating cells and could serve as a target for the development of drugs which inhibit cell division. Alpha-difluoromethylornithine (DFMO), an irreversible inhibitor of ODC, has been shown to inhibit cell proliferation and in fact appears to be selectively cytotoxic *in vitro* for some tumours such as small cell carcinoma of lung. DFMO has been shown to inhibit experimental metastasis (Sunkara and Rosenberger, 1987) but its possible clinical use remains to be evaluated.

1.7.22 Intracellular signals

Many observations have been reported on shifts in the levels of intracellular signalling molecules, such as the cyclic nucleotides and calcium, within tumours. Most of these shifts are associated with the higher activity of tumour cells when compared to normal cells (e.g. their commonly higher mitotic index) and they represent quantitative rather than qualitative differences. Nevertheless it would seem likely that the expression of the transformed phenotype is intimately associated with a range of intracellular signals, although exactly how they manifest the recognizable changes seen on transformation is by no means certain.

Three major intracellular signalling pathways have been shown to exist in a variety of cell types:

(a) The tyrosine kinase receptor pathway.
(b) The cyclic adenosine monophosphate (cAMP)-dependent pathway.
(c) The phosphatidylinositol pathway.

All three pathways are activated following the binding of an appropriate ligand to its receptor. The paradigm for activation of the tyrosine kinase receptor pathway is EGF interaction with its receptor on 3T3 cells, that for the cAMP-dependent pathway is the binding of adrenaline to beta receptors in muscle cells, while that for the phosphatidylinositol pathway is the binding of platelet-derived growth factor (PDGF) to its receptor also on 3T3 cells. Many of the biochemical events associated with intracellular signalling may be illustrated by the response of cells to growth factors (Fig. 1.4). Generally speaking, the response to a growth factor involves seven stages:

(a) The activation of a receptor.
(b) Transduction of the signal to the cytoplasmic side of the cell.
(c) Amplification of the signal.
(d) Transfer of the signal to the nucleus via secondary messengers. It should be noted here that evidence is mounting for the additional presence of nuclear receptors for a range of growth factors including EGF and PDGF, but the functional significance of this location is not yet clear (reviewed in Burwen and Jones, 1987).
(e) Activation of mRNA synthesis (transcription).
(f) The production of crucial proteins on cytoplasmic ribosomes (translation).
(g) DNA synthesis and subsequent division.

Although the full sequence of events for all three pathways is far from clear, a good deal is known about the responses of mouse 3T3 cells to PDGF, which acts via an inositol phospholipid pathway (reviewed in Cochran, 1985; Abdel-Latif, 1986; Rozengurt, 1986; Berridge, 1987). As far as PDGF is concerned, binding to its receptor (a transmembrane glycoprotein) leads to phosphorylation of receptor tyrosine residues, redistribution of the receptor–ligand complex within the plane of the membrane, and endocytosis. PDGF binding also induces hydrolysis of membrane polyphosphoinositides such as phosphatidylinositol 4,5-bisphosphate (PtdIns $4,5P_2$) with the production of diacylglycerol (DAG) and inositol triphosphate (Ins $1,4,5,P_3$). This event is brought about by the membrane bound enzyme PtdIns $4,5P_2$ phosphodiesterase (PDE) which thus acts as the amplifying system, and both DAG and Ins $1,4,5,P_3$ then act as intracellular messengers. During transduction, the message seems to be conveyed from the receptor to the enzyme via a G protein whose activity is regulated by guanosine triphosphate (GTP) in a manner similar to that described below for the cAMP-dependent pathway (reviewed in Abdel-Latif, 1986).

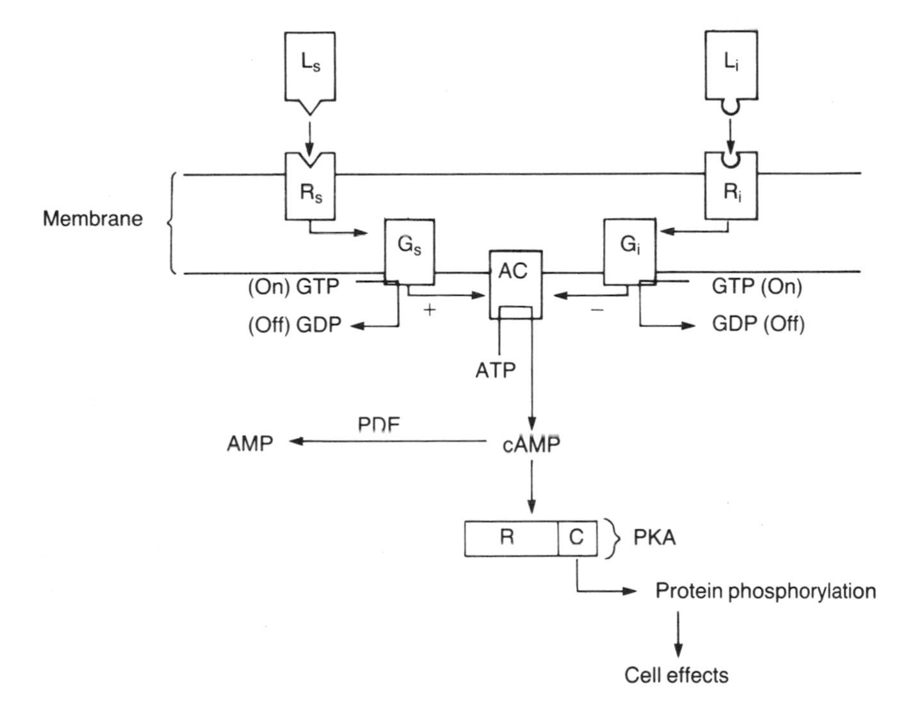

Figure 1.4(a) Intracellular signalling pathways. The cAMP-dependent pathway (modified after Berridge, 1985). The cAMP-dependent pathway has stimulatory (s) and inhibitory (i) arms. Transduction of the signal, initiated after receptor–ligand binding, is mediated via stimulatory and inhibitory guanosine nucleotide-binding G proteins. These proteins are active when GTP is bound, but become inactive when it is hydrolyzed to GDP. Abbreviations: L_s: stimulatory ligand; L_i: inhibitory ligand; R_s: receptor for L_s; R_i: receptor for L_i; G_s: stimulatory G protein; G_i: inhibitory G protein; AC: adenylate cyclase; GTP: guanosine triphosphate; GDP: guanosine diphosphate; ATP: adenosine triphosphate; cAMP: cyclic 3′,5′-adenosine monophosphate; PDE: phosphodiesterase; AMP: adenosine monophosphate; R: receptor sub-unit of PKA; C: catalytic sub-unit of PKA; PKA: cAMP-dependent protein kinase.

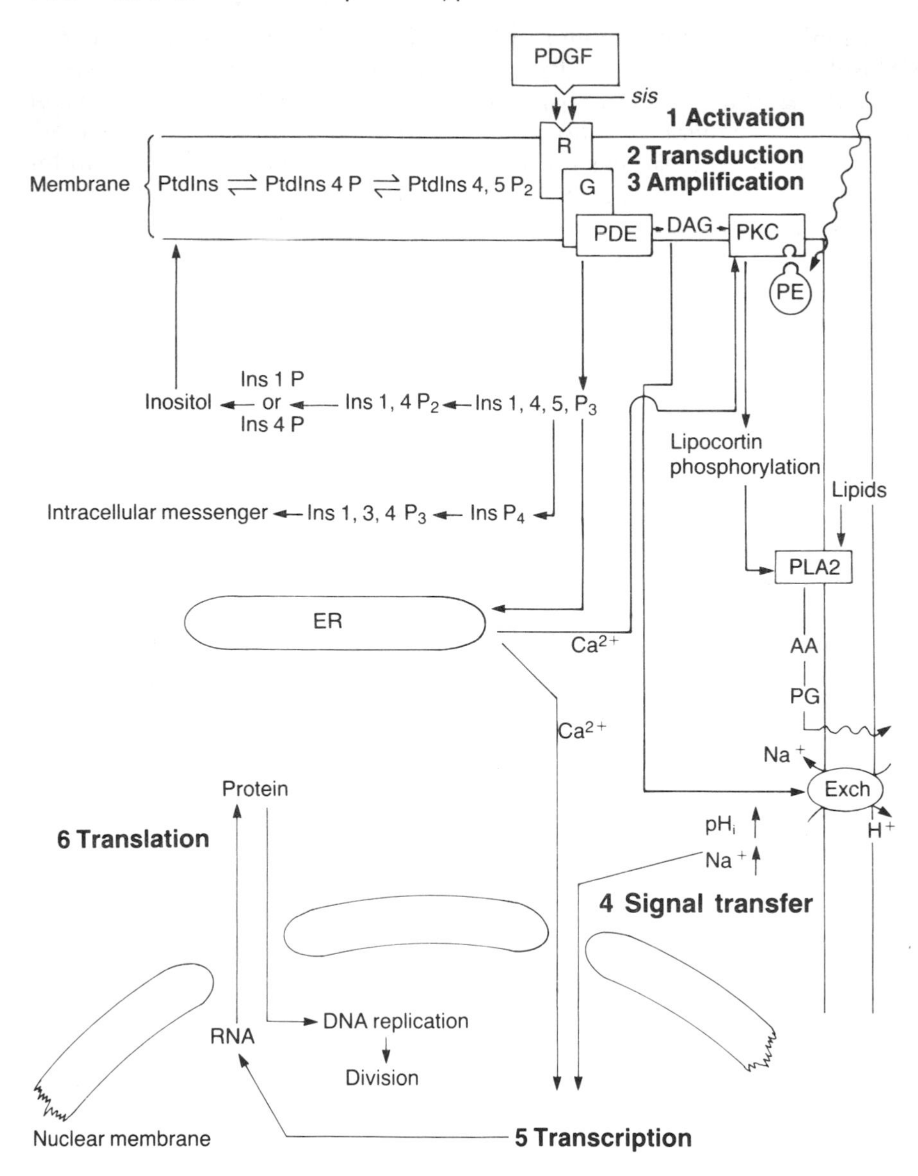

Figure 1.4(b) (opposite) The phosphatidylinositol pathway (modified after Berridge, 1987). Note that the effects of the Na^+/H^+ antiporter leading to increased pH_i may occur only under conditions where bicarbonate is low, and that under more physiological conditions a decrease in pH_i may actually result largely as a consequence of Na^+-independent Cl^-/HCO_3^- exchange. Abbreviations: PtdIns: phosphatidylinositol; $PtdIns_4P$: phosphatidylinositol 4-phosphate; $PtdIns_{4,5}P_2$: phosphatidylinositol 4,5-bisphosphate; $Ins1,4,5P_3$: inositol 1,4,5-triphosphate; $Ins1,4P_2$: inositol 1,4-diphosphate; Ins1P: inositol 1-phosphate; Ins4P: inositol

DAG, in the presence of Ca^{2+} activates protein kinase C (PKC) which stimulates the Na^+/H^+ antiport system increasing intracellular pH (pH_i) whilst also promoting the influx of Na^+. The Na^+ influx stimulates Na^+/K^+ pump activity with a consequent increase in K^+ influx. DAG may be metabolized into arachadonic acid (AA) which in turn could yield eicosanoids such as the prostaglandins (PGs) with numerous cellular effects including influence on cyclic nucleotide levels. The release of arachidonic acid involves the enzyme phospholipase A_2 (PLA_2), whose activity is inhibited by lipocortin. The inhibitory activity of this protein is blocked following phosphorylation via the DAG/PKC pathway. In an alternative metabolic pathway, DAG is conserved for re-use.

It is worth noting here that PKC activity can be inhibited by sphingosine and lysosphingolipids, which are the *N*-deacylation breakdown products of the sphingolipids (section 1.7.5). Cell surface gangliosides are an obvious source of these breakdown products. Providing the lytic products can traverse the plasma membrane (since the parent molecules reside in the outer lipid leaflet), then the possibility exists that they may function as intracellular messengers counteracting the effects of DAG (reviewed in Hannun and Bell, 1989).

Although the DAG-induced increase in intracellular pH is associated with protein and DNA synthesis, the activation of glycolysis, and the assembly of microtubules important in cytokinesis, its precise role in cell division is not clear. Perona and Serrano (1988) have recently induced the expression of yeast plasma membrane H^+-ATPase in NIH 3T3 and monkey Vero fibroblasts via a plasmid intermediate carrying the appropriate cDNA. Cells expressing the enzyme which showed an increase in pH_i were able to grow in sloppy agar and formed tumours in nude mice. The interpretation of this and other similar observations is complicated by a report that under physiological conditions certain growth factors (arginine vasopressin) may actually induce a decrease in pH_i (Ganz *et al.*, 1989). The original work by Moolenaar and colleagues (1983, 1984) linking the growth factor induced increase in

---

4-phosphate; InsP4: inositol tetrabis-phosphate; Ins1,3,4P_3: inositol 1,3,4-triphosphate; PDGF: platelet-derived growth factor; R: growth factor receptor; G: G protein; PDE: PtdIns4,5P_2 phosphodiesterase; DAG: diacylglycerol; PKC: protein kinase C; PE: phorbol ester; PLA_2: phospholipase A_2; AA: arachidonic acid; PG: prostaglandin; Exch: Na^+/H^+ antiport exchanger; ER: endoplasmic reticulum.

Consult the text for further details. See Fig. 5.9 for further information concerning the metabolism of arachidonic acid.

pH_i with stimulation of the Na^+/H^+ antiporter via PKC activation was actually done in the absence of bicarbonate (HCO_3^-) and this may account for the results obtained. In the presence of HCO_3^-, it appears that pH_i is regulated not so much by the Na^+/H^+ antiporter as by Na^+-dependent and Na^+-independent Cl^-/HCO_3^- exchange systems, with the Na^+-dependent system predominating. The alkali extruding effects of this system will tend towards cytoplasmic acidification. Tannock and Rotin (1989) make the interesting point that HCO_3^- levels may be relatively low in the acidic regions of tumours, which thus might leave scope for a pH_i regulatory role for the Na^+/H^+ antiporter *in vivo*.

The Ins 1,4,5,P_3 generated following PDGF binding to its receptor is thought to mediate the mobilization of Ca^{2+} from stores in the endoplasmic reticulum (ER) and these ions may then cooperate in the activation of PKC and other Ca^{2+}-dependent cellular processes. It is postulated that Ins 1,4,5,P_3, which is water soluble, may diffuse from the plasma membrane to the endoplasmic reticulum where it may bind to a specific receptor which leads to Ca^{2+} mobilization. The increased cytosolic Ca^{2+} may also shift the activity of the membrane bound Ca^{2+} pump promoting the Ca^{2+} efflux which is detectable within seconds of growth factor binding to 3T3 cells. Ins 1,4,5,P_3 may be metabolized in two ways, one involving sequential dephosphorylation to yield inositol, and one via phosphorylation to yield inositol 1,3,4,5-tetrakisphosphate ($InsP_4$) which is then dephosphorylated, yielding first Ins 1,3,4,P_3. There is some indication that Ins 1,3,4,P_3 may also act as an intracellular messenger.

Cytosolic Ca^{2+} is regulated primarily by the calcium binding protein **calmodulin**. This is a single chain, 17 kD polypeptide of 148 amino acids with four high affinity (Kd 10^{-6}M) Ca^{2+} binding sites. The cellular effects of Ca^{2+}/calmodulin are remarkably broad, extending from influence over a wide range of enzymatic activities to effects on cell behavioural patterns such as adhesion and motility. Calmodulin synthesis occurs in cells at the G_1/S interface (reviewed in Veigl, Vanaman, and Sedwick, 1984). Despite the importance of calmodulin in regulating cell growth, its synthesis at the G_1/S interface is not unique in that it occurs in concert with that of many cellular proteins. Various cell types seem to differ in their requirements for calmodulin, but in part this may reflect technical difficulties associated with quantifying intracellular levels of this protein. Furthermore, subtle changes in the cytoplasmic position of calmodulin in different cell types may have profound effects on cellular responses. Such difficulties notwithstanding, numerous reports have suggested that calmodulin levels may be higher in transformed cells than in their normal counterparts. In many studies these differences remain even when the data on calmodulin levels in normal and transformed cells are corrected for total cell protein content. Thus RSV-transformed NRK cells and SV40-transformed 3T3 cells both show a two- to three-fold increase in calmodulin levels per mg protein relative to their normal counterparts (Chafouleas *et al.*, 1980, 1981). Confirmation, for some cell types at least, that the increase in calmodulin seen in transformed cells relative to normal cells is not a simple function of increased division rates in the former has been gained from a comparison of leukaemic and normal lymphocytes (Takemoto and Jilta, 1983). Thus both dividing and non-dividing (thymidine blocked) leukaemic lymphocytes display increased calmodulin levels relative to normal lymphocytes. However, other studies such as those by Connor *et al.* (1983) using various temperature-sensitive mutants have failed to show any relationship between calmodulin content and transformation, and thus generalizations are unlikely to hold across the spectrum of tumours.

Another Ca^{2+}-binding protein known as **oncomodulin** has been identified in tumour cells from both rats and humans (MacManus, Whitfield and Stewart, 1984). Since its presence has also been reported in extra-embryonic tissues but never in normal adult tissues, oncomodulin is a good candidate for an oncofetal antigen (section 1.7.5.1). Co-transfection of primary mouse kidney cells or Rat-1 cells with an oncomodulin-containing plasmid, however, failed to result in any obvious changes in growth properties of the cells and thus the significance of the molecule in transformation is unclear (Mes-Masson *et al.*, 1989).

Recent studies on the effects of PDGF on 3T3 cells have also shown that this growth factor stimulates cAMP production, and that a sustained increase in this cyclic nucleotide acts as a second signal for cell division. In fact, the stimulatory effects of PDGF on cAMP levels in 3T3 cells appear to be indirect, occurring via E-series prostaglandins which escape from the cell and stimulate cAMP via their own receptors. Earlier work on the cellular effects of cAMP, either indirectly by inhibiting phosphodiesterase activity (Burk, 1968) or directly by adding the nucleotide itself (Ryan and Heidrick, 1968), suggested that cAMP inhibited cell division and that some transformed cells had relatively low cAMP levels (reviewed in Pastan, Johnson and Anderson, 1975). This simplistic view is complicated by studies such as those of Gottesman and colleagues (1984), for example, who have provided evidence indicating that whereas the growth of normal CHO cells is inhibited by cAMP, that of RSV-transformed variants is promoted. Nevertheless, in contrast to the early view that cAMP inhibited cell division, Goldberg *et al.* (1974) suggested that cGMP stimulated cell division: they thus put forward their yin–yang hypothesis in which division is thought to be regulated by the ratio of the two cyclic nucleotides. This too now appears to be an oversimplification of the process, and indeed many of the early results were probably due to lengthy stimulation with non-physiological concentrations of nucleotides. Of interest here is the observation that guanylate cyclase appears to be activated in association with the phosphatidylinositol pathway, leading to an increase in cGMP. Exactly how this activation is brought about is not yet clear, and although cGMP activates a specific kinase how this molecule then influences cellular events remains to be determined.

Although there is some convergence in their general modes of action, initially the cAMP-dependent pathway proceeds differently to the phosphatidylinositol pathway described above. Ligands involved in the cAMP-dependent pathway either stimulate or inhibit adenylate cyclase activity. Signal transduction involves two different G proteins, so called because they bind guanosine nucleotides. One G protein (G_i) inhibits adenylate cyclase activity while the other (G_s) stimulates it. The interaction between ligand and receptor stimulates binding of GTP to the appropriate G protein. This leads to the release of a subunit that interacts with adenylate cyclase, either stimulating or inhibiting its activity as the case may be. Control is effected through the transducing system being switched off by GTPase-mediated hydrolysis of the bound GTP to GDP, which is thus the nucleotide that binds to G proteins when the receptor is in an inactive state. Following stimulation, adenylate cyclase acts as the amplifying system generating cAMP. This binds to a cAMP-dependent kinase liberating a catalytic component which then phosphorylates target proteins that effect cellular responses appropriate to the original activating ligand.

Ligands which act via the cAMP-dependent pathway generally influence cellular responses other than division. Adrenaline, for example, stimulates a variety of responses depending on cell type. These include glycogen breakdown, cell contraction, fluid secretion and so on. This is not to say that cAMP is without effect on cell division, but it must be admitted that its role in this process still remains somewhat enigmatic. In fact, it appears that cAMP can have positive or negative effects on cell growth dependent upon dose, cell type, and *in vitro* growth conditions. It is even possible that some cells may be able to do without cAMP mediation altogether: cAMP exerts its effects on cells via two cAMP-dependent protein kinases, but mutant CHO cells containing defective versions of these enzymes are nevertheless capable of dividing normally (reviewed in Gottesman and Fleischmann, 1986).

The major effects of cAMP are thought to result from the activation of cAMP-dependent protein kinases, the general notion being (as seen above) that phosphorylation/dephosphorylation events can exert regulatory control over enzymatic processes. Of course the regulation of enzymatic activity can also be influenced by phosphorylation-independent steps such as feedback inhibition, gene induction or repression, and transcriptional and translational associated events, and many molecular details remain to be clarified.

The roles played by the prostaglandins are diverse, but they contribute significantly to platelet aggregation, either inhibiting (e.g. PGI_2) or stimulating (e.g. TXA_2) the strength of the response. Platelet interactions may influence blood borne tumour spread, and the significance of this in metastasis (along with a postulated anti-metastatic role for PGI_2) is described in section 5.7.2.2. Although prostaglandin content has been shown to be elevated in a number of tumours, their precise involvement in tumorigenesis is not clear. Inhibition of prostaglandin synthesis by indomethacin has been shown to inhibit the growth of a some murine tumours, while exogenous PGE derivatives apparently inhibit the growth of the B16 murine melanoma both *in vitro* and *in vivo*, and also enhance melanogenesis (reviewed in Cho-Chung, 1982). As the intracellular signalling story unwinds it is proving to be so complex that it would be premature to reach any meaningful conclusions on only one aspect when several interrelated events are involved.

Thus although the ways in which cells respond to growth factors are becoming clearer, there is still some uncertainty associated with many of the molecular events involved. According to Rozengurt and Mendoza (1985), only two of a wide range of growth factors they have studied can independently induce DNA synthesis in 3T3 cells: these are PDGF and various bombesin derivatives. All other growth factors they have studied (including insulin, cAMP, PGE_1 and EGF) only stimulate cell division in association with other growth factors. This synergistic requirement for almost all growth factors seems to indicate that there is a need for cooperation between different molecular pathways within the cell. Since any given cellular response may be subject to a number of molecular controls, it is conceivable that different cell types may well utilize different intracellular signalling pathways in their responses to the same stimulatory molecule. Despite this possibility, most cells are likely to work to the same basic principles, and although not all cell types will respond to the same stimulatory molecule, those that do respond are likely to utilize similar intracellular mechanisms. Indeed, even if a cell does not respond to PDGF, for example, its response to other mitogens will almost certainly involve many of the molecular pathways currently being elaborated for 3T3 cells. It must be emphasized, however, that not all growth factors act via the inositol phospholipids: EGF and insulin, for example, appear to act via different pathways. Thus in many responsive cell types EGF induces changes in pH_i and Ca^{2+} with minimal effect on inositol phospholipid breakdown, and unlike the responses of most growth factors it is dependent on an external (rather than internal) Ca^{2+} supply. As we have seen, the receptors for EGF and insulin are both tyrosine kinases, and it seems that the phosphorylating activities of these molecules initiate the intracellular responses.

The intracellular signalling pathways referred to above are influenced by tumour associated events in a number of ways:

(a) Tumour promoters such as the phorbol esters (PE) appear able to directly activate PKC, which thus serves as a cell receptor for these agents. The exact mechanism by which PKC activation promotes carcinogenesis, however, is not clear.
(b) The signalling pathway may be usurped by *onc* gene products because of their homology with the growth factor or its receptor as described previously.
(c) PDGF and other growth factors induce expression of the cellular *onc* genes c-*fos* and c-*myc*, again as described earlier in this chapter. Exactly what the products of these *onc* genes achieve is unclear, but they appear to be involved in nuclear events associated with signalling.
(d) Finally, adenylate cyclase activity and the level of the second messenger cAMP may be influenced more directly by some *onc* gene products such as $p21^{ras}$, the product of the Harvey and Kirsten murine sarcoma viruses. This protein binds to guanosine triphosphate (GTP) and functions as a G protein as described above. Interestingly, $p21^{ras}$ is also a GTPase which might at first be thought to lead to decreased cAMP production since this enzyme would hydrolyze GTP and switch off the action of the G protein on adenylate cyclase. However, the story is more complicated (as is usually the case!) since the product of the transforming *ras* gene is less active as a GTPase than the product of its normal cellular counterpart (c-*ras*) which usually does not transform cells. A decrease in GTPase activity of c-*ras* can be induced by site specific mutation (involving as little as a single amino acid substitution) and this too can lead to transformation (section 1.7.2). It seems that the decreased GTPase activity of transforming *ras* gene products leads to a stimulation of adenylate cyclase, but how this change actually contributes

to oncogenesis is unknown. Mutated *ras* genes with lowered GTPase activity have been found with high incidence in several human tumours including those of the pancreas (90% incidence), colon (59%), thyroid (50%) and lung (30%), and they have also been found in about 30% of patients with AML (reviewed in Bos, 1985). It should be borne in mind, however, that not all human tumours have been shown to contain mutated *ras* genes.

Microinjection of $p21^{ras}$ induces DNA synthesis (Feramisco *et al.*, 1984) and anti-$p21^{ras}$ antibodies (also injected into the cells) block serum and PDGF/EGF stimulation of DNA synthesis (Mulcahy, Smith and Stacey, 1985). These results confirm that $p21^{ras}$ is involved in the responses of cells to growth factor signals, but since some tumours do not contain mutated *ras* genes it would seem that several avenues exist for perturbation of cell growth in a way which leads to transformation.

1.7.23 Glycolysis and cell respiration

Many normal cells display what is referred to as the Pasteur effect i.e. the presence of oxygen inhibits the glycolytic pathway in which glucose is broken down relatively inefficiently to yield ATP, with lactic acid as an end product. Some malignant tumour cells, however, display a high rate of glycolysis, even in the presence of adequate oxygen tensions (the Warburg effect) and this may lead to the accumulation of lactic acid. Warburg (1930) implied that the high level of aerobic glycolysis shown by tumour cells was due to irreversible damage of the respiration system by carcinogens, but there is no evidence that tumour cells characteristically have impaired respiration. Furthermore, many normal cells can display high rates of aerobic glycolysis and other studies actually have shown certain tumours to have a normal Pasteur effect (Weinhouse, 1955). Burk, Woods and Hunter (1967) have suggested a correlation between glycolysis and growth rate but this is unlikely to be the preserve of tumour cells.

Although the original proposal of Warburg (1930) is no longer tenable, the centres of large tumours are often necrotic and it is possible to envisage that some tumour cells might survive better than normal cells in such an anaerobic environment if they could utilize the glycolytic pathway more effectively. What effects this might have on tumour development and progression, however, remain speculative.

In a review of oxidative metabolism in tumours, Pedersen (1978) compares the mitochondrial content of a range of tumours with that of normal tissue. Most of the tumours studied showed a relative decrease in mitochondrial content, as did fetal liver indicating a possible oncofetal relationship for this trait. This rather startling observation requires more detailed analysis, however, particularly in relation to mitochondrial function.

1.7.24 Anchorage independence

Unlike non-transformed cells, many transformed cells are able to grow in an anchorage-independent mode in soft agar (Macpherson and Montagnier, 1964). In experimental systems such as the C3H mouse $10T^1/_2$ fibroblastic cell line, treatment with chemical carcinogens leads to the development of cells which grow in soft agar and form tumours on injection into syngeneic mice (Mondal, Brankow and Heidelberger, 1976). In their detailed *in vitro* study of a number of rat liver epithelial cell lines, Montesano *et al.* (1977) found good correlation between the ability to grow in soft agar and tumorigenic potential. Of 15 types tested, all the tumorigenic cell lines formed colonies in soft agar whereas none of the non-tumorigenic lines did, but investigations in other systems question the universality of this observation. Thus some normal cells such as lymphocytes, for example, can grow in suspension and not all tumour cells are necessarily anchorage-independent (Stanbridge and Wilkinson, 1980).

1.7.25 Cell adhesion

The suggestion that malignant cells might be *less* adhesive than their normal counterparts was raised originally by Ludford (1932) and Cowdry (1940). Other early work in which the force of cell detachment was measured suggested that malignant cells had reduced adhesion (Coman, 1944; 1953) and it was proposed that this would aid cell release from the primary tumour during metastasis. However, detachment of cells may result from membrane rupture rather than the breaking of adhesive bonds *per se*, and thus results based on this technique may actually be a measure of membrane strength rather than intercellular adhesiveness. The notion that tumour cells might be poorly adhesive when compared to normal cells has gained some support from the observations of Criborn *et al.* (1974), who found that aspirates from needle biopsies of malignant tissues often contained more free cells than did aspirates from benign tumours. Indeed, surgeons experienced in the needle biopsy of breast lesions can be remarkably accurate in distinguishing benign from malignant tumours simply by the ‘feel’ of resistance to aspiration. However, tumours may contain many enzymes which could break otherwise strong adhesive bonds between cells or between cells and the connective tissue matrix, and they could also contain significant necrotic portions in which the cells are no longer adherent to each other. Thus resistance to aspiration may provide a measure of adhesive strength only in relation to the condition of the tumour at the time of biopsy. Nevertheless, Murray *et al.* (1980) have reported that spontaneously transformed adult murine connective tissue fibroblasts are less adhesive to a range of collagens than are their normal counterparts and tumour cell-connective tissue interactions may well play a role in malignancy.

Further support for the ideas of Ludford (1932) and Cowdry (1940) has come from the studies of Bubenik *et al.* (1979) who used a completely different assay system in order to compare normal and malignant cell adhesion. The approach employed was based on counting the number of latex particles adherent to cells suspended in a medium lacking any protein source, including serum. Although carried out under extremely artificial conditions, these authors nevertheless found that high malignancy correlated with low cell-surface adhesiveness.

Other studies using cell–cell attachment methods such as cell aggregation (Cassiman and Bernfield, 1975), the aggregate cell capture assay (Dorsey and Roth, 1973) or the collecting lawn technique (Walther, Ohman and Roseman, 1973), rather than cell detachment methods, have shown virally transformed cells to have *higher* adhesiveness than non-transformed cells. In one study by Wright *et al.* (1977), a range of transformed cells (with one exception) were found to aggregate more than non-transformed cells in medium lacking Ca^{2+}, Mg^{2+} and serum. Interestingly, the exception related to Balb/c 3T3 clone A31 cells which appear to be 'partially transformed' in that they can form tumours in experimental animals when injected attached to glass beads (Boone, 1975). It should be emphasized here that cell detachment and cell attachment are fundamentally different phenomena, and that one is not simply the opposite of the other. Indeed, considerable confusion has resulted from the use of the term 'adhesion' to describe cell behaviour based on both attachment and detachment phenomena. It has been claimed that detachment is likely to be important in the spread of a malignant tumour from its primary, while attachment is likely to be important in lodgement within blood or lymphatic vessels prior to extravasation. Intuitively, this might appear to be the case providing allowance is made for differences in the cells involved. Thus increased tumour cell detachment from an essentially **homotypic** primary might enhance metastatic behaviour, as might increased **heterotypic** attachment (perhaps to endothelial cells).

The term **homotypic attachment** refers to the adhesion of like cells to like cells, whereas **heterotypic attachment** refers to the adhesion of two unlike cells. It can be appreciated that another dimension in complexity is added when the nature of the cells involved in attachment or detachment is taken into account. Thus although homotypic detachment might aid spread from the primary, attachment to form both homotypic and heterotypic clumps could also enhance metastasis by physical entrapment in small vessels. Yet it can be argued equally as forcibly that decreased heterotypic attachment might allow tumour cells to avoid host cell mediated cytotoxicity, thereby rather perversely aiding tumour growth and spread! Attention has already been drawn to the fact that tumours actually represent a mixed bag of cells, possibly containing macrophages, other defence cells and fibroblasts in addition to connective tissue components, blood vessels and so on.

It should be clear from the above that the contribution of adhesiveness to metastatic outcome is far from being straightforward. It is by no means certain that all invasive tumour cells will display similar adhesive behaviour, and nor does it follow that benign tumour cells will necessarily differ in their adhesiveness from malignant tumour cells. The role of cell adhesion in invasion and metastasis is considered in detail in Chapter 5, but from a general point of view the concept that tumour cells are less homotypically adhesive than their normal counterparts is unlikely to be universally true. This conclusion is enforced by the heterogenity of tumours, with the possibility that some cells may be more adhesive than others within the same tumour mass. Finally, we should note that because of the complexity of the metastatic process, it is unlikely that proposals based solely on adhesive changes will adequately explain tumour spread.

1.7.26 Phenotypic instability and tumour heterogeneity

In accordance with the theory of progression put forward by Foulds (1975), it seems that a single tumour may contain cells which are heterogeneic in nature. Both human and animal tumours have been shown to be heterogeneic with respect to a number of features including their immunogenicity, growth kinetics, motility, invasiveness, metastatic ability, radiosensitivity, karyotype, metabolic capacity, membrane receptors and drug sensitivity (reviewed in Poste and Fidler, 1980). This heterogeneity is thought to be due to the emergence of variant cells, and it can occur irrespective of whether the original tumour was monoclonal or polyclonal in nature (Nowell, 1976). It should be noted, however, that normal tissue cells also display heterogeneity and thus this phenomenon is not an exclusive attribute of tumour cells. Nevertheless, it appears that tumour cells may display quantitatively more heterogeneity than normal cells, although a good deal of variation is to be expected among different tumour types (reviewed in Nicolson, 1982a, 1984b, 1988). It should not be overlooked that heterogeneity exists not only for tumour cells within a primary mass (Prehn, 1970; Pimm and Baldwin, 1977; Pimm, Embleton and Baldwin, 1980), but also for tumour cells in metastases. Studies of both primary and metastatic lesions have provided the basis for establishing the concept of zonal heterogeneity, namely that tumours may have detectable areas within them which are noticeably different from each other. Finally, it should be noted that cells in secondary lesions can differ in their behaviour, drug responsiveness, antigenic properties and other biochemical characteristics from cells in the primary which seeded them (Sugarbaker and Cohen, 1972).

When tumour cells are cloned, an often rapid diversification of their progeny occurs. Polyclonal populations, on the other hand, appear to be more phenotypically stable since the co-cultivation of different clones leads to a reduction in the extent of phenotypic change (Poste, Doll and Fidler, 1981). These results imply some form of interaction between mixed cell populations which leads to phenotypic stabilization, although exactly how this is brought about is not known. Miller and Heppner (1979) and Miller *et al.* (1980) have shown that subpopulations within mouse mammary tumours can exert regulatory constraints on each other which affect immunogenicity and drug sensitivity. An important point to be made here is that *in vitro* cloning will not necessarily result in a cell line with uniform properties. In fact, by removing the stabilizing effect of a mixed population it may actually generate diversity. Clinically, it may be argued that certain drugs could perturb the tumour cell population thereby reducing stabilization and promoting the development of variant cells. Much more research, however, remains to be done in this area. Nevertheless, the phenotypic instability of tumour cells may have profound clinical consequences. It could lead to the rapid generation of variant cells with altered metastatic ability, or altered susceptibility to drugs and irradiation.

What processes are responsible for generating tumour heterogeneity? The selection pressure promoting variability can either be natural (e.g. cell mediated cytotoxicity) or artifical (e.g. drug induced). As explained in section 1.5.3, genetic events could generate changes in the phenotype but the rate of change is apparently not fast enough to explain the rapid diversification shown by some tumours. According to

Bosslet and Schirrmacher (1981), ESb cells isolated from lung or liver metastases are lysed by cytotoxic T lymphocytes whereas cells from spleen metastases are not. Since the authors could not find cytotoxic T cell-resistant ESb cells in the parental population by cloning techniques, they suggested that the resistant variants were induced by the splenic environment and were not selected from a pre-existing population. In the light of rapid phenotypic change (occurring within three cell generations) under both *in vivo* and *in vitro* conditions, Schirrmacher (1980) proposed that preformed genetic programmes might become activated by inductive signals from the microenvironment. Epigenetic events such as DNA methylation and transcriptional and translational control processes may be capable of generating rapid diversity. Neri and Nicolson (1981) have coined the term **phenotypic drift** to describe the generation of heterogeneity via epigenetic processes.

Intuitively, metastatic heterogeneity is likely to have the same basis as other forms of tumour heterogeneity. Cifone and Fidler (1981) have shown that highly metastatic variants of murine fibrosarcomas have higher spontaneous mutation rates to 6-thioguanine and ouabain resistance than do poorly metastatic variants. Other workers, using different experimental systems, however, have been unable to show any correlation between spontaneous mutation rate and metastasis, although a correlation has been noted with induced mutation rate (Yamashina and Heppner, 1985). Furthermore, Elmore, Kakunaga and Barrett (1983) were unable to find any differences in spontaneous mutation rate between normal and chemically transformed human fibroblasts and thus there may not be any simple relationship between mutational ability and malignancy. Since tumour variants may arise at a rate faster than can be explained by genetic mutation (section 1.5.3), epigenetic events such as changes in DNA methylation may be involved in the generation of metastatic variants. Altered patterns of DNA methylation can be somatically inherited, but the full significance of this possibility in generating metastatic variants is not yet clear. Another source of variability is suggested by the studies of Kerbel *et al.* (1983) who showed that shifts from a non-metastatic to a metastatic phenotype in the MDAY-D2 murine tumour system are due to fusion with host cells followed by chromosomal segregation. Finally, various sorts of signals emanating from other tumour cells or from host cells and even from the extracellular matrix (which is known to influence cell differentiation) might affect heterogeneic development.

It would seem that there may be considerable differences between tumours in how variants arise, and any given tumour might in fact

Table 1.8 Postulated mechanisms underlying progression

1. Genetic instability e.g. high DNA mutation rate
2. Activation of preformed genetic programmes leading to a rapid phenotypic shift
3. Epigenetic change e.g. alteration in levels of DNA methylation
4. Intercellular signals from other tumour cells (e.g. clonal destabilization) or from host cells (e.g. hormonal effects)
5. Environmental signals from the extracellular matrix (e.g. fibronectin, laminin)
6. Hybridization with somatic cells and subsequent chromosome segregation

employ a range of strategies during progression to yield the observable variants. A summary of possible mechanisms important in progression is provided in Table 1.8.

1.7.27 The nature of metastatic cells

A problem of key interest to experimental pathologists concerns the nature of the cells which give rise to metastases. Two conflicting hypotheses have been proposed:

(a) *The specialist model of metastasis*

The central notion of this hypothesis is that metastasis is a non-random process, requiring a specialized subset of cells which pre-exists in the tumour mass (having developed in accord with the concept of tumour progression).

We have seen that not all cells within a malignant tumour appear to have the same metastatic potential, and thus heterogeneity exists for this particular phenotypic trait. Koch (1939) provided the first evidence that this might be the case when he successfully isolated a metastatic subline of Ehrlich's carcinoma by serially transplanting lymph node metastases. In other pioneering studies on metastatic potential, Klein (1955) found that tumour cells adapted for ascites growth spread preferentially to the lungs. On the basis that the change was stable and heritable she proposed that these cells pre-existed in the original tumour as a particular subpopulation which was able to grow better in ascites form than the other subpopulations. Zeidman (1965) subsequently showed that cells from B16 melanoma lung nodules formed more colonies when harvested and re-injected into host mice than did the original parental cells, and he suggested that a particular type of mutant cell was required for metastasis to occur. Fidler and Kripke (1977) were later to confirm that metastatic subpopulations exist in tumours before metastasis occurs, and that although some of these subpopulations have a high metastatic potential others may be incapable of metastasizing. This was done using a variation of the fluctuation test devised originally by Luria and Delbruck (1943) to distinguish between selection and adaptation during bacterial growth. Briefly, Fidler and Kripke (1977) prepared a cell suspension from a subcutaneously growing B16 tumour and divided it into two aliquots. The first aliquot was tested for its ability to colonize the lungs after tail vein injection in syngeneic mice without further selection, whereas the second aliquot was used to isolate 17 clones. Each of these clones was then tested separately for their lung colonizing ability. Whereas the original suspension produced similar numbers of tumours in different animals, each of the clones differed markedly in their lung colonizing potential. Attempts were made in this study to show that although clones of tumour cells are highly variable, the differences detected between clones were unlikely to have been generated as a consequence of the cloning process itself.

(b) *The stochastic model of metastasis*

The basic premise of this model is that the likelihood of a maligant cell forming a metastasis is entirely random and dependent on local events (Weiss, 1979b; Weiss, Holmes and Ward, 1983). In this model there is no subset of specialist metastatic cells pre-existing before metastatic lesions develop. Instead, cells are thought to be recruited randomly into a transient metastatic pool.

Now if metastatic cells do pre-exist as a stable subpopulation then, on the average, cells from metastases might be thought to metastasize better than cells from the primary since the metastatic cells could be selected for. That this seems to be the case in some studies (as cited above) but not others originally provoked the development of the stochastic model (reviewed in Weiss, 1983). In such a model, both the primary and its metastases should give rise to similar numbers of metastases, providing a large enough survey is undertaken. Unfortunately, with the available data it is difficult to resolve between the two hypotheses. It may also be that components of both apply, and thus that the most appropriate model would contain aspects of the two outlined above. One major problem with trying to resolve between these two hypotheses is that malignant cells may diverge as the metastases develop. Young and Hill (1986), for example, have reported that B16 lung colonies contain more metastatic cells (determined by cloning) than the parental tumour early in their growth, but the proportion declines with time as the metastases diversify.

Regardless of whether metastatic variants pre-exist in a stable pool or move in and out of a transient pool, at any one time metastatic cells seem to represent only a small fraction of the total tumour cell population. Consequently, research aimed at developing anti-metastatic regimens may require the isolation of these variant cells in order to avoid confusion from the majority of non-metastatic cells.

1.8 THE SIGNIFICANCE OF THE IMMUNE SYSTEM IN TUMOUR DEVELOPMENT: IMMUNOSURVEILLANCE

The concept of immunological surveillance (Ehrlich, 1908; Thomas, 1959; Burnet, 1970) has as its basis the idea that tumour incidence is limited by immunological defence mechanisms of the host. Thomas (1959), in an attempt to explain the origins of transplantation immunity, argued that it could not have evolved in response to grafted tissues and that it was more likely to have developed as a natural defence mechanism against tumours. The immunosurveillance concept is widely accepted since, amongst other attributes, it seems to provide an explanation for the age incidence of cancer and for the association between tumour growth and immunodeficiency which has often been reported in clinical studies. However, it is worth reiterating at this point that not all tumours are antigenic and indeed most spontaneous tumours in both humans and mice may be poorly antigenic, if at all. This leads us to seriously question the significance of the immune response in limiting tumour establishment, growth and spread.

If the immune system does limit tumour incidence, then it might be argued that immunosuppressed animals and patients should display a higher incidence of tumours than normal. Although immunosuppressed individuals may be more prone to getting cancer, they usually display a narrow range of neoplasms, particularly lymphomas, and it is consequently difficult to determine any generalized causal relationship. Athymic nude mice (which lack functional T cell responses) do not have an abnormally high incidence of spontaneous tumours (Stutman, 1978), but this is probably because they have natural killer (NK) cells capable of tumour cell destruction. The extent to which we hold rigidly to the concept of immunosurveillance depends very much on the plasticity of the tumour cell phenotype and on how we envisage tumour cell destruction to occur. Certainly, there is often no correlation between tumour cell destruction under *in vitro* conditions and tumour rejection *in vivo* (Baldwin and Embleton, 1974).

1.8.1 Immune stimulation–inhibition

It seems that in different experimenal systems, antitumour immune responses can either reduce or enhance the metastatic process, and these contradictory results led Prehn (1972) to propose his theory of immune stimulation–inhibition. According to this concept, the immune response during early stages of tumour development (when tumours are often weakly antigenic) may be stimulatory, becoming inhibitory only at later stages of development when the tumours are more antigenic. Although this is an attractive hypothesis, it is doubtful that this can fully explain the observed spectrum of anti-tumour immune effects.

1.8.2 Immune effector mechanisms

The body has several different ways by which it may potentially rid itself of a transformed cell. These include the production of soluble factors such as antibody and complement, and the activation of cytotoxic cells. Generally speaking, complement-mediated cytotoxic responses directed against tumour cells are ineffective *in vivo* although they may result in considerable cell killing under *in vitro* conditions (Algire, Weaver and Prehn, 1954). Effector cells for cell-mediated cytotoxic reactions include cytotoxic T cells, natural killer (NK) cells, lymphocyte-activated killer (LAK) cells, neutrophils (PMNs), and macrophages.

The tumour-directed cytotoxic activity of macrophages requires macrophage activation by a lymphokine known as macrophage activating factor (MAF). MAF is released from specifically sensitized T lymphocytes, but its action on macrophages is relatively non-specific. This two step mechanism of macrophage activation forms the basis of BCG tumour therapy. BCG is an attenuated form of *Mycobacterium tuberculosis* which was isolated by two French scientists, Calmette and Guerin; hence the acronym BCG (*bacille Calmette Guerin*). Its main clinical use is in immunizing individuals against tuberculosis, but because it is a potent activator of macrophages (as well as of B cells and T cells) it may be able to promote anti-tumour activity. Thus the direct injection of BCG into a tumour within a specifically sensitized patient should induce an intense inflammatory response, the release of MAF, non-specific activation of local macrophages, and lysis of the neoplastic cells. Unfortunately, the clinical application of this technique is rather limited, and in fact results from clinical trials have been rather unimpressive.

Macrophage activation can also be induced by the tetrapeptide

tuftsin (thr-lys-pro-arg), which is a cleavage product of the γ-globulin component leukokinin (reviewed in Fridkin and Najjar, 1989). Tuftsin enhances a number of macrophage activities including phagocytosis and tumour-directed cytotoxicity, the latter being augmented through the enhancing effects of tuftsin on TNF production (section 6.4).

A possible anti-tumour role for human PMNs was first postulated by Bubenik *et al.* (1970), and Pickaver and colleagues (1972) later provided both *in vivo* and *in vitro* experimental evidence for the anti-tumour activity of rat PMNs. In the *in vivo* study, rats were injected i.p. with beef heart infusion broth in order to activate a PMN response and this was followed 4 h later by the i.p. injection of 10 WBP1(A) syngeneic ascites tumour cells. When rats treated in the same way were left for two days before tumour cell injection, the inflammatory cell population consisted predominantly of macrophages and no inhibitory effect on tumour cell take was noted. Inflammatory reactions *in vivo* usually involve a mixture of defence cell types (not to mention a plethora of their products) and thus it is difficult to reach a conclusion on the precise cellular basis of any observed anti-tumour activity. Additional studies by Fisher and Saffer (1978) using a murine system, however, have at least confirmed the *in vitro* tumour cytotoxic activity of PMNs. These authors also showed that PMN anti-tumour activity was specific and inhibitable by some type of blocking factor which was present in the serum of tumour-bearing animals.

Cytotoxic T cells (Table 1.9) were once thought to be the most active cell type against tumours. However, it now seems that while they may afford protection against virally induced tumours they are not so effective against spontaneous or chemically induced tumours. NK cells probably represent a diverse group: in humans, the predominant cell type is represented by a population of $CD3^-$, $CD16^+$, Leu19 $(NKH\text{-}1)^+$ lymphocytes with distinct morphology giving rise to the name large granular lymphocyte (LGL). In mice, the predominant cell with NK activity is a $Ly1^-$, $Ly2^-$ cell expressing asialo-GM_1. Functionally, the important aspects of NK cell anti-tumour cytotoxic activity are that it is independent of antigenic sensitization, that it has no specific memory component, and that it is independent of the major histocompatibility complex (MHC). NK cell behaviour is not solely related to anti-tumour activity since they appear to have diverse roles which include involvement in microbial infections and other diseases. It is important to note here that clinical evidence for anti-tumour activity of NK cells in humans is rather limited. Furthermore, immunohistochemical studies suggest that NK cells are relatively scarce in solid human tumours. Patients with ascites tumours,

Table 1.9 Characteristics of the principal human cytotoxic effector cells*

Character	*Cell type*		
	NK cell	LAK cell	Cytotoxic T cell
(a) Determinants			
CD3	–	+	+
CD8	–	+	+
CD16	+	+	–
Leu7	+	–	–
Leu11	+	–	–
Leu19	+	+	–
(b) Properties			
Development time	spontaneous	2–3 days	5–7 days
Stimulus	none	IL-2	antigen

* Data compiled from various sources. Not all cells of any particular group may express every determinant. Some cytotoxic T cells, for example, may be CD8 negative. Whilst it is convenient to categorize cells, it may be argued that this is artificial and that the various types might really represent a continuum of functionally overlapping cells.

however, have shown some response to treatment with OK432, a streptococcal preparation thought to augment NK cell activity (Uchida and Micksche, 1983). It may be that NK cells are less active against solid tumours than ascites and leukaemic type tumours. There is considerable experimental evidence that NK cells may be active against metastatic cells in the blood (section 2.16).

Another cytotoxic effector cell type, characterized by activation in response to certain lymphokines including interleukin-2 (IL-2), has recently been identified. Effector cells activated by IL-2 are apparently derived from a phenotypically diverse pool (Damle, Doyle and Bradley, 1986), but there is some evidence for the existence of a category of lymphokine-activated killer (LAK) cells separate from other cytotoxic cells (Table 1.9). LAK cells are able to destroy many tumour cell lines as well as a wide range of cells from naturally occurring tumours. Since they do not destroy normal tissue cells, this has encouraged early clinical trials using IL-2 adoptive immunotherapy (see section 6.4). Experiments with mice have shown that injection with LAK or with LAK plus IL-2 can reduce or eliminate established tumours (Mazumder and Rosenberg, 1984). How IL-2 works to activate LAK cells is not clear. It has been postulated that it may work in a non-receptor mediated manner, perhaps by direct intercalation of the hydrophobic IL-2 molecule into the plasma membrane of the LAK precursor cell. LAK activation is remarkably sensitive to hydrocortisone (Grimm,

Muul and Wilson, 1985), and there is also evidence that some tumours may secrete immunosuppressive factors which can inhibit LAK activation.

1.8.3 Tumour escape mechanisms

If immunological surveillance is operative *in vivo* it might be argued that tumours and their metastases would never develop. That they can and do has lead to the suggestion of numerous and ingenious mechanisms by which transformed cells may avoid immune destruction (reviewed in Woodruff, 1980; Siegel, 1985). In fact, by the time of diagnosis many cancer patients have demonstrable humoral and cellular immunity directed against their tumour cells but these still survive and grow. How is this achieved? At least five possibilities have been suggested:

(a) *Immunogenicity*

Novel antigens on the surface of the tumour cells may be absent, weakly immunogenic, variable, or masked so that the immune system is not properly activated. Hewitt, Blake and Walder (1976) have suggested that many tumours which arise spontaneously lack demonstrable TATAs (tumour associated transplantation antigens) and hence they are unlikely to elicit an appropriate immune response. Prehn (1975) has shown that tumour immunogenicity is a function of carcinogen concentration, which might offer a possible explanation for the lack of TATAs on naturally occurring spontaneous tumours. Perhaps not surprizingly, there is also some evidence which suggests that the immunogenicity which develops following exposure to chemical carcinogens *in vitro* is a function of cell culture density (Carbone and Parmiani, 1975).

Some tumours are known to be able to modulate their antigenic nature in response to the immune status of the host. Thus thymic leukaemia cells expressing the TL antigen (Boyse and Old, 1969) lose it from the cell surface when they are injected into immune hosts (see section 1.7.6). There is some evidence that the products of the E1a locus in cells transformed by adenoviruses may inhibit transcription of the genes for Class 1 MHC molecules, thereby enabling them to escape detection by T lymphocytes (Schrier *et al.*, 1983; Tanaka *et al.*, 1985). On the other hand, it seems that reduced MHC expression could open tumour cells to enhanced NK cell mediated killing, and to what extent these two process balance out *in vivo* is uncertain. Antigen masking may be brought about by the presence of quantities of substances such as sialomucin, for example, which could obscure the antigenic nature of the tumour cell surface.

(b) *Immunosuppression*

Some tumours may secrete novel immunosuppressive molecules which could act on the immune system at various levels, perhaps shifting the suppressor/helper cell ratio in favour of immune inhibition thereby reducing the effectiveness of the surveillance system. A rather elegant demonstration of the role of T suppressor cells in the emergence of malignant tumours was performed by Fisher and Kripke (1978). These authors showed that chronic exposure of mice to UV light induced highly antigenic fibrosarcomas which were rapidly rejected. Intermittent doses of UV radiation, on the other hand, rendered mice tolerant to fibrosarcoma implants through the induction or activation of T suppressor cells. Further evidence on the importance of immunosuppression comes from clinical studies. Immunosuppression is often extensive in patients with Hodgkin's disease, for example, even though they may only have a single affected lymph node. As mentioned previously, immunosuppressed patients (as a consequence of organ transplantation, for example) have a relatively high incidence of lymphomas, but the reasons for this selectivity are unknown. It has been suggested that many lymphomas have a viral origin and that tumours of this type, which may normally be particularly sensitive to T cell mediated immunosurveillance, are able to capitalize on the induced immunosuppressive state. Alternatively, the immunosuppressive drugs may have an oncogenic effect on immune cells. Many viruses implicated in tumorigenesis are immunosuppressive, although the precise mechanisms by which immunosuppression is achieved are not known.

(c) *Immunoblocking*

Blocking of the immune response may be induced by the antigen, its antibodies, or by immune complexes. In antigen blocking, antigens released from the tumour cell surface may compete with whole tumour cells as the target for effector cells thereby reducing their effectiveness (Currie and Basham, 1972). Antibody blocking might result from a screen of essentially ineffective anti-tumour antibodies restricting the access of cytotoxic cells to their targets. Similarly, immune complexes deposited on either the tumour cell (via the antigen) or the cytotoxic cell (via its receptor) might hinder the recognition step necessary for lysis. Although both antibodies and immune complexes appear to be able to block *in vitro* cell mediated immunity against animal and human tumour cells (Hellstrom and Hellstrom, 1969, 1971), the precise mechanisms by which this is achieved remain speculative and the significance of their roles *in vivo* is open to question.

(d) *Immunoselection*

The fundamental heterogeneity of tumours may provide the variability necessary for the selection of cell variants which are resistant to the immune defence mechanisms of the body (section 1.7.26).

(e) *Sneak through*

This refers to a critical mass phenomenon. If the first transformed cell can survive a few divisions before the immune response is fully developed, then the tumour may reach a critical mass beyond which it is less likely to be destroyed by the immune system (Old, 1962).

The possibility of immunological escape does not invalidate the concept of immunosurveillance, for many tumours may be destroyed by effector systems before they are detected clinically. However, it does suggest that immunosurveillance (if it operates) is not fully effective, and clinically speaking it would be of considerable importance to know whether this effectiveness could be improved by some form of immunotherapy.

Largely on the basis of experimental studies we can summarize the position as follows:

(a) Cancer cells may be immunogenic, but the extent to which immune phenomena limit tumour growth and metastasis in the natural situation is uncertain.
(b) The growth of some tumours may be suppressed by host immunity, while in others it may be enhanced.
(c) The pattern of stimulation–inhibition may change during the growth of a tumour, with growth being stimulated during early development and inhibited later in development.
(d) Similarly, the growth of weakly antigenic tumours may be stimulated while that of strongly antigenic tumours may be inhibited.
(e) The destruction of tumours is largely host cell mediated, particularly by NK cells, LAK cells, cytotoxic T lymphocytes, macrophages and neutrophils.
(f) Tumour-directed cell-mediated cytotoxic activity may be either antibody dependent or antibody independent.

(g) Direct complement-mediated tumour cell lysis does not seem to play a significant role *in vivo*.
(h) It is difficult to prove conclusively that immunosurveillance operates *in vivo*. In fact, its popularity stems largely from the intuitive feeling that it seems to ring true.

1.8.4 Bystander lysis

When mice are treated *in vivo* with specific cytotoxic T cells, variant tumour cells which fail to express the appropriate antigens may soon be generated. This problem may be overcome by the adoptive transfer of non-specific killer cells such as NK cells and LAK cells. Recently, Shiohara *et al.* (1987 a and b) have presented evidence in support of a possible role for bystander lysis in tumour eradication. These authors suggest that artificially generated autoreactive T cells (directed against self I-A gene products) can be activated by class II MHC specific syngeneic stimulator cells (such as macrophages which are found within tumours) to release lymphotoxin and interferon-γ which act synergistically to lyse nearby neoplastic cells. Both syngeneic class II MHC positive target cells (lipopolysaccharide stimulated lymphocytes) and class II MHC negative B16 malignant melanoma variants can be killed in this way, and the intratumoural injection of autoreactive T cells can prolong survival of mice bearing the B16 melanoma. Furthermore the i.v. injection of autoreactive T cells (but not control T cells) immediately after the injection of B16 cells can markedly reduce the number of lung colonies that develop. It should be noted, however, that this latter treatment was relatively inefficient in that at least ten times more autoreactive T cells than melanoma cells were required before a significant reduction of lung colonies was observed. Bystander lysis of tumour cells is clearly of interest as a possible mechanism to be harnessed in tumour therapy, but there is no evidence as yet, however, that autoreactive T cells actually populate spontaneous tumours *in vivo*.

1.8.5 Leukocyte adherence inhibition

Leukocyte adherence inhibition (LAI) is an *in vitro* assay developed by Halliday and Miller (1972) for assessing cell immunity. LAI is employed predominantly as a means for detecting sensitization to tumour antigens, although it works equally well with other antigens, such as myelin basic protein, expressed by non-neoplastic cells. LAI can be studied in various ways, but the simplest involves culture in a horizontal glass tube of a tumour extract mixed with autologous buffy coat leukocytes in a serum free medium (Grosser and Thomson, 1975). After 2 h of culture, the tube is stood vertical and a sample of non-adherent cells taken and quantified. This non-adherent cell count is compared with that determined after culture of the same cells with an extract prepared from an unrelated tumour from a different patient or with an extract of normal tissue prepared from the same organ type which yielded the original tumour extract. The key to the effect is that LAI is pronounced only when leukocytes from patients with cancer of a particular organ are cultured with extracts of cancers from that same organ. Not all leukocytes become non-adherent in LAI assays. In fact, it seems that only about 30% do so and these include PMNs, monocytes and $CD4^+$ (inducer/helper) and $CD8^+$ (suppressor/cytotoxic) T lymhocytes. It should be noted that although LAI is highly dependent on a cell-mediated immune response to organ specific cancer antigens, it is also a function of the protein content of the assay medium (typically 20 μg ml^{-1}) and tumour burden. Patients with heavy tumour burdens give less positive LAI results than those with small tumour burdens, and problems thus arise with false negative observations.

The molecular and cellular events involved in LAI are still under study, but it seems that the cancer extracts contain organ-specific antigens which activate responsive leukocytes. In allogeneic LAI, the responsive cells appear to be predominantly monocytes which act in an antibody-dependent manner (Marti *et al.*, 1976), whereas in autologous LAI the response is mediated primarily by T cells (Shenouda *et al.*, 1984). It would seem likely, however, that there is some interaction between the various cell types in both the allogeneic and autologous responses. Recent studies on the nature of the organ specific antigen involved in LAI associated with colorectal cancer has implicated a role for a 40 kD cell-surface glycoprotein (Artigas *et al.*, 1986). A similar glycoprotein was found to be involved in LAI associated with lung cancer. Interestingly, at least some of these organ specific antigens found in cancer extracts are also expressed by fetal organs, indicating that they may be another class of oncofetal antigens (section 1.7.5.1).

Some progress has been made towards understanding how interaction between the organ specific antigen of LAI and its leukocyte receptor induces adherence inhibition. Thus interaction of specifically reactive cells with organ-specific antigens has been shown to result in the release of a number of cytokines which cause bystander leukocytes to detach from the substrate. Many of these cytokines are arachidonic acid metabolites, foremost amongst which with respect to LAI is the leukotriene LTB_4 (section 5.7.2.2). This leukotriene is a known chemotactic substance, and in fact a number of such agents including *N*-formyl-met-leu-phe (FMLP), C5a (a fragment of the fifth component of complement) and interleukin 1 (IL-1) can induce LAI (Thomson, 1984). Although exactly how these agents induce LAI is not clear, they may work through thromboxane intermediates induced in the bystander cells. The *in vivo* generation of mediators of inflammation such as the leukotrienes may assist the host in combating cancer by encouraging reactive leukocytes to invade the tumour mass, but the precise significance of this remains to be evaluated fully.

1.9 THE INDUCTION OF TUMOUR DORMANCY

Dormancy is a state in which tumour cells persist in the host without obvious growth. An apparent resting state will exist if the rate of tumour cell production is equal to the rate of tumour cell destruction. In this case, the emergence of a formerly dormant tumour may be due to some alteration in the host (perhaps in its immune status, for example) shifting the cell division/cell death ratio in favour of division. Dormancy will also result if tumour cells temporarily stop cycling by entering into the G_0 phase of the cell cycle, and this is the sense in which the term is used here. In this case, the appearance of a formerly dormant tumour may be due to some event (such as supply of an essential growth factor) which induces non-cycling tumour cells to divide.

At least four mechanisms have been proposed to account for this phenomenon:

(a) *Lack of vascularization*

Dormancy may result from the lack of vascularization, possibly as a consequence of nutrient deprivation and/or the build up of toxic metabolites. In angiogenic experiments, tumours stop growing when they are about 3 mm in diameter unless they become vascularized (Folkman, 1985). In some experimental studies, however, dormancy is thought to be associated with single cells, and thus the avascular state may not explain all aspects of this phenomenon.

(b) *Microenvironmental effects*

Dormancy may also result from the trapping of cells in a particular environment, perhaps remote from host defence cells and not conducive to growth. Warren (1976) has suggested that during the process of extravasation tumour cells may become trapped between the endothelial cells of blood vessels and their underlying basement membrane, and that this location could foster dormancy. Given the vicinity of a blood vessel in this location, it is difficult to ascribe dormancy under these conditions to nutrient deprivation or metabolic waste concentration. In this example, some change in the environment (such as trauma) may offer the cells a more favourable environment and growth could then proceed. Of interest here are the experiments of Fisher and Fisher (1959) in which they injected rats via the portal vein with various doses of Walker 256 carcinosarcoma cells. At low doses (down to as few as 50 cells per animal) liver metastases were not common, but they could be induced by subsequent trauma such as that resulting from laparotomy and liver manipulation. Walker 256 cells are poorly immunogenic, since they may be implanted allogeneically, and this suggests that the immune system may not be primarily involved in the induction of dormancy in these experiments.

(c) *Supply of growth factors and hormones*

Dormancy may result from the lack of appropriate growth conditions. Some malignancies have a constitutive dependency on certain hormones, for example, and if these are lacking (perhaps as a consequence of age or another disease) then a tumour may not be able to develop.

(d) *Presence of defence cells*

The basis of this mechanism is that dormant tumour cells are maintained as such by the defence systems of the body. Preliminary evidence for this possibility emerged from the studies of Eccles and Alexander (1975), who were able to show that dormant tumour cells in the lungs of experimental animals (from which the primaries had been removed) could resume growth following immunosuppression by such mechanisms as whole body irradiation. Wheelock and colleagues (reviewed in Wheelock, Weinhold and Levich, 1981) explored this observation further using a model system based on L5178Y lymphoma cells in DBA/2 mice. They injected mice i.p. with mitomycin-C-treated L5178Y cells, which were incapable of growth but were otherwise metabolically active. The mice were then challenged with fully viable lymphoma cells, and it was found that these immunized mice lived considerably longer than control, unimmunized ones. Some immunized mice remained clinically normal for a period of months, but ascitic tumours suddenly developed. That even clinically normal, immunized mice actually harboured tumour cells was indicated by the *in vitro* growth of cells obtained from their spleens and peritoneal cavities. It appeared that cytotoxic T cells working synergistically with macrophages were responsible for the dormant state, but this effect slowly diminished with time and the tumour cells were then able to become established. Viewed from this aspect, tumour dormancy is not so much a function of the tumour cell as a consequence of limitations in the host immune defence systems. However, given the range of possibilities available, it would be unwise to narrow down dormancy to a single causative event.

2 The invasive and metastatic behaviour of malignant cells

2.1 INVASION AND METASTASIS *IN VIVO*

We have seen that the behaviour of tumours allows us to categorize them into two general types, benign and malignant. Benign tumours do not invade surrounding tissues although they will displace them as they increase in size. Malignant tumours, on the other hand, frequently invade and destroy local tissues and, furthermore, they often spread or metastasize to give rise to secondaries remote from the original or primary tumour (Willis, 1973). Local invasion is distinct from but part of the process of metastasis. Some invasive tumours rarely metastasize (e.g. basal cell carcinoma of the skin), but all metastases involve an invasive step. Furthermore, not all cells within a tumour may have the same propensity to metastasize, and this provides us with an example of tumour heterogeneity as discussed in section 1.7.26.

Three cellular properties may be identified which are of importance in invasion, namely proliferation, motility and lytic action. Although there is no generalized scheme which adequately explains the invasive behaviour of all malignant tumours, on the basis of histopathological data Strauli (1980) has proposed two general patterns of invasive behaviour:

(a) *Type 1 invasion*

This type (often called **invasive growth**) is characterized by malignant cells which invade normal surrounding tissue by penetration *en masse*. It is often shown by tumours of epithelial origin (the carcinomas), and it may be associated with a more or less general pattern of lytic activity in the immediate vicinity of the tumour. Proliferation rather than locomotion is the main agent of invasion: after division two cells have to adapt to a site previously occupied only by one and this growth pressure leads to invasion.

(b) *Type 2 invasion*

This type of invasion results from the migration of individual cells or cell islands, and lytic activity tends to be restricted to the microenvironments of the invasive cells. Although locomotion contributes to this type of invasion, Strauli (1980) believes that proliferative activity is mainly responsible. Invading cells emerging from the tumour mass migrate only a small distance before dividing again, and this division contributes to cellular relocation. Interestingly, another model put forward by Gabbert *et al.* (1985) stresses the importance of cellular dedifferentiation in mobilizing tumour cells from the main mass. Type 2 invasion is shown predominantly by tumours of mesenchymal origin (the sarcomas) and by malignant melanomas.

2.2 RESISTANCE TO INVASION

Not all tissues are invaded by malignant tumours to the same degree. Numerous clinical observations, for example, suggest that uncalcified cartilage is relatively resistant to invasion by malignant tumour cells. Other studies have also indicated that cartilage is resistant to blood vessel invasion, except at one particular developmental stage (reviewed in Kuettner and Pauli, 1981; Pauli and Kuettner, 1984). Other tissues which are relatively resistant to tumour invasion include structures such as the heart valves, the cornea, and the lens and it is noteworthy that each of these also lacks an intrinsic blood supply. The basis of this resistance is still as yet uncertain, but with respect to cartilage it appears to be related to both its physical nature and its chemical content. Type II collagen (which is specific for cartilage), for example, is generally more resistant to destruction than other types of collagen such as types I and III. Furthermore, the presence of proteoglycans in cartilage may have an anti-invasive effect. According to Mikuni-Takagaki and Gross (1981), the unusual ability of the Yoshida rat sarcoma to invade cartilage may be related to its ability to degrade proteoglycans. Other studies have shown that the anti-invasive character of cartilage is associated in part with the presence of two major classes of inhibitory substances:

(a) *Proteinase inhibitors*

This category of anti-invasive components includes a serine proteinase inhibitor (7 kD) resembling the commercially available trypsin inhibitor trasylol (aprotinin), a collagenase inhibitor (22 kD), and a cysteine proteinase inhibitor (13 kD) active against cathepsin B and papain. In association with the structural properties of cartilage, these inhibitors are likely to severely restrict the ability of most tumour cells to actively invade cartilage.

(b) *An endothelial growth inhibitor*

This inhibitor is not yet fully characterized, but in growing cultures of endothelial cells it has been reported to exert both cytostatic and cytotoxic effects. These dual effects may possibly be a consequence of incomplete purification of the growth inhibitory fraction. The active principal is apparently without effect on control cell populations such as human foreskin fibroblasts and rat urinary bladder epithelial cells.

2.3 THE PROBLEM OF METASTASIS

Metastasis is the major problem which faces the cancer clinician since it is the spread of malignant tumours which prevents successful treatment by simple excision of the primary growth. It is often the case that metastases have already been seeded by the time a patient presents with a malignant tumour. In many cases these metastases, perhaps composed of as little as one or a few cells, may not be immediately obvious even with modern mechanisms of detection. The presence of undetectable metastases compounds the clinical problem, for it is difficult to treat what cannot be seen.

The heterogeneic nature of malignant cells provides the potential for widespread metastasis. It also provides the basis for the development of treatment-resistant variants. Since metastatic lesions may progress differently from their primary tumours, the effective treatment of metatstatic disease presents a formidable problem.

2.4 THE METASTATIC PROCESS: GENERAL PRINCIPLES

There are three main routes by which a tumour may metastasize in the body (Willis, 1973): (a) via the blood; (b) via the lymphatic system; (c) by surface implantation. Each of these will be considered further in the following sections.

2.4.1 Haematogenic (blood) spread

Many experimental systems are based on tumour cell spread through the blood, and this mechanism will therefore be considered in some detail later (section 2.14). Clinically, sarcomas appear to spread more via the blood system than do the carcinomas, which predominantly spread via the lymphatic system. We shall soon see that these options are not mutually exclusive, and in any case a number of tumours such as melanomas, for example, seem to spread equally well by either route (section 2.5). Briefly, during haematogenic spread tumour cells gain access to the blood (usually via the venous circulation) in which they may be transported to a distant site. Within the blood, tumour cells may interact with blood components such as lymphocytes, platelets, and fibrin (derived from fibrinogen). Although some of these interactions may lead to tumour cell destruction, they may also promote tumour cell lodgement in small blood vessels through clump formation. Once lodged within a vessel, the tumour cell may extravasate into the tissue parenchyma where it can divide to establish a secondary growth. A diagrammatic summary of possible events occurring during haematogenic spread is provided in Fig. 2.1.

Reviews of metastasis with some emphasis on haematogenic spread are provided by Wood, Holyoke and Yardley (1966), Fidler, Gersten and Hart (1978), Roos and Dingemans (1979), Poste and Fidler (1980), Hart and Fidler (1980b, 1981), Nicolson (1982 a and b), Schirrmacher (1985), Killion and Fidler (1989) and Hart, Good and Wilson (1989).

Figure 2.1 A schematic overview of blood-borne metastasis (after Nicolson and Brunson, 1977). The figure illustrates the formation and haematogenic spread of a melanoma such as the B16. Numerous alternative scenarios are equally likely (see text for details). The process begins with a transforming event (1) which allows local growth (2) and invasion of malignant cells. These ultimately break through the basement membrane and invade blood vessels (3), some probably newly developed under angiogenic influences from the tumour. Within the vascular system the malignant cells may interact with host components such as platelets and NK cells (4). Such interactions could either promote subsequent lodgement of the malignant cells or they could result in their death. Once in the blood the malignant cells are carried to the heart (5), entrance to the vascular system usually having been gained on the venous side. The figure illustrates malignant cell lodgement in the lungs (6), a common occurrence for the B16F10 variant specifically selected for this trait. Exactly how lodgement occurs is unclear. A cluster of tumour cells is depicted in the figure in association with platelets and fibrin. The malignant cells are presumed to have bound selectively to determinants on the surface of the endothelium. Following lodgement the malignant cells extravasate (7) with enzymatic dissolution of the basement membrane, and under appropriate conditions the cells divide to form visible metastatic lesions (8).

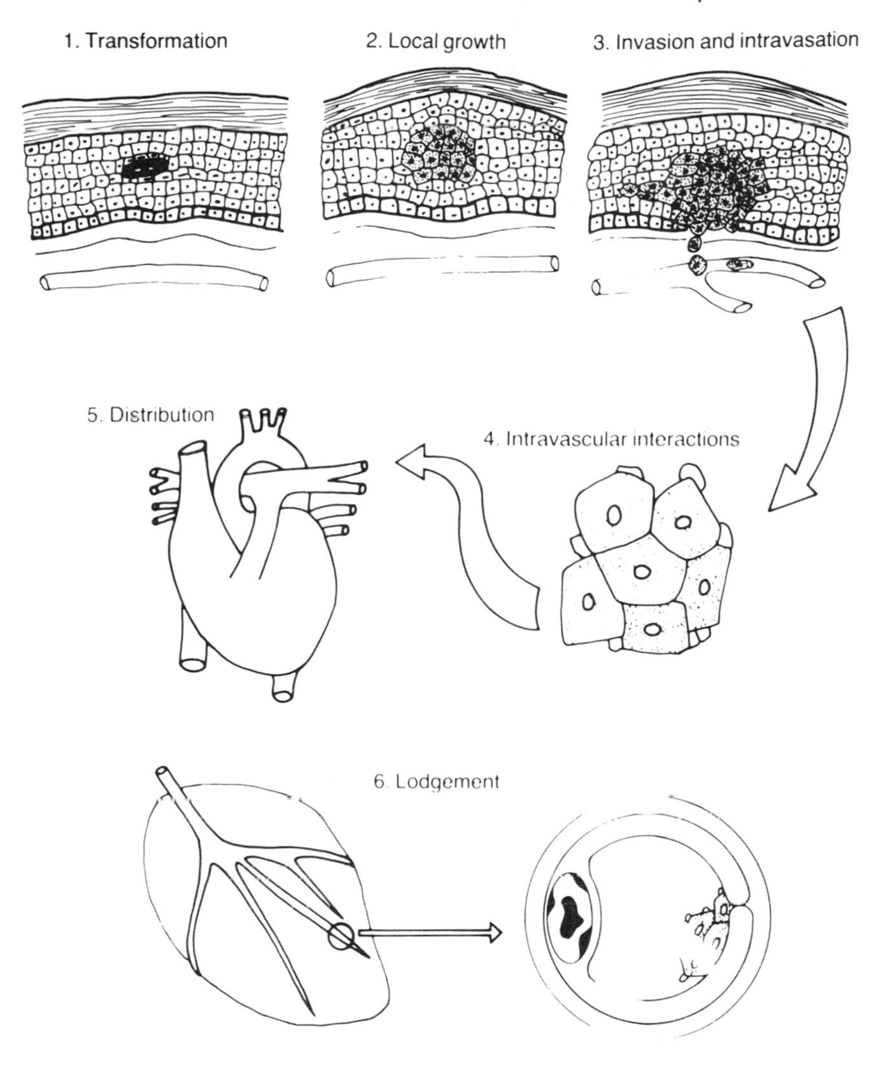
1. Transformation
2. Local growth
3. Invasion and intravasation
4. Intravascular interactions
5. Distribution
6. Lodgement

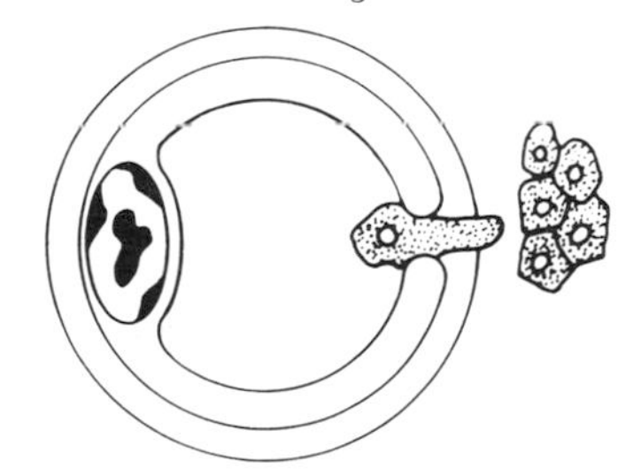
7. Extravasation and growth

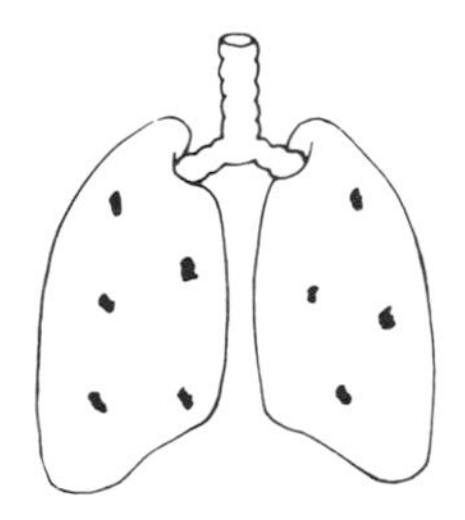
8. Metastases apparent

2.4.2 Lymphatic spread

Since tumours do not contain lymphatic vessels (Swabb, Wei and Gullino, 1974), their main port of entry into the lymphatic system is probably via lymph capillaries present near the edge of the tumour mass. The blind-ended nature of these capillaries and their lack (or the minimal presence) of a basement membrane is presumed to make it relatively easy for an invading tumour cell to gain entrance to this system. As described later, there is evidence of fluid flow from the tumour mass to nearby lymph capillaries, and this may passively carry or otherwise facilitate the transport of tumour cells into the vicinity of lymph vessels. Within lymph vessels tumour cells may exist singly or as small clumps or they may actually grow as a column within the lumen of the vessel before fragmenting. Tumour cells usually spread via the lymph vessels to lymph nodes where they may lodge in the subcapsular sinus. The extent to which a lymph node may act as a barrier for tumour spread and as a focus for tumour cell destruction by the immune system is still undecided. Using an *in vivo* experimental system based on the rabbit V_2 carcinoma, it has been shown that tumour cells injected into the afferent lymph vessels of the popliteal node (Fisher and Fisher, 1966) or the mesenteric nodes (Madden and Gyure, 1968) may be found in the respective efferent lymph within minutes after injection. Earlier studies by Zeidman and Buss (1954), however, had suggested that the lymph nodes might act at least as temporary barriers to spread. From the efferent lymph, tumour cells may spread via the thoracic duct into the blood system. Tumour cells not destroyed within the lymph node may colonize it and use it as a base for the seeding of further metastases. With respect to breast cancer, Berg *et al.* (1973) have proposed that the presence of enlarged but non-tumourous regional nodes may offer better prognosis than the presence of smaller ones (which are presumably poorly reactive). Once the nodes are colonized by tumour cells, however, the prognosis is likely to be markedly different.

The complex nature of the lymphatic system and its interactions with the blood system suggest that spread will soon come to involve both. There are several routes by which tumour cells initially spreading via the lymph may gain access to the blood system. These include intravasation of intranodal blood vessels, and transfer via connections between the lymphatic system and the venous blood stream such as the thoracic duct. Tumour cells initially spreading via the blood may also gain access to the lymphatic system, particularly within the structure of the lymph nodes.

2.4.3 Surface implantation

Cancer cells may seed onto cell surfaces such as those lining the pleural and peritoneal cavities. Carcinoma of the bronchus, for example, may invade into the pleural cavity allowing tumour cells to settle and implant in the pleura and form secondaries. Similarly, cells from ovarian carcinomas may detach and implant in the linings of the peritoneal cavity. The escape of malignant cells into the body cavities may give rise to ascites tumours in which the cells grow essentially in suspension, both singly and in clumps of various sizes. Ascites tumours usually invoke haemorrhagic and inflammatory reactions and they are associated with the accumulation of large volumes of fluid which may often be fatal because of pressure effects on major organs.

2.5 THE SPREAD OF CARCINOMAS AND SARCOMAS

In certain tumours one of the three major routes of dissemination may be favoured over the other two. There is some clinical evidence to support the notion that carcinomas spread via the lymphatics (as judged by the presence of metastases in lymph nodes) while sarcomas spread preferentially via the blood stream. There are exceptions to this generalization, but in numerous cases its validity is upheld. Several points may be relevant:

(a) Metastases in lymph nodes may occasionally derive from the nodal blood supply rather than from the afferent lymph. Thus the differences between the spread of sarcomas and carcinomas may not be as marked as it seems at first sight.
(b) The environment within a lymph node may only be suitable for the growth of particular types of tumours such as the carcinomas.
(c) Some tumours such as the sarcomas may find it easier to invade blood vessels than others.
(d) There may be differences in blood vessel structure within different tumours. The vascular clefts seen particularly in sarcomas, for example, are not lined by endothelial cells thus possibly promoting spread via the blood stream.
(e) Some tumours such as the sarcomas may survive more readily in the blood stream than others.
(f) Since tumours do not contain lymphatic vessels, spread via this route must involve prior detachment from the tumour mass and migration through tissue space. Carcinomas may be more successful in carrying out these steps than sarcomas.
(g) Because of the complex interrelationships which exist between the blood and lymphatic systems within the body, it is unlikely that

the dominance of one route over the others will last beyond the intital stages of spread.

It seems that the route of intitial spread is probably governed by anatomical considerations, although clearly the tumour cells must have the requisite properties in order to metastasize. The two types of invasive mechanisms proposed by Strauli (1980) and discussed in section 2.1 will certainly influence the patterns of spread, but their precise contribution is uncertain.

2.6 THE BLOOD SYSTEM

Before considering the metastatic process in more detail, it is appropriate to examine the structure of the major systems which are utilized during tumour cell spread. We should begin by taking a simple look at the heart which is a four chambered muscular pump, with two chambers (an atrium and a ventricle) in each half. The right atrium collects blood from the veins that drain all the parts of the body other than the lungs. This blood is passed into the right ventricle which then pumps it through the rich capillary beds of the lungs where gaseous exchange takes place. The oxygenated blood now passes via the pulmonary veins to the left atrium and thence to the left ventricle which pumps it into the aorta for distribution. There are thus two circulatory systems, the pulmonary system in which blood is pumped by the right ventricle to the lungs, and the systemic system in which blood is pumped by the left ventricle to the rest of the body. From the aorta, the blood courses through arteries of narrowing diameter finally emptying into capillary beds which then drain into the venous system. It should be noted here that not all of the blood must pass through capillaries *per se* since some may traverse the system by metarterioles or various other types of shunts (see below). The blood from all of the veins in the systemic circulation eventually reaches the right atrium via either the inferior or the superior vena cava. Tumour cells arising from malignant primaries gain access to the vena cavae either through smaller vessels such as venules which drain into them, or via the lymphatic system which also ultimately drains into the venous blood system. Such cells pass rapidly through the heart before encountering their first capillary bed in the lungs. As might be expected, there are a number of additional anatomical factors determining tumour spread. The details of these may be gained from the study of any one of a number of relevant textbooks in anatomy, though it is worth pointing out the following:

(a) Metastases from certain tumours (such as carcinoma of the prostate) are found distributed in the axial skeleton. These result from the presence of the internal vertebral venous plexus, a longtitudinal arrangement of valveless veins on each side of the spinal canal. The plexus receives its tributaries from the basivertebral veins which drain the red marrow of the vertebrae and pass on either side of the spinal column to connect with the longtitudinal veins. When the inferior vena cava cannot cope with a sudden flush of blood due to a sharp increase in intra-abdominal pressure (as might occur during coughing, for example), the plexus acts as a bypass for pelvic and abdominal blood which courses into the posterior intercostal veins and thence into the superior vena cava. Tumour cells from neighbouring organs may be drawn along with this venous blood into the spinal column where they may develop into metastases.

(b) The blood supply to the liver is both venous and arterial in nature, being carried by the portal vein (from the intestines, spleen, pancreas and gallbladder) and the hepatic artery respectively. Blood from both vessels courses through the liver in capillary-like structures referred to as sinusoids before draining into the inferior vena cava. Tumour cells which gain entrance into the hepatic portal system, perhaps from intestinal tumours, will thus encounter the sinusoidal bed of the liver before they reach the lungs.

Now it might be supposed that the distribution of metastases from a primary tumour would depend on its location in relation to the anatomical arrangements referred to above. However, there appear to be a large number of anomalies in the distribution of metastases which leads one to consider the possibility that more than just anatomical aspects are involved in the spread of tumours. Tumour cells which leave the left ventricle, for example, might be disseminated according to the pattern of arterial blood flow but this is not necessarily the case. The patterns of tumour spread will be discussed further in section 2.14.

2.7 THE STRUCTURE OF BLOOD VESSELS

Although the structure of each class of blood vessel is fairly consistent allowing classification by histological means, the actual arrangement of the microcirculation varies from organ to organ so that the picture presented here (Fig. 2.2) is very much a generalized one. The reader should refer to review papers by Rhodin (1967, 1968) and by Ryan (1986), and to classic textbooks on histology.

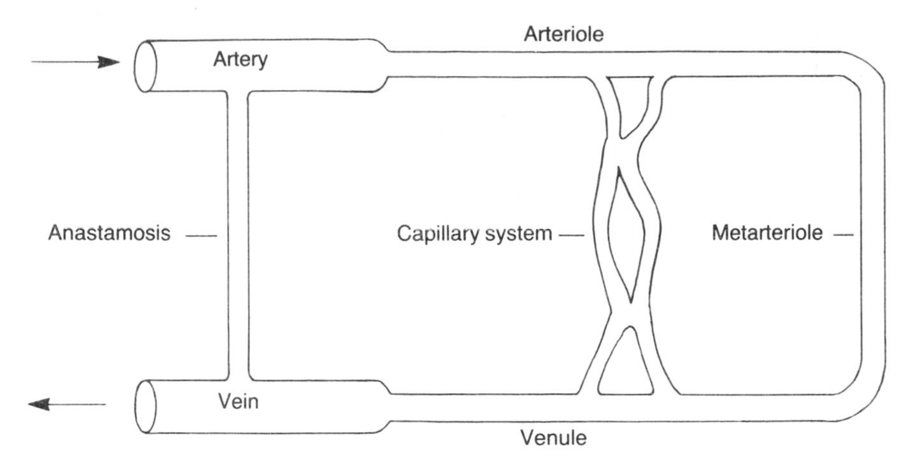

Figure 2.2 Generalized blood vessel relationships. The figure illustrates schematically the relationships between the arterial and venous systems of the body.

2.7.1 The arterial system

There are three principal types of arteries in the systemic circulation:

(a) *Elastic arteries*

These maintain the pressure of the ventricle between contractions.

(b) *Muscular arteries*

These regulate the blood supply to particular parts of the body.

(c) *Arterioles*

These act as pressure reduction valves before delivering the blood to a capillary bed. **Metarterioles** connect arterioles directly with venules (although some authorities use the term differently). In many tissues and organs, metarterioles may in fact represent the preferred channel for blood flow between the arterial and venous systems.

Each type of artery is of the same general stucture with the walls being composed of three coats known as the tunica intima (the innermost), the tunica media, and the tunica adventitia. The volume each coat occupies within the arterial wall varies both within and between each vessel type.

(1) *The tunica intima*

The luminal aspect of the tunica intima is composed of an endothelial lining and its associated basement membrane which either sits directly on an elastic lamina (the internal elastic lamina) or intercellular tissue (composed largely of smooth muscle cells, fibroblasts, collagen and some elastin) is interposed between the two. The endothelial cells may not be smooth and indeed they may have projections on their luminal surface.

(2) *The tunica media*

This coat is composed predominantly of elastin in elastic arteries and smooth muscle in muscular arteries and arterioles. The outer border of the tunica media is composed of another elastic lamina (the external elastic lamina), but this may be missing in some arterioles.

(3) *The tunica adventitia*

In the muscular arteries and the arterioles this consists mostly of elastic fibres with some collagen, but in the elastic arteries it is relatively thin. The tunica adventitia in elastic and muscular arteries is supplied with blood vessels (vasa vasorum) and lymphatics, as may be the outer part of the tunica media. The tunica intima, however, is not supplied with such vessels.

As arterioles (max. diam. 100 μm) become smaller their coats are lost, such that a precapillary arteriole may consist of nothing more than an attenuated endothelium sitting on a thin layer of discontinuous smooth muscle with precious little connective tissue. The last muscle cell before the capillary network is referred to as the **pre-capillary sphincter**.

2.7.2 Arteriovenous anastomoses

Arteriovenous anastomoses are composed of vessels which arise as sidebranches of arteries and which connect directly to veins. That such connections might exist was indicated by experiments in which glass spheres (too large to pass through capillaries) nevertheless managed to reach the venous circulation after arterial injection (Prinzmetal *et al.*, 1948). As might be expected, the structure of anastomosing vessels varies along their length, from being somewhat like an artery to being somewhat like a vein. In the middle region anastomoses have a thickened muscular wall supplied with sympathetic nerves, which presumably allows them to adjust the local flow of blood. Arterio-venous anastomoses are particularly common in body extremities (such as the tips of fingers and toes) where, by adjusting blood flow, they may help in temperature regulation.

2.7.3 Capillaries

From the general viewpoint presented here, capillaries arise from arterioles and metarterioles and they are largely called into play when blood flow (regulated by smooth muscle cells in the arterioles) increases. There are essentially two types of capillaries, continuous (which contain closely apposed cells with recognizable junctional complexes) and discontinuous (with patent intercellular channels). The lumen of a capillary is less than 10 μm in diameter and it is typically the size of an erythrocyte. It is lined by a single layer of endothelial cells which themselves sit on a thin basement membrane. In some organs the capillary basement membrane may be substantial as in the brain, for example, where it may contribute to the blood–brain barrier. In other organs, such as the kidney, the capillary basement membrane may be incomplete. Cells known as pericytes are found in association with the basement membrane; they may be contractile and thus assist in the regulation of blood flow through the microcirculation. Since the endothelial cells themselves also contain contractile proteins, they too may assist in regulating blood flow.

2.7.4 The venous system

The blood from capillaries is collected by venules which are slightly larger vessels lined by an endothelium that sits on a thin basement membrane. The venules pass into small veins whose walls now contain smooth muscle cells. Larger veins have the three coats described for arteries, but the tunica media is generally less developed and contains a significantly smaller quantity of smooth muscle. There is also generally a smaller amount of elastin in veins. Some veins of the lower limbs tend to have thicker and more muscular walls because of hydrostatic pressure (from the column of blood in the body above them) and because they often run a superficial course which does not give them significant structural support. Because the pressure within veins is not as great as that within arteries, their walls may contain more extensive lymphatic vessels.

2.8 FUNCTIONAL PROPERTIES OF THE VASCULAR ENDOTHELIUM

The metabolic activities of endothelial cells are remarkably diverse, but events of particular interest to the spread of tumour cells predominantly involve the luminal surface. The luminal aspect of the endothelial cell surface (particularly in the lung microcirculation) is covered by microprojections which increase the surface area, and it is rich in calveolae (pits with a thin covering diaphragm) which seem to contain many of the endothelial enzymes (reviewed in Ryan, 1986). Experiments in which isolated rat lungs were perfused with 5–10 μm microbeads showed that endothelial cells were phagocytic. Although phagocytic behaviour may occur *in vivo*, the physiological significance of this observation is not clear. External to the cell membrane is a fuzzy coat, the **glycocalyx**, which contains numerous components including adsorbed plasma proteins and proteoglycans and which may be up to 60 μm thick. The glycocalyx probably covers many surface receptors, it may act as a molecular sieve for plasma proteins, and it probably represents the surface which normally comes into contact with circulating cells in the blood, including cancer cells. The passage of blood through the vascular system is aided by the non-thrombogenic nature of the endothelial cell surface. This is provided by the presence of heparin-like proteoglycans in the glycocalyx, by surface ADPase activity, by the secretion of plasminogen activator, by mechanisms for the clearance of circulating thrombin, and by the production and release of prostacyclin (PGI_2). The antithrombogenic nature of the endothelial cell surface can be readily reversed by the actions of certain agents such as endotoxin (bacterial lipopolysaccharide), which induces the cells to express procoagulant (tissue factor) activity and to secrete an inhibitor of plasminogen activator.

Finally, it is worth noting that a stationary boundary layer may exist immediately adjacent to the endothelium. Such a layer arises at the boundary between a flowing fluid (such as blood) and a solid surface (such as the endothelium) because the contacting fluid can bind to the surface itself thereby yielding zero fluid velocity in this region. The thickness of the stationary boundary layer depends on a number of properties including fluid viscosity and vessel diameter. Although it must be penetrated for a malignant cell to extravasate from the blood, the boundary layer is unlikely to prove to be an effective physical barrier to metastasis. The boundary layer could feasibly restrict access to membrane located determinants, but its role here is likely to be secondary to the potential energy barrier which arises between cells of similar (negative) charge.

2.9 HAEMODYNAMICS

The flow of a fluid such as blood in a vessel may be streamlined or turbulent in nature, subject to local eddies, to pulsatile and retrograde flow, and even to stasis. A thorough analysis of blood flow would be extremely complicated and will not be attempted here. There are some features of flow, however, which may be relevant to the metastatic process. Before we discuss these we must first consider two important terms used in fluid dynamics:

(a) *Shear rate*

Imagine a fluid flowing over a solid surface such that all its elements travel in the same direction with a velocity proportional to the perpendicular distance from the surface. The velocity gradient which is established is known as the shear rate.

(b) *Shear stress*

Since neighbouring layers within such a fluid travel at different rates, they must flow over each other. This induces a shearing stress which is tangential to the surface and proportional to the shear rate.

2.9.1 The pattern of fluid flow in vessels

The flow of blood in a vessel may be likened to the flow of a somewhat viscous fluid in a pipe. Most biophysical models take blood as a Newtonian fluid (in which the shear rate is proportional to the shear stress) and they assume the flow to be Poiseuille or laminar (in which cylinders of fluid flow over each other, with maximum velocity at the vessel axis). The assumptions in these models should be noted carefully, since the situation described is rather ideal. Measurements have been made of vessel velocity profiles using experimental systems in which model vessels bifurcate symmetrically. After the bifurcation, the fluid velocity is greater nearer the inside wall of a daughter vessel than its outside wall, but this returns to normal some distance downstream. Thus just after the bifurcation there are areas of relatively low wall shear stress and, in theory at least, these may represent advantageous sites for metastatic cell adhesion. There is experimental evidence based on model systems that cells can spend relatively long times in such areas and this may encourage adhesive interaction and subsequent extravasation. In the venous system, the flow is reversed with the blood from two daughter vessels flowing into the parent and the velocity profiles are different.

Within blood vessels such as arterioles, there is a general tendency for the cells to assume an axial position, although this will change as a function of a number of properties including cell size, rigidity and number, the distribution of fluid shear rate within the vessel, fluid viscosity and flow rate. The tendency for axial cell flow creates a more or less cell free layer of relatively slow moving plasma, about 2–4 μm thick. Cells which are going to migrate from the blood within vessels which display axial flow clearly must breach this cell-free layer adjacent to the vessel wall. However, the blood is not exclusively divided into an axial, cellular component and a peripheral cell-free component, and there is considerable evidence that cells can enter the so-called cell-free layer. One possible mechanism by which cells might breach the cell-free layer is described in section 2.12.1. Note that axial flow does not exist in capillaries, since the vessel diameter is often similar to or smaller than that of the cells which pass through it.

2.9.2 The effects of flow on cells

Cells within the blood rotate as a function of shear rate differences over their surfaces and, depending on their rigidity and the forces involved, their shape may deform. As seen in experiments with erythrocytes, shear stresses induce cellular deformation but very high shear stresses (around 200 Nm^{-2}) can cause their lysis. Normal erythrocytes (biconcave discs about 8 μm × 3 μm) can pass through pores as small as 3 μm diameter, but they are damaged when forced through smaller ones.

The passage of blood cells through small vessels is aided by the glycocalyx on the endothelial surface which acts as a lubricating layer, and this is also likely to assist in the transport of tumour cells. Thus whether tumour cells survive passage through small vessels or not will depend both on this lubricating layer and on the deformable nature of the vessel and of the tumour cell itself. Sato and Suzuki (1976) assessed the deformability of four types of rat ascites tumour cells by measuring the negative pressure necessary to suck cells from each tumour type into micropipettes. After obtaining an assessment of the deformability of each tumour type in this way, cells were then injected into host rats and recovered after one passage through the lungs. These authors found a correlation between relative deformability and survival after transpulmonary passage, with more viable cells passing through the lungs for the more deformable tumour types.

2.10 THE EXTRACELLULAR MATRIX

The extracellular matrix (ECM) comprises both the interstitial stroma and basement membranes. Generally speaking it is composed of collagen, proteoglycans and various glycoproteins such as laminin, fibronectin and possibly vitronectin (Tables 2.1. and 2.2.). Although recent research has unveiled a number of collagen types, three are of principal interest here: collagen types I and III make a significant contribution to the interstitial stroma, while type IV collagen is a major component of the basement membrane. Digestion of these collagenous components of the ECM is likely to facilitate the invasion and extravasation of malignant cells. Enzymes which break down the interstitial collagens, however, are not effective against the type IV collagen of the basement mebrane. The degradation of this collagen involves a specific neutral metalloproteinase (type IV collagenase) which is known to be released by metastatic tumour cells (Liotta, Thorgeirrson and Garbisa, 1982).

Most of the proteoglycans (including hyaluronic acid) are thought to contribute to the ECM. Considerable research has centred on the role

Table 2.1 Major non-collagenous components of the extracellular matrix

Component	$M_r \times 10^{-3}$ *D*	*Location and possible role*
(a) Protein/glycoproteins		
Laminin	1000	basement membrane (bm); cell adhesion
Fibronectin	440	vessel walls, bm; cell adhesion
Nidogen (entactin)	150	bm; binds to laminin
Elastin	72[1]	large arteries, skin, lung; provides elastic support
(b) Proteoglycans		
Heparan sulphate	130–350[2]	various forms. Bm, cell surface; role in adhesion and molecular filtration
Chondroitin sulphate	30–200[2]	various forms. Cartilage, aorta, skin, muscle, bone and bm; possible role in adhesion
Dermatan sulphate	40–500[2]	various forms. Skin, tendon, aorta
Keratan sulphate	30–40[2]	various forms. Cartilage, cornea
Hyaluronic acid	500–5000	various forms. Vitreous, cartilage; tissue volume and shape

1. Subunit M_r. The subunits cross-link to form the elastin fibre.
2. Approximate core protein M_r. Proteoglycans consist of a protein core (of variable M_r) to which are attached the glycosaminoglycan side-chains and *N*– and *O*–linked oligosaccharides.

Table 2.2 Major collagen types

Type	Form	Location
I	triple helix fibril	bone, tendon, ligament, cornea
II	triple helix fibril	hyalin cartilage
III	triple helix fibril	dermis
IV	triple helix network	basement membrane (bm)
V	triple helix fibril	hyaline cartilage, bone, cornea, bm
VI	?	blood vessels
VII	?	epithelial/mesenchyme anchoring fibril
VIII	?	endothelium, mesenchyme
IX	?	cartilage
X	?	cartilage

of heparan sulphate proteoglycan which is believed to be degraded during tumour cell invasion (Kramer, Vogel and Nicolson, 1982). Laminin is a large molecule (approx. M_r 1 × 10^6 kD, under non-reducing conditions) composed of 3 × 200 kD subunits and 1 × 400 kD unit arranged in the general shape of a cross. Like fibronectin, it seems to be composed of a number of domains which subserve numerous functions including collagen binding, heparin binding and cell binding. Fibronectin is present both in the plasma and in the ECM. It is has an unreduced M_r of about 450 kD and is composed of two disulphide bridged chains. Like laminin, it contains functional domains for the binding of heparin, collagen and cells. Vitronectin, which occurs as 65 kD and 75 kD forms, is a plasma glycoprotein which also localizes in the interstitial stroma (Hayman *et al.*, 1983) and appears to be immunologically related to epibolin (Stenn, 1981 a and b).

Other, possibly more specific components of the ECM include epinectin, a 70 kD protein released from the matrix secreted by A431 carcinoma cells following their solubilization with deoxycholate and treatment of the residue with 6 M urea. Enenstein and Furcht (1984) have shown that epinectin is active in promoting the substrate attachment of this cell type. The roles played by other ECM components such as the proteoglycans, collagen, fibronectin, vitronectin and laminin in so far as they might influence adhesion and metastasis are discussed in detail in Chapter 5. It seems, however, that interaction with ECM components confers more information than that to do with adhesion alone. Indeed, since the ECM is a major part of the immediate environment of tumour cells, as such it is likely to exert considerable influence over their growth and behaviour. There is evidence that the ECM exerts an influence on differentiation. Pigmented retinal epithelial cells, for example, can be induced to transdifferentiate into neurons by laminin and this suggests that the ECM may provide a signal for induction (Reh, Nagy and Gretton, 1987). The ECM may also exert an effect as a competence inducing factor, acting in association with various growth factors and hormones (section 1.7.1). Most tumours appear to lack or have altered ECMs, and this may significantly influence tumour growth, invasion and metastasis, although it must be admitted that its contribution remains to be evaluated fully.

2.10.1 Structure of the basement membrane

The basement membrane is a trilaminar, acellular sheet-like structure. It is thought to have two main functions: to provide anchorage and elastic support for the overlying cells, and to act as a barrier for filtration. During development, however, it plays a key role in many morphogenetic events such as branching of the salivary glands, and it is also crucial to the establishment and maintenance of epithelial cell polarity. The basement membrane also serves as a barrier to malignant cell invasion, though the extent to which this is achieved may vary for different tumours and for the basement membranes of different tissues. Transition from a benign epithelium to a carcinoma is generally associated with the loss or disorganization of the basement membrane. Exactly how this is brought about is not clear, and both perturbed synthesis and increased degradation through the release of proteolytic enzymes may be involved. A basement membrane is found around non-invasive carcinoma *in situ*, but focal defects may be observed at the ultrastructural level. Such minute defects may provide a first indication of progression to the invasive stage.

The idealized basement membrane is composed of three layers (reviewed in Kefalides, Alper and Clark, 1979; Timpl and Dziadek, 1986):

(a) *The lamina lucida*

This layer is about 10–50 nm thick and it lies immediately external to the basal plasma membrane of the overlying cells. It is occasionally crossed by fine filaments (10–15 nm in diameter) which extend between the overlying cells and the underlying lamina densa.

(b) *The lamina densa*

This layer is about 20–300 nm thick and is composed of fine filaments in an amorphous matrix. The term **basal lamina** has been used to refer to the lamina densa.

(c) *The lamina fibroreticularis*

This layer is usually discontinuous and contains reticular and anchoring fibrils. Microfibrils (15–20 nm diameter) within the lamina fibroreticularis may connect collagen fibres to the lamina densa (Ichumura and Hashimoto, 1984), and the series of interconnections which exists will almost certainly serve to strengthen the overall structure. Interconnections reaching from the cells through the basement membrane into the underlying connective tissue, whether direct or indirect, will also serve to anchor the overlying cells in place. Such a network of interconnections may extend from the hemidesmosomes of certain epithelial cells as described elsewhere (section 5.4.1).

Recently the suggestion has been made that the basement membrane may be a homogeneous structure and that separation occurs on fixation (Goldberg and Escaig-Haye, 1986). Indeed, the lamina lucida may represent a space normally occupied by the glycocalyx of the cell surface and the lamina fibroreticularis may be a product of underlying connective tissue cells. In these still rather controversial terms, the basement membrane proper is represented only by the lamina densa (basal lamina).

The basement membrane is composed of diverse components including collagen (type IV is specific for basement membranes), fibronectin, laminin, proteoglycans, and nidogen (Laurie *et al.*, 1982, 1984), the quantities of which appear to vary depending on the source of the basement membrane. Type V collagen usually has a pericellular location, but it has been reported to be present in some basement membranes. According to Laurie, Leblond and Martin (1982), most of these components are localized in the lamina densa with an integrated distribution. Type IV collagen (M_r 550–600 kD) is a triple-stranded molecule composed of two different polypeptide chains. It contains about 10–12% carbohydrate and it is able to form a meshwork, looking somewhat like molecular grade chicken wire. *In vitro* experiments indicate that type IV collagen does not aggregate when incubated alone in solution, but it does do so when incubated with equimolar laminin (Kleinman *et al.*, 1983). This aggregation is potentiated by heparan sulphate proteoglycan, but fibronectin is without effect. These studies indicate that at least some of the major macromolecular constituents of the basement membrane interact in a highly specific manner, and it would seem likely that these interactions govern the nature and *in situ* deposition of this structure. Indeed, laminin appears to be deposited first in the developing basement membrane and this would suggest a key role for this molecule.

Interestingly, fibronectin is found more in newly synthesized basment membrane than in older basement membrane, but its precise role in basement membrane development remains uncertain. Specific details relating to the various molecules which contribute to the basement membrane will be discussed in subsequent sections.

The basement membrane is thought not to be a rigidly stable structure since there is evidence that it can undergo continuous breakdown and synthesis. By measuring the rate of release of radiolabelled components, Nakajima *et al.* (1984), for example, found that the subcellular matrix of bovine aortic endothelial cells solubilized spontaneously *in vitro* at a rate equivalent to 0.3%–0.4% per hour. It should be emphasized here that, despite its use in many experimental studies, there is some uncertainty as to how closely the substratum-attached ECM deposited by cells in culture actually resembles *in vivo* basement membrane. Although in a few cases both ultrastructural and biochemical studies suggest a close resemblance, there is usually considerable variation from culture to culture and for this reason the material deposited *in vitro* will be referred to as matrix rather than basement membrane *per se*. Another source of basement membrane material much used in experimental studies is the mouse Engelbrecht-Holm-Swarm (EHS) sarcoma. This tumour produces basement membrane in considerable quantities, with type IV collagen, laminin, and heparan sulphate present in a molar ratio of about 10:10:1. Unfortunately, similar quantitative studies have not been made on basement membrane material from normal tissues, and thus to what extent the EHS basement membrane is typical remains uncertain.

It seems likely on both ultrastructural and biochemical grounds that different types of the basement membrane exist, and there is every reason to believe that some tumour cells can penetrate certain basement membranes but not others. Kramer and Vogel (1984), for example, found that B16F1, B16F10 and B16-BL6 melanoma cells could not penetrate the matrix deposited *in vitro* by the mouse endodermal cell line PF HR9, whereas other evidence from both *in vitro* and *in vivo* studies suggest that these B16 melanoma cells can penetrate through a wide range of matrices. The B16-BL6 cell line, for example, was actually selected for its ability to penetrate the bladder wall *in vitro* (Hart, 1979; Poste *et al.*, 1980). That some basement membranes can act as effective barriers to tumour cells as they extravasate from blood vessels has been confirmed ultrastucturally, and indeed there are numerous reports of tumour cells lodged between endothelial cells and the underlying basement membrane. Here they may remain for weeks or longer (possibly in a dormant state) before migrating further, or they may go on to divide and invade destructively *en masse*. Not all basement membranes are continuous structures. That of the intestinal villi, for example, contains numerous fenestrations ranging from 0.5–5 μm in diameter (Komuro, 1985). It has been argued that these fenestrations may facilitate lymphocyte traffic across the gut wall, but there is no evidence that they markedly influence the metastatic process. The role of the basement membrane in metastasis is discussed further in section 5.7.4

2.11 THE LYMPHATIC SYSTEM

The lymphatic system draws fluid from the interstitial space into the bloodstream and (in conjunction with the blood system) it also transports lymphocytes around the body. Lymphatic capillaries drain from the tissues into lymph nodes and thence into larger vessels which have the three recognizable coats of blood vessels. All lymphatic vessels have an endothelial lining. Lymphatic capillaries, however, seem to be without a true basement membrane, although they are usually associated with a small amount of connective tissue. In larger lymphatic vessels, the tunica intima contains elastic fibres, while the tunica media contains variable amounts of smooth muscle in a connective tissue framework which also contains some elastic fibres. The outermost coat, the tunica adventitia, is usually reasonably well developed and it includes connective tissue and small blood vessels. Small lymph vessels do not progressively unite with one another to form larger and larger vessels; instead they form unions and branch again before finally draining into larger vessels which ultimately empty into the venous system. The overall drainage pattern is not simple, however, and connections may exist between lymph vessels, between lymph nodes, and between lymph vessels and various parts of the venous system.

Lymphatic fluid (lymph) enters lymph nodes from tissue spaces via afferent lymphatic vessels. It percolates through the substance of the node exiting in efferent lymphatic vessels which ultimately join up with an appropriate collecting duct that feeds into the venous system as described above. Lymphocytes enter lymph nodes from the blood stream via specialized endothelial cells of post-capillary venules within the node. Instead of being rather flat, as is normally the case, the endothelial cells of these venules are somewhat columnar so that they are often referred to as high endothelial venules (HEVs). Lymphocytes bind to these specialized cells and extravasate from the blood to enter the node. Within the node, B cells and T cells migrate to specific areas (the primary follicles and the paracortex respectively). From their specific areas, the lymphocytes progress to the efferent lymphatic vessels from where, like the lymph, they ultimately reach the blood system. Lymphocytes, of course, do more than just circulate between the blood and the lymph, and they may enter the tissue spaces to combat infections and attempt to deal with tumour cells. The extravasation of lymphocytes into lymph nodes is discussed in more detail in section 2.12.2.

Unlike haematogenic spread in which tumour cells lodge in blood vessels, lymphatic spread is thought to occur primarily via trapping in lymph nodes, although cords of cells may permeate along a vessel. Tumour cells which penetrate afferent lymphatic vessels may lodge and grow in the appropriate lymph node and perhaps act as a generating site for the seeding of further metastases. Alternatively, tumour cells may pass straight through to lodge in another more proximal node ('skip metastases'), or they may reach the venous system and thence be distributed by the blood. Occasionally, retrograde tumour spread may be seen as a consequence of obstruction of the efferent vessels and reverse flow. The penetration of lymphatic vessels by tumour cells is likely to be aided by the lack of a distinct basement membrane and by the low pressure within.

2.12 THE EXTRAVASATION OF LEUKOCYTES: GENERAL PRINCIPLES

Malignant cells are not the only cell type which extravasate from blood vessels and invade the tissue spaces. Normal white blood cells such as PMNs, monocytes and lymphocytes may also emigrate from the blood system. The emigration displayed by erythrocytes, however, is fundamentally different since although these cells may reach the tissues (as in haemorrhage, for example) they do so passively rather than actively. Erythrocytes have been observed to escape from the blood system through gaps in vessel walls (resulting from physical trauma or after another cell has migrated through), and they are presumably forced out in response to pressure differences.

2.12.1 Leukocyte extravasation during inflammation

Many scientists have looked for associations between the ways in which leukocytes and malignant cells extravasate from blood vessels. Indeed, common sense would suggest that some parts of these processes are likely to be similar, if not the same. As will become clear later (section 2.13), however, there are crucial differences between the extravasation processes of leukocytes and some malignant cells which make them not strictly comparable. Nevertheless, there may be some insight to be gained by closer examination of the mechanisms involved in leukocyte extravasation.

The first cells which emigrate from the blood into the tissues during an acute inflammatory reaction are the polymorphonuclear neutrophils (PMNs), followed later by monocytes (which become wandering macrophages) and lymphocytes. Indeed, PMNs characterize the acute inflammatory reaction whereas macrophages and lymphocytes are more characteristic of chronic inflammation. It is not known what governs this different temporal behaviour, although it is presumed to lie in local parameters such as changes in chemotactic factors and other cellular signals (including lymphokines and monokines) as well as in the cells themselves. Interestingly, it has been claimed that monocytes cannot invade collagen gels whereas PMNs and lymphocytes can (Lackie *et al.*, 1985), pointing to significant differences in invasive and motile behaviour between these blood cell types.

An understanding of the extravasation of leukocytes during inflammation requires a basic knowledge of the inflammmatory reaction itself. For convenience more than anything else, many experimental studies have been made on artificially induced acute inflammatory reactions. In response to an inflammatory stimulus such as a burn, immediate arteriolar constriction usually leads to a transient blanching of the injured tissue. This is quickly followed by relaxation of the arterioles so that, along with the capillaries and venules, they become engorged with rapidly flowing blood. Probably aided by increased hydrostatic pressure, protein-rich fluid leaks from the engorged vessels, particularly the capillaries and venules. Although the escaping fluid is rich in proteins, more fluid is actually lost than proteins and the viscosity of the remaining blood thus increases. This increase in viscosity contributes to a slowing of the blood which may allow erythrocytes to form rouleaux, thereby further increasing blood viscosity and enhancing the possibility of stasis.

Experimental evidence suggests that there are two phases in which vascular permeability is increased:

(a) *The early phase*

This phase shows a rapid onset and lasts about 30 min. Leakage is seen predominantly from venules, possibly as a consequence of histamine release from tissue mast cells.

(b) *The late phase*

This begins about an hour after injury, is prolonged, and appears to involve both capillaries and venules.

Since PMNs can be seen to extravasate from vessels within minutes of tissue damage, fluid changes associated with the early phase (involving the venules) are likely to be important in initial cell emigration. Within the venules, changes in flow rate and viscosity appear to be associated with disruption of normal axial flow and leukocytes (particularly PMNs) make temporary and rolling contacts with the endothelium of the vessel wall. This process is known as **margination** or **pavementation** and, as described below, it is likely to be a normal phenomenon which is markedly increased under inflammatory conditions. Indeed, there is abundant evidence that cytokines such as TNFα and IL-1 promote PMN adhesion to the endothelium, although another cytokine (TGF$_\beta$) apparently inhibts the process (Gamble and Vadas, 1988). Be that as it may, the length of contact time during margination increases until eventually some adherent cells are able to extend pseudopodia which probe the junctional areas between endothelial cells. In due course the junctions are disrupted and the cells leave the blood to enter the tissue space by passing through the basement membrane associated with the vessel.

Observations suggest that PMNs and other leukocytes extravasate largely from venules and not from capillaries or arterioles. Presumably the venules are the location of optimal physico-chemical changes which are necessary for extravasation, but the exact nature of these changes is still somewhat obscure. It may be, however, that they are related in part to venule leakiness and to changes in hydrostatic pressure. Since PMNs have an average diameter greater than that of most capillaries, they can be expected to distort as they enter and pass through a capillary network. In fact, PMNs probably represent a temporary obstruction (of the order of seconds) at capillary openings. Furthermore, since PMNs move along the capillaries more slowly than RBCs there is a tendency for a train of erythrocytes to build up behind. According to the ideas of Bagge and colleagues (1983), once on the venular side the PMNs are pushed against the endothelial wall by the faster moving train of RBCs behind. If the conditions are suitable, PMNs displaced from the axial flow will then be able to adhere to the lining endothelial cells and subsequently emigrate from the venules. Not all adherent cells appear to extravasate, and this is presumably the source of the two pools of PMNs classically described in the literature. These two pools, referred to as the circulating and marginal pools, are roughly equal in size, although it must be admitted that to what extent the processes described above actually influence pool size is uncertain. Cells of the circulating pool flow in the axial stream whereas cells in the marginal pool roll along the endothelial surface. Direct observation suggests that these marginal cells probe the endothelial cell surface;

they may form temporary adhesions with this surface, and they may even move back along the endothelium against the normal flow of blood. This behaviour is apparently displayed by the marginal cells under normal conditions, and it presumably places them in a position to respond to subtle changes in their environment, perhaps by extravasation towards a mild inflammatory stimulus. According to Clark and Clark (1935, 1936), cells of the marginal pool respond to local changes in the endothelium since cells which temporarily adhere to one part of the endothelium do not have a tendency to stick elsewhere and, furthermore, particular patches of the endothelium are sticky for more than one circulating leukocyte. Under conditions of tissue damage (as is often seen experimentally) or in the more severe cases of inflammation, changes induced in the blood and its vessels (as described above) may lead to increasingly marked perturbation of axial flow, thereby bringing more circulating PMNs into the marginal pool. These ideas, in principle at least, may also apply to malignant cells circulating within the blood. Thus tumour cells may be able to join the marginated pool under normal or inflammatory conditions and thereby find themselves in a suitable position for extravasation. As will become clear in due course, however, there is evidence that simple mechanical processes alone cannot fully explain the observed patterns of metastatic distribution.

Studies of PMN migration across endothelial or epithelial cell monolayers *in vitro* suggest that PMNs penetrate such barriers only in the presence of a gradient of some chemotactic signal (Evans *et al.*, 1983). A certain amount of vessel leakiness may be necessary for this signal to reach the luminal aspect of the vessel wall since the junctional complex between endothelial cells could act as an effective barrier to all but the smallest molecules and ions. Although precise details are unkown, chemotactic molecules may bind to endothelial cells so that they are not swept away in the blood or they may induce changes in the endothelial cells leading to the expression of other molecules which are chemotactic or adhesive in nature. The direct secretion of PMN chemoattractants by endothelial cells has been shown using agents such as angiotensin II, a molecule which forms part of the renin–angiotensin system regulating blood pressure. Recognition of adhesive and chemoattractant molecules eventually leads to the binding and extravasation of PMNs. Note that activated PMNs may be induced to degranulate, thereby releasing additional chemotactic factors and relaying the chemotactic signal. Interestingly, it has been supposed in the past that chemoattractants probably serve a dual role, both increasing PMN adhesion to the endothelium and promoting directed locomotion. Recent evidence suggests that different molecules are probably involved in each of these processes, however, since chemoattractants inhibit directed locomotion at concentrations where they stimulate adhesion (Charo *et al.*, 1986). Studies by Bevilacqua and colleagues (1987, 1989) suggest that inflammatory cytokines such as interleukin-1 (IL-1) and bacterial endotoxin induce a cell surface glycoprotein of M_r 115 kD on human endothelial cells. Expression of this molecule, known as the endothelial leukocyte adhesion molecule-1 (ELAM-1), correlates with the increased adhesiveness of PMNs to cytokine-activated endothelium. The intercellular adhesion molecule ICAM-1, which is involved in lymphocyte adhesion as the putative receptor for LFA-1 (section 2.12.2), is also inducible on endothelial cells by cytokines and under such circumstances it appears to promote the adhesiveness of both PMNs and lymphocytes. Whereas the expression of ELAM-1 is induced relatively rapidly (peaking within 2-4 h) that of ICAM-1 is more prolonged (>48 h), and this had led to the suggestion that the former may act primarily in acute inflammation while the latter might operate under more chronic circumstances.

The majority of studies suggest that PMNs emigrate from blood vessels by passing between the endothelial cells in the junctional regions, although studies of PMN extravasation in the cerebral microcirculation suggest two other pathways (Faustmann and Dermietzel, 1985). Thus PMNs crossing the blood–brain barrier may use either an intracellular route (referred to as **emperipolesis**) in which the PMN seems to be 'phagocytosed' by the endothelial cell, or another route involving the formation of a pore in the endothelial cell (possibly in the region where the PMN first becomes adherent). Pore formation has been proposed as the mechanism by which leukocytes escape from the bone marrow, and it may be that leukocytes use different pathways to traverse endothelia in different organs.

2.12.2 The extravasation of normal and malignant lymphocytes

Lymphocytes circulate or traffic from the blood to the lymph, passing through venules modified by the presence of high (cuboidal to columnar) endothelial cells (section 2.11). They also extravasate from the blood to enter inflammatory sites, typically after the emigration of other leukocytes such as PMNs. Ultrastructural studies of rat lymph nodes by Anderson and Anderson (1976) suggested that microvilli extend from lymphocytes to make contact with characteristic pits on the surfaces of high endothelial venule (HEV) cells which are not seen on normal endothelial cells. Binding was typically near junctional regions, and the cells then passed across the endothelial barrier via an intercellular route. Hypertonic fixation, which distorts normal cellular configurations, showed localized adhesive contact sites which contained fibrillar material when stained with alcian blue (a stain commonly used for demonstrating the presence of negatively charged molecules such as the proteoglycans). Earlier, Marchesi and Gowans (1964) had suggested that lymphocytes emigrated through the endothelial cell cytoplasm rather than between the cells, and it may be that lymphocytes (like PMNs) can use either intra- or intercellular routes depending on the nature of the endothelium. Interspecies differences may also exist. Localized antigen has no effect on the pattern of circulation of stimulated B lymphocytes. The pattern appears instead to be determined by interaction with the endothelium, possibly through complementary receptors as suggested originally by Gesner and Ginsburg (1964) and as described more fully below. Nevertheless, virgin (non-activated) lymphocytes traffic widely throughout the body whereas activated lymphocytes display a more restricted migratory pattern, thus indicating that antigen activation is not without an effect.

Stamper and Woodruff (1976, 1977) developed an experimental approach for studying the adhesion of lymphocytes to glutaraldehyde-fixed high endothelial cells. Briefly, these authors seeded frozen sections of lymph nodes (warmed to about 7°C) with lymphocytes, washed away non-adherent cells, and then counted the number remaining specifically bound to HEVs. Interestingly, glutaraldehyde fixation of the lymphocytes but not the HEVs abrogated adhesion suggesting a passive role for the endothelium, and lymphocytes bound at 7°C detached at physiological temperature. It was proposed that this reversibility of adhesion might help transmigration of the lymphocytes across the endothelium, although how they would bind physiologically in the first place is not exactly clear. *In vitro* binding is in fact dependent upon agitation of the preparation, and it may be that rheological events also influence binding and detachment *in vivo*. Butcher, Scollay and Weissman (1979) used an adaptation of this technique (with unfixed tissue) to show that there was an inverse, exponential relationship between the adhesion of lymphocytes from various animals to mouse HEVs and the evolutionary distance separating the lymphocyte donor from the mouse. On the presumption that proteins mediated the adhesive interaction, a linear rate of amino acid mutation was considered to underlie the exponential decline in adhesiveness with evolutionary separation. Since these rather heady and speculative days, a good deal of research has been directed towards characterizing the biochemical nature of the molecules involved.

It seems that several mechanisms probably exist for the adhesion of lymphocytes to the endothelium. Thus mouse Peyer's patch and lymph node lymphocytes bind preferentially to HEVs in frozen sections of Peyer's patches and peripheral lymph nodes respectively (Butcher *et al.*, 1980), while lymph node and Peyer's patch-specific B lymphoblastoid cells fail to bind to HEVs in the inflamed synovium (Jalkanen *et al.*, 1986 a and b). Peyer's patches are accumulations of lymphoid cells in the intestinal sub-mucosa, particularly in the ileum where they develop (rather intriguingly) opposite the site of attachment of the mesentery. It should be noted here that HEVs from mesenteric lymph nodes appear not to be so discriminating in their adhesions with lymphocytes, since both Peyer's patch and peripheral lymph node lymphocytes apparently bind equally to these structures. To a degree, these patterns of adhesion reflect the circulation or 'homing' profiles of lymphocytes *in vivo*, since experiments have shown that lymphocytes from intestinal and peripheral lymphatic tissue recirculate through Peyer's patches and peripheral nodes respectively. Furthermore, gut-derived lymphoblasts circulate on i.v. injection to fetal intestine grafted under the kidney capsule (Guy-Grand *et al.*, 1974), an experiment reminiscent of the selective colonization of ectopic tissue by tumour cells (Kinsey, 1960; Sugarbaker *et al.*, 1971). In line with studies by Auerbach and colleagues on the adhesion of tumour cells to the endothelium (Alby and Auerbach, 1984; Auerbach *et al.*, 1985), it seems that endothelial cells may also express tissue-specific determinants for lymphocyte adhesion, although the nature of these determinants remains to be elucidated fully. Immature lymphocytes, such as thymocytes, apparently bind poorly to HEV whereas mature, circulating lymphocytes usually bind well. According to Evans and Davies (1977), transitory changes in cell adhesiveness may be responsible for cells leaving one organ and migrating to another. Their data suggest that immmature thymocytes might fail to leave their organ of origin because of high homotypic adhesiveness and that changes in adhesion might be required before intravasation can occur.

Woodruff and her colleagues (1977) determined that rat lymphocyte-HEV interaction was Ca^{2+}-dependent and required viable lymphocytes whose binding was sensitive to trypsin but not to neuraminidase treatment. Furthermore, binding was found to be independent of cytoskeletal microtubules although it was sensitive to microfilament disrupting agents such as cytochalasin B. The role of the endothelium in this interaction has been explored in mice by Rosen and colleagues (1985), who found that neuraminidase treatment of fixed HEVs from lymph nodes (but not Peyer's patches) abrogated lymphocyte adhesion. The suggestion was thus made that sialic acid on endothelial cells from peripheral lymph node HEVs was involved in lymphocyte adhesion. The results of these two groups, although using different species, suggest that different molecular entities may be involved on lymphocytes and the endothelium and they also highlight intertissue differences. Since lymphocyte adhesion to both peripheral lymph node and Peyer's patch HEVs can be inhibited by periodate oxidation of the HEVs, it would seem that adhesion to HEVs of both tissues is carbohydrate dependent (with only peripheral lymph node interaction involving sialic acid). Interestingly, exogenous sialic acid does not competitively inhibit lymphocyte binding to lymph node HEVs, even though its removal does so. In earlier work Stoolman and Rosen (1983) had shown that L-fucose, D-mannose and fucoidin (a sulphated polysaccharide) inhibited rat lymphocyte adhesion to cervical lymph node HEVs. Fucoidin was the most active of the three molecules as an inhibitor (ID-50 at about 10^{-8}M), and pretreatment experiments suggested that it exerted its effects on the lymphocyte rather than on the HEVs. An ID-50 refers to the dose of reagent necessary to give 50% inhibition of an observed response.

In a later publication, Stoolman, Tenforde and Rosen (1984) showed that mannose-6-phosphate, the related fructose-1-phosphate, and a phosphomannan core polysaccharide from yeast were also potent inhibitors of lymphocyte-HEV adhesive interaction. Inhibitory activity did not correlate with negative charge (as determined by cell electrophoresis), but was enhanced by increased ionic strength. These studies consistently point in the direction of carbohydrate involvement in lymphocyte-HEV adhesion, but further research is required before the molecular details are clear. At least in so far as the larger inhibitory molecules (fucoidin and the phosphomannan core polysaccharide) are concerned, the possibility exists that they may manifest their effects via steric hindrance rather than by direct interaction with specific molecules involved in adhesion.

In another approach directed towards understanding the molecular events involved in lymphocyte-HEV adhesion, Gallatin, Weissman and Butcher (1983) generated monoclonal antibodies (McAbs) against a peripheral lymph node HEV-binding mouse B cell lymphoma line (38C-13). Specific McAbs were selected by screening against lymphocytes which bound to Peyer's patch HEVs, and one (referred to as MEL-14) was isolated which interacted with mouse lymphocytes and blocked their binding to lymph node HEVs but not to Peyer's patch HEVs. Furthermore, the treatment of mesenteric lymphocytes with MEL-14 prior to tail vein injection *in vivo* inhibited their migration to peripheral lymph nodes. Detergent extraction of cell surface molecules, followed by affinity chromatography on bound MEL-14 and electrophoresis under reducing conditions, suggested that the antigen recognized by MEL-14 had an M_r of about 80 kD in normal mesenteric lymph node lymphocytes and about 92 kD in the original B lymphoma cells. Although encouraging, these results are not definitive proof that the antigen recognized by MEL-14 is directly involved in mediating lymphocyte-peripheral lymph node HEV adhesion. More recently, Siegelman *et al.* (1986) have shown that the MEL-14 antigen (now called $gp90^{MEL-14}$) is composed of a glycosylated core polypeptide modified by the ubiquitination of side chains. The actual antigenic determinant recognized by MEL-14 is thought to reside in a small part of the C-terminal region of ubiquitin where it binds to the peptide side chain (St John *et al.*, 1986), and it has been proposed that this region is also involved in binding of $gp90^{MEL-14}$ to its receptor on peripheral lymph node HEVs. Ubiquitin is a small polypeptide of M_r 8451 D that is widespread in the animal kingdom (as is reflected in its name). There is some doubt about its physiological role(s), but it may be involved in chromosomal organization and function since it is linked to a histone, and it also appears to serve as a label for proteins which are to be degraded. Siegelman *et al.* (1986) have suggested that a lymphocyte homing receptor is likely to be rapidly turned over, since once a lymphocyte has bound to and crossed the endothelium, receptor degradation may foster de-adhesion and block the possibility of reverse migration. Such ideas remain speculative, however. It should be noted that $gp90^{MEL-14}$ appears not to be the only ubiquitinated cell surface glycoprotein, which raises questions as to how it might exert its specificity. Interestingly, Hall (1985) has proposed that effete lymphocytes may be cleared from the circulation by binding to HEVs, after which they enter the parenchyma of the lymph node to be ultimately destroyed by dendritic cells. Thus the possibility exists that ubiquitinated surface molecules may represent labels for sick cells. It may be of some interest here that in some animal species ubiquitinated proteins appear in response to heat shock treatment.

More recent studies utilizing a cDNA clone of the core glycoprotein have identified three other possible recognition sites on $gp90^{MEL-14}$, namely an N-terminal located lectin-like domain (which is arranged in tandem with an EGF-like domain containing an LFA-1 homologous region), and two repeated complement regulatory protein domains (Siegelman, van de Rijn and Weissman, 1989). The presence of a lectin-like domain is especially significant, given the observations of Stoolman, Tenforde and Rosen (1984) indicating that lymphocyte adhesion to lymph node HEVs is inhibitable by sugars such as L-fucose and D-mannose. Interestingly, the domain structure of $gp90^{MEL-14}$ is very similar to that of ELAM-1, the cell surface molecule expressed on cytokine-activated endothelial cells which is adhesive for PMNs (section 2.12.1).

It should be recalled that lymphocyte-HEV adhesive interaction involves two cells with either similar or different receptors, and indeed more than one receptor type may be involved on each cell. If gp90^{MEL-14} is the receptor on mouse (and possibly human) lymphocytes which mediates binding to peripheral lymph node HEVs, what is the nature of the complementary receptor on the endothelium (the so-called **vascular addressin**)? We are ignorant at present of the nature of this proposed receptor on peripheral lymph node HEVs (PLN-HEVs), but Streeter and colleagues (1988) and Nakache *et al.* (1989) have recently identified a 58–66 kD endothelial cell surface molecule involved in lymphocyte binding to Peyers's patches. Pretreatment of mice with a McAb directed against the endothelial molecule blocks lymphocyte homing to Peyer's patches almost completely (97%) but is without significant effect on homing to peripheral lymph nodes (PLNs). Lymphocyte homing to mesenteric lymph nodes is partially blocked (37%) which is in keeping with the obervation that HEVs in these nodes bind both PLN and Peyer's patch specific lymphocytes and lymphoma cells. Independent studies by Holzmann *et al.* (1989) have identified a 160/130 kD heterodimer (LPAM-1) on murine lymphocytes which seems to be involved in their traffic to Peyer's patch HEVs. The larger α chain of this molecule is virtually identical to that in human VLA4, a member of the integrin family of adhesion receptor molecules (section 5.8). At this stage it is not known if the 58–66 kD vascular addressin in Peyer's patch HEVs is the endothelial receptor for LPAM-1 lymphocyte Peyers patch HEV adhesion molecule-1.

Using a similar antibody based approach but with rats rather than mice, Woodruff and her colleagues (reviewed in Woodruff and Clarke, 1987) have presented evidence indicating the possible involvement of multiple lymphocyte determinants (135 kD, 63 kD, and 40 kD) in the adhesion of B cells and T cells to peripheral lymph node HEVs (Rasmussen *et al.*, 1985), and an 80 kD determinant in lymphocyte adhesion to Peyer's patch HEVs (Chin *et al.*, 1986).

Progress in humans has not been so rapid as that in the mouse and the rat, but recent studies have identified involvement of a family of molecules centred around a gp90^{MEL-14}-related receptor. Although originally prepared against mouse lymphoma cells, MEL-14 cross-reacts with human lymphocytes and inhibits their binding to PLN-HEVs (Jalkanen *et al.*, 1986b; 1987). The antigen recognized on human cells is a glycoprotein of M_r 85–95 kD which is also recognized by an antibody known as Hermes-1, although this fails to block lymphocyte adhesion. Recent evidence implicates the pan-leukocyte antigen leu-8 as the human equivalent of $gp90^{MEL-14}$ (Camerini *et al.*, 1989). Another McAb (Hermes-3) prepared against the Hermes-1 antigen has inhibitory activity, blocking lymphocyte adhesion to HEVs in the appendix. Based on the observed cross-reactions, it has been suggested that a family of related glycoproteins influences lymphocyte-HEV adhesion, with specificity of interaction for particular HEV targets possibly residing in the oligosaccharide side chains. Woodruff and her colleagues have identified another human lymphocyte determinant involved in PLN-HEV traffic which is recognized by the McAb 3.A.7 (Table 2.3), but it relationship (if any) with the $gp90^{MEL-14}$ family is not yet clear (reviewed in Woodruff and Clarke, 1987).

In parallel with the studies described above, other research groups have been exploring the role played in lymphocyte adhesion by another family of molecules collectively referred to as the leukocyte adherence-related proteins. This avenue of research began with the preparation of McAbs specific for murine leukocyte antigens, and it has yielded interesting results which add an extra dimension to the problem of how lymphocytes adhere to other cells, including those of the endothelium. The leukocyte adherence-related proteins are discussed in more detail in section 5.7.2.4, but it is worth noting here that one of these (LFA-1) has been implicated in the adhesion of lymphocytes to endothelial cells (Pals *et al.*, 1988). Unlike the other molecules involved in lymphocyte-HEV adhesion, however, LFA-1 appears to function as

Table 2.3 Molecules implicated in lymphocyte–endothelial cell adhesion

Species	*Defining McAb*	*M_r (kD)*	*Location*	*Function*
Mouse	MEL–14	90	lymphocyte	PLN traffic
	anti–LPAM–1	160,130	lymphocyte	PP traffic
	anti–LFA–1	180,95	lymphocyte	non-specific adhesion?
	MECA–367	58–66	PP–HEV	PP traffic
	MECA–79	?	PLN–HEV	PLN traffic
Rat	A.11	135,63,40	lymphocyte	PLN traffic
	1.B.2	80	lymphocyte	PP traffic
Man	MEL–14	90	lymphocyte	PLN traffic; appendix
	anti–LPAM–1	150,130*	lymphocyte	GALT traffic?
	Hermes–3	90	lymphocyte	GALT traffic
	3.A.7	?	lymphocyte	PLN traffic
	anti–LFA–1	180,95	lymphocyte	non-specific adhesion?

* Equivalent to VLA4.

Abbreviations: GALT: gut associated lymphoid tissue (appendix); HEV: high endothelial cell venule; PLN: peripheral lymph node; PP: Peyer's patch.

a non-organ-specific adhesion molecule. It has been suggested that LFA-1 may be more important in lymphocyte binding to stimulated rather than to unstimulated endothelial cells, but as yet the situation is not completely clear. Another possibility is that LFA-1 may have a stabilizing role allowing more specific forms of adhesion to develop. It is evident from the foregoing that the adhesions of lymphocytes are rather complex, involving a spectrum of cell surface determinants. Some of these are likely to have more or less restricted roles, such as in cytotoxicity (section 5.7.2.4) or in binding to specific HEVs, whereas others may have a broader scope. A summary of the various molecules implicated in lymphocyte adhesion to endothelial cells (as defined by their respective McAbs) is provided in Table 2.3.

Although circulation between the blood and the lymph is probably utilized by normal lymphocytes as a means of gaining access to tissues throughout the body, the process can be usurped by lymphoma cells. The profile of lymphoma cell arrest can be altered dramatically by treatment with neuraminidase or inhibitors of glycosylation such as swainsonine and 1-deoxynojirimycin, but the significance of these results in the light of the molecular mechanisms discussed above remains to be explored. Swainsonine inhibits mannosidase II leading to the production of hybrid *N*-linked sugars, whereas 1-deoxynojirimycin is a glucosidase inhibitor which results in the accumulation of high mannose type *N*-linked sugars.

MEL-14 was originally raised against a murine B cell lymphoma, but EL-4 T lymphoma variants expressing high $gp90^{MEL-14}$ levels have been selected using fluorescence flow cytometry followed by expansion *in vitro*. The selection for high levels of $gp90^{MEL-14}$ expression was found to cosegregate with the ability to bind to peripheral lymph node (but not Peyer's patch) HEVs. Some clonal lymphoma variants of either B or T cell origin can in fact show virtually exclusive specificity for either peripheral lymph node or Peyer's patch HEVs, although others may recognize neither HEV type. Thus although some lymphomas may utilize $gp90^{MEL-14}$ as a binding site during spread *in vivo*, it would seem that other molecular systems probably also exist. To what extent these involve the leukocyte adherence-related proteins referred to above has yet to be evaluated fully, but Roos and Roossien (1987) have recently provided evidence that a McAb directed against LFA-1 can inhibit murine MB6A lymphosarcoma cell adhesion to and invasion of rat hepatocytes *in vitro*. MB6A cells metastasize to the liver and spleen and the authors suggest that liver metastasis at least might involve LFA-1 mediated adhesion. This presupposes that the LFA-1 receptor or counterstructure is present on hepatocytes and that interaction with hepatocytes rather than with endothelial cells is an important aspect of selective metastasis. ICAM-1, which may be equivalent to the LFA-1 counterstructure in humans (Rothlein *et al.*, 1986) is apparently not present on human hepatocytes (Dustin *et al.*, 1987) which suggests either that human lymphosarcoma cells may not utilize LFA-1 in liver metastasis or that other counterstructures are involved.

2.12.3 Leukocytes and contact inhibition

The behaviour of PMNs, monocytes and lymphocytes has been analysed after they have been allowed to come into contact with fibroblasts. In such experiments, all of these blood cell types show some failure to display contact inhibition of locomotion (see section 1.7.4). Similarly, there is some evidence that PMNs, monocytes and lymphocytes are not contact inhibited by endothelial cells. This seems to make some sense since leukocytes are invasive cells and the ability to migrate between endothelial cells is crucial during their extravasation from blood vessels. However, although the failure to display contact inhibitory behaviour may assist in extravasation, the phenomenon is likely to involve many other behavioural aspects of cells, including chemotaxis. Although PMNs can move on the surfaces of fibroblasts and endothelial cells, whether they do so or not appears to be dependent on the adhesiveness of the rest of the substrate. Thus, when the adhesiveness of the normal substrate is poor, the PMNs tend to migrate over the surface of the other cells (Lackie and De Bono, 1977). The possibility that similar changes relating to contact inhibition may be important in tumour cell invasion is discussed in section 4.1.

2.13 ESSENTIAL STEPS IN THE METASTATIC PROCESS

The metastatic process has been likened to a reaction cascade in which individual steps must be accomplished before the sequence can continue. It is convenient to concentrate here on haematogenic spread, for which we may delineate the following steps in the metastatic process:

(a) Local invasion
(b) Detachment from the primary
(c) Invasion of a blood vessel
(d) Transport within the blood system
(e) Lodgement at a distant site
(f) Extravasation
(g) Growth

It should be obvious from some of the foregoing discussions that a strictly sequential cascade analogy may be a little far-fetched. Invasion of a blood vessel, for example, may occur before detachment from the primary and growth may occur before extravasation is complete. The cascade analogy is primarily one of convenience in that it divides the metastatic process into a number of steps which can be studied independently of each other, thereby simplifying the analysis of what is, in reality, a very complex phenomenon. In addition, it is important to make the point that if metastasis can be represented by a cascade process then any step in the sequence is potentially rate limiting. For example, a cell may achieve all metastatic steps up to extravasation only to fail to grow at the new site. Clearly, an invasive cell that cannot grow at a distant site is not a metastatic cell. This possibility suggests that single steps in the metastatic cascade should not be studied completely in isolation and, as will become clear in the following chapters, continued reference should be made towards parallel testing of *in vitro* and *in vivo* behaviour.

2.13.1 Local invasion

As used here, the term **local invasion** refers to the penetration of host tissue immmediately adjacent to a neoplasm. When discussing local invasion, it is of considerable importance to appreciate that although it is a dynamic process, it is not necessarily continuous in time. Not surprisingly, reliable correlations are particularly difficult to achieve when dynamic structures or processes are under consideration. Indeed, in many studies relating particular histopathological features to invasion, it is not clear whether active invasion was taking place in the fragment of tissue examined at the particular time of tissue preparation.

Although invasion usually takes place in the extracellular space, there are some carcinomas which are thought to be capable of intracellular invasion through the cytoplasm of striated muscle fibres, in what is a rather bizarre example of emperipolesis. The extent of local invasion is probably the outcome of three processes (growth, motility and tissue destruction) although a fourth (differentiation) may prove to be important:

(a) *Growth*

We might expect that a tumour will invade locally as it grows and exerts pressure on surrounding tissues (Eaves, 1973). Growth will, of course, result in an outward expansion of the tumour mass, but we use the term invasion in relation to malignant cell behaviour to refer to more than just spread by growth. In some cases, malignant invasion may have a destructive component, although a destructive aspect to benign tumour growth may be introduced through pressure occlusion of blood vessels vital for nutrient supply and waste removal. There is in fact a poor correlation between rapid growth and invasiveness since benign tumours, such as fibroadenomas, may grow rapidly but never invade. Conversely, malignant tumours such as carcinoma of the breast may grow slowly but actively invade. Furthermore, tumour invasion in experimental systems can be demonstrated in the presence of agents that inhibit proliferation (Mareel and De Mets, 1984). That invasion is more than just growth can also be tested *in vitro* by seeding malignant tissue onto normal tissue. The malignant cells are seen to invade the normal tissue in a very short time during which period growth pressure would be minimal (Easty and Easty, 1974). It should not be forgotten here that the degree of tissue oedema associated with a tumour correlates with its grade of malignancy, and that oedema at the tumour periphery may open up tissue spaces providing access for the tumour cells. Because tumours are often richly supplied with relatively leaky blood vessels they have a high interstitial fluid volume. This interstitial fluid does not drain away within the tumour mass because tumours lack lymphatic vessels. Instead, there seems to be a streaming of the fluid towards the periphery of the tumour mass and into the adjacent tissue. This fluid streaming has been measured at speeds of up to 25 μm sec^{-1} in mouse mammary tumours by Reinhold (1971), and it weakens local tissue cohesiveness while at the same time carrying tumour cells passively towards the draining lymphatic vessels.

(b) *Motility*

It might seem at first sight that cell motility is required for invasive cancer spread. However, this need not always be the case since displacement (without locomotion) occurs when cells divide. Nevertheless, examination of tissue sections shows that some malignant cells may be detached from the tumour mass (Strauli and Weiss, 1977). Although this may have occurred during the preparation process, the observation is highly suggestive of locomotion having occurred. *In vitro* experiments in which cell behaviour can be observed and filmed provide strong evidence for a role of motility in the invasion of some tumours. More significantly, Wood (1958) filmed the movement of malignant cells *in vivo* as they emigrated from blood vessels growing in a rabbit ear chamber.

However, although claims have been made for a correlation between increased motility and tumorigenic potential (Gail and Boone, 1971), on balance there seems to be no such general relationship between motility and invasive potential. Thus the inhibition of cell motility in *in vitro* experiments prevents tumour cell invasion in some (Mareel and De Brabander, 1978 a and b) but not all tumour systems (Easty and Easty, 1974).

(c) *Tissue destruction*

Local spread often occurs along a plane of least resistance such as that offered by loose connective tissue or between two fasciae. During the process of invasion, however, malignant cells are often seen to break through natural boundaries or barriers such as the basement membrane, thereby gaining access to the blood system. In some cases, structures such as organ capsules may remain essentially intact while the malignant tumour invades around them or breaks through only in places. The extent of local tissue destruction by an invading tumour is variable, ranging from microscopic (seen in initial tissue weakening) to gross. Tissue destruction will result from the summation of a number of processes including the obstruction of the blood supply (ischaemic necrosis) perhaps by pressure of the growing tumour, by the production of tumour-associated cytolytic products, by indirect ('bystander') activity as a result of host defence actions, and by the activation of normal host cells to produce lytic enzymes.

The weakening of tissue structures by the secretion of enzymes will clearly contribute to the extent of tumour spread. According to Strauch (1972), areas of human breast tumours rich in collagenolytic activity coincide with areas of the tumours with morphological evidence for invasion. However, although some malignant tumours secrete abundant lytic enzymes, others secrete little if any at all (section 1.7.17). Alternatively, malignant cells which fail to secrete lytic enzymes may act on normal host cells to induce their secretion. The invasion of bone by tumour cells, for example, appears to be mediated by the enzymatic activity of osteoclasts, which are locally activated by products of tumour cells. Other studies by Biswas (1984) have shown that human fibroblasts can synthesize and secrete high levels of collagenase when they are co-cultured with human tumour cell lines.

(d) *Differentiation and invasion*

Clinical studies suggest that poorly differentiated tumours are more invasive and that they are generally associated with a poor prognosis. Gabbert and colleagues (1985) have used dimethylhydrazine-induced rat colonic carcinomas to explore the relationship between differentiation and invasiveness. These authors found that cells at the invasion front of both differentiated and poorly-differentiated colonic tumours underwent further dedifferentiation, and that this correlated with their detachment from the tumour mass. According to the hypothesis put forward in this study, the first and essential step of tumour invasion is tumour cell dedifferentiation, which mobilizes tumour cells out of the main tumour bulk. It is suggested that this is a temporary and reversible phenomenon, so that observation at any one time need not show detached, dedifferentiated cells at the invasion front. The extent of dedifferentiation that occurs at the invasion front is clearly dependent on the state of differentiation of the tumour bulk. For poorly differentiated masses, the change usually involves loss of intercellular junctions and a decrease in tumour associated basement membrane. For well differentiated tumour masses, on the other hand, the changes can be quite considerable. This hypothesis warrants further study, since at the moment its applicability to carcinomas in general is not clear, let alone its applicability to other sorts of invasive tumours.

2.13.2 Detachment from the primary

Detachment from the primary may occur prior to blood vessel invasion or after invasion has already taken place. Experimentally, it has been shown that the appearance of tumour cells in the blood correlates with the onset of tumour vascularization and that their concentration is related to the number and diameter of blood vessels present in the tumour (Liotta, Kleinerman and Saidel, 1974, 1976). These authors used a model system based on femoral implantation of the T241 murine fibrosarcoma. At an appropriate time after implantation, the tumour-bearing leg was removed and the tumour vascular bed was hooked into a perfusion circuit which allowed collection of the venous effluent. Tumour cell release per day from the implanted tumour was quite prodigious, increasing from an estimated 10^3 on day 5 to 5×10^5 on day 15. Although no data are available on how this relates to the size of the primary, Liotta, Saidel and Kleinerman (1976) have correlated this release rate with the rate of accumulation of metastases in the lungs. Interestingly, tumour cells collected in the venous drainage solubilized basement membrane to a greater extent than did cells taken from the primary. Other studies by Butler and Gullino (1975) in the rat indicated that 2–4 g tumours of the MTW9 mammary carcinoma could release cells at a rate of $3–4 \times 10^6$ cells per gram of tumour tissue per day, although it was not reported exactly how many of these cells were viable.

Tumour cells which detach from the primary may do so in two

ways: as single cells, or as small clumps. It is not known whether any particular tumour favours one form of detachment over the other. Liotta, Kleinerman and Saidel (1974, 1976) have shown experimentally that tumour trauma or massage can lead to an increase in the release of tumour cells and clumps in the venous effluent. Furthermore, these authors were able to confirm that the injection of tumour cells in clumps of 6–7 cells results in more lung colonies than does the injection of a similar number of single cells. Available evidence indicates that cell clumps of a number of tumours (including the T241 fibrosarcoma) survive longer in blood vessels of the lungs than do single cells, and Liotta, Kleinerman and Saidel (1974) have also shown that there is a direct relationship between clump size distribution and blood vessel diameter for this fibrosarcoma.

What mechanism underlies detachment from the primary? If malignant cells were poorly adherent to each other then it would not be too difficult to envisage them breaking away from the primary in response to a number of events such as physical stress from a cough or knock, or from the shearing effects of flowing blood. Not surprisingly, it has been appreciated for some time that the trauma of surgery can enhance tumour cell detachment leading to what has been referred to as **iatrogenic dissemination.** Even though the presence of tumour cells in the blood does not ensure metastasis, their presence in the circulation is a prerequisite for haematogenic spread. As a consequence, it makes common sense to limit the possibilities of this type of spread through the adoption of appropriate preventative measures during surgery. Detachment prior to gaining access to a blood vessel would be enhanced if the tumour was not encompassed in a distinct capsule or if the local tissue was damaged by secreted enzymes. There is some evidence that fibrinolytic activity might promote tumour cell detachment. Griffiths and Salsbury (1965), for example, were able to increase the release of tumour cells from perfused human malignant tumours of the gut by adding plasmin to the perfusing fluid. Tumour cells which are adjacent to necrotic zones within the tumour mass may be poorly adherent and this too may permit relatively easy detachment (Weiss, 1977).

The idea that malignant cells within a primary might be poorly adherent to each other was suggested a number of years ago from the work of Coman and colleagues (Coman, 1944; McCutcheon, Coman and Moore, 1948). This is actually difficult to prove *in vivo* since cells can adhere to other types of cells or to different types of substrates to different degrees. In most studies, detachment is measured by gaining some estimate of the force required to prise one cell off another or off a deposited substrate such as collagen. When the cell finally breaks free, it is difficult to be sure if this is due to the breakage of adhesive bonds or to rupture of the cell membrane. In a study of the adhesiveness of Balb/c 3T3 cells, Elvin and Evans (1983) suggested that adhesiveness varies as a consequence of the cell cycle. If the actual extent of adhesiveness of a cell is not constant at all times, then it is possible to construe a mechanism by which cells may detach from time to time from an otherwise coherent mass in much the same way as malignant cells are seen to do, but whether cell cycle changes actualy underlie tumour cell detachment or not remains speculative.

2.13.3 Blood vessel invasion

Numerous types of blood vessels are found within a tumour, including open channels which do not have a complete endothelial lining (Warren, 1979). Tumour cells may also invade blood vessels adjacent to and even some distance remote from the tumour itself, and thus blood vessel invasion is not restricted to intra-tumoural vessels alone. Not surprisingly, there is evidence that different types of vessels are invaded by tumour cells to different extents. Arteries, for example, are seldom invaded, presumably because of their thick walls and the fibromuscular nature of their coats (section 2.7.1). The limited invasion of arteries may also be due to their relatively high internal hydrostatic pressure. It has been shown that tumours can invade arteries deprived of their normal blood flow, but such experimental manipulations may induce destructive changes in the arterial wall thereby inadvertently promoting invasion.

Tumour cells may invade blood vessels in two principal ways:

(a) *As single cells with minimal damage.* In many cases of blood vessel invasion by single cells, it would seem that the invading cell must first migrate to come into contact with the basement membrane in the vessel wall. The distance travelled is probably highly variable, and whether this migratory process is random or in response to some chemotactic signal emanating from the direction of the blood vessel is unknown, although there is evidence that tumour cells can respond chemotactically (section 5.11.1). Invasion of blood vessels without significant damage may be facilitated by vessels which have incomplete endothelial linings or basement membranes. New blood vessels develop and invade tumours in response to angiogenic signals and the delicate nature of their walls may render them easily penetrable by tumour cells. The centres of sizeable tumours are usually necrotic and the presence of toxic products diffusing from the tumour core may also lead to blood vessel damage or leakiness thus increasing the possibility of infiltration.

Warren (1980) studied the invasion of blood vessels by melanoma cells following tumour implantation into the hamster cheekpouch. He found that irregular, thin walled 'giant capillaries' frequently grew at the edge of such implants and that these were readily invaded by the melanoma cells. In this study, single melanoma cells were found to first breach the basement membrane around the vessel, but whether the breach was found or made by the tumour cell could not be discerned. A thinning of the endothelium then occurred, possibly reflecting retraction or rupture, and the tumour cell then inserted a blunt process into the lumen of the vessel. Again, it was difficult to discern whether the cell process forced the gap or the gap opened in advance of the process. However, single tumour cells found their way into the circulation and, after hanging onto the vessel wall by a tail, they finally let go to float off in the blood. In an ultrastructural study of invasion in the rat, De Bruyn and Cho (1979) showed that penetration of blood vessels by allogeneic myelogenous leukaemic cells involved the formation of a temporary migration pore through the cytoplasm of the endothelial cell. Three or four cells were able to use the same migration pore before it closed off to seal the blood vessel without any overt signs of vessel damage. This type of intravasation is very similar if not identical to that shown by normal myeloid cells (and leukaemic cells) within the bone marrow. The obvious conclusion here is that whereas some malignant cells display drastic changes from normal behavioural patterns in order to invade blood vessels, others may utilize features of their normal behavioural repertoire.

(b) *As clumps of cells, usually in association with blood vessel damage.* As a tumour grows it may gradually obliterate vessels with which it comes into contact. Destruction of the endothelial wall may occur with the local formation of a thrombus. Organization and malignant infiltration of the thrombus may then allow the tumour cells to grow along the vessel as a cord and/or to be swept away in the blood either as clumps of tumour cells or as thrombic fragments, but also as single cells. The extents to which lumen obliteration and thrombus formation occur are variable, and in some cases tumour cells may invade the blood vessel without recourse to such events.

The time course of appearance in the blood of murine fibrosarcoma cells after intramuscular implantation was monitored by Liotta and his colleagues (1974, 1976). They reported the presence of circulating tumour cells after about four days and this was coincident with the first appearance of blood vessels in the tumour. In fact, the number of circulating tumour cells (particularly as clumps) correlated to an extent with the density of tumour blood vessels and also with the incidence of lung metastases. In so far as tumours which spread by the blood are concerned, it might seem that metastasis depends upon tumour vascularization. The universality of this conclusion remains to be tested rigorously, however, since as explained above tumour cells also have the possibility of access to extra-tumoral blood vessels.

2.13.4 Transport

Once they have gained access to the blood system, tumour cells may be swept away with the possibility of eventually colonizing a remote site. It should be noted that the presence of even large numbers of tumour cells within the blood does not guarantee metastasis formation, and indeed such a situation does not necessarily indicate a poor prognosis (Salsbury, 1975). Within the blood, the tumour cells may interact with host components such as lymphocytes, monocytes and platelets. This may lead to the formation of larger clumps incorporating tumour cells and, as described below, this could enhance the possibility of metastasis. Indeed, once in the circulation there is a good correlation between size of the tumour cell clump, rate of arrest, and likelihood of secondary tumour formation (Fidler, 1973b; Liotta, Kleinerman and Saidel, 1974, 1976). Conversely, host cells which adhere to tumour cells in the circulation may be defence cells which could initiate tumour cell destruction and thereby decrease the possibility of a metastasis developing. The extent to which each of these conflicting processes influences metastasis remains to be established.

2.13.5 Lodgement

Most tumour cells, other than those of haematopoietic origin, appear to be arrested in the first capillary bed they encounter, but this does not necessarily guarantee the subsequent growth of a secondary. In experimental studies, tumour cells (either singly or in clumps) are seen to lodge in small vessels particularly arterioles and capillaries, and occasionally venules. It may be that the shearing forces are too great in larger vessels, such as arteries, for lodgement to take place. Alternatively, conditions of flow in conjunction with the size of the vessel lumen may be such that the tumour cells fail to come into appropriate contact with the vessel wall. Whether a tumour cell lodges in an arteriole or a capillary or a venule may have profound consequences on the metastatic outcome, but unfortunately almost nothing is known about how entrapment in different vessels influences metastasis. The walls of arterioles might present an effective barrier to extravasation, and it has been claimed that this might be one explanation for the infrequent finding of metastases in tissues such as skeletal muscle.

Shape changes may occur soon after lodgement and many arrested tumour cells show a loss of microvilli on their surface.

During the process of lodgement within a blood vessel a tumour cell will usually interact either directly or indirectly (via a platelet, white blood cell, or fibrin) with one or other of two surfaces, the basement membrane or the endothelium:

(a) *Interaction with the basement membrane.* As a consequence of normal processes, the endothelium is subjected to minor damage possibly leading to cell death. In such cases, the dead cell is rapidly extruded from the monolayer by its neighbours which then spread out so as to keep the endothelial barrier intact. Cell division may then be activated to replace the lost cell. In more severe cases, as is often seen in pathological circumstances, endothelial damage may reveal components of the underlying basement membrane. Circulating platelets (present in the blood at $1.5–3.0 \times 10^5$ mm^3) rapidly adhere to the exposed basement membrane where they undergo morphological and biochemical changes. These changes promote further platelet adherence and aggregation which leads to the development of an occluding platelet plug. Endothelial injury also stimulates blood coagulation and this leads to the formation of fibrin which adds to the platelet plug thereby forming a thrombus. Excessive blood vessel leakiness and the extravasation of leukocytes are not necessarily associated with platelet plug formation, indicating that the basement membrane may not be significantly exposed. When the basement membrane does become exposed, the following question must be asked: can a circulating tumour cell find the exposed basement membrane before the platelets? Although entirely conceivable, given the vast number of circulating platelets this would seem unlikely unless the damage was extensive. One possible exception might include certain haematopoietic malignancies where large numbers of circulating malignant cells may exist.

There is some evidence that tumour cells can selectively lodge in areas of endothelial damage, particularly if this is severe. Cotmore and Carter (1973) found more metastases in the livers of animals implanted with a lymphoma after treatment with the detergent Triton WR 1339 when compared to controls. Apparently the detergent destroyed the sinusoidal lining of the liver and the damaged endothelium promoted extensive metastasis. Under less severe conditions, lodgement involving indirect interaction via a platelet plug or thrombus might seem to be favoured over direct interaction between the tumour cell and the basement membrane. As will become clear later, however, there is little ultrastructural evidence that either platelets or fibrin mediate the attachment of most metastatic cells to either the endothelium or to subendothelial components.

One other possibility should also be considered here: a moderate sized tumour embolus might physically impact in a blood vessel whose lumen is smaller in diameter than the embolus itself. This form of mechanical entrapment or microembolism (see Malik, 1983) might abrade endothelial cells from the lumen wall thus allowing the embolus to come into direct contact with the basement membrane. It must be admitted, however, that there is no direct evidence in support of this phenomenon. Alternatively, by altering the pattern of local pulmonary bloodflow, a microembolism might induce local changes which could influence lodgement nearby without recourse to physical abrasion. This possibility would be enhanced if vasoconstricting agents or chemotactic substances for tumour cells are released, but the situation is rather hypothetical. Nevertheless, vascular damage is likely to influence metastatic outcome and this it may do either directly (e.g. by allowing escape from the blood system) or indirectly via the clotting, fibrinolytic, kinin and complement cascade systems. The various chemical mediators involved in these cascade systems have diverse functions including vasoconstriction and vasodilation with alterations in vascular permeability, fibrin deposition and breakdown, platelet aggregation and aggregation-inhibition, and the release of chemotactic factors for both leukocytes and tumour cells. In fact, all of these activities may influence tumour cell lodgement and extravasation, thereby exerting an influence on the metastatic process. The complexity of possible interactions is all too evident, and in general we are obliged to admit that precisely how the events outlined above will influence the spread of any given malignant tumour is not clear.

(b) *Interaction with the endothelium.* Histological observations of many tumours have rarely shown platelets or fibrin mediating between a presumably lodged tumour cell and the endothelium. This suggests some form of direct interaction based on recognition between the endothelium and the tumour cell, but it should be clear that lodgement may also result from simple mechanical trapping. Furthermore, it is known that the associations of both platelets and fibrin may be transient, and thus histological observations may be misleading. Nevertheless, studies with W256 carcinosarcoma cells have shown that cells within small blood vessels are associated with an intact endothelium for at least the first 3 h after injection and that extravasation involving vessel damage usually occurrs between 12 and 36 h after injection (Chew, Josephson and Wallace, 1976).

Wood and his colleagues made direct visual observations of the lodgement and extravasation of V_2 carcinoma cells using the rabbit ear chamber (Wood, Holyoke and Yardley, 1961; Wood, Baker and Marzocchi, 1967 a and b). Following injection of tumour cells into the central artery of the rabbit ear (which may have induced pressure effects), they observed adhesion of cells to the capillary endothelium which was then followed by thrombus formation around the lodged cells. Within minutes, but stretching to hours in some cases, the underlying endothelium was perturbed. Leukocytes migrated out in this zone and they were followed by tumour cells from about 2.5 to 72 h after injection.

The studies of Wood's group and of Chew and his colleagues leave no doubt that different tumours can follow different lodgement and extravasation kinetics. Futhermore, not all tumour cells must of necessity lodge in the first capillary bed they meet, and thus we may envisage four possible outcomes for tumour cells in the blood system: (a) the cells lodge and form metastases; (b) the cells lodge and become dormant, giving rise to metastases at a later date; (c) the cells lodge but do not survive; (d) the cells fail to lodge. Each one of these represents a possible scenario for every tumour cell which escapes from a primary into the blood system. Since different tumours can metastasize to different degrees and display a biased pattern of spread, it would seem that the process is not entirely random.

(a) *The cells lodge and go on to form metastases*. The arrest of a malignant cell or clump within a blood vessel at a site remote from the primary may occur in three ways, which are not necessarily mutually exclusive:

1. Mechanical entrapment. This is a purely physical process in which a cell or a clump of cells is jammed in a vessel whose diameter is smaller than that of the lodged particle. This would suggest that clumps would have a greater chance of lodging than single cells and therefore of giving rise to metastases, and as we have already seen this has been shown experimentally to be the case (Fidler, 1973b; Liotta, Kleinerman and Saidel, 1976). Although size is likely to be important in mechanical arrest, Zeidman (1961) has also shown that lodgement is a function of embolic deformability. That single tumour cells as well as clumps can occur in the blood of human cancer patients was shown by Moore, Sandberg and Watne (1960) who found clumps of up to 40 cells to be present. Tumour cell clumps may offer a degree of protection to some of the enclosed cells and this could contribute to the reasons why larger emboli are more likely to give rise to metastases. Mechanical trapping may be related to endothelial injury (see above), and Fisher *et al.* (1967) have shown that circulating tumour emboli tend to localize at sites of tissue damage. Fidler and Zeidman (1972) among others have suggested that tissue damage occurring after X-irradiation of the host may explain the increased lung colonization of tumour cells seen

following i.v. injection after this procedure. This observation is not without clinical relevance, since it suggests that sub-lethal injury induced by X-irradiation may actually enhance the possibility of a metastasis developing. Interestingly, a more recent publication has proposed that the main effects of X-irradiation on lung colonization may actually be manifested through perturbation of host defence cells rather than through blood vessel damage, even though the latter still occurs (Kawano *et al.*, 1986).

Mechanical entrapment by plugging of a blood vessel may lead to stasis and this could induce clot formation. Indeed, some tumours (but certainly not all) may be associated with the adherence of platelets and the deposition of fibrin, even if only temporarily. Clot formation may be enhanced by the secretion of procoagulant substances from the lodged cells, and the temporary nature of the clot could be modulated by the secretion of fibrinolytic substances such as plasminogen activator. The role of platelets in metastasis will be considered in more detail later (section 5.7.22).

2. Specific adhesion. This form of adhesion implies specific recognition between the malignant cell and the wall of the blood vessel in which it is to become lodged. The most obvious basis of specific recognition lies in the molecular content of the surfaces involved. However, there appears to be no evidence that tumour cells adhere to one cell type through an exclusive mechanism and thus specific adhesion remains speculative as the basis for lodgement.

3. Selective adhesion. In this form of adhesion, lodgement results from molecular interactions which are not exclusive to the adhering surfaces. Although lodgement can undoubtedly result from mechanical entrapment, there is now considerable evidence that some tumour cells may adhere selectively to blood vessel endothelium (Alby and Auerbach, 1984). The molecular nature of this interaction is not yet clear, but selective adhesion has been proposed as a basis for the observation that certain tumours are seen to metastasize more frequently to particular organs. Metastatic patterns, however, also reflect anatomical factors such as the pattern of venous drainage and it would seem highly unlikely that any single physical property or behavioural trait underlies the process of metastasis to particular organs.

Cells which successfully lodge within blood vessels and succeed in forming metastases can themselves give rise to further metastases by repetition of the process (section 2.14.9). Metastases which act as relay stations for the seeding of further metastases are referred to as generalizing sites (section 2.14.10).

(b) *The cells lodge and become dormant*. In certain instances, metastatic cells may exist in a resting or dormant state to become active at a much later date (section 1.8.6). Although of some clinical significance, it must be admitted that little is known about dormancy, particularly with regards to its mechanisms of induction and its frequency in different tumours. Interestingly, a form of dormancy has been reported for normal cells by Taptiklis (1968, 1969). This author reported that i.v. injected thyroid cells lodged in tissues such as the lung, but failed to grow until the host animals were made thyroxine deficient after which they could divide and develop into typical acini at the ectopic site.

(c) *The cells lodge but do not survive to form metastases.* Irrespective of the underlying mechanism, most (but not all) cells lodge somewhere within the blood system (if only temporarily) and most of these apparently die, at least as judged from experimental studies. According to Fidler (1970), for example, most tumour cells injected into the tail vein of an unprimed mouse die within 24 h and less than 1% survive to develop into metastases. What is the likely cause of this extensive cellular destruction? The answer is as yet unknown, but classical immune responses involving sensitized T cells might be excluded because of the short timescale involved in experimental systems in which tumour cells are injected directly into the venous system. However, if the immune system of experimental animals is primed before injection of viable tumour cells, then in some cases more cells appear to be destroyed (Proctor, Auclair, and Rudenstam, 1976). Other host cells, such as monocytes and PMNs, may destroy tumour cells through antibody dependent mechanisms, but these too would seem unlikely given the timescale involved in unprimed animals. Some cells, particularly natural killer (NK) cells can utilize antibody independent mechanisms for destroying cancer cells and thus they do not need prior priming; however, whether there is enough of these cells locally to be rapidly effective is uncertain. Since tumour cells, particularly those injected directly into the venous sytem, are subjected to considerable shearing forces in the small blood vessels of the first organ they reach, it would seem likely that much of the cell death is due to these physical effects.

In a few cases, tumour cells may lodge and grow only to spontaneously regress at some later stage. There are some reported clinical cases of spontaneous regression of malignant tumours, but such examples are extremely rare. Presumably the status of the body's defence systems and anatomical considerations including non-vascularization are important in the induction of spontaneous regression.

(d) *The cells fail to lodge and pass through the first organ they contact.* A small percentage of tumour cells in the blood appear able to bypass the destructive events associated with the capillary network of the first organ they encounter. These cells may be particularly deformable such that they are not destroyed by shearing forces within the capillary system, or they may be shunted through anastamoses or the metarteriole system thereby avoiding destruction (section 2.7). The ability to metastasize is correlated with deformability as is the ability to remain viable after one passage through the lungs (Sato and Suzuki, 1976), but the extent to which shunts are used in passaging through organs with extensive capillary beds is not clear. Viable cells which pass through the first capillary bed may lodge in subsequent beds or ultimately they may be destroyed. Note here that many cells which appear to pass through particular capillary beds may in fact be relayed from generalizing sites as described in more detail later.

2.13.6 Extravasation

The exit of tumour cells from the circulation would appear to be dependent on prior lodgement within a blood vessel, except in those relatively rare situations where haemorrhage (usually for some other reason) has occurred. Extravasation itself may take place in a number of ways:

(a) The cells may divide and/or pile up within the lumen of the blood vessel and invade outwards *en masse* (Chew *et al.*, 1976). In this situation the tumour cells are thought to extravasate through destruction of the blood vessel rather than by active migration.

(b) Single cells may migrate between endothelial cells either destructively or non-destructively. As noted above, observations on the extravasation of V2 carcinoma cells from blood vessels growing in the rabbit ear chamber indicated that many of the cells emigrated out in the wake of extravasating leukocytes (Wood, Holyoke and Yardley, 1961; Wood, Baker and Marzocchi, 1967a), but this appears not to be necessary for many other tumours. In studies in our laboratory we have found that co-incubation of PMNs with B16 melanoma cells in chambers lined with MDCK epithelial cells failed to induce tumour cell invasion. This proved to be the case even when the PMNs were stimulated to penetrate the epithelial barrier by a gradient of the chemoattractant FMLP.

Sindelar, Tralka and Ketcham (1975), working with a murine fibrosarcoma, noted extravasation through lung endothelium within about 24 h. In this example, the tumour cells extravasated in the junctional region between apposed endothelial cells, but it was not possible to confirm that they did not take advantage of some previously occurring defect in the endothelial junctional zone. Endothelial gaps are thought to occur naturally in some organs or tissues, such as the venules of the rat diaphragm where intercellular gaps of about 6 nm width have been described (Simionescu, Simionescu and Palade, 1978). Temporary gaps may also appear as a result of normal wear and tear. Some gaps may be too small to be capitalized upon by tumour cells within the bloodstream, however, while others may exist for too short a time, and thus there need not necessarily be any relationship between natural gap occurrence and extravasation. Other endothelial gaps may be induced by the tumour, either as a result of the production of lytic or toxic substances inducing endothelial cell degeneration or retraction, or by active tumour cell probing. The emigration of AH7974 rat ascites hepatoma cells from brain capillaries following their injection into the carotid artery has been studied by Kawaguchi, Tobai and Nakamura (1982). These authors reported that AH7974 cells inserted cytoplasmic projections into the endothelial cells inducing their fragmentation and thus exposure of the underlying basement membrane. Small pores (0.07–1.8 μm diam.) were then formed in the basement membrane through which the cytoplasmic projections were thrust, eventually to be followed by translocation of the entire cell. Penetration of the basement membrane is presumably aided by the production and release of specific enzymes such as type IV collagenase. Tumour cells which produce cytoplasmic extensions as they invade may have such enzymes concentrated at the tips of these processes.

(c) Single cells may leave the blood vessel by passing through (rather than between) the lining endothelial cells (Dingemans *et al.*, 1978). This is essentially the reverse of the process by which many normal and malignant haematopoietic cells leave the bone marrow to enter the circulation.

A review of the time sequence of extravasation reported for various tumours after i.v. injection is included in a study by Crissman *et al.* (1985). Generally speaking, platelets may or may not be seen associated with tumour cells in capillaries: if they are involved it is usually only temporary since they generally disappear within 24–48 h. Tumour-induced platelet aggregation can begin within 2 min of the injection of appropriate cells into recipient animals, and this usually leads to thrombus development over the next 4 h. The presence of this thrombus and the associated tumour cell may lead to cessation of blood flow, and this condition may persist until the thrombus is dissipated about 24 h after injection. Although platelets are aggregated by some tumour cells, in most instances they appear not to mediate the interaction between the tumour cells and the endothelium. After tumour cell arrest, breaching of the endothelium may then occur, in many cases involving retraction of the lining cells. Some reports suggest that breaching may occur within 30 min, although others suggest about 6 h as being more typical. Penetration of the basement membrane takes about 2–3 days for most tumours, although Wood (1958) reported that the V2 squamous cell carcinoma could reach a perivascular location within 3 h. Regardless of how a tumour cell may penetrate the endothelial lining of a blood vessel, invasion of the basement membrane need not follow immediately. Thus tumour cells can remain dormant or actively divide within the relatively protected environment between the endothelium and the underlying basement membrane. Alternatively, the endothelium may not extend to recover the tumour cells adherent to the basement membrane, in which case intravascular division could proceed to plug the vessel and ultimately break through the basement membrane.

In summary, although the process of extravasation has many generalized features, malignant cells appear to be able to use a variety of mechanisms to emigrate from the vascular system. A diagrammatic perspective of the key steps involved is provided in Fig. 2.3.

2.13.7 Growth

For tumour cells that extravasate by migration, the metastatic process is still not yet complete since they must now grow at the secondary site. Generally speaking, metastasis is an inefficient process since fewer than 1% of the cells which gain access to the vascular system may successfully grow to form metastases (Fidler, 1970). Tumour growth is dependent upon a number of factors relating to the nature of its environment (the soil) and to the nature of the tumour itself (the seed).

Figure 2.3 A schematic overview of key events involved in tumour cell lodgement and their extravasation from blood vessels. The scenario depicted in this figure refers to the generalized processes involved in the lodgement and extravasation of three murine tumour cell lines (3LL carcinoma, B16a melanoma and 16c mammary adenocarcinoma) as envisaged by Crissman *et al.* (1988). Malignant cells are thought to lodge in small blood vessels by direct interaction with the endothelium (1). An impacted malignant cell takes on an activated form with pseudopodial extensions, and platelet aggregation and fibrin deposition ensue (2). Next, the malignant cell is envisaged to induce endothelial retraction (3) thereby gaining access to the basement membrane. The thrombus now dissipates allowing the re-establishment of blood flow (4). This is followed by intravascular proliferation (5) and extravasation with dissolution of the basement membrane (6).

1. Lodgement

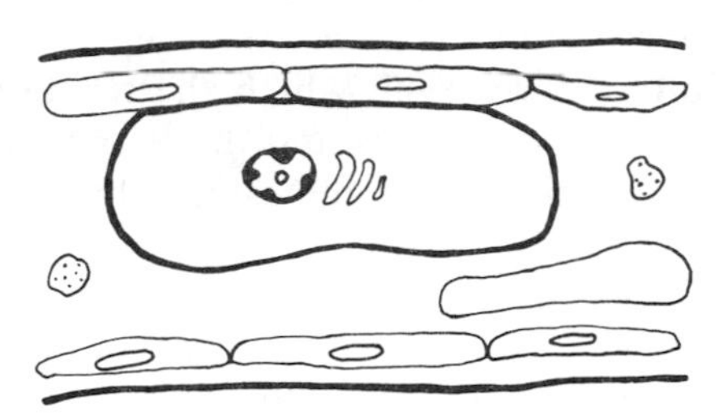

2. Platelet aggregation

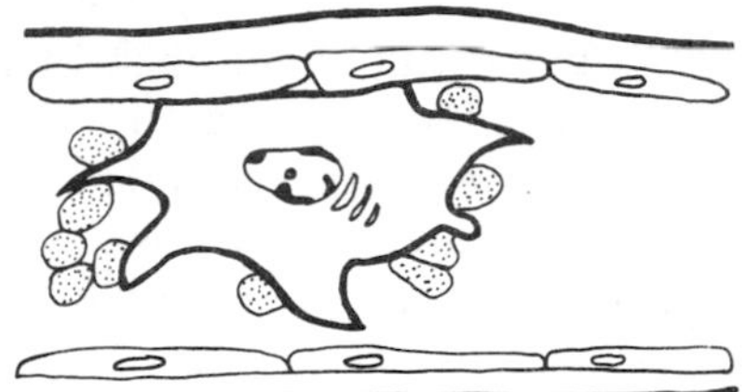

3. Endothelial displacement

4. Thrombus dissipation

5. Intravascular proliferation

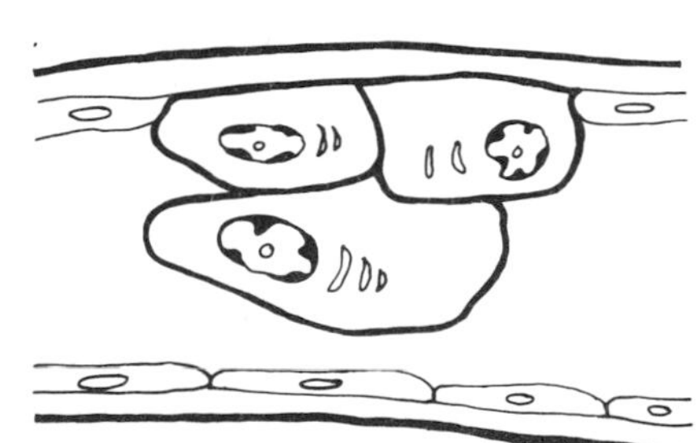

6. Penetration of basement membrane

The factors involved are now thought to include resistance to host defence mechanisms of humoral and cellular nature, and responsiveness to or requirement for particular growth factors. Secondary growth may begin within the lumen of the blood vessel in which the tumour cell is lodged, or between the lining endothelial cells and the basement membrane, or after complete extravasation. The metastasis formed may then proceed to metastasize itself.

In a review of their studies with about 60 rat ascites hepatomas, Kawaguchi and Nakamura (1986) describe the arrest and proliferation of cells from two lines (AH7974 and AH7974F) following their injection into the common carotid artery of host animals. After intra-arterial injection in this manner, the AH7974 cells colonize the brain parenchyma, meninges and choroid plexus but not the liver, whereas the AH7974F cells colonize the liver with little involvement of the brain. Histological examination of these organs after injection of tumour cells showed significant arrest of the AH7974 cells in brain capillaries but not in the liver sinusoids. Many cells were then lost from the brain capillaries over the next 6 h, but the cells remaining invaded the parenchyma to proliferate (with a peak at 72 h) and form tumour colonies. AH7974F cells also showed a high rate of initial lodgement in the brain followed by rapid cell loss, but in this case little cell division took place in the brain and thus few brain colonies developed. AH7974F cells which lodged in the liver sinusoids, on the other hand, initiated rapid proliferation and invaded the parenchyma. These results suggest that the organ microenvironment influences the eventual fate of arrested tumour cells, since cells which cannot divide in certain locations cannot form metastatic lesions there. It should also be remembered that the endothelial lining of the liver sinusoids is not complete and that it lacks a basement membrane, whereas the same structure in brain capillaries is complete with a relatively thick basement membrane and a lining of both pericytes and astrocytes. On purely structural grounds it might be thought that tumour cells would colonize the liver in preference to the brain. That this is not necessarily the case suggests that other factors such as cell growth after lodgement are involved.

Tumour cells may grow at different rates in different parts of the same organ or tissue. Hewitt (1953) reported that sarcmoma 37 cells grew faster when injected subcutaneouly into the groins of recipient mice than when injected into their axillae. Auerbach, Morrissey and Sidky (1978) showed that the C755 mammary tumour grew at increasing rates in the dorsolateral lumbar region, the midventral region, and the lateral and dorsal thoracic regions respectively following intradermal injection into host mice. In a comprehensive review of regional differences in the growth of tumours, Auerbach and Auerbach (1982) provided evidence in support of the greater growth of a variety of tumour cells when they were injected intradermally or subcutaneously into cranial regions of the lateral trunk rather than more caudal regions. These effects were seen in male and female hosts of various mouse strains, in both normal and nude mice, and in both normal and irradiated animals. Such effects are likely to extend to the growth of metastatic lesions in these sites, although conclusive *in vivo* evidence is lacking.

What phenomena might underlie these observations of regional differences in tumour growth rates? The majority of available evidence (both indirect and direct in nature) points towards differences in the microcirculation as being the principal source of variation. At room temperature, for example, there is a 0.5°C decline in mouse skin temperature along the trunk in the anterior to posterior direction which almost certainly reflects vascular differences. In other studies, thermosensitive colour crystals have highlighted both anteroposterior and dorsoventral temperature gradients. More directly, measurement of local blood flow in unanaesthetized rats has shown that capillary blood flow is higher anteriorly than posteriorly. It thus seems likely that regional differences in the vascular system are responsible for the observed differences in tumour growth. Although not without intrinsic interest, the observation of regional differences in tumour growth emphasizes the need for precision in experimental studies based on techniques employing tumour transplantation.

As we shall see in section 2.14.6, some tumour cell types display a tendency to spread selectively to particular organs. This trait most likely reflects a multitude of tumour cell activities, including selective adhesion and growth. Selective growth is likely to be primarily mediated by the presence of organ-specific paracrine growth factors which act (possibly selectively) on certain tumour cell types. Liver metastasizing RAW117-H10 cells, for example, are stimulated to grow in liver conditioned medium to a greater extent than are their parental cells, which metastasize to this organ only poorly (Nicolson, 1987b). Although not all metastatic variants apparently behave in this way (Hart, 1982 a and b), responsiveness to paracrine growth factors is obviously not without its advantages in the successful establishment of a metastatic lesion.

2.14 THE NATURE AND PATTERN OF TUMOUR SPREAD

2.14.1 Metastatic potential

It is clear that not all malignant tumours have the same metastatic potential. Basal cell carcinoma of skin, for example, never gives rise to metastases even though it may be highly invasive. Small cell carcinoma of lung, on the other hand, has high metastatic potential and almost all patients have multiple organ involvement at the time of presentation. Little is known at the present time as to why malignant tumours should differ so widely in their metastatic potential, but it may relate to the nature of the original cell type and to the nature of the transforming agent. Despite these difficulties, we do have some insight into the factors determining whether any particular tumour will develop metastases.

Interestingly, evidence is available which suggests that the extent of invasion and/or metastasis in experimental systems is dependent on the site of injection of malignant cells (reviewed in Meyvisch, 1983). In experiments performed by Gao *et al.* (1984), for example, cells from a human oesophageal cancer line formed invasive tumours when injected i.p. into nude mice, but encapsulated tumours after s.c. injection. To what extent this might be involved in spontaneous *in vivo* metastasis, however, is uncertain.

2.14.2 Angiogenesis

We have seen that tumour cells (and also some normal cells) can secrete angiogeneic factors which encourage the directional growth of new blood vessels (section 1. 7.20). Angiogenesis appears to be a programmed response on behalf of endothelial cells, and it may be summarized as a number of sequential steps (Folkman, 1985):

(a) New vessels usually originate from small venules or capillaries, beginning with the local degradation of basement membrane.
(b) Endothelial cells migrate through the degraded basement membrane and align to form sprouts.
(c) A lumen then appears, probably by reshaping of the cell or possibly by alignment and fusion of vacuoles.
(d) The sprout elongates by migration and division. Endothelial cells about halfway along the sprout begin to divide, whereas those at the tip migrate but do not divide.
(e) Capillary loops now form as sprouts anastomose. Luminal flow

becomes apparent, and a basement membrane is formed to complete the process.

Folkman has argued that tumour growth is dependent on neovascularization (reviewed in Folkman, 1985). Indeed, solid tumour growth is markedly limited in the absence of angiogenesis. Tumours implanted in the avascular anterior chamber of the rabbit eye float in the aqueous humor and reach a volume of about 0.5 mm^3 after 14 days. When implanted into the vascular iris, however, such tumours can reach 330 mm^3 over a similar period. With time, vascular tumours grow to compress their blood vessels and, as flow within them ceases, the tumour becomes necrotic. Necrosis is usually centrally located, presumably because it is here where the effects of compression will be the greatest, but also because cells at the periphery of a tumour can undergo replenishment of nutrients and removal of waste products by diffusion.

In section 2.13.3 we saw that Liotta and his colleagues (1974, 1976) have provided evidence for a correlation between the formation of tumour blood vessels and the appearance of maligant cells in the blood. Folkman (1985) has also argued that metastasis is influenced by angiogenesis on the grounds that avascular tumours have a low probability of metastasizing compared to their vascularized counterparts. Relatively thin melanomas, for example, have a good prognosis since they reside entirely above the basement membrane, are avascular, and rarely metastasize (Seigler and Setter, 1977; Srivastava *et al.*, 1986)). Since tumours lack lymphatic vessels (Swabb, Wei and Gullino, 1974), this could compound the poor metastatic behaviour of avascular tumours. It would seem that the dependency of metastatic tumours on angiogenesis is likely to be a function of the process by which invasion occurs. Some tumours may shed single cells from the periphery which could invade lymphatics and/or small blood vessels outside the bulk of the tumour mass: this possibility might explain the different metastatic profiles of some tumours. It is generally accepted that most malignant tumours may have seeded metastases before they became clinically observable. Whether this occurred in a prevascular state or not is uncertain, but the possibility is not entirely out of the question.

One final point of considerable importance is to recall that angiogenesis has two dimensions to it. Thus although angiogenesis is necessary for tumour growth and may promote metastasis, it also delivers immune cells, macrophages and humoral factors direct to the tumour and in this respect it may represent an important arm of the body's defence system against malignant growth.

2.14.3 The effects of time and tumour size

Clinical evidence supports the idea that for many common human tumours the probability of metastasis increases with increasing size of the primary (Sugarbaker, 1981). Many advantages thus lie in the early detection of malignant tumours, and techniques for doing so are continuously being developed. Most tumours would be expected to increase in size with time, but time also adds complexity to the pattern of spread. Thus a primary in the gut may give rise to a secondary metastasis in the liver which could then act as a generalizing site for a tertiary in the lungs (Bross and Blumenson, 1976). This too may act as a generalizing site giving rise to quarternary metastases which, as a consequence of arterial dissemination, may be widespread in the systemic organs. Thus the pattern of spread for the same malignant tumour may be expected to be different in patients in whom the disease is diagnosed early and in patients who have died from the disease after it has become widespread, with considerable variation in between. As will be discussed below, some tumours metastasize to particular organs with a relatively high frequency. For many tumours, particularly those that spread via generalizing sites, this pattern may not be fully expressed for some time and indeed, death of the patient may inhibit it from being expressed at all.

Finally, several reports such as that by Sugarbaker and Ketcham (1977) have indicated that the primary tumour can exert an inhibitory effect on the growth of its secondaries, but how this is brought about is not clear.

2.14.4 The effects of cell number and viability

By definition, tumour cells which metastasize via the haematogenic route must be found in the blood at some stage of the process. With other things being equal, it might seem obvious that the frequency of metastasis to a particular organ would bear some relationship to the number of cells which reach it via the blood. However, it is well known that the mere presence of tumour cells in the blood does not guarantee the establishment of a metastasis, and other things clearly are not equal since blood supply to an organ does not always closely correlate with the frequency with which it develops metastatic lesions. That large numbers of malignant cells may be released into the blood without evidence of metastases developing provides further evidence for the inefficiency of metastasis.

How many cells are required for the successful establishment of a metastasis? This question has two parts, one relating to the primary which seeds metastatic cells, and the other to the cells which establish the secondary. It is not known how many cells are necessary in the primary to seed a successful metastasis. Nor is it really known whether a single cell can establish a metastasis, although the likelihood that this may be the case, however, can be deduced from tumour transplant experiments in which different numbers of tumour cells are inoculated into recipient animals (Hewitt, Blake and Walder, 1976). A single tumour cell may transmit leukaemia from one host to another, but this is clearly not identical with metastasis. Perhaps not surprisingly, the ability for tumours to take from low cell numbers depends on the tumour under study and on the techniques employed.

Hart, Talmadge and Fidler (1983) have shown that the numbers of lung colonies which form after tail vein injection are not linearly correlated with cell input. From their studies they have suggested that comparisons based on a single inoculum dose cannot be used to test the colonizing capacity of different tumours. Yet in many experimental cases it is useful to be able to compare the metastatic ability of different tumour cell types. According to Hart and colleagues (1983), the injection of a range of viable tumour cells mixed with a constant number of X-irradiated tumour cells provides less variable results (in terms of the number of cells required to form each tumour nodule) and these may be used in establishing a baseline for comparing colonizing activity. Using this approach, these authors showed that on the average about 8× more K-1735 mouse melanoma cells were required to form a lung colony than K-1735-M4 cells which were originally selected for increased metastatic capacity. It is not known how the simultaneous injection of X-irradiated cells actually affects the metastatic outcome, although the suggestion has been made that their bulk may serve to dampen the effect of host defence systems on the non-irradiated cells. Other studies have shown that the injection of heat killed AH7974 cells into the carotid artery of rats before the injection of other viable tumour cells by the same route leads to an enhancement of brain tumour colonization by the second set. Presumably, the injection of heat treated dead cells induces some conditions which are conducive to colonization, although there is no indication of what precisely these may be. Note that the effects of cells treated with X-irradiation are not necessarily the same as those killed by heat, and the mechanism of treatment may thus have some bearing on the experimental outcome. It is known that tissue injury can lead to metastasis formation (possibly as a result of enhanced extravasastion through damaged vessels). Furthermore, hepatomas such as the AH7974 cell line which colonize the brain can induce cerebral oedema, focal tissue softening and the local development of granulation tissue, all of which may assist in colonization. Whether the products of heat treated dead cells can induce tissue damage thereby facilitating colonization by subsequently injected viable tumour cells is not known.

2.14.5 Cell cycle dependency

Cells change both biochemically and morphologically as they proceed through the division cycle (section 1.2.1). Suzuki *et al.* (1977) tested the cell cycle dependency of metastasis using a model system based on lung colonization by synchronized mouse fibrosarcoma cells. They found that lung colony formation after i.v. injection was cell cycle dependent in the order $S > G_2 >> G_1$. Unfortunately, it is not known as yet what cell cycle related parameters are responsible for these reported differences. It could be, for example, a function of differences in size, cell surface determinants, or adhesion, all of which alter as cells pass through the division cycle.

2.14.6 Preferred patterns of spread

(a) *Clinical evidence*

There is a considerable amount of clinical evidence that certain types of tumours have a propensity to metastasize to particular organs (Willis, 1973). Two main explanations, known as the anatomical and the seed and soil hypotheses of metastatic spread, have been put foward to explain this evidence. In simple terms, the anatomical hypothesis argues that conditions of distribution are important while the seed and soil hypothesis argues more that conditions of growth are important. Historically, many researchers have tended to align themselves with one or the other of these hypotheses. However, there are no sound reasons for considering the two hypotheses to be mutually exclusive and, as will become clear, many tumours probably spread using a combination of the methods proposed in the anatomical and seed and soil hypotheses.

(b) *Experimental evidence*

Reasonably convincing evidence for preferred metastatic spread to the lungs was provided by some ingenious experiments designed by Kinsey (1960). In earlier experiments, it had been shown that Cloudman melanoma cells colonized only the lungs of host mice after intravenous or intra-arterial injection (Kinsey and Smith, 1959). Kinsey (1960) explored this preferred pattern of spread further by comparing the ability of the melanoma cells to colonize grafted lung tissue with that of other tissue grafts or sham controls. In essence, one thigh of an experimental mouse was implanted by trochar with a lung fragment while the other thigh was implanted with another organ fragment or it was sham treated by injection of the trochar only. After 10 days, by which time the implants had become vascularized (only liver grafts were rejected), the mice were inoculated with melanoma cells via the tail vein, tail artery or the left ventricle. Regardless of the route of inoculation, tumours formed only in the lungs of the recipient animals or in lung grafts. It might be argued that the other grafts were from organs that did not allow growth of the melanoma cells, but this appeared not to be the case since direct injection of most of the organs studied allowed tumour growth. Of course, direct injection probably involves different numbers of cells and tissue destruction, which limits the usefulness of these control experiments. Similar sorts of studies were carried out by Sugarbaker, Cohen and Ketcham (1971) using cells from a 3-methylcholanthrene-induced sarcoma and by Hart and Fidler (1980a) using the B16 melanoma. Parabiotic experiments to examine selective organ colonization were performed by Fidler and Nicolson (1977). These authors reported that B16F1 cells colonized non-injected parabiont partners at a lower frequency than their injected partners and, somewhat surprisingly, that B16F10 cells failed to colonize the uninjected parabionts, even though they recirculated from injected to uninjected partners. One possible explanation for these results is that a specific lung-colonizing sub-population of cells was selected out in the lung capillary bed of the injected host, but because of the anomaly between the distribution of B16F1 and B16F10 cells in the uninjected parabionts this remains rather speculative.

In an earlier set of experiments, Schour and Faraci (1974) used tumour cells which preferentially metastasized to the spleen in recipient mice. When these authors splenectomized the tumour bearing mouse of a parabiotic pair, they found that increased metastases developed in the spleen of the other mouse. Pilgrim (1971) had earlier shown, however, that when otherwise normal mice were splenectomized and inoculated with the same tumour cells, other organs developed metastases. Furthermore, if tumour bearing mice were splenectomized in order for them to live longer, metastases soon became apparent in organs other than the spleen. Thus it would seem that for some tumours at least there is a selective rather than an exclusive preference for metastasizing to a particular organ.

2.14.7 The anatomical or mechanical hypothesis

The idea that anatomical considerations might be responsible for preferred patterns of spread was championed by Ewing (1928), but the theory has its origins in a number of papers published last century (see Sugarbaker, 1981).

Many carcinomas which metastasize via lymphatic channels colonize regional lymph nodes. This may not seem a particularly startling observation, but it is evidence that patterns of drainage are crucial in the metastatic process (Lindberg, 1972). Indeed, unusual patterns of metastasis may sometimes be traced at autopsy to lymphatic blockage and the rerouting of lymph (containing tumour cells) through collateral channels. Further evidence along these lines confirms that anatomical relationships are important in the metastatic process. In lymph-borne metastasis we might expect to see the profile of spread reflect the pattern of lymphatic drainage, and in blood-borne metastasis we might expect to see secondaries develop initially in relation to the pattern of venous drainage. Thus some squamous cell cancers of the head and neck (particularly those of moderate metastatic potential) show preferential lymph node metastasis, and the liver (which represents the first capillary bed encountered) is the dominant site for blood-borne metastases from cancer of the colon. The pattern of metastatic distribution is not always so clear-cut, however. It may be confused, for example, by aberrant connections (some possibly unknown) and by the presence of lymphaticovenous shunts. The metastatic colonization of bone by a number of tumours including nasopharyngeal cancer and cancers of the breast and prostate might at first sight be thought to be independent of anatomical considerations. It seems clear, however, that in many cases such metastases can be ascribed to the presence of the internal vertebral venous plexus as described in section 2.6. Furthermore, distribution patterns are likely to increase in complexity with the age of the tumour since metastases may act as relay stations for the seeding of further metastases, and this too will present a confusing picture.

According to the anatomical theory, tumour cells which reach the left ventricle are likely to be distributed in relation to arterial supply. That this may not always be the case suggests that the pattern of metastatic distribution may be governed by more than simple anatomical considerations. Blood-borne metastases are rare in the muscle and gut, for example, even though between them they take more than half of the cardiac output. Nevertheless, for many human cancers the location of initial metastases appears to be determined by anatomical considerations. Thus the first organ encountered by circulating cells is a frequent site of metastatic development.

2.14.8 The seed and soil hypothesis

This hypothesis developed primarily from the autopsy studies of Paget (1889). He observed that breast carcinoma preferentially metastasized to the liver when compared to carcinoma of the uterus, and that both tumours metastasized poorly to the spleen. His interpretation was that the tumour cell (or seed) required a particular organ (or soil) for growth. Numerous other reports confirm that some tumours metastasize in a pattern which seemingly bears no direct relationship to anatomical considerations. Clear cell carcinoma of the kidney, for example, frequently metastasizes to the thyroid gland and follicular carcinoma of the thyroid frequently metastasizes to bone.

2.14.9 The metastasis of metastases

There is no doubt that metastases can themselves give rise to metastases and indeed, the phenomenon was reported by clinicians in the last century. It is conceivable that the metastases of some types of tumours may do this better than others, but the appropriate data are lacking. Hoover and Ketcham (1975) provided experimental evidence in support of the ablity for metastases to metastasize through the use of parabiotic experiments in mice. These authors used three different types of tumours which metastasized to the lungs after injection into the footpad. After lung metastases had formed, the tumour-bearing limb was amputated and each experimental animal was parabiosed via the skin with a normal animal. In time, about 80% of the guest animals in the parabiotic pair developed lung metastases, presumably from the lung metastases present in the host parabiont.

2.14.10 Generalizing sites

In haematogenic spread, the capillary bed of the first organ encountered is often (but not always) the site of most metastases. This organ may then act as a relay or generalizing site for the seeding of further metastases which may display a particular pattern of spread irrespective of simple anatomical considerations. The liver and/or the lungs are the main generalizing sites for a number of adenocarcinomas, yet breast adenocarcinoma still has a propensity to metastasize to the brain and endocrine organs when compared to other adenocarcinomas (Viadana and Au, 1975). In lymphatic spread, the draining lymph nodes may act as generalizing sites for the establishment of further metastases.

In experimental studies with the M5076 tumour (a murine reticulum cell sarcoma of ovarian origin), Hart and colleagues (1981) showed that it metastasized preferentially to the peritoneal viscera regardless of the site of tumour inoculation. Intravenous injection of radiolabelled cells showed that they were initially trapped in the lungs and that they were retained there for 3–4 days. Subsequently, the cells dislodged and recirculated to become arrested again, but this time primarily in the liver where they grew to form overt tumour nodules. One interesting aspect of this metastatic behaviour was that although the cells lodged and survived in the lungs, they did not form obvious nodules in these organs. Thus arrest and retention in the lungs do not necessarily correlate with subsequent tumour growth. In fact, more M5076 cells are retained for longer in the lungs than are B16F10 melanoma cells, even though the latter have actually been selected for their lung colonizing ability (Fidler, 1973a). Clearly, the nature of the soil would seem to be crucial for tumour growth, even if only for some. Finally, the results bring another dimension to the study of generalizing sites, since it would seem that such sites are not necessarily associated with overt tumour growth.

2.14.11 A combined hypothesis for metastatic spread

We have seen that there are two aspects to the often repeated observation that metastases are not found randomly in the body:

(a) Not all parts of the body are supplied with tumour cells in the same quantities, suggesting an anatomical influence.
(b) Some sites within the body may be more conducive for metastatic growth than others, suggesting an influence of the soil on the growth of the seed.

The combined hypothesis argues that both of these aspects are likely to be important in the distributiom of metastases, although their effects may vary for different tumours or at different stages of spread. Put simply, it suggests that malignant tumours spread along definable anatomical pathways (some of which may not be readily discernible), but that metastases only form if both the tumour cells and the site of metastatic development have properties conducive to growth.

Much of the clinical data on the spread of tumours are interpretable in terms of a combined hypothesis rather than in the exclusive terms of either the anatomical or seed and soil hypotheses. Thus although many primary tumours spread to generalizing sites as a consequence of anatomical considerations (Viadana, Bross and Pickren, 1978), they will only grow if each site provides an environment appropriate to the needs of the metastatic cell. Similarly, further spread from generalizing sites is dependent on a mixture of anatomical and growth components. Thus, although major generalizing sites occupy strategic anatomical locations, they must also provide an appropriate environment for growth. Indeed, the dual properties of strategic location and ability to support metastatic growth are characteristic of generalizing sites.

2.14.12 Protected sites

The lungs and the liver represent sites in which metastases are commonly found. An explanation for the fact that more metastases are found in the liver than might be expected is largely due to its having two blood supplies, arising from the systemic and portal systems. In so far as the lungs are concerned, we know that all of the blood ultimately passes through the fine capillary structure of this organ and thus it is perhaps not surprizing that it is a common site for metastatic growth. However, other organs, such as the kidneys, are poor sites for metastasis even though they receive a considerable portion of arterial blood. Like the kidneys, skeletal and cardiac muscle appear to represent protected sites in that they are poorly colonized by metastatic tumours in relation to their arterial bood supply. We have seen, however, that tumours do not spread only via the arterial system since they may also use venous and lymphatic routes and they may even spread by surface implantation. The possibility of multiple routes of spread complicates any mechanical explanation of the failure for tumours to grow in certain organs and renders interpretation of clinical data difficult. Furthermore, delivery of tumour cells is only one aspect of tumour spread. Thus cells may fail to lodge or may be destroyed depending on the nature and extent of the blood circulation in the organ involved. The architecture of some organs and tissues, for example, may be such that tumour cells are destroyed in the microcirculation for haemodynamic reasons before they have the opportunity to metastasize.

Although mechanical aspects are important in the establishment of metastases it is clear that environmental conditions must also be favourable for growth. Certain organs and tissues may either lack appropriate growth factors or they may contain certain toxic components (or both). Organs and tissues may indeed vary in their abilities to defend themselves against tumour growth. Herberman, Nunn and Lavrin (1975), for example, have shown that different sites for tumour growth have different natural killer cell activities. It is not known, however, whether or not this cytotoxic activity correlates with metastatic patterns.

The spleen is often thought of as an organ that is relatively resistant to metastasis, but this appears to be a misconception based on the fact that many splenic metastases are not observable on gross autopsy and require more detailed histological observaton for detection. Excluding problems of delivery and detection, there appear to be no simple reasons to explain why some organs appear to be relatively resistant to metastatic growth. Possible reasons put forward include the lack of appropriate growth factors, the presence of toxic factors and/or host defence cells, and various mechanical and physical characteristics of the particular organs.

It would seem from the foregoing that there is no simple explanation for the distribution pattern of metastases displayed by a number of tumours. Although some tumours do tend to spread almost solely in accord with anatomical considerations, for many others which display site-selective metastatic behaviour we are obliged to consider the possibility that a multitude of events might be involved. A summary of

Table 2.4 Processes likely to contribute to selective metastasis

1. Anatomical considerations. The ability to colonize selectively a particular organ cannot be expressed if tumour cells are not delivered to it in a viable state.
2. Selective attraction of circulating metastatic cells e.g. chemotaxis.
3. Selective recognition e.g. specific cell surface determinants allowing tumour cell–organ cell recognition. Such recognition may trigger a number of secondary pathways (e.g. paracrine growth factor release) which could be conducive to metastatic establishment.
4. Selective adhesion between the circulating tumour cell and components of the target organ:
   (a) the endothelium;
   (b) its basement membrane;
   (c) the organ stroma;
   (d) parenchymal cells.
5. Promotion of organ invasion by adherent cells e.g. chemotaxis; haptotaxis.
6. Inhibition of invasion by the production of enzyme inhibitors e.g. cartilage.
7. Physical and mechanical attributes of particular tissues e.g. dense collagenous structures may be poorly invaded.
8. Attributes characteristic of particular tumour cells e.g deformability; motility; hormonal requirements.
9. Viability:
   (a) toxic factor production by particular organs;
   (b) presence of host defence cells.
10. Growth:
   (a) production of chalones inhibiting growth (and possibly contributing to dormancy);
   (b) production of paracrine growth factors acting selectively on particular tumour cells;
   (c) availablility of nutrients and processes for waste removal e.g. angiogenic potential.

possible processes which might contribute to site selective metastasis is provided in Table 2.4. Many of these processes will be referred to repeatedly in subsequent chapters, and these should be read for further details.

2.14.13 The clonal origin of metastases

There is some evidence which suggests that metastases (like their primaries) are clonal in origin. Talmadge, Wolman and Fidler (1982), for example, irradiated K-1735 mouse melanoma cells, injected them s.c. into syngeneic C3H mice, and examined the metastases which formed in the lungs. They found that the cells within each metastatic lesion all exhibited the same radiation induced chromosomal abnormalities, although of 21 metastases studied, 10 could not be evaluated because they lacked radiation induced chromosomal markers. Apart from the low detection rate, the main disadvantage of this approach is that chromosomal instability might present a false picture of the origin of metastatic lesions. Poste *et al.* (1982b) used cells with different drug-induced phenotypes to test for the clonal origin of metastases. These authors found that fewer lesions which developed after s.c. injection of drug-marked B16BL6 cells were clonal in origin when compared to cells injected intravenously. This difference was suggested to be due to the routes of inoculation since i.v. injection involves single cells in suspension, whereas spread from an s.c. site may involve the intravasation of heterogeneic clumps of tumour cells. More recently, Talmadge and Zbar (1986) re-examined the possible clonal origin of metastases using B16BL6 cells genetically marked by transfection with a bacterial plasmid. Two differently transfected B16BL6 sublines were mixed together in equal proportions and injected into the footpads of recipient mice. The primary tumours were excised 28 days after inoculation, and individual lung metastases were isolated after another 28 days. Cells from these metastases were expanded in nude mice in order to get enough material for subsequent DNA extraction and restriction enzyme mapping. It was found that the mixed footpad tumours always contained the genetic markers of both sublines, whereas virtually all (13/15) of the metastases only contained a single marker (thereby confirming their clonal origin). Although many experiments suggest that tumour cell clumps are more likely to form metastases than single cells, the data above suggest that metastases in fact result largely from single cells. Three possibilities might resolve this paradox:

(a) Tumour cells circulate preferentially *in vivo* as single cells rather than as clumps.

(b) Although multicellular clumps may lodge in vessels, in most cases only one cell may survive to form a metastasis.
(c) Finally, the clumps of cells which detach from the primary may represent a 'local clone' i.e. they are all daughters of the same parent cell and thus a tumour is populated by zones of related cells.

Which of these three possibilities is most likely to explain the *in vivo* situation has yet to be explored fully.

2.14.14 The biochemistry of metastasis

The isolation of tumour cell variants with enhanced metastatic or colonizing capabilities stimulated active research directed towards gaining an understanding of the biochemistry of metastasis, the simple premise being that there must be something different between low and high metastasizing variants. Many of these studies have been based on the B16 melanoma (section 3.4), but Shearman, Gallatin and Longnecker (1980) carried out an early study using liver colonizing variants of a Marek's disease lymphoma cell line. Marek's disease (MD) is a naturally occurring, Herpes virus-induced T cell lymphoma of chickens which has been used as a model for Burkitt's lymphoma in humans. Shearman and his colleagues were able to derive a monoclonal antibody against an antigen on MD cells whose expression correlated with increased liver colonizing ability following the i.v. injection of cells into 11 day old chick embryos. Liver foci resulting from a variant line with enhanced liver colonizing ability were significantly reduced by preincubation with the antibody, whereas invasion of the chorioallantoic membrane (CAM) was not inhibited. Expression of the antigen was deemed necessary but not sufficient for liver colonization, but further analysis of its nature is required before the full significance of these observations can be appreciated.

Sinha and Goldenberg (1974) have reported that neuraminidase treatment of W256 carcinosarcoma cells can alter the distribution of i.v. injected cells between the lungs and the liver, but the precise determinants involved are not clear and in any case not all tumours they tested behaved in the same way. Although perhaps not correlating with distribution to particular organs, cell surface sialylation appears to correlate with metastatic expression, at least in so far as the tumour cell lines studied by Yogeeswaran and Salk (1981) are concerned. These authors, working with a wide range of tumours (including the B16 melanoma) found spontaneous metastatic potential to correlate particularly well with the degree of sialylation of cell surface Gal and

GalNAc residues, but further characterization was not attempted. Bolscher *et al.* (1986) studied the surface glycopeptide content of four pairs of normal and transformed cells after proteolytic digestion of membrane glycoproteins. The general conclusion of these authors is that increased carbohydrate complexity and increased sialic acid content are associated with malignancy and, furthermore, that invasive behaviour is also related to altered cell surface carbohydrate content. In yet other studies, this time with the 13762NF rat mammary adenocarcinoma system, Steck and Nicolson (1983) have shown that the presence of a large sialogalactoprotein (gp580) correlates with metastatic potential. However, separate studies with B16 variants have failed to show a correlation between either colonization or spontaneous metastatic potential and sialylation (Yogeeswaran, 1980; Raz *et al.*, 1980). Furthermore, Weiss, Fisher and Fisher (1974) were unable to detect any differences in either lung localization (5 min) or retention (60 min) between neuraminidase treated and control W256 carcinosarcoma cells when they were injected into the femoral or portal veins of rats. More recently, the ability of R13762 cells to metastasize after injection into the mammary fat pad of experimental rats has been correlated with the presence of fucosylgangliosides (Wright *et al.*, 1988). In previous studies from this group it had been reported that cells with metastatic potential displayed a nuclear magnetic resonance (NMR) signal characterized by a long, transverse relaxation time. This distinctive signal was identified as being due to the presence of fucosylgangliosides and indeed, the experimental treatment of R13762 cells with fucosidae was found to reduce metastatic potential without affecting their tumorigenicity.

Studies by Dennis and colleagues (1987) with the highly metastatic DBA/2 mouse tumour cell line MDAY-D2 failed to show any correlation between metastatic potential and increased sialylation of a specific plasma membrane glycoprotein (gp130). These authors were able to show, however, that increased β1-6 linked branching of complex-type, aspargine-linked oligosaccharides correlated with metastatic potential in this cell line and two others. Glycosylation mutants were selected from the metastatic MDAY-D2 cell line using leukoagglutinin (L-phytohaemagglutinin or L-PHA) which binds to β1-6 branched complex-type oligosaccharides. Cells deficient in these particular structures (probably because of a deficiency in GlcNAc transferase V activity) remained highly tumorigenic but displayed poor metastatic behaviour as assessed by the relatively low development of nodules in the liver after subcutaneous injection into syngeneic mice. Following biochemical analysis, the plasma membrane glycoprotein gp130 in these poorly metastatic cells was found to be deficient in L-

PHA binding. Conversely, the acquisition of metastatic potential by the normally non-metastatic mouse mammary carcinoma line SP1 correlated with the expression of L-PHA binding gp130. Precisely how these changes might relate functionally to metastatic potential remains to be clarified, as does the role of gp130 itself in metastasis.

Sialic acid may serve a more indirect role in tumour invasion and metastasis by masking certain sugar groups, particularly Gal and GalNAc. Collard *et al.* (1986) found that sialic acid masked the sugar receptors for various agglutinins and that this correlated with the invasive and metastatic ability of hybrids formed from activated T cells and cells from a non-invasive T lymphoma (BW5417). These authors prepared mouse T cell hybrids by fusing activated T cells with BW5147 cells. Recloning of one particular hybrid (TAM 4D1) gave rise to subclones which differed markedly in their invasive potential as evaluated by their ability to infiltrate rat hepatocytes in culture. When the subclones were tested for reactivity with soyabean agglutinin (SBA), an inverse relationship was found between its reactivity and invasiveness. Thus the more invasive the subclone, the less receptor for SBA was available for binding on its surface. Two other lectins, peanut agglutinin (PNA) and *Ricinus communis* agglutinin (RCA), seemed to discriminate between the subclones in a similar manner to SBA. Since SBA is specific for GalNAc residues, PNA for Gal-GalNAc, and RCA for Gal and Gal-GlcNAc, it would seem that these sugars might be involved in the discrimination process, if not in the different invasive abilities of the subclones. Further work showed in fact that the decreased binding of the three lectins to invasive hybrids was due to the masking of the receptor sites by sialic acid. When the sialic acid was removed from highly invasive hybrids by neuraminidase treatment, the lectin receptors were exposed and the invasive potential of the cells was reduced in a reversible manner. Thus sialylation of certain sugar residues (predominantly Gal and GalNAc) may promote the invasive potential of tumour cells. No effects were found, however, when hybrids treated in a similar manner were tested for changes in metastatic ability by intravenous injection. This difference may perhaps be attributable to the reversibility of the change in conjunction with the longer nature of the *in vivo* experiment.

Masking of sugar residues by sialic acid has also been implicated in the metastatic ability of the mouse L5178Y lymphoma cell variants Eb and Esb (Cheingsong-Popov *et al.*, 1983). These authors found that cells of the highly metastatic variant ESb (which displays considerable liver involvement) formed more rosettes *in vitro* with mouse hepatocytes than did those of the poorly metastatic variant Eb. After treatment with neuraminidase, however, the Eb cells formed considerable hepatocyte rosettes. The formation of rosettes between ESb cells and hepatocytes was inhibited by 100 mM Gal or 100 mM GalNAC suggesting an involvement of these sugars in rosette formation. The authors suggested that hepatocytes contain a lectin-like molecule which recognizes Gal and GalNAC residues and that these are unmasked on Eb cells following neuraminidase treatment. Earlier, Schirrmacher *et al.* (1982) had reported that receptors for some lectins (such as SBA) expressed on Eb cells were sialylated on ESb cells and, conversely, that receptors for other lectins (such as PNA) expressed on Esb cells were sialylated on Eb cells. It should be noted that both ESb and Eb cells contain similar amounts of neuraminidase accessible sialic acid and thus a simple quantitative variation is unlikely to account for the observed differences in hepatocyte rosette formation. Since the liver metastasizing variant forms more hepatocyte rosettes than the poor metastasizing variant it is tempting to correlate this behaviour with organ-selective metastatic ability, but the overall picture is likely to be more complex.

Using the RAW117 lymphosarcoma system, Reading and colleagues (1980 a and b) have shown that the loss of con A receptors correlates with increasing metastatic potential to the liver, whereas the loss of WGA receptors correlates with decreased metastatic potential to the same organ. The parental RAW117 cell line was originally induced *in vitro* from spleen cells of Balb/c mice infected with Abelson leukaemia virus. Reading *et al.* (1980 a and b) speculated that their results with con A selection might have been due to the loss of a 70 kD glycoprotein equivalent to the major Abelson viral envelope glycoprotein gp70. If this proves to be the case, the increased metastatic ability of cells selected for non-adherence to con A might be due to nothing else other than the ability of such cells to avoid host defence systems because they are probably deficient in gp70 (which binds to the lectin).

Although of considerable interest, not one of these studies really allows us to reach any general conclusions on the biochemistry of metastasis. Virtually all have explored the surface biochemistry of malignant cells using classical biochemical techniques, and it may be that significant progress will only now be made following the advent of new methods in molecular biology. Further discussion on possible biochemical correlates with metastasis is made in section 3.4.7.2 with respect to the B16 malignant melanoma.

2.15 THE ROLE OF ENZYMES IN INVASION AND METASTASIS

There is considerable evidence supporting a role for enzymes during tumour cell invasion. It is generally thought that such enzymes may degrade the local tissues through which tumour cells invade, and that they may also be involved in penetration through the basement membrane during the extravasation of tumour cells from blood vessels (Sloane and Honn, 1984; Dano *et al.*, 1985). It is important, however, to raise several questions at this stage:

(a) Do *all* tumour cells need to breakdown connective tissue structures in order to invade?
(b) Do different tissues require different strategies of invasion?
(c) Are tumours heterogeneic in their ability to produce particular enzymes? This heterogeneity could be expressed between different types of tumours (e.g. carcinomas and sarcomas) and between different cells within a particular tumour.
(d) Do tumour cells produce enzymes endogenously, or can some tumours utilize enzymes produced by normal or activated host cells?
(e) Is there any universal correlation between endogenous and/or exogenous ezymatic activity and tumour cell invasion?
(f) We have seen earlier that invasion is a critical step in metastasis, but is there any correlation between enzyme activity and metastatic (rather than invasive) potential?
(g) Finally, how do enzyme inhibitors influence metastatic invasion?

These and related questions will be discussed in the following sections.

2.15.1 Proteolytic enzymes

Proteolytic enzymes (proteinases or proteases) are enzymes which hydrolyze peptide bonds. As summarized in Table 2.5, there are four major classes of proteinases, three of which appear to be particularly important in invasion and metastasis. Although a variety of proteases are likely to contribute to invasion, those principally involved include the serine protease plasminogen activator (Ossowski and Reich, 1983 a and b), the cysteine protease cathepsin B (Sloane *et al.*, 1986), and the metalloproteinase collagenase, of which there appear to be specific types (Liotta *et al.*, 1979). With respect to invasion, the precise location of protease activity may be of considerable importance since some enzymes may be less effective if they are generally secreted and allowed to diffuse over a wide range. Few attempts have been made to examine the location of tumour proteases, but Tokes, Sorgenti and

Table 2.5 Classification of the major proteinases

Type	*Examples*	*Range*	*Inhibitor*
Aspartic	pepsin	pH 2–7	diazoketones
Serine	plasmin plasminogen activator thrombin elastase trypsin	pH 7–9	fluorophosphates
Cysteine	cathepsin B	pH 3–8	N–ethyl–maleimide
Metalloproteinase	collagenases stromelysin	pH 7–9	TIMP and metal chelators

Okigaki (1977) explored this problem by making ingenious use of protease substrates immobilized on beads. In this way, tumour cells could be kept at a distance or brought into contact with the various substrates. The general conclusion reached by these authors for a variety of rat liver epithelial tumours was that both cell surface-associated and secreted enzymes were higher in the tumour cells than in normal cells. It should be noted that studies based on tissue homogenates cannot fully evaluate the enzyme contribution of contaminating normal host cells, and nor can they differentiate between cytoplasmic, cell surface and released enzymes.

(a) *Plasminogen activator (PA)*

This enzyme generates plasmin (itself an enzyme with broad substrate specificity) from plasminogen. Plasmin is principally responsible for degrading fibrin, but it can also degrade various glycoprotein and proteoglycan components of basement membranes. Fibrinogen degradation products resulting from plasmin activity are found in plasma, and at least one of these has an immunosuppressive effect (Girmann *et al.*, 1976). To what extent a tumour that produces PA can capitalize on this effect, however, is uncertain. Plasmin also activates latent type IV collagenase, thereby influencing basement membrane degradation which is likely to be important in some examples of metastasis. Available evidence suggests that there are at least two types of PA:

1. The urokinase type (u-PA) with an M_r of about 54 kD, which is particularly involved in tissue degradation by both normal and malignant cells.
2. The tissue type (t-PA) with an M_r around 70 kD, which is a key enzyme in activating fibrinolysis.

Cell surface receptors for both PA forms have been identified, and evidence exists confirming that the bound PAs retain their enzymatic activity, thus offering a mechanism for localization of action (reviewed in Laiho and Keski-Oja, 1989).

Fibrinolytic activity due to PA production has been correlated with experimental metastatic potential in B16 mouse melanoma cells (Wang *et al.*, 1980). An earlier report failed to show any significant difference in fibrinolytic activity between B16F1 and B16F10 cells, but this may have been due to using one tenth as many cells in the assay system (Nicolson *et al.*, 1976). According to Mignatti *et al.* (1986), cells of the B16 melanoma express u-PA on their surfaces, and invasion of human amnion by B16-BL6 can be blocked by anti-urokinase antibodies, as well as by inhibitors of plasmin and collagenase. These results are in contrast to those of Ostrowski *et al.* (1986), that inhibition of u-PA was without effect on lung colonization by B16F10 cells.

The production of PA does not always correlate with transformation or metastatic potential. Thus Talmadge, Starkey and Stanford (1981) found no relationship between PA production and metastatic potential in a rat hepatocarcinoma system. Furthermore, many normal cells (such as activated macrophages) can show high levels of PA activity. Studies with fused tumour cells and normal cells have shown that hybrids that lack metastatic potential express low levels of PA activity (Ramshaw *et al.*, 1983). However, as the hybrids reverted to the metastatic phenotype during culture *in vitro* no change was found in the level of PA activity. Within the limits of this experimental system, these results suggest that high levels of PA production may not be essential for metastasis. Finally, Chen and Buchanan (1975) have reported that some tumour cells secrete enzymes which result in fibrinolysis independently of plasminogen. Note that problems of interpretation of much of the existing data focus on the possible secretion of single chain PA forms (since these are not active in most routine assays) and the simultaneous production of PA inhibitors.

(b) *Cathepsin B*

This cysteine proteinase is found primarily in lysosomes, but some tumour cells also secrete it in large amounts. Cathepsin B, which is active near pH 7, degrades basement membrane glycoproteins and proteoglycans as well as collagen types I–IV. Cathepsin B appears also to be able to activate latent procollagenase, and cathepsin B-like activity has been correlated with metastatic potential in B16 melanoma cells (Sloane and Honn, 1984). Ostrowski *et al.* (1986), however, found that inhibitors of cathepsin B released via implanted minipumps were without effect on lung colonization by B16F10 cells. In their minipump experiments with the cathepsin B inhibitor leupeptin, for example, Ostrowski and colleagues (1986) found they were able to maintain a steady state *in vivo* concentration of inhibitor that was 1000× greater than its K_i, but even this concentration was still without effect on colonization. Persky *et al.* (1986) have also recently questioned the role of serine and cysteine proteases in invasion by tumour cells. These authors found that high, effective concentrations of the enzyme inhibitors leupeptin (acetyl-leu-leu-argininal) and glu-gly-argininal failed to inhibit the invasion of epithelial-stripped amnion by B16F10 cells. The stripped amnion was still essentially intact as judged by morphological studies and by the failure of non-invasive cells or inert beads to penetrate it.

(c) *The collagenases*

Collagens of various sorts are found in connective tissues and basement membranes (section 2.10) and a corresponding range of collagenases may be involved in their degradation. According to the studies of Liotta and colleagues (1979, 1980), some tumour cells secrete a type IV collagenase (which is specific for the collagen found in basement membranes), and the amount secreted correlates with metastatic potential in model systems such as the B16 melanoma. Furthermore, tumour cells which escape from the primary tumour into the venous blood flow also show increased levels of collagenase activity relative to the cells which remain behind (Liotta et al., 1977). Other studies by Dresden, Heilman and Schmidt (1972) suggest, however, that although normal tissues rarely produce significant quantities of collagenase, not all human tumours necessarily do so. According to Salo *et al.* (1985) fibroblasts secrete type IV collagenase, while Mainardi *et al.* (1980) and others have shown that broad spectrum proteases (such as leukocyte elastase) produced by non-malignant cells can also degrade type IV collagen.

In so far as B16F10 cells are concerned, the results above suggest that the collagenases (particularly type IV) may be more important than serine and cysteine proteases in the invasion of basement membranes. Other evidence, however, leads us to conclude that different tumours may invade through tissues by different means. Zucker *et al.* (1985), for example, studied the invasion of rat urinary bladder by Walker 256 carcinosarcoma cells and found that although W256 cells produced both serine proteases and metalloproteinases, it was the trypsin-like serine proteases which appeared to be more important in this system. Two ingenious assays were used by Chen and Chen (1987) to show that degradation of fibronectin containing substrates by RSV transformed chick fibroblasts (but not normal cells) required two cell-surface-bound enzymes: a 120 kD serine protease and a 150 kD metalloproteinase. In the first assay, cells were grown on a fibronectin substrate conjugated with rhodamine and cross-linked to an underlying gelatin film. Discrete patches lacking flourescence were seen beneath the transformed cells, and the appearance of these correlated with the release of fibronectin into the surounding medium. In the second assay, based on a technique described by Heussen and Dowdle (1980), membrane fractions were electophoresed in SDS-PAGE copolymerized with fibronectin. After staining, the membrane-located enzymes active in digesting fibronectin could be identified by transparent bands in the gel. Other reports have confirmed the importance of fibronectin degrading enzymes in invasion (Jones and De Clerck, 1980; Fairbairn *et al.*, 1985).

In keeping with these observations that different tumours may employ different enzymes during invasion is the report from Lowe and Issacs (1984) that rat prostatic cancer cell lines of different metastatic potential possess numerous proteinases and that no single enzyme activity correlates with metastatic potential. Bogenmann *et al.* (1983) studied the invasion and destruction of rat R22 smooth muscle cell (SMC) multilayers and their underlying matrix by cells from various primary paediatric neoplasms, but they could find no simple correlation between the extent of degradation and invasive ability. These authors found that invasion by some tumours (e.g. yolk sac carcinoma) could proceed with little degradation of the SMC matrix and without destruction of SMCs, while other tumours utilized either or both of these processes during invasion. Invasion by some tumours in this study (e.g. Ewing's sarcoma) was also found to occur in the absence of extensive cell division, confirming that the pressure of growth is not a prerequsite for invasion. Using a similar system, Jones and De Clerck (1980) reported that the matrix produced by SMCs was extensively digested by only one (HT1080 fibrosarcoma) of four human tumour cell lines they tested. Interestingly, the elastolytic and collagenolytic activities of HT1080 cells required direct contact with the SMC matrix suggesting that, at least for this cell type, these enzymes are probably localized to the plasma membrane. The degradation of elastin and collagen by HT1080 cells was accelerated by the hydrolysis of glycoproteins via plasmin activation, an observation suggesting that glycoproteins may restrict the access of the enzymes to some of the fibrillar components of the matrix. This possibility also provides an alternative role for the function of tumour-secreted PA.

We must conclude that although proteolytic enzymes can be utilized in the metastatic process, it seems unlikely that a specific proteinase is used by all tumours. Furthermore, the endogenous production of lytic enzymes appears not to be an absolute requirement for invasion. As mentioned below, host connective tissue cells such as fibroblasts may be activated *in vivo* by the tumour cells or their products to produce numerous enzymes which could then facilitate tumour cell invasion. Host leukocytes and macrophages (which are often associated with tumours and could be activated by tumour cells) can also induce fibroblasts to secrete collagenases.

2.15.2 Glycosidases

The basement membrane produced by endothelial cells is rich in collagen, other proteins such as fibronectin and laminin, and proteoglycans. The major glycosaminoglycan of the endothelial matrix produced by bovine aortic endothelial cells *in vitro* is heparan sulphate (>70%) with smaller amounts of chondroitin 6-sulphate and chondroitin 4-sulphate (Kramer, Vogel and Nicolson, 1982). These authors found that B16F1 melanoma cells, which invade the subendothelial matrix, partially degrade some of its glycoproteins and proteoglycans, particularly fibronectin and heparan sulphate (HS) proteoglycan. Degradation was shown to be independent of serum plasminogen and probably involved two different enzymes, one of which has turned out to be a novel endo-β-D-glucuronidase (Nakajima *et al.*, 1984). This enzyme (approx. M_r 96 kD) cleaves β-D-glucuronosyl-*N*-acetylglucosaminyl linkages in HS, and has been called **heparanase** to distinguish it from **heparitinase** which can also cleave HS, but with different site-specificity. The optimal pH for heparanase activity is about 5.6, but apparently it retains significant activity under physiological conditions.

According to Nakajima *et al.* (1983), B16F10 cells solubilize HS at a faster rate than B16F1 cells, and this may be related to their increased lung colonizing ability. More recently, however, Kramer and Vogel (1984) found that although B16F1, B16F10 and B16BL6 cells all solubilize HS there is no difference in the rate or extent to which they do this. The crucial difference between these two studies may lie in the nature of the extracellular matrix studied: whereas the matrix used by Nakajima *et al.* (1983) was derived from lung, that used by Kramer and Vogel (1984) was derived from PF HR9 cells (a differentiated embryonal carcinoma). The PF HR9 matrix is a continuous structure (about 5 μm thick) of fine fibrillar material deposited on the substrate during cell culture. Its recognizable components include laminin, type IV collagen, the sulphated glycoprotein entactin (Carlin *et al.*, 1981), HS proteoglycan and only trace amounts of fibronectin. Melanoma cells were found to adhere to denuded PF HR9 matrix but they were unable to penetrate it suggesting some fundamental difference from the matrix produced by endothelial cells. Furthermore, although melanoma cells could digest heparan sulphate within the PF HR9 matrix there was no detectable digestion of matrix proteins such as type IV collagen and fibronectin. These results suggest that macromolecules within different matrices may be masked or protected in various ways from particular enzymes: collagen, for example, could be protected by proteoglycans or by the presence of inter-molecular crosslinks. They also suggest that the production of enzymes in itself may not be enough to assure invasion.

2.15.3 Serum enzyme levels and metastasis

A relationship between the activity of several glycosyltransferases and metastatic behaviour has been demonstrated by Chattergee and Kim (1977, 1978) and Chattergee (1979). According to these authors, who worked with a metastatic rat mammary carcinoma, metastasizing tumour homogenates had increased fucosyltransferease and galactosyltransferase activities, but decreased sialyltransferase activity. When serum levels were considered, however, rats with growing, highly metastatic tumours had increased sialyltransferase and galactosyltransferase activities. Similarly, Capel *et al.* (1979) found that the serum level of galactosyltransferase correlated with tumour burden in cyclophosphamide treated mice bearing Lewis lung carcinoma metastases. The possibility of an equivalent serum marker for metastatic disease in humans is currently being evaluated.

2.15.4 Modulation of enzyme production and activity

Although some tumours may secrete collagenases directly, others may release cell products which activate normal fibroblasts to secrete collagenase (Biswas, 1982). Thus solubilization of collagenous components during tumour cell invasion may be brought about either directly or indirectly. The collagenase inhibitor TIMP (Tissue Inhibitor of Metallo-Proteinases) has been shown to inhibit the invasion of M5076 mouse reticulum sarcoma cells using the amnion experimental system, indicating that collagenases are important in invasion (Thorgeirsson *et al.*, 1982). Similarly, invasion was inhibited by a cartilage extract (which includes both metalloproteinase and serine proteinase inhibitors), but not by soybean trypsin inhibitor (which blocks serine proteinases) or by a non-enzymatic, control protein such as BSA. In most experiments of this type, however, the cellular source of the collagenase is not known: is it tumour derived or host derived? Support for the concept that tumour cells may employ host collagenases in invasion comes from the studies of Hicks, Ward and Reynolds (1984) who used cell lines from a mouse fibrosarcoma model selected for different invasive potential after s.c. injection. Invasive variants of the fibrosarcoma apparently do not secrete collagenases directly, but they do show reduced TIMP secretion. From their observations, the authors suggest that the ability of at least some tumours to invade may depend on a balance between the level of TIMP secreted by the tumour (and possibly also normal host cells) and the level of localized collagenase secretion by host cells. TIMP, a glycoprotein of M_r 28.5 kD, has been sequenced (Docherty *et al.*, 1985) and it appears to be identical to the serum component β-1 anti-collagenase. Its secretion is enhanced by the cytokine interleukin-1 (IL-1) produced by monocytes.

In a more recent study, Khokha and colleagues (1989) have confirmed that low levels of TIMP production are associated with invasiveness in at least some cell types. These authors transfected 3T3 cells with TIMP antisense RNA and found that resulting clones which produced low levels of TIMP were not only more invasive in the amnion assay (section 4.8.5), but they were also tumorigenic and metastatic in nude mice after i.v. or s.c. inoculation. How decreased TIMP expression influences transformation is not clear, but it may act indirectly via perturbation of the local microenvironment thereby disrupting signals crucial for maintaining the normal differentiated state. It should be recalled here that established cell lines such as 3T3 show many indications of already being partially transformed.

Easty *et al.* (1986) studied a group of 10 cell lines (derived from human squamous carcinomas of the tongue and larynx) for their abilities to produce and secrete collagenase, elastase, PA, and cathepsin B-like enzymes, but they could find no simple correlation between enzyme activity and invasive potential. They did find a degree of synergistic activity, however, between tumour cells and host cells. Thus some tumour cells acted on host fibroblasts to induce type I collagenase release, and some acted on osteoclasts to promote bone resorption. This modulation of normal homeostatic processes by tumour cells is likely to be of considerable importance in tumour pathogenesis and it must be borne in mind as a major limitation of *in vitro* studies.

Other studies by Jones and his colleagues (1981) indicate that host endothelium may have considerable effects on tumour cell proteolytic activity. These authors found that the hydrolysis of rat smooth muscle multilayers by human HT 1080 fibrosarcoma cells was considerably retarded in the presence of bovine aortic endothelial cells, even at a ratio as low as 1 endothelial cell to 25 tumour cells. Endothelial cells secrete type 1 PA inhibitor (PAI-1), a 54 kD protein, but it is not known whether this was the effective inhibitor in the studies above. PAI-1 acts on both t-PA and u-PA, whereas PAI-2 (secreted by the placenta and monocytes) acts more rapidly on u-PA. Interestingly, TGFα has been reported to stimulate PAI-1 secretion, and it also negatively regulates the production of both u-PA and t-PA by fibroblasts, whereas TGFα and EGF enhance PA productin. The major serum inhibitor of u-PA is α2-macroglobulin, but because of its size (restricting diffusion from the blood) it is probably not effective in tissue compartments.

Since plasmin hydrolyzes fibrin, a role for it in breaking down fibrin meshworks (possibly aiding tumour spread by freeing enmeshed cells) has been postulated. On the other hand, a cocoon of fibrin may protect tumour cells from the cytolytic activity of host defence cells, and thus the appropriate modulation of both fibrin formation and breakdown may be advantageous to the invading tumour cell. Since inhibitors of PA have been reported in human tumours (Naito *et al.*, 1981), the production of both PA and an inhibitor may underlie the ability of some tumour cells to modulate the extent of fibrin deposition around them. It should be noted here, however, that not all tumour cells are necessarily associated with a fibrin coat.

Clearly, there are a considerable number of interactive processes in action during the pathogenesis of tumours, the full impact of which remains to be evaluated. With respect to tumour invasion, enzyme production is likely to be advantageous in penetrating through the tissues and basement membranes. A range of enzymes probably act in concert during the invasive process, and their activities are no doubt modulated by local levels of enzyme inhibitors. Many enzymes are likely to act close to the invading cell surface, since the tip of an extending pseudopodium is probably a crucial site for invasion at the cellular level. This has led to the proposal that the degradative enzymes of invasion may actually be cell surface bound, and a receptor for u-PA has already been identified (Stoppelli *et al.*, 1985). Such a location could conceivably promote the possibility of invasion by concentrating enzymatic activity at the appropriate location, by reducing the effects of inhibitors, and by providing a suitable environment for maximizing the activation of proenzymic forms.

2.16 METASTASIS AND IMMUNE STATUS

Several authors have reported that the metastatic ability of tumour cells may be a function of the immune status of the host. Eisenbach, Segal and Feldman (1985), for example, found that clones of the Lewis lung carcinoma (3LL) and of the T10 sarcoma which varied in their metastatic capacity nevertheless all metastasized in irradiated hosts or nude mice. At least for the cells and the conditions employed in this study, even apparently non-metastatic clones may have the ability to invade the circulation and extravasate at a remote site; that they do not do so in normal hosts probably reflects the activity of a functioning immune system. This is an important experimental observation which should not be overlooked. We shall see later that a standard approach when searching for genes which might control metastatic behaviour has been to test for metastasis in nude mice (section 6.1). However, not all tumours may metastasize more in immunodeficient animals. The formation of lung colonies following tail vein injection of B16 melanoma cells, for example, is greater in normal animals than in experimentally immunosuppressed animals (Fidler and Nicolson, 1978), and yet at least as good in three-week-old nude mice which lack both T cells and mature NK cells (Hart, Talmadge and Fidler, 1983). In the study by Hart and colleagues (1983) it was reported that K1735 tumour cells produced significantly more lung colonies in young nude mice than in adult sygeneic mice. It should be noted that many studies with nude mice (including the one above) have used allogeneic combinations since the nude genotype is not always available in the appropriate genetic background. The significance of allogeneic pairing in nude mice is not properly understood, although since they lack T lymphocytes a full immune response cannot be generated.

Experiments with radiolabelled cells have suggested that the increased formation of colonies in relatively immunosuppressed mice by some tumours is not due to enhanced trapping in lung capillaries, but rather to increased survival of the tumour cells (Hanna and Fidler, 1980) although even this does not guarantee the growth of overt metastases (Hart *et al.*, 1981). Following adoptive transfer in mice, normal spleen cells have been shown to have an anti-metastatic effect (as judged by inhibition of organ colonization) that is attributable to their NK cell content since this effect is lost when they are treated with anti-NK cell antibodies and complement (Hanna, 1982 a and b).

Activation of NK cells in three-week-old mice by i.p. injection of polyinosinic-polycytidilic acid (poly I:C) prior to tumour challenge can reduce the extent of lung coloniztion. Poly I:C is thought to exert its effects by inducing interferon production and release which then activates pre-NK cells present in the young mice. It appears that the effector cells in these experiments exert their functions relatively shortly (12–24 h) after tumour cell injection, and the suggestion has been made that NK cells may destroy tumour cells while they are still within the circulation (Riccardi *et al.*, 1979).

When adult mice are treated with a single dose of cyclophosphamide about four days before the i.v. injection of tumour cells, there is a marked but reversible enhancement of both lung tumour colonies and extrapulmonary colonies compared to controls (Hanna and Fidler, 1980). The effects of cyclophosphamide correlate with the kinetics of inactivation and recovery of NK cells following a single injection of the drug. Furthermore, the adoptive transfer of normal spleen cells to cyclophosphamide treated mice 24 h before tumour inoculation can reverse the metastatic enhancement. That this effect was due to the transfer of NK cells was shown by fractionation experiments in which spleen cells, depleted of other cell types such as macrophages, T cells and B cells, were still able to reverse metastatic enhancement in cyclophosphamide treated mice.

In a relatively early report, Gorelik *et al.* (1979) showed that cells obtained from a metastatic lesion of the 3LL carcinoma were more resistant to NK cell mediated cytotoxicity *in vitro* than cells from the primary tumour. Later, Gorelik, Feldman and Segal (1982) explored the role of NK cells in inhibiting metastasis further by selecting NK-cell-resistant 3LL variants *in vitro*. Although these authors found that increasing NK cell resistance paralleled enhanced lung colonizing ability, other studies indicate that NK cell resistance alone is insufficient to explain tumour growth and metastasis. Thus Hanna and Fidler (1981), for example, have found that metastases arising from the UV-2237 fibrosarcoma express a level of NK cell sensitivity comparable to that shown by their parent cells. Interestingly, these authors were able to select UV-2237 cells for resistance to NK cell mediated cytotoxicity, although their susceptibility to activated macrophages was unchanged. It was found that the resulting cell variants produced more lung colonies than did the parent cell line when injected i.v. into syngeneic recipients. Thus resistance to NK cell cytotoxicity can enhance the metasatic potential of tumour cells, but it is not necessary for metastasis to occur. Presumably, some NK sensitive cells are able to escape from being destroyed by NK cells and these can then give rise to metastases.

It is worth noting here that immunoselection has been proposed as a possible basis for metastatic spread (Fogel *et al.*, 1979). These authors put forward this proposal on the basis of two observations made with the 3LL Lewis lung carcinoma:

(a) Antigenic differences can exist within a given tumour.
(b) Differences in surface antigens can exist between a primary and its metastases.

The idea behind this proposal is that immunoselection of antigenic variants in the primary population can lead to the generation of cells with increased metastatic capacity. Although an intriguing concept, there is little hard evidence that immunoselection drives the metastatic process. Indeed, metastatic variants can be generated by a mechanism faster than that based on immunoselection (section 1.5.3).

Some reports have indicated that there may be a relationship between macrophage content and metastasis, with more metastatic tumours having a lower macrophage content (Eccles and Alexander, 1974). Support for this possible relationship between macrophage content and metastasis has come from the studies of Wood and Gillespie (1975). These authors found that the removal of macrophages from a non-metastasizing tumour resulted in metastases being able to develop after subsequent transplantation. There also appears to be a correlation between antigen shedding and metastasis, which may be related to the phenomenon of tumour escape. Studies by Kim (1970) and Kim *et al.* (1975) on a number of rat mammary tumours, for example, have shown that highly metastatic types can shed more antigens than poorly metastatic types. This conclusion was supported by the studies of Alexander (1974) and Currie and Alexander (1974) working with methylcholanthrene-induced rat sarcomas who showed that the shed antigens from metastatic cells inhibited lymphocyte mediated cytotoxicity.

3 *In vivo models for the study of invasion and metastasis*

3.1 GENERAL FEATURES OF MODEL SYSTEMS

Given the ethical constraints which apply to experimentation on humans, it is not surprising that a number of experimental models based on both cell and tissue culture and animal systems have been developed for research into the mechanisms underlying tumour cell spread. The main benefit of any model system is that it allows a complex, difficult or otherwise experimentally unapproachable problem to be studied, usually by modifiying, adapting or simplifying the particular problem in some way. Although such systems should be developed and operated under conditions designed to reflect the true situation as closely as possible, as simplified analogies, model systems can only approximate the true *in vivo* situation. It should be clear that some models can do this more realistically than others. Furthermore, the experimental analysis of particularly complex problems is likely to generate more than one model system, some of which may be designed to study only one particular aspect of the overall problem. It follows that due care needs to be employed in the interpretation of results obtained from model systems, particularly with repect to the limitations of the model and to any assumptions on which it may be based. Generally speaking, when complex processes such as metastasis are being studied it usually pays to have available a number of assays, some of which are complementary and others of which are designed to examine different aspects of the entire process.

There are a number of features which characterize a good model system, the hallmark of which is its correlative or predictive significance: can results based on a model system predict the *in vivo* outcome? In using model systems to assess the effects of drugs on invasive behaviour, for example, it is pertinent to remember that markedly different pharmacokinetics exist *in vivo* and that this seriously restricts the predictive significance of model systems in studies of this type. A good model need not be based on the human system to provide useful information, although clinical relevance may be increased if human cells and tissues are employed. Ideally, a model system should approximate the *in vivo* situation in some definable way so that its limitations, for example, are clear. It should be relatively simple, it must be accessible experimentally under defined and controlled conditions, and it should yield quantifiable results relatively quickly. It is of little value to employ a model system, no matter how intricate or impressive its design, if it fails to confer some advantage over existing approaches either *in vitro* or *in vivo* in nature. Finally, it is worth recalling the paradox that relates to all model systems: the more simple they become, the less likely they are to yield data of predictive significance.

Two types of model systems are available for the study of most clinical problems which are usually not accessible to experimental manipulation because of ethical constraints. The first type of model system we shall examine is largely *in vivo* based, and it makes use of the widespread availabilty of highly inbred experimental animals such as the laboratory mouse. The second type of model system is exclusively *in vitro* based, but it nevertheless may provide ground for the formulation of ideas of clinical relevance. Both model systems rely considerably on the use of transplantable tumour cell lines.

3.2 TRANSPLANTABLE TUMOUR CELL LINES

Studies of human tumours and their metastases are more or less restricted to their clinical history before, during and after treatment, and to their histopathology following biopsy or surgical removal. Given the ethical constraints associated with clinical research, it might be argued that the best way to study metastasis would be to wait for or induce primaries in experimental animals and then attempt to observe the behaviour of metastatic cells as secondaries develop. This is a commendable approach which has been adopted by a number of research groups, but it is subject to a good deal of variation. With chemical carcinogens, for example, each tumour will be different and there is no guarantee as to at what time and to what extent metastases will develop. Thus, naturally occurring or induced animal tumours offer little advantage over human tumours, other than that multiple biopsies may be performed and ethical considerations (although certainly not absent) can be less demanding. In an attempt to provide a degree of consistency in experimental studies, most research groups have turned towards tumour cell lines which can be propagated *in vitro* and which have a more predictable metastatic behaviour. The other major advantages of using cell lines in such work include the availability of the cells for world-wide studies and the possibility of growing the cells in sufficient quantities for detailed biochemical and molecular studies.

The use of cell lines does not significantly restrict the nature of the experiments which can be undertaken, but it does require caution if results are to be extrapolated to naturally occuring primaries and their metastases. Many cell lines, for example, may have altered considerably from when they were first isolated and indeed the actual act of growing them *in vitro* may lead to considerable changes in itself. Serial transplantation often leads to alteration in tumour immunogenicity, with the highly transplanted tumour bearing little relationship immunologically to the original primary, and if metastases are transplanted highly metastatic variants may soon emerge. Although efforts can be made to overcome these drawbacks, it should not be forgotten that all model systems represent analogies with inherent limitations.

In working with transplantable tumours it is essential to monitor the transplanting process itself. Transplantation is achieved using either of two principal methods: by inoculation with cells derived from lines propagated *in vitro*, or by the serial transplantation of tumours from animal to animal with or without an intervening *in vitro* step. *In vitro* passaging is likely to lead to changes in the original tumour, but these changes may become stable over an extended period of time. When tumours are transplanted as fragments from animal to animal it is possible to accidentally select for or against certain phenotypic characteristics. Transfer of pigmented fragments of the B16 melanoma, for example, produces uniformly black tumours while trasplantation of amelanotic fragments yields white tumours. Maintenance of the diversity of the original tumour would seem to demand the serial transfer of cell suspensions prepared from all zones of the tumour, although there is no need to pass through an *in vitro* growth stage (Fidler and Hart, 1981).

The use of cell lines short circuits the need to induce a primary, but for many experiments this still leaves a good deal of variation. The host is usually a healthy individual lacking the factors which would have contributed to the original development of the tumour, and the cell line is unlikely to contain the full range of cell types present in the primary. The process of inoculation is also likely to affect metastatic outcome and it is known that the site of inoculation can influence the subsequent growth and behaviour of tumour cells.

3.3 'ARTIFICIAL' AND 'SPONTANEOUS' METASTASIS

In models based on 'spontaneous' metastasis, tumour cells are injected (usually subcutaneously) into selected sites such as the footpad or the ear, and the development of metastases in specific organs is subsequently followed. Wood and his colleagues (1967a), for example, injected the Lewis C150 carcinoma subcutaneously into host animals and observed subsequent events which led to the establishment of lesions in the lungs. The injected tumour cells first invaded the venous system and then spread in the blood to lodge primarily in capillaries (but also in arterioles) within the lungs. The tumour cells became associated with a delicate meshwork of fibrin and platelets (with some leukocytes) which presumably aided in their entrapment. Intravascular growth after lodgement was followed by extravasation and subsequent establishment of the secondary.

It should be made clear in the first instance that not all malignant cells display similar behaviour following s.c. injection. Even with this caveat, however, this is hardly a perfect model of the natural progression of a tumour. Nevertheless, as an experimental approach it does have advantages in that it provides some control over the development of metastases which is lacking in truly natural circumstances. Since it usually takes some time for metastases to develop in these types of model system, it is occasionally necessary to prolong the life of the host which would otherwise die or be seriously perturbed by the effects of the primary. This is usually achieved by surgical removal of the primary through, for example, resection of the lower part of the limb or removal of the ear.

In order to avoid complications associated with surgery, some *in vivo* models of metastasis, particularly those concerned with haematogenic spread, have gone one step further by introducing cells directly into the blood stream. This offers a very useful experimental situation since tumour cells can be injected into either the venous system or to the pulmonary or systemic branches of the arterial system. In this way the effects of various organs or capillary beds on metastatic outcome, for example, can be studied experimentally. However, studies in which tumour cells are introduced directly into the blood stream do not model fully all of the steps in the metastatic cascade, and for this reason many research groups refer to this experimental system as 'artificial metastasis'. The terms **artificial metastasis** and **colonization** are used throughout this text to refer to the process by which tumours develop following direct injection into the blood or lymphatic systems, while the term spontaneous metastasis refers to tumour development following injection into tissues. As will be expanded upon below, there is some doubt that tissue injection at different sites (e.g. the flank or footpad) is an equivalent process. The term **metastasis** is used to refer to the general process of tumour spread, be it from a naturally occurring or induced malignancy, or after the injection of malignant cells by any route. Again, as will be discussed in more detail later, it seems that colonizing ability is not necessarily predictive of spontaneous metastatic behaviour.

In normal metastatic situations the host will have one or more primaries which will affect not only its defence systems but also its general physiological condition, and this is likely to have considerable impact on the metastatic outcome. It is known, for example, that primary tumours can induce repression of metastatic growth. Thus direct injection of tumour cells into the blood system represents an acute model of what is for most tumours a chronic situation. Many tumours progress *in vivo* from a non-invasive state to an invasive and metastatic state, and this progression is lacking when tumour cells already endowed with metastatic capability are employed experimentlly. Direct injection into the bloodstream also avoids the early steps in metastasis, namely local invasion, escape from the primary, and penetration of blood vessel walls. Furthermore, the direct injection into the blood system of tens of thousands of tumour cells suspended in as much as 0.2 ml of physiological saline is a traumatic experience for the experimental animal, particularly if it is a small rodent such as the mouse. Fidler, Gersten and Riggs (1977) showed that the i.v. injection into syngeneic mice of 50–100 000 B16 cells induced a rapid drop in circulating leukocytes (**leukopenia**) which lasted less than 30 min. This leukopenia was followed by a more prolonged increase in circulating leukocytes (**leukocytosis**). The initial leukopenia induced by B16 cells was seen whether the injected tumour cells were alive or dead, but it was not so marked if normal embryonic cells were injected. Histological observation showed that the tumour-induced leukopenia was due to the accumulation of leukocytes in the lungs. A drop in circulating platelets (**thrombocytopenia**) associated with their lodgement in the lungs also occurs immediately after the i.v. injection of B16 tumour cells (Gasic *et al*., 1973). The significance of these observations on lung colonization is not fully understood, but it is clear that caution is required when extrapolating from the lung colonization technique because of the effects of direct i.v. injection on blood components. Interestingly, Fidler, Gersten and Riggs (1977) showed that B16 immunized mice which can reject an s.c. tumour challenge nevertheless allow tumours to grow in the lungs following i.v. injection. This observation indicates that there are marked differences between the two experimental techniques. Clearly, different types of models are required to explore the various aspects of the metastatic cascade properly.

It is known that the number of lung colonies which result from i.v. injection is not linearly correlated with cell input, but when viable and X-irradiated tumour cells are injected together, there is less deviation in the number of viable injected cells required to form each resulting colony (Hart, Talmadge and Fidler, 1983). The effects of s.c. injection of either viable or dead B16 melanoma cells, or B16 melanoma cell products, on lung colonization after i.v. injection of viable tumour cells are somewhat different than the effects of simultaneous i.v. injection described above. According to Kalish and Brady (1980), this process results in an increase in the incidence of lung colonies. Presumably this process of faciltation acts through effects on the host's immune system, which actually may be altered to either enhance or inhibit tumour growth (section 1.8.1).

Stackpole (1981) used B16 melanoma cells to test whether the direct injection of tumour cells into the bloodstream was predictive of spontaneous metastatic potential from subcutaneous transplants. Syngeneic mice were injected at different sites with B16F10 cells and an area around the injection site was surgically removed after 1 h. Three weeks later, lung colonies were found in mice injected intravenously (i.v.) in the lateral tail vein, intramuscularly (i.m.) in the footpad, intradermally (i.d.) in the external ear, and subcutaneously (s.c.) in the thigh. No lung colonies were found, however, after injection s.c. into the abdominal flank. This particular site had relatively few microscopically visible blood vessels, suggesting that the colonies which formed at the other sites could have resulted from gaining direct access to damaged blood vessels (as is obviously the case with tail vein injection). This is an important observation, since i.m and s.c. inoculation are thought by many researchers to model metastasis more closely than i.v. injection: in fact, all routes of inoculation could conceivably result in direct access of tumour cells to the blood, although the percentage of cells gaining access will obviously be less after i.m or s.c. injection than after i.v. injection.

When the B16F1, B16F10 and B16BL6 variants were tested for colonizing ability (following tail vein injection) and metastasizing ability (following s.c. injection into the abdominal flank), their colonizing ability could be ranked B16F10>B16BL6>>B16F1 (Stackpole, 1981). However, there was no difference in their metastasizing ability, with all three lines metastasizing poorly. If these variant lines are poorly metastatic (or perhaps even not metastatic at all when tested under rigorous experimental conditions), then many studies based upon them and purporting to relate to metastatic behaviour may be misleading. Although the B16 variants referred to above are poorly metastatic following s.c. injection into the flank, Stackpole (1981) was able to obtain other variants which metastasized under these experimental conditions. He was able to confirm with these cells that there was no apparent relationship between the ability to colonize the lungs and the ability to metastasize there. It is important to recognize here that colonization refers to the ability to form lung tumours after tail vein injection, whereas metastasis refers to the ability to form lung tumours after s.c. injection specifically into the flank, a procedure claimed by Stackpole (1981) to test for true metastatic ability. Perhaps not surprisingly, Stackpole (1981) also showed that the standard of surgical removal of tissue at the inoculum site affected the metastatic outcome, with careless surgical excision yielding relatively more lung colonies than careful excision.

Bresalier and colleagues (1987) recently have described a novel model system for studying the spontaneous metastasis of colon cancer. These authors used the 51B cell line derived from a 1,2 dimethylhydrazine induced colon carcinoma of Balb/c mice. Cells from this line were injected into the caecal wall of syngeneic mice and cells from metastatic lesions were isolated and reinjected into the caeca of other host mice. This cycling process was repeated five times to yield variant tumour lines (LiM-1 to LiM-5) with generally enhanced ability to spontaneously metastasize to the liver after caecal injection. Bresalier's group (1987) then subjected the metastatic variants to a battery of *in vitro* tests designed to explore various phenotypic properties possibly associated with metastasis. They found that the ability of the 51B tumour cell variants to invade a reconstituted basement membrane and to secrete type IV collagenase was directly proportional to their metastatic ability. Furthermore, the liver metastasizing variants showed chemotactic behaviour in a Boyden chamber system towards liver extracts (as compared to extracts from brain or lung) and the variant cells adhered more to hepatic sinusoidal endothelial cells relative to hepatocytes. Each one of these behavioural characteristics is likely to contribute to the outcome of the metastatic process *in vivo* and thus could conceivably explain the liver metastasizing ability of the selected cell variants. This possibility will be discussed in more detail later in association with the B16 malignant melanoma, since numerous laboratories have attempted to correlate various features of the *in vitro* behaviour of B16 cells with their organ colonizing ability (section 3.4).

Studies with other model systems, such as the TS/A murine adenocarcinoma (Nanni *et al.*, 1986) have confirmed that there is no correlation between colonizing ability and spontaneous metastatic capacity. The TS/A cell line contains high and low spontaneously metastatic populations, clones of which have epithelioid (E) or fibroblastic (F) morphology respectively. Despite the existence of such clones in the TS/A population, cyclic selection procedures failed to yield cells which displayed spontaneously metastatic behaviour in excess of that shown by the parental cells. The selection procedure involved s.c. injection of parental TS/A cells into the right hind leg followed by excision of nodules from the lungs, *in vitro* expansion, and repeated injection for a total of 10 such cycles. These results are clearly different from those reported by Bresalier *et al.* (1987) described above, indicating how different experimental model systems initially designed to test the same phenomenon (spontaneous metastasis) can yield markedly different results. Nanni *et al.* (1986) also used an alternative selection procedure based on i.v. injection into the tail vein which was very similar to that employed by Fidler (1973a) in his selection of the B16F1 and B16F10 melanoma variants. Using this selection procedure, variant TS/A cells were obtained which showed enhanced lung colonizing ability after one cycle. Further cyclic selection (up to 10×) failed to improve the ability of these cells to colonize the lungs. More importantly, such cells failed to display increased spontaneous metastatic ability even though they showed enhanced colonizing ability, and thus the two behavioural patterns appear to be independent of each other to some degree. It would seem that the cyclic procedures used by Fidler (1973a) and by others emphasize selective phenomena, and while this does not necessarily denigrate from their use as model systems it does caution us on any extrapolations we may wish to make. Since the B16F1 and B16F10 variants have been selected using colonization procedures, since there is no correlation between spontaneous metastasis and colonization, and since these variants metastasize poorly in any case, it is perhaps not surprising that many studies attempting to explain aspects of metastasis using these cells have failed to meet their expectations. Nevertheless, as we shall see in the following section, there is little doubt that the multitude of studies based on this experimental system have had considerable impact in the general area of experimental cancer biology.

3.4 THE B16 MALIGNANT MELANOMA: A MURINE MODEL SYSTEM FOR THE STUDY OF INVASION AND METASTASIS

As discussed above, many aspects of the growth and spread of tumours can be studied *in vivo* using spontaneous, autochthonous tumours, but their relatively low incidence and their variability from individual to individual serve to restrict their practical use. Transplantable tumours, on the other hand, offer a more repeatable system for experimental analysis and it is this feature which has made them the system of choice in many studies. There are a large number of transplantable tumours available, some with special characteristics which make them suitable for particular experimental situations. It would not be possible to discuss all the transplantable tumours in detail and we shall concentrate here instead on one particular tumour, the B16 malignant melanoma of C57Bl6 mice.

3.4.1 Advantages of the B16 melanoma

The key advantages of the B16 melanoma as a model system relate to those typical for other transplantable tumours: it is easy to work with, and it offers repeatability of experiments under definable conditions (Hart, 1982b). The tumour, which arose spontaneously in a C57Bl/6J mouse, was first described by the Jackson Laboratory in 1954 (see Green, 1968). Such a tumour, which has outlived its original host by over 18 generations, may not seem to represent the model system of choice since the original tumour is likely to have undergone considerable change over this period of time. Hu and Lesney (1964) succeeded in establishing two cell lines from the B16 melanoma, but as will be seen, it was the selection of the so-called 'metastatic variants' by Fidler (1973a) which resulted in the boost of popularity of this particular transplantable tumour over many others.

3.4.2 Applicability

Like all animal model systems, the question immediately arises as to whether the B16 malignant melanoma is an appropriate model system for studying either specific aspects of human melanomas or the more general features of human malignant tumours, or both. Even after three decades of research with the B16 melanoma, this question cannot be answered satisfactorily since many properties of human tumours cannot be adequately studied in the clinical situation. Although it is clear that the B16 melanoma has generated a level of research activity and hypothesis which renders it one of the most thoroughly studied tumour systems, we thus still remain uncertain as to whether human malignancies can be described in a parallel manner. As will be seen, the accumulation of knowledge on one system does not necessarily render it any more relevant than what it was initially.

3.4.3 Nature of the host

Many aspects of tumour growth are likely to be affected by the nature of the host, some of which are referred to below using the B16 melanoma as an example.

3.4.3.1 *Effects of immune status and stress*

The immune response against a tumour is likely to vary as a function of several parameters including the age, sex, and health of the host, and the actual site of tumour growth. Environmental stress can influence both the immune response and tumour growth, possibly through effects mediated in some way by corticosteroids (Riley, 1981), which are generally considered to be immunosuppressive. Other substance such as the endorphins and enkephalins, for example, are also known to be released under stressful conditions. There is considerable evidence that these peptides can influence *in vitro* immune functioning through the enhancement of natural killer (NK) cell activity (Kay, Allen and Morley, 1984), and in this way they may counteract the more deleterious effects of the corticosteroids. Murgo (1985) has shown that methionine-enkephalin (up to 50 μg per mouse per day s.c.) can markedly inhibit local tumour growth of B16BL6 cells following their s.c. injection into the flanks of syngeneic mice. Interestingly, the course of treatment had to begin after tumour inoculation, since met-enkephalin can enhance tumour growth if injected before the malignant cells. This specific example illustrates almost classically a key problem which emerges when studying tumours, namely that aspects of their behaviour can change markedly with relatively subtle alterations in experimental conditions. Great care indeed is required if misleading interpretations are to be avoided, and this is a point that the reader should continually bear in mind. Although they affect immune function, exactly how agents such as met-enkephalin may alter the course of malignancy is not clear. Indeed, the endogenous opioids can influence the release of a range of hormones *in vivo* thereby affecting many processes likely to impact upon tumour development.

The interferons (IFNs) represent a group of cellular proteins which can influence various aspects of immune activity including NK cell and macrophage functions. Additionally, some IFNs may be directly cytotoxic for certain tumour cells. There are three main catgories of IFNs (IFN-α, IFN-β and IFN-γ) whose effects on tumour development appear to differ. Sakurai *et al.* (1986) have shown that previous i.p. inoculation of recombinant murine IFN-β can significantly inhibit lung colonization by i.v. injected B16 melanoma cells. These authors showed that the IFN-β used in their study was not directly cytotoxic to B16 melanoma cells, but that it augmented NK cell activity. The inhibitory effect of IFN-β on B16 lung colonization was reduced by pretreatment of experimental animals with carrageenan, an agent known to be cytotoxic for macrophages. This latter result suggests that macrophages too may be involved in the tumour colonization inhibitory activity of IFN-β (see below). It should be noted, however, that in this study the IFN-β was added before the tumour cells, which themselves were injected directly into the blood stream and such conditions are far removed from the clinical situation.

Considerable interest, both experimentally and clinically, is currently being expressed in the therapeutic use of interleukin-2 (IL-2), which is now available in large quantities in recombinant form. IL-2, originally designated as T cell growth factor (TCGF), is a 15 kD glycoprotein secreted by helper T cells which acts primarily as a second signal for the proliferation of antigen primed T lymphocytes. It seems that IL-2 actually has more diverse effects than first thought, including stimulation of the cytotoxic activity of both NK cells and lymphokine-activated killer (LAK) cells, although this situation remains to be clarified (section 1.8.2). Saijo *et al.* (1986) have demonstrated recently that treatment of mice with IL-2 can reduce the extent of lung colonization by B16 melanoma cells. They injected syngeneic mice with 5×10^5 B16 cells, and 24 hr later they began daily s.c. injections of high doses of human recombinant IL-2 (5 or 10×10^4 IU per injection). The outcome of these experiments was that fewer tumour colonies were found in the livers and lungs of experimental animals than controls. Apart from the high doses that were necessary, recombinant IL-2 was far from curative since no experimental mice were actually without lung colonies. These studies need to be extended to include spontaneous metastases of a wider range of tumours before they yield information which may be of interest to clinical studies. Nevertheless, as demonstrated earlier by Warner and Dennert (1982), NK cells can markedly inhibit lung colonization by B16 melanoma cells and, as shown by Riccardi *et al.* (1979), this activity can be manifested in the circulation. The strategy adopted by Warner and Dennert (1982) was to adoptively transfer a cloned cell line with NK cell activity into NK deficient syngeneic mice. Under these conditions, the lung colonizing ability of the B16 melanoma cells was markedly inhibited relative to controls. In other experiments, Hanna (1980) had shown that the experimental metastatic capacity of B16 cells was enhanced in three-week-old nude mice (which are low in NK cells) relative to six-week-old nude mice (which are relatively high in NK cell activity). The point to be emphasized here, however, is that not all tumour cell types are equally susceptible to NK cell mediated cytotoxic effects. Indeed, cells within any given tumour seem to vary in NK cell sensitivity and it is entirely possible that NK resistant phenotypes might be more likely to produce metastatic lesions.

Other cytotoxic effector systems involving PMNs and macrophages, for example, also influence tumour development (section 1.8.2). PMN involvement in the inhibition of lung colony formation has been suggested by the studies of Glaves (1983), although their i.v. injection may actually lead to an increase in colony formation (Starkey *et al.*, 1984b). In a rather unusual experiment, Ishikawa *et al.* (1986) induced a marked granulocytosis in C57Bl6 mice by subcutaneously transplanting a syngeneic fibrosarcoma and subsequently challenged the recipients i.v. with B16 melanoma cells. It was found that the lung-colonizing ability of B16 cells was enhanced in these granulocytic mice, but some caution is necessary in interpreting these results because of the nature of the model system employed. Macrophages are commonly found in primary tumours and their metastases, where they may have a dual role depending on their state of activation. Normal macrophages appear to show only limited cytotoxic acivity against tumour cells, and they may in fact release factors conducive to the growth of certain tumours. Once activated or 'armed' by lymphokine containing supernatants from sensitized T cells, however, macrophages display heightened anti-tumour activity. Fidler (1975 a and b) reported that release of the lymphokine macrophage activating factor (MAF) from lymphocytes of mice bearing the B16 melanoma progressivly declined after tumour inoculation. This decline in MAF release as detected by *in vitro* activity assays was tumour specific, was clearly a function of tumour mass, and presumably lessens the anti-tumour defence systems of the host. Conversely, it has been reported that the mass of a primary tumour can be related to the degree of inhibition of metastatic growth (Sugarbaker and Ketcham, 1977). Of course, it is not really known to what extent these phenomena reflect local defence activities, let alone macrophage involvement, and this illustrates again our considerable ignorance in this particular area. When activated macrophages were injected i.v. into host mice 48 h after the injection of B16 melanoma cells, a significant reduction in the development of lung colonies was noted (Fidler, 1974 a and b). Similar effects were not seen after i.p. injection of the activated macrophages, possibly indicating a need for cell contact between the macrophages and their tumour cell targets.

B16 variant cells have been selected for resistance to a variety of cytotoxic processes including drugs, antibodies, T cells and NK cells, but all variants apparently show some susceptibility towards macrophage mediated lysis, at least in *in vitro* assay systems (Fidler, 1980). Perhaps not surprisingly in the light of these studies, activation of macrophages is often associated with increased resistance to tumour growth (Bast *et al.*, 1975) and conversely, the impairment of macrophage function may lead to the promotion of tumour growth (Keller, 1976). Fidler and his colleagues have shown that the incorporation of muramyl dipeptide (MDP) or its derivative muramyl tripeptide (MTP) into liposomes (multilamellar lipid vesicles) can activate the tumouricidal activity of macrophages (Sone and Fidler, 1981; Key *et al.*, 1982). MDP was originally isolated from *Mycobacterium* but its structure is now known, and in fact both macrophage activating peptides are available in synthetic form. Key *et al.* (1982) incorporated MTP conjugated to phosphatidylethanolamine (MTP-PE) into liposomes formed from phosphatidylcholine and phosphatidylethanolamine (7:3 molar ratio). When these modified liposomes were repeatedly injected (twice weekly for four weeks) into mice bearing spontaneous B16BL6 metastases, 70% of experimental animals were alive and disease-free after 180 days, whereas only 10% of control animals were still alive. The use of fluorescently labelled liposomes showed that these vesicles were first phagocytosed by circulating monocytes and alveolar macrophages, which subsequently extravasated into the organ parenchyma. Within 24 h of i.v. injection of labelled liposomes, 5% of macrophages obtained from lung metastases were shown to contain phagocytosed liposomes. Such macrophages were actively tumoricidal only when the liposomes incorporated MTP-PE. It would seem from these results that there is a strong correlation between the liposomal activation of macrophages and tumour regression, but the clinical potential of this possible form of therapy remains to be fully evaluated. Perhaps an indication of the potential of macrophage therapy in tumour treatment comes from studies which show that although variants of the B16 melanoma can be selected for resistance to lymphocyte (Fidler, Gersten and Budmen, 1976) or NK cell (Hanna and Fidler, 1981) mediated cytotoxicity, it has not so far been possible to select for macrophage-resistant variants even though the B16 line is presumably heterogeneous in this respect. There are at least two possible explanations for the difficulty in selecting macrophage-resistant tumour cell variants. The first relates to the fact that macrophages can release a barrage of cytotoxic effector molecules and the chance of developing resistance to all of these is correspondingly low. The second possibility relates to the way macrophages detect the presence of tumour cells. Although macrophages can sense and respond to altered proteins and carbohydrates on cell surfaces, Fidler and his colleagues have suggested that changes in tumour cell membrane phospholipid expression might be more important in the recognition of altered-self by macrophages (reviewed in Fidler and Schroit, 1988). In their view, this possibility explains experimental observations that macrophages recognize and destroy tumour cells seemingly regardless of their different phenotypic characteristics. It is also consistent with the data on the uptake of liposomes by macrophages referred to above. Interestingly enough, this possibility may also explain the process underlying the macrophage-based removal of effete red blood cells. In healthy erythrocytes, phosphatidylserine is restricted to the inner lipid leaflet of the membrane. Perturbation of this arrangement through cell damage or as a consequence of ageing might promote erythrocyte removal by the reticuloendothelial system. Indeed, the expression by tumour cells of negatively charged phospholipids such as phosphatidylserine seems to be more important in directing macrophage cytotoxicity than do changes in the expression of positive or neutral phospholipids. The nature of the postulated link between transformation and altered phospholipid expression leading to macrophage-directed tumour cell destruction is unknown.

The role of lymphocytes and immunity in general in tumour growth and metastasis has been controversial since the experiments of Prehn (1972) in which he demonstrated that immunity could either stimulate or inhibit tumour growth. Fidler, Gersten and Riggs (1977) have shown that more B16 cells arrest in the lungs of mice sensitized to this tumour than they do in the lungs of normal mice. The numbers of resulting lung colonies, however, were usually greater in normal mice relative to sensitized animals, presumably reflecting heightened reactivity in the latter. Earlier experiments by Fidler (1974 a and b) had shown that X-irradiation and thymectomy tended to inhibit the formation of lung colonies and this effect could be abrogated by treatment with lymphocytes derived from tumour-bearing animals. At first sight, it might be thought that immune suppression of this type would lead to increased lung colony formation, but the opposite proved to be the case. It is perhaps worth recalling here that it was in an attempt to explain the somewhat confusing results of immune manipulation that Prehn (1972) originally formulated his hypothesis. Fidler (1974 a and b) also reported that injection of B16 cells with relatively high numbers of syngeneic lymphocytes yielded more lung colonies than tumour cells injected alone or with fewer lymphocytes. One possible interpretation of these latter results is that heterotypic tumour–lymphocyte emboli enhance the possibility of lung colonies developing, possibly through increased arrest within the narrow lung capillaries. Essentially similar results were found when heterotypic *nu*/+ mice were treated with large numbers of lymphocytes prior to tumour inoculation, but not in their homotypic (*nu*/*nu*) nude littermates (Fidler, Gersten and Riggs, 1977). Since both *nu*/*nu* and *nu*/+ lymphocyte treated mice displayed similar profiles of initial cell arrest, it is clear that this attribute neither correlates with nor predicts lung colonization. In order to put these results into some sort of perspective, it is important to recall that we are dealing here with a very artificial system based on the i.v. injection of *in vitro* cultured cells, often into experimentally manipulated animals. To what extent many of the attributes tested actually relate to spontaneous *in vivo* metastasis can only be alluded to.

Irrespective of the immune status of the host, Fidler, Gersten and Riggs (1977) found that B16F10 cells always adhered and survived more in the lungs of host animals than did B16F1 cells. Interestingly, mice which could reject the B16 melanoma after s.c. inoculation failed to reject the same tumour if it was given intravenously. Like many other of the observations reported here, this result highlights the complexity of tumour pathogenesis, and it indicates that more than immunity alone may be necessary for the suppression of tumour growth *in vivo*.

Allogeneic experiments, in which tumour cells are injected into histoincompatible recipients, would seem to confuse what is already a very complex field. Indeed, it was the early use of histoincompatible hosts which resulted in a considerable misunderstanding of the immune response against tumours and we shall not pursue this avenue further. However, there is some evidence that the MHC (major histocompatibility complex) influences tumour growth and metastasis. Haywood and McKhann (1971), using murine sarcoma cells, were amongst the first researchers to show that the MHC can influence metastasis. According to their study, tumour cells expressing high MHC antigen levels had a tendency to spread to the lungs when compared to tumour cells with low MHC expression. Now the B16 melanoma is of H-2^b haplotype and it usually expresses MHC class I antigens (K, D and L), whereas MHC class II antigens appear to be lacking (Heath and Boyle, 1985). It is possible, however, to select for variants with high or low MHC expression which originate from the same tumour mass, and this was accomplished by Taniguchi, Karre and Klein (1985) using a fluorescent activated cell sorter (FACS). The low MHC variant (H-2^-) is deficient in lung colonizing ability when compared to the high MHC variant (H-2^+), and experiments have shown that this is largely due to their increased susceptibility to NK cell mediated lysis (Kawano *et al*., 1986). It follows from these studies that the B16 melanoma is heterogeneous with respect to MHC antigen expression, although precisely how this correlates with the metastatic capability of the various cell lines selected by Fidler (1973a) and others is not certain. One reason for this dilemma is that whereas NK cells attack H-2^- variants, H-2^+ types are subject to attack by LAK cells and cytotoxic T cells, and thus the eventual outcome will reflect the relative efficiencies of the different processes. In this respect, Toshitani *et al.* (1987) have provided evidence that the inhibition of lung colonization by H-2^- B16 cells in normal hosts is NK-dependent, whereas inhibition of lung colonization by H-2^+ B16 cells in NK-depeleted hosts is dependent on the presence of LAK cells. It might be expected from the above that either high or low levels of H-2 antigens would correlate with reduced metastatic potential. Eisenbach and colleagues (1983, 1984) have shown metastatic potential to be correlated with the expression of an imbalance in K and D products (low H-$2K^b$ and high H-$2D^b$), but the precise mechanisms involved as well as the generality of this observation are currently not clear. Interestingly, down regulation of MHC gene products is associated with transformation by a number of oncogenes including *myc* and *ras*, but while protecting against one host defence mechanism this process would open up the transformed cell to attack by another.

3.4.3.2 *Effects of age and sex*

The age of recipient animals has an effect on the course of metastatic disease. Generally speaking, the onset of sexual maturation in animals is associated with a decline in immune function, particularly in T cell mediated processes (Makinodan and Kay, 1980). This decline is presumed to account at least in part for the increasing susceptibiity of older animals to infection, autoimmunity and tumour development. However, in experiments designed to test the effects of age on B16 tumour growth and host survival, Ershler *et al.* (1984) revealed an apparent paradox. These authors found that tumour growth after s.c. injection into the flank of B16F10 cells was actually slower in old mice (24 month) than in younger mice (3 month). Furthermore, after i.v. injection of B16F1 cells the number of lung colonies was less and host survival was greater in the old mice. There are several possibilities which might help explain this paradox. It is known, for example, that the B16 melanoma is only weakly immunogenic and thus any decline in host immunity with age may not relate to tumour growth in this system. This implies that some host factor other than T cell mediated host immunity (e.g. NK cell activity) is involved in controlling tumour growth. Another possibility is that defects of the vascular bed, often seen in older animals (Rockwell, 1981), may limit tumour growth by restricting the amount of tumour vascularization. Indeed, Ershler *et al.* (1984) reported that the tumours removed from the old animals were less vascularized than those from the young ones, but it must be admitted that which (if any) of the explanations above may be more likely is unknown.

The studies of Hanna (1980) referred to earlier in which young (3 wk) syngeneic mice were found to yield more lung colonies after B16 injection than older ones (6 wk) are thought to reflect age dependent changes in the host defence system, here the development of NK cell competence. Nevertheless, other changes (perhaps hormonally related) cannot be rigorously excluded.

Of interest here is the work of Pierce and his colleagues which has shown that fewer tumours than expected develop in the skin of the back following the injection of unselected B16 melanoma cells into this site in ten-day-old mouse fetuses. Similar results were found when melanoma cells were injected into the skin of the developing limb bud of fourteen-day-old mouse fetuses (Gerschenson *et al.*, 1986). The inhibitory effects on tumour growth at each specific site were noted at specific embryonic ages which correlated with the time of arrival into each site of migrating neural crest cells. Conditioned medium from cultured embryonic limb buds was found to be cytotoxic for B16 melanoma cells, an effect now attributable to the generation of lethal products such as H_2O_2 during polyamine oxidation (Parchment and Pierce, 1989). In conditioned medium a large proportion of the polyamine oxidase which catalyzes the oxidation reaction stems from the serum supplement, but a similar enzymatic activity has been shown to be present in limb bud homogenates. Saunders (1966) has assembled evidence in support of considerable cell death during limb development, particularly between the digits as these are formed from the elongating limb bud, and it has been argued by Parchment and Pierce (1989) that this could be a consequence of endogenous limb bud polyamine oxidase activity. Indeed, the presence of the enzyme could conceivably lead to selective or localized cytotoxic activity and this behaviour could account for the destruction of melanoma cells in both the developing limb and back skin, but further evidence in support of the hypothesis is required. It should be emphasized that as yet there is no evidence that invading neural crest cells are actually the prime source of the developmentally regulated endogenous polyamine oxidase.

Clinical data suggest that melanomas grow more slowly in females and this seems to be the case with the B16 murine melanoma system. According to Proctor *et al.* (1976) the B16 melanoma grows more slowly and produces fewer metastases when implanted into the hindlegs of female C57Bl6 mice compared to male mice. At least as far as melanomas are concerned, it would seem from these observations that different hormone levels found in males and females affect the outcome of the malignant process, but other studies (involving the s.c. injection of B16F10 cells into the flank) failed to show any effects of sex (Simon and Ershler, 1985). Hart (1982 a and b) actually found that female mice developed *more* lung tumour colonies than male mice following tumour cell injection via the lateral tail vein. Furthermore, he also reported that castration of male mice resulted in an increase of lung colony number in excess of that shown by normal females. These results suggest that hormonal effects on tumour growth may be influenced to some degree by the route of tumour implantation (see below).

3.4.3.3 *Site effects*

The site of tumour implantation seems to contribute to the malignant process (Auerbach *et al.*, 1978). The reasons for this are not immediately clear, but they presumably reflect differences in the blood supply to particular sites and the nature of the local environment (e.g. the abundance of host defence cells).

There is also a site effect on metastasis which is reflected in the tendency for certain tumours to metastasize to particular organs. This possibility is discussed more fully in section 2.14.6, and seems likely to result from a multitude of events chief among which are selective adhesion and growth. With respect to the B16 melanoma, it seems that some variants adhere specifically to particular organs (section 3.4.8.2), while others respond by increased growth in the presence of extracts from particular organs. Both of these phenomena correlate to a degree with the colonizing abilities of B16 melanoma variants. In terms of selective growth, Nicolson and Dulski (1986) have reported that lung colonizing B16F10 cells and ovary colonizing B16-O10 are stimulated by lung- and ovary-conditioned media respectively. Earlier work on the B16F10 variant reported by Hart (1982 a and b), however, was less definitive.

3.4.3.4 *Effects of health and diet*

Rous (1914), of viral fame, was one of the first researchers to note that tumour growth may be slower in underfed animals. Since that time several workers have shown that experimental animals fed a calorie restricted diet (usually 60–85% of the normal intake) may have a more active immune system, develop fewer spontaneous tumours, and live longer than control animals (Jose and Good, 1973; Weindruch and Walford, 1982). Ershler, Berman and Moore (1986) have shown that local tumour growth of the B16F10 melanoma is slower in calorie-restricted mice compared to control animals, whereas lung colonization of B16F1 cells is actually enhanced. While at first sight these results might seem extraordinary (even though different B16 variants were used for the local growth and colonization experiments), they probably only reflect the observation that local tumour growth and lung colonization represent different facets of malignancy.

Various nutritional studies have implicated an inverse relationship between high cholesterol diet and metastasis in experimental animals, and there is some evidence that cancer incidence may be elevated in humans with low serum cholesterol levels (Sorlie and Feinleib, 1982). Studies with B16F10 cells have shown that their binding to bovine endothelial cell monolayers can be inhibited by treatment of the monolayers with agents which inhibit cholesterol sysnthesis (Ramachandran *et al.*, 1986) although the clinical significance of these results remains to be evaluated.

3.4.3.5 *Effects of trauma and surgical intervention*

It is well known that tumour massage and other forms of trauma can enhance the possibilities of metastatic development. Weiss *et al.* (1982) developed a simple roller technique to reproducibly massage i.m. implanted B16 tumours. They reported that massage increased the metastatic capability of growing tumours of unselected B16 cells as well as those of the B16BL6 invasive variant, but not those of either the B16F10 or the B16F10LR6 variants. Unfortunately, the significance of this intriguing difference between B16 variants is not clear.

Mead and her colleagues (1985) have reported on an experimental study in which they examined the effects of wide excision of the primary (amputation below the knee) and wide excision plus prophylactic lymphadenectomy (removal of femoral and popliteal lymph nodes) following injection into the footpad of either B16F1, B16F10 or B16Bl6 cells. There is some evidence that human melanoma patients with relatively thin primaries may be cured by wide excision, that patients with primaries of intermediate thickness have a relatively low risk of blood-borne metastases after wide excision plus prophylactic lymphadenectomy, but that patients with relatively thick primaries gain little therapeutic benefit from these surgical processes (Balch *et al.*, 1979). In their experiments with the B16 melanoma in mice, Mead *et al.* (1985) noted for all three tumour variants that excision and lymphadenectomy conferred a significant survival advantage over excision alone, but unfortunately this was not simply related to size of the primary. This latter result may be attributable in part to the small number of experimental animals used in this report and it is to be hoped that the study will be extended since this will give us further insight into the clinical relevance of the B16 melanoma as a model system. Additionally, it is important to note that the growth of an experimentally injected melanoma cell is significantly different from that exhibited by a naturally ocurring, spontaneous melanoma. Nevertheless, histological analysis of the excised nodes generally indicated increased involvement with increasing size of the primary. Furthermore, there appeared to be greater involvement of the nodes in animals bearing the B16BL6 line than in those bearing either the B16F1 or B16F10 lines. As will be seen in due course, the B16BL6 melanoma line was originally selected for its invasive capability (section 3.4.5.3).

3.4.4 Mechanisms and patterns of B16 spread

Many variants of the B16 melanoma appear to metastasize poorly, and of those that do metastasize a number show considerable organ selectivity. This is in stark contrast to the clinical situation, since human malignant melanoma may metastasize widely and it usually shows a relatively small degree of organ selectivity (typically to the lungs and brain). This would seem to raise serious doubts about the validity of using the B16 malignant melanoma as a model system for studying human malignancy. However, most studies with the B16 melanoma have not been extended to include the full history of progression of this experimental tumour. Alterman, Fornabaio and Stackpole (1985) carried out a thorough study of the progression of the B16 melanoma using slow (G3.5) and fast growing (G3.12) metastatic variants inoculated s.c. into the flanks of syngeneic recipients. These metastatic variants were derived from a clone (G3) of tumorigenic but non-colonizing and non-metastatic 'null' B16 cells (Stackpole, Alterman, and Fornabaio, 1985) which underwent phenotytpic diversion *in vitro* (G3.5) and *in vivo* (G3.12). Most mice that died from disseminated disease after s.c. inoculation of these two variants did so as a consequence of respiratory failure due to excessive lung metastasis. Clinically relevant extrapulmonary metastases (defined as metastases which were numerous or large, and/or likely to contribute to discomfort or death) were apparent, however, in 42% of the mice that died from G3.5 metastases and in 19% of those that died from G3.12 metastases. The kidneys, adrenal glands and ovaries were the most common extrapulmonary sites for metastases, and such metastases were only apparent in mice with lung metastases. According to Alterman and her colleagues (1985), the pattern of spread following B16 injection into the abdominal flank begins with dissemination to the lungs. Involvement of the draining lymph node (the ipsilateral axillary lymph node) is next apparent, and finally spread occurs to a variety of systemic sites. This pattern of spread suggests that the lungs act as a generalizing site for the spread of this tumour (see section 2.14.10). However, because the frequency of draining node metastases was reduced following surgical removal of the tumour which grew at the site of inoculation, it seems that spread to the node may be independent of spread to the lungs. The later appearance of metastases in the draining node might thus result from a different mechanism of spread, perhaps via the lymphatics rather than the blood. Interestingly, the number of visible lung metastases increased progressively following excision of the tumour which formed at the site of s.c. inoculation. An apparent inhibitory effect of the primary tumour on metastatic growth has been noted in several studies, although previous work by Stackpole (1981) suggests that the trauma of surgery may contribute to this situation. The general conclusion of the study by Alterman, Fornabaio and Stackpole (1985) is that progression of the B16 melanoma can result in widespread metastasis to all organs including the brain (commonly involved in human melanoma), although some organs such as the liver and spleen are relatively poorly affected. Since human melanoma can metastasize fairly frequently to the liver, it would seem likely that there are subtle differences between the metastatic capacities of the two tumours. This of course is not surprising, but the demonstration that the B16 melanoma can progress to systemic involvement like its human equivalent is generally supportive of its use as an experimental model system.

The process of arrest and extravasation of B16 melanoma cells has been examined microscopically by a number of groups. Crissman *et al.* (1985), for example, used both light and electron microscopy to study the process of extravasation following tail vein injection of a highly metastatic amelanotic variant of the B16 melanoma (B16a). These authors found that B16a cells in the pulmonary vasculature 10 min after injection were associated with platelets and fibrin as well as with occasional leukocytes, although there was never evidence of cytotoxic inflammatory cell reaction. Many of the platelets associated with the tumour cells in the capillaries were apparently activated (as judged by degranulation). Individual tumour cells were large enough to completely occlude capillaries, giving the appearance of expanding these blood vessels, and they appeared to make direct contact with the endothelium rather than through a bridge of platelets or fibrin. The extent of associated thrombus reached a peak 4 h after tumour cell injection, but then the amount decreased to be almost absent by 48 h. Thrombus formation could lead to the cessation of blood flow, but this was restored as the thrombus was broken down. Within 4 h of injection, some tumour cells had displaced part of the endothelium so that they came into direct contact with the basement membrane. Mitosis was apparent in these tumour cells within 24 h. Penetration of the basement membrane by tumour cell processes was first evident at 72 h, and dissolution of this structure progressed over the course of the study (120 h). Extravasation of the tumour cells was presumed to result from their proliferation in association with the progressive destruction of the basement membrane since there was no evidence of active migration (although this would be impossible to assess reliably).

An ultrastructural study of the invasion of the brain (from day 3 to death) by the B16 melanoma has been made by Kawaguchi and colleagues (1985) using B16-B14b cells. These cells were selected from the parental B16F1 line for their brain-colonizing ability following intra-arterial injection into the common carotid artery. A major route for invasion of the brain parenchyma was along the blood vessels, and tumour cells were frequently found adherent to the basal side of the basement membrane having displaced the perivascular astrocytes. Tumour cells which invaded nervous tissue produced numerous cell extensions which appeared to fragment and engulf nervous elements. Although this study suggests that invasion of the brain by B16-B14b cells may be promoted by their destruction of nervous tissue, the destruction was localized and no significant signs of tissue necrosis were evident.

Further discussion of the mechanisms involved in tumour cell spread, particularly with respect to arrest and extravasation, is provided in later sections. Suffice to say here that there is no single mode of arrest or extravasation which applies to all metastatic tumours. Indeed, there is every likelihood that a single tumour can capitalize on a multitude of processes during the course of metastatic spread.

3.4.5 B16 melanoma variants

Several variants of the B16 melanoma cell line which metastasize preferentially to particular organs now exist. The term 'metastasize' has been used rather loosely in many publications concerning the B16 melanoma, however, since the majority of variant lines have in fact been obtained by colonizing methods. As we have seen, i.v. injection models only part of the metastatic pathway, and in keeping with the comments above we shall refer to the 'metastatic' lesions which develop using such methods as **colonies**.

3.4.5.1 *Lung-colonizing variants*

Fidler (1973a), selected the first B16 melanoma variants as part of a study designed to test whether metastases were derived randomly from a single population of tumour cells, or non-randomly from tumour cells which possessed particular properties. In the original work, the B16 was first adapted to grow in tissue culture and several clones were then isolated (Fidler, 1970; Fidler and Zeidman, 1972). During the subsequent selection process (Fidler, 1973a), cells were detached from a confluent monolayer of one clone by treatment with 0.25% trypsin and a single cell suspension of 50 000 viable cells in 0.25 ml salt solution was then injected into the tail veins of recipient C57Bl/6J mice. Several B16 colonies were removed from the lungs of these mice three weeks later and isolated cells were grown *in vitro*. Some of the pigmented colonies which grew from these cells were then passaged and allowed to grow to confluence. Cells from these cultures were then prepared as above and 25 000 in 0.25 ml salt solution were this time injected into recipient mice. Originally, this procedure was repeated five times, but since then two variants have become established in research, namely the B16F1 and the B16F10 lines with 1 and 10 pasages through the lung respectively.

How is it possible to select for variant cell lines which display increased colonizing or spontaneous metastatic ability to particular organs such as the lungs? If pre-existing subpopulations of cells exist within a tumour and some of these have the propensity to metastasize to the lungs, then repeated selection as performed by Fidler (1973a) may lead to the emergence of lines with particular metastatic characteristics. If this interpretation is correct (see section 1.7.27) then the obvious question to ask next is whether there is any simple explanation for the different colonizing abilities of B16F1 and B16F10 cells? The situation may be summarized as follows:

(a) The differences between the lung-colonizing abilities of B16F1 and B16F10 cells are probably not due to the nature of the injection site (usually the tail vein in colonization experiments). For many tumours, it seems that the location of the first capillary bed encountered is a major contributor to the site of metastasis. Injection of W256 rat carcinosarcoma cells into the portal vein, for example, leads to colonization of the liver whereas injection into the tail vein leads to colonization of the lungs (Griffiths and Salsbury, 1963). After i.v. injection more B16F10 cells colonize the lungs than B16F1 cells, but this appears not to be related to the fact that the lungs represent the first major capillary bed encountered since this difference remains even if the B16 cells are injected directly into the left ventricle of the heart (Fidler and Nicolson, 1976). Tumour cells injected into the left ventricle are widely dispersed and other capillary beds are likely to be met before such cells reach the lungs. The general conclusion from these studies is supported to a degree by those of Hart and Fidler (1980a), who examined the growth of B16F10 tumour colonies in experimental mice bearing organ tissue grafts. They found that i.v. injected B16F10 cells could colonize ectopic lung tissue (in 20/28 experiments) and ectopic ovarian tissue (7/10), whereas there was little colonization of ectopic kidney tissue (4/28). This pattern of results using ectopic grafts is suggestive of a degree of organ selectivity, but the mechanisms remain unclear. It should be noted here, however, that other studies have cast some doubt on the selective ability of B16F10 cells to colonize the lungs at the expense of other organs. Thus Roos and Dingemans (1979) found that the direct injection of B16F10 cells into the portal veins of experimental mice resulted in extensive liver colonization. Close examination of such experimental animals revealed that most of the tumours had proliferated intravascularly and that they had not extravasated into the organ stroma (Dingemans and Van den Bergh Weerman, 1980). The nature of many of these liver lesions appears at first sight to be different from that of lesions which grow in the lungs, but the significance of this point is somewhat obscure since given time invasion of the liver stroma will almost certainly take place. B16F10 cells can in fact invaginate into liver cells and this is seen both *in vivo* (although relatively rarely) and *in vitro* (Roos, 1980).

(b) Although more B16F10 cells lodge in the lungs after i.v. injection than do B16F1 cells, this may not explain their colonizing differences. Fidler and Nicolson (1976) radiolabelled B16F1 and B16F10 cells with ^{125}IUDR (^{125}I-5-iodo-2′-deoxyuridine, a thymidine analogue incorporated into DNA) and injected them separately either into the tail veins (i.v.) or into the hearts (i.c.) of syngeneic mice. Radioactivity within the lungs (an estimate of the cells present) was then counted 2 min after injection. Following i.v. injection, 64% of B16F1 cells and 99% of B16F10 cells were estimated to be lodged in the lungs after this time interval. At the same time interval after i.c. injection, about 30% of both cell types were located in the lungs. It should be noted that the detection of radiolabelled tumour cells is rather crude in comparison to the small numbers of cells which give rise to metastases (probably less than 1% of those injected), but it would seem that the extent of cell arrest does not always provide a good estimate of subsequent growth. Thus although there was no difference in initial arrest in the lungs between the two cell types after i.c. injection, the B16F10 cells still went on to form more visible colonies than the B16F1 cells.

In experiments with parabiotic mice, relatively more B16F10 cells reached the circulation of the non-injected partner than did B16F1 cells. Most of these cells failed to be arrested in the lungs, however, and consequently did not give rise to tumour colonies (Fidler and Nicolson, 1977). There are several possible interpretations of these results, but one suggests that malignant cells capable of adhering to lung endothelium were selected out as they passed through the lungs of the injected partner prior to crossing between the parabionts.

(c) The ability of B16 melanoma cells to colonize the lungs is a function of cell culture density *in vitro*. Bosmann and Lione (1974), found that B16 cells derived from more confluent cultures (4×10^4 cells cm^{-2}) have higher colonization rates than cells from sparse cultures (8×10^2 cells cm^{-2}). This effect of cell culture density, however, does not explain the differences between the low and high lung-colonizing variants. Rather confusingly, a study almost 10 years later using the B16F1 and B16F10 variants (Gilbert and Gordon, 1983) showed lung-colonizing ability to decrease with time in culture. This leads one to question the long-term stability of some phenotypic characteristics of the B16 melanoma.

(d) Lung-colonizing variants cannot be selected simply by their ability to grow in this organ. Nicolson and Custead (1982) reported experiments in which they tested whether the ability to grow in the lungs could give rise to cells with enhanced colonizing potential. They grew B16F1 cells on microbeads (120–180 μm diam) and then injected the beads and adherent cells into recipient animals via the tail vein. The beads lodged mechanically in the lungs and the tumour colonies which grew were isolated and cultured *in vitro*. These cells were then grown on more beads and the procedure repeated nine times to yield B16F1A9 cells, which were then compared for their lung-colonizing ability against B16F10 cells which Fidler (1973a) had selected from B16F1 cells by nine further passages involving i.v. injection and subsequent *in vitro* expansion of lung colonies. Like their parental B16F1 cells, and in contrast to B16F10 cells, the B16F1A9 cells colonized the lungs poorly following tail vein injection even though they were selected for growth in this organ. Similarly, B16F1A9 cells metastasized poorly after s.c. injection, implying that the ability of tumour cells to grow in the lungs is insufficient to select for cells capable of lung colonization or metastasis. Essentially similar experiments have been performed by direct intracerebral inoculation of B16 melanoma cells. It proved impossible to select brain colonizing variants by this technique, although such variants could be obtained by the repeated selection of cells which grew as colonies in the brain following i.v. injection. Although selection for growth in a particular organ does not necessarily result in the emergence of cells with selectivly enhanced colonizing ability, it should nevertheless be clear that growth is an absolute requirement for metastatic development. It is thus not surprizing that some groups have reported that certain B16 melanoma variants display a degree of selectivity in their growth responsiveness to organ extracts and that this correlates with their colonizing abilities (Nicolson and Dulski, 1986).

(e) Both B16F1 and B16F10 cells arrest and survive better in the lungs of normal mice than immunosuppressed mice, and their colonizing differential is maintained in both sets of animals. More lung colonies are found when either B16F1 or B16F10 cells are injected into normal mice compared to immunosuppressed mice (X-irradiated and thymectomized or nude), and this is reflected in cell trapping and survival as determined by following the kinetics of injected, radiolabelled cells (Fidler and Nicolson, 1978). Since both B16F1 and B16F10 cells aggregate more or less similarly with organ cells prepared from either normal or nude mice, it would seem that the poor colonization of immunosuppressed mice may not be due to differences in interaction with target organ cells. It is clear that the presence of host lymphocytes somehow or other affects both initial lodgement of cells in the lung and their subsequent growth: the mechanisms involved are by no means understood, but we might speculate that heterotypic aggregation with host lymphocytes could promote lodgement, and that subsequent growth (at least for weakly immunogenic tumours) could be enhanced by the secretion of growth factors from lymphocytes. Indeed, B16F10 cells do adhere to host lymphocytes at a higher rate than do B16F1 cells, but since differences between B16F1 and B16F10 cells still exist in immunosuppressed animals, it would seem that interaction with lymphocytes cannot fully explain their different colonizing abilities.

(f) B16F1 and B16F10 cells display similar NK cell sensitivity. Talmadge and colleagues (1980) were able to select NK cell-sensitive B16 cells by *in vitro* culture, and they found that such variants grew more slowly and produced fewer metastases in normal mice than in NK-deficient mice. No differences have been found, however, in the susceptibility of B16F1 and B16F10 cells to NK cell mediated cytotoxicity: both cell types show only modest but similar sensitivity as tested *in vitro* (Hanna and Fidler, 1981), and thus NK cell responsiveness probably does not underly their differences in lung-colonizing ability (reviewed in Hanna, 1982a).

(g) Compared to B16F1 cells, B16F10 cells may have increased resistance to cytolysis by activated macrophages. This possibility is somewhat controversial since Fidler (1978b) could not show any significant differences between B16 variants in their responses to activated macrophages. Miner *et al.* (1983), on the other hand, reported that both B16F10 cells and brain-colonizing B16-B14b variants were less susceptible to killing by activated macrophages. It seems that studies based on macrophage cytotoxicity may depend on the method of their activation and on the macrophage to target-cell ratio; since the latter cannot be quantitated or manipulated easily *in vivo*, the full significance of these studies is difficult to evaluate.

(h) Cell division rates and colonization. The doubling times of B16F1 and B16F10 cells *in vitro* seem to vary from laboratory to laboratory: values of 22 h for B16F1 cells and 17 h for B16F10 cells were reported by Fidler, Gersten and Budmen (1976). Despite differences in doubling time, it would seem unlikely that this parameter can explain the differences in lung-colonizing ability for the B16F1 and B16F10 cell lines.

(i) Lectin binding sites. B16F10 cells have fewer wheat germ agglutinin (WGA) and soybean agglutinin (SBA) binding sites but more concanavalin A (con A) binding sites when compared to poor lung-colonizing variants such as the B16LR6. The relevance of this in metastasis is not clear, however, since studies with other tumour cell lines, such as RAW117, have shown that the loss of con A binding sites has a better correlation with lung-colonizing potential (Reading *et al.*, 1980a).

(j) Surface charge distribution. Raz and colleagues (1980) used cationic ferritin to label cell surface anionic sites on glutaraldehyde-fixed B16 melanoma variants, and then they determined the distribution of the charged groups by electronmicroscopy. According to these authors, the lung-colonizing B16F10 variant contains a greater percentage (41%) of cells with clustered anionic sites than do either the B16F1 (21%) or B16F10 lymphocyte resistant (28%) variants (both of which colonize the lungs poorly following tail vein injection). The general applicability of these observations is uncertain, however, since both tumour cells and non-transformed cells have been shown to have anionic sites clustered on microvilli (Weiss and Subjeck, 1974 a and b; Grinnell, Tobleman and Hackenbrock, 1975).

(k) B16F10 cells are more invasive than B16F1 cells. Nicolson and his colleagues (1977) compared the invasive properties of B16F1 and B16F10 cells using the chick chorioallantoic membrane (CAM) as a model system (section 4.8.6). They found that B16F10 cells could invade the CAM within 12 h, whereas B16F1 cells required considerably longer. Similar studies have been made using blood vessels and the bladder wall (Hart, 1979; Poste *et al.*, 1980), as well as the amnion (Liotta *et al.*, 1980). Although invasion is probably an important aspect of metastasis, these studies do not explain why B16F10 cells have greater ability to preferentially colonize the lungs.

(l) The differences in lung-colonizing ability between B16F1 and B16F10 cells seem not to be due to differences in suceptibility to the high O_2 tensions present in lung capillaries. Because of the delicate structure and location of lung capillaries, their O_2 tension is in equilibrium with that of the atmosphere (about 18%). This O_2 tension is 3–6× higher than that typically found in interstitial tissue and is toxic for many cell types. In unpublished work from my laboratory, however, we have found that both B16F1 and B16F10 cells can adapt rapidly to high O_2 tensions and there are no consistent differences in their susceptibility towards O_2 toxicity.

It might seem from the above that we are at a loss to explain convincingly the differences in lung-colonizing ability between B16F1 and B16F10 cells. This is indeed true, although an explanation may be forthcoming from more detailed biochemical and molecular genetic studies. We should note that many of our current techniques are rather insensitive, particularly if we are searching for quantitative rather than qualitative differences. Some of the differences between B16F1 and B16F10 cells may be cumulative or potentiating, and thus studying any one in isolation may not be particularly illuminating. As will become clearer later, B16F10 cells adhere more to lung cells *in vitro* than do B16F1 cells, and this selective adhesion perhaps coupled to other slight differences (such as enhanced growth rate or decreased cytotoxic susceptibility) might be enough to result in more B16F10 lung colonies. Other features which might correlate with selective B16 metastasis are discussed further in section 3.4.7.

Since the original selection studies of Fidler (1973a) which resulted in the B16F1 and B16F10 variants, several other selection procedures have been adopted, including selection based on homotypic aggregation performance, substrate adherence or detachment, and the ability to penetrate living cellular barriers, extracellular matrices (ECMs), or synthetic structures such as filters.

3.4.5.2 *Brain-colonizing variants*

Nicolson and Brunson (1977) and Brunson, Beattie and Nicolson (1978) selected B16F1 melanoma cells for their propensity to colonize the brain after intracardiac (i.c.) injection. The rare brain colonies which developed were removed, expanded *in vitro*, and subjected again to i.c. injection essentially as described by Fidler (1973a). This technique was performed for four selection cycles, after which the authors resorted to intravenous (i.v.) rather than i.c. injection procedures. After seven selection steps, tumours were found to form predominantly in either of two sites, namely the meninges of the dorsal cerebrum or in the vicinity of the rhinal fissure (between the olfactory bulb and the cerebral cortex). After three further selection steps, the variant colonizing the meninges was designated B16-B10b, while that colonizing the rhinal fissure was designated B16-B10n. When injected i.v., each of these established variants showed a high degree of selective site colonization reflecting the restricted nature of the selection processes. The colonization pattern, however, was not absolutely specific in that tumour colonies also developed occasionally at other sites. It is of interest here to note that although B16-B10n cells have qualitatively similar cell surface determinants to that expressed by their parent cells (as judged by lactoperoxidase-catalysed cell surface iodination), they have quantitatively more of two particular proteins (95 and 100 kD as judged by PAGE). Exactly how this difference relates to their selective patterns of colonization, however, is far from clear.

Other procedures have been attempted in order to see whether brain-colonizing cells can be obtained by selection for growth in this organ after direct implantation rather than i.v. inoculation (Brunson and Nicolson, 1980). After 10 selection steps for brain adaptation the resulting cells were still no more effective in their colonizing potential than the parental variant. Thus, as we have already seen with lung-colonizing cells, the ability to grow in an organ is not sufficient alone to ensure colonization. The procedure employed in selecting cells with enhanced colonization potential presumably involves a number of steps which, by acting in concert, result in successful colony establishment.

3.4.5.3 *Ovary-colonizing variants*

The selection of B16 melanoma cells with increased preference for colonization of the ovaries after i.v. injection was reported by Nicolson, Brunson and Fidler (1978) and Brunson and Nicolson (1980). After one selection step, the cells gave rise to ovarian colonies in experimental animals with a frequency of about 15%, but after 10 steps the cells (designated B16-O10) formed ovarian tumours with a frequency nearer 80%. The B16-O10 variant appears not to be as site-selective as the lung and brain variants in that it forms colonies at a number of locations. Analysis of B16-O10 cell surface proteins by iodination in conjunction with PAGE has shown the increased exposure of 140 kD and 150 kD components relative to that on their parental cells but, as with the brain colonizing variants, the significance of this is uncertain.

3.4.5.4 *Liver-colonizing variants*

Liver-colonizing variants of B16F1 and B16F10 cells were selected by Tao *et al.* (1979). Their procedure was similar to that developed originally by Fidler (1973a), but it involved direct injection into a vein within the omentum of the small intestine (which drains into the portal vein of the liver), and a total of eight selection steps. After i.v. injection, both L8-F1 and L8-F10 cells formed tumours in the livers of syngeneic mice with a higher frequency (100% incidence) than did control B16F1 (4%) or B16F10 (0%) cells. Since the lungs contain the first capillary bed to be met by i.v. inoculated tumour cells, the liver colonizing variants could have resulted either from selection for increased detachment from the lungs (using it as a generalizing site) or for increased liver preference *per se*. Tao *et al.* (1979) attempted to test between these two possibilities by i.c. (left ventricular) injection of L8-F1 and L8-F10 cells. Both variants displayed higher colonization incidences in the liver than in the lungs, but this is hardly proof that the authors had indeed selected for a liver-colonizing preference.

Tao and colleagues (1979) also examined the pattern of blood flow after i.v. or i.a. inoculation by determining the distribution at 19 days of ^{113}Sn-labelled Sephadex microspheres injected immediately after the tumour cells. After i.v. injection, 92% of the radioactivity was associated with the lungs, 7% with the liver, 1% with the spleen and 0.07% with the kidneys. After i.a. injection, the distribution pattern was considerably different: lungs (11%), liver (13%), spleen (9%), and kidneys (67%). Even though the spleen received a reasonable proportion of the blood supply it was never the site of tumour growth, and similarly the kidneys were rarely colonized even though they retained a high proportion of the injected beads. These results support the notion that colonization does not necessarily relate to the proportion of injected cells carried to a particular organ, but we should reflect on the problems of using Sephadex microspheres to analyze blood distribution.

Interestingly, the authors reported that i.a. injection always yielded significantly more liver colonies than did i.v. injection, and they also found that the lung colonizing ability of B16F10 cells was dependent on the route of inoculation: i.v. injection yielded more lung colonies (mean 25) than liver colonies (0), whereas i.a. injection yielded more liver colonies (45) than lung colonies (4). Essentially similar data were obtained for the B16F1 variant: i.v. injection yielded a mean of 37 colonies in the lungs and none in the liver, wheras i.a. injection yielded a mean of two colonies in the lungs and 44 in the liver. These results indicate that B16 cells may be non-specifically trapped in the first capillary network they reach, but they need to be compared and contrasted with those of Fidler and Nicolson (1976) where an essentially opposite conclusion was reached (section 2.14.6). These authors reported that both B16F1 cells and B16F10 cells formed equivalent numbers of lung colonies after i.v. (9 +/− 3 and 79 +/− 16 respectively) or i.a. (10 +/− 2 and 83 +/− 10) injection. It is undoubtedly of some significance that the report by Tao *et al.* (1979) is one of few where the B16F1 and B16F10 variants did not behave in the manner predicted originally by Fidler (1973a).

3.4.5.5 *Invasive variants*

Fidler (1975a) reported that B16F10 cells were more capable of metastasizing than B16F1 cells, although Stackpole (1981) could find no differences in this ability between the two variants using a supposedly more discerning experimental system. Other studies by Nicolson *et al.* (1977) provided evidence that B16F10 cells could invade the chick CAM more successfully than B16F1 cells, but it was Hart (1979), however, who was the first to succeed in selecting a variant of the B16 melanoma using a procedure based on invasive ability. He achieved this by injecting B16F10 cells into the urinary bladder of male C57Bl6 mice via the vas deferens (a route not without considerable anatomical obstacles). The bladder was then ligated, excised and cultured on semi-solid agar. Melanoma cells which migrated through the bladder wall were collected from the agar, expanded *in vitro* and recycled through the same process a total of six times to yield a variant which he designated the B16BL6. Relative to its parental line, the B16BL6 was found to produce less plasminogen activator, it was less motile, and it was more resistant to trypsin mediated detachment from tissue culture surfaces. Interestingly, both the B16F10 and B16BL6 variants adhered equally to bovine aortic endothelial cells even though the latter formed considerably more metastases after intramuscular (i.m.) injection than did the former in similar circumstances. It should be pointed out here also that the difference in motility between the B16F10 and B16BL6 variants was very minor, amounting to less than 10 cell diameters after 4 days (in favour of the B16F10). Such variation could well be accounted for by the slight difference in growth rate between the two lines, with the B16F10 variant dividing marginally faster. No qualitative differences in cell surface proteins were found between B16F10 and B16BL6 cells after lactoperoxidase iodination. Although some quantitative variation was detected, the significance of this remains uncertain.

In their study of B16 melanoma metastatic inefficiency, Weiss *et al.* (1982) reported that whereas both the B16F10 and the B16BL6 lines formed lung colonies in all the mice tested after i.v. inoculation, the incidence dropped to 36% for B16F10 cells after i.m. injection but remained at 100% for the B16BL6 variant. As shown in Table 3.1, the median number of tumours per animal was considerably higher after i.v. or i.m. inoculation with B16BL6 cells relative to that for B16F10 cells. Unselected B16 cells colonized fewer animals than either B16BL6 or B16F10 cells, and they metastasized about as efficiently as B16F10 cells. These results from an independent group provide a clear indication that the selection procedure yielding B16BL6 cells has resulted in a variant with increased metastatic capacity rather than just increased colonizing ability, providing we accept that the i.m. injection schedule actually does test for metastasis (section 3.3).

Tullberg and Burger (1985) succeeded in isolating B16 variants with increased spontaneous metastatic capacity (after i.m. or s.c. injection

Table 3.1 Formation of lung tumours by different B16 variants*

Variant	*i.v. colonization*		*i.m. metastasis*	
	incidence (N)	*median (range)*	*incidence (N)*	*median (range)*
B16BL6	100% (12)	240 (77–352)	100% (25)	6 (1–51)
B16F10	100% (45)	103 (5–400)	36% (64)	0 (0–19)
B16F10LR6	83% (41)	3 (0–152)	50% (68)	1 (0–10)
Parent	93% (30)	4 (0–69)	40% (30)	0 (0–4)

* Data modified from Weiss *et al.* (1982).

into the leg or flank respectively) by selecting and cloning B16F1 cells which were capable of migrating through 2 μm diameter pore size chemically modified Nuclepore filters. Filter migration did not always correlate with metastatic potential, however, since at least one clone showed greater penetration than the B16F1 parent although its metastatic potential was equivalent.

3.4.5.6 *Lymphocyte-resistant variants*

Lymphocyte-resistant variants were selected from both the B16F1 and B16F10 cell lines by Fidler, Gersten and Budmen (1976). After 6 passages involving *in vitro* culture with syngeneic lymphocytes from mice immunized with the B16 melanoma, these authors obtained cells (B16F1LR6 and B16F10LR6) which were resistant to lymphocyte mediated cytotoxicity. The resistant variants had normal doubling times *in vitro* and they showed normal growth patterns following s.c. injection, but each variant showed reduced lung colonization relative to its parental line. An independent study by Weiss *et al.* (1982) showed that whereas B16F10 cells colonized 100% of animals after i.v. injection, B16F10LR6 cells colonized only 83%. As shown in Table 3.1, there were marked differences between the two variants in the median number of colonies established per animal, with the B16F10 cells forming considerably more. At first, this result might appear somewhat surprising since by escaping lymphocyte mediated toxicity it could be argued that lymphocyte-resistant cells should give rise to more colonies rather than less. However, studies with nude mice suggest that lymphocyte mediated cytotoxic reactions may not be too important in limiting tumour growth. Indeed, experiments with radiolabelled cells indicate that lymphocyte resistance does not seem to confer any advantage in terms of cell survival following lodgement in the lungs. Weiss *et al.* (1982), however, reported that B16F10LR6 cells were more sensitive to the presence of normal, syngeneic heparinized blood as judged by their *in vitro* plating efficiency. Furthermore, we have noted that a range of host cells including PMNs, macrophages, and NK cells can lyse tumour cells, and thus the contribution of host cell-mediated cytotoxicity to the differences in colonizing potential between B16F10 and B16F10LR6 cells is rather difficult to judge.

Other studies with radiolabelled cells have shown that fewer cells from lymphocyte-resistant variants are trapped in the lungs 2 min after tail vein injection when compared to control cells. These results imply that the selection procedure has generated cells with altered lodgement properties, possibly as a result of failing to form embolic clusters with lymphocytes (section 3.4.8.2). According to the studies of Fidler and

Bucana (1977), immunized lymphocytes cluster more around B16F10 cells than they do around B16F10LR6 cells.

Syngeneic mice immunized with B16F10 cells are protected against challenge with B16F10 cells, whereas mice immunized with B16F10LR6 cells are not protected against B16F10 cell challenge. Furthermore, although lymphocyte-resistant variants are not destroyed by immunized syngeneic lymphocytes, they are susceptible to cytotoxicity by immunized allogeneic lymphocytes and immunized syngeneic (or allogeneic) macrophages (Fidler and Bucana, 1977). These results suggest that B16F10LR6 are poorly immunogeneic although they still express MHC antigens involved in cell lysis.

3.4.5.7 *Lectin-resistant variants*

Lectin-resistant variants of the B16 melanoma were first selected by Tao and Burger (1977) from B16F1 cells using toxic concentrations (100 μg ml^{-1}) of the plant lectin wheat germ agglutinin (WGA). Colonies which grew *in vitro* under toxic concentrations of the lectin were repeatedly selected to yield cells which survived progressively better in WGA but which also showed a progressive decline in both metastasizing ability (after i.p. injection) and in lung colonizing ability (after tail vein injection). It should be noted here that the control cells in this study were the parental B16F1 cells, which are known to be relatively poor metastasizers and poor colonizers.

3.4.6 B16 tumour heterogeneity

Phenotypic heterogeneity appears to be expressed by all tumours despite the clonal origins of many. The general consensus is that new variants are generated as the clonal population multiplies. When a malignant tumour metastasizes, the secondary lesions may also be clonal in origin and the process of diversification begins again.

3.4.6.1 *Three major behavioural subpopulations*

We have seen that not all cells within a B16 tumour are likely to have equal metastasizing potential. Thus Tao and Burger (1977), for example, were able to isolate non-metastasizing variants from the B16F1 malignant melanoma by selecting for resistance to toxic concentrations of WGA, and Fidler and Kripke (1977) successfully isolated a number of clones with different metastatic capabilities from the unselected B16 parent line. According to Stackpole (1981, 1983) the B16 melanoma is comprised of three basic cell subpopulations:

(a) Spontaneously metastasizing cells
(b) Colonizing cells
(c) 'Null' cells

The spontaneously metastasizing and colonizing cell types are separable by appropriate assay systems such as s.c. injection into the flank and i.v. injection into the tail vein respectively (section 3.3). The 'null' cell type has neither colonizing nor spontaneous metastatic ability, but it still retains tumorigenic potential. 'Null' cells appear to be able to convert into either spontaneously metastatic or colonizing cells, and spontaneously metastatic cells can become colonizers or even revert to 'null' cells, but colonizers are apparently end cells (at least in so far as this aspect of phenotypic progression is concerned). These three possible subpopulations clearly contribute to the heterogeneity of the B16 melanoma, with shifts in their proportions likely to yield cell lines of different colonizing and spontaneously metastasizing ability.

3.4.6.2 *Clonal interactions*

Tumour cell lines, such as the B16 malignant melanoma, may be cloned *in vitro* and the clones tested for their metastatic ability (Fidler and Kripke, 1977). The general conclusion from such studies is that malignant tumours contain clones with different metastatic capabilities, including, as indicated above, some which cannot metastasize. It has been suggested that the possibility of metastasis might increase as a consequence of the selective (non-random) growth of clones with enhanced metastatic potential. Talmadge and Fidler (1982) examined three B16 variants (B16F1, B16F10, and B16BL6) for cells with increased metastatic potential by taking cells from metastases of these lines which had arisen after footpad injection and re-injecting them into different mice by the same route. They found that cells derived from B16F1 and B16F10 metastases formed more metastases after one selection *in vivo* than did the parental lines. Cells derived from B16BL6 metastases, however, failed to show any increase in metastatic potential when assayed by this technique. When tested by i.v. inoculation, cells from B16F1 metastases again produced many more lung colonies than the parental line. B16F10 metastasis-derived cells failed to show any increase in colonizing potential, however, and B16BL6 derived cells showed only a relatively moderate increase. These results suggest that populations of tumour cells selected for lung colonization (B16F10) or invasion (B16BL6) cannot be further selected for each particular metastatic phenotype, although they may be selectable for other phenotypes. It would seem that both the B16F10 and the B16BL6 cell lines are performing more or less at peak ability for their respective phenotypes, yet both populations probably contain non-metastasizing and non-colonizing cells and further selection should be possible. Hill and colleagues (1984), for example, have reported that both B16F1 and B16F10 clones contain few lung colonizing cells. Interestingly, these authors were able to show that B16F10 cells generated colonizing variants at a higher rate than B16F1 cells and they suggested that this may underly the different lung colonizing ability of the B16F1 and B16F10 lines. If B16F10 cells have reached the maximum rate of generating lung colonizing variants, then this may explain why further selection of this cells line for lung colonizing potential has not been possible. Such possibilities, however, remain speculative and serve only to remind us how ignorant we are of many aspects of tumour cell biology.

When B16 melanoma clones were cultured individually or with two to three other clones, relatively rapid generation of variant cells with altered metastatic potential was noted. When the same clones were cultured with six or more other clones, however, the overall metastatic phenotype remained stable (Poste, Doll and Fidler, 1981). It was suggested from this study that heterogeneous cell lines such as the B16F1 and B16F10 maintain a stable pattern of heterogeneity both *in vitro* and *in vivo* through some form of recognition between their clonal subpopulations. Interestingly, phenotypic destabilization on cloning was not seen for all the phenotypic properties studied (such as cell surface glycolipids and proteins, fibronectin content, and plasminogen activator production) indicating that metastatic potential may be independent of major changes in these parameters. Clinically, it has been suggested that the therapeutic elimination of subpopulations of cells within a tumour may destroy the stabilizing equilibrium, leading to enhanced phenotypic instability and the generation of possibly more metastatic, drug-resistant subpopulations. It should be noted that some tumour lines have been shown to rapidly diversify *in vivo* while parallel *in vitro* cultures of the same cells remained stable (Talmadge *et al.*, 1979). Since diversification *in vivo* is likely to be influenced by the host's defence systems, it would be surprising if all tumours displayed similar diversification characteristics to the B16 melanoma.

3.4.7 Correlates with selective B16 metastasis

We have seen that under certain circumstances B16 cell variants may be selected which show a propensity to colonize or metastasize particular organs. What is the basis of this selectivity? Generally speaking, features of both the host and the tumour probably influence selective metastasis, and we may identify four particular processes which are likely to be involved:

(a) More tumour cells could be delivered to those organs which show preferential metastases.
(b) Tumour cells delivered in the blood could adhere preferentially to the endothelium lining vessels in certain organs.
(c) Tumour cells could extravasate preferentialy into particular organs.
(d) Tumour cells could grow preferentially in particular organs. Note that there are several aspects to growth: the cells may respond positively to growth stimulatory factors; alternatively, they may fail to repond to chalones; the recipient organ may be relatively deficient in anti-tumour defence cells; or, some other feature of the organ (such as tissue architecture or blood supply) may be directly or indirectly supportive of growth.

3.4.7.1 *Anatomical correlates*

In experimental studies with the B16 murine malignant melanoma, Weiss *et al.* (1984) examined the relationships between the percentage of cardiac output delivered to various tissues and organs, the arrest of malignant cells in these tissues and organs after delivery via the arterial blood, and the subsequent incidence of tumours. The distribution of cardiac output was measured using radiolabelled microspheres, the number of cells arrested was calculated following the left ventricular injection of radiolabelled B16 melanoma cells, and tumour growth in each organ or tissue was examined after 3 weeks. Two important findings were made in this research study:

(a) A good correlation was found between arterial blood supply and malignant cell arrest in all of the organs and tissues studied except for the lungs and liver, where cell arrest was greater than expected in terms of the percentage of cardiac output they received.
(b) No correlation was found, however, between malignant cell arrest and subsequent tumour growth.

Thus although delivery of tumour cells may be largely a function of anatomical considerations, whether the cells form metastases or not appears to be dependent on other properties which enable the cells to be retained and grow in the organ to which they have been delivered. Based on this evidence it would seem that both of the major hypotheses proposed to account for the development of metastatic patterns, namely the seed and soil and the mechanical hypotheses (section 2.14.11), contribute to the outcome of tumour cell spread. Interestingly, Weiss *et al.* (1984) were able to obtain some correlation between malignant cell arrest and growth when they divided their results into three groups, but there are no apparent biological reasons for making such groupings.

According to Weiss *et al.* (1984) B16 cell arrest is greater in the lungs and liver than would be expected in terms of the percentage of cardiac output to these organs. These results suggest that some selectivity of interaction may exist between B16 cells and these organs. This selectivity may result from the physical nature of the vascular beds in these organs, or it could conceivably be due to adhesive interaction involving particular cell surface determinants on the tumour cells and the endothelium (section 3.4.8.2).

Interestingly, Dingemans, van Spronsen and Thunnissen (1985) found that although B16F10 derived cells initially lodged randomly within the lungs and the liver, there was a distinct preference for the lesions which grew to be located near the surfaces of these organs. The cells which arrested centrally in these organs were presumed to have been eliminated more efficiently than those which lodged peripherally, although they could conceivably have been in a dormant state and thus remained undetected. The processes responsible for this non-random growth of metastatic lesions are not understood, but differences in the microenvironment affected by variation in the blood flow, oxygenation, nutrient supply, waste removal and presence of immune effector cells may be important. Unfortunately, the differential role these processes might play in selective metastasis is not understood.

3.4.7.2 *Biochemical correlates*

It is generally suspected that recognition events mediated at the cell surface are likely to be involved in the preferential lung colonizing ability of B16F10 cells relative to B16F1 cells. A rather elegant demonstration of this possibility was provided by the microvesicle fusion experiments of Poste and Nicolson (1980). These authors took advantage of the fact that B16 cells spontaneously shed membrane vesicles into the culture medium. Fusion of B16F10 vesicles with B16F1 cells reversibly increased the lung colonizing ability of these cells, and this was associated with an increase in rapid arrest (2 min after i.v. injection) of the fused cells in the lung compared to that for normal B16F1 cells. In contrast, the reciprocal fusion experiment did not alter the ability of B16F10 cells to colonize the lungs. The use of I^{125}-labelled vesicles indicated that the newly introduced surface determinants persisted for a maximum of about 24 h in recipient cells. These results would suggest that relatively early events after injection are probably important in selective organ colonization. Such events are likely to include melanoma cell adhesion to the lung endothelium, although the experiments are far from being definitive. Confirmation that membrane integration was achieved during cell-vesicle fusion was obtained by fusing cytotoxic lymphocyte resistant cells with membrane vesicles from non-resistant cells. In this situation, the resistant cells gained sensitivity to lymphocyte mediated cytotoxicity thus confirming the integration of plasma membrane. More recently, it has been shown that microvesicles are derived from particular membrane domains with regard to enzyme activities and lipid composition (Schroeder and Gardiner, 1984). Microvesicles thus are not representative of the entire plasma membrane, although it would seem that they contain at least some determinants which effect lung colonization.

Other studies in which the cell surface has been subjected to enzymatic modification support its involvement in metastasis. Fidler (1978 a and b), for example, found that trypsin treatment of B16 melanoma cells reduced their lung colonizing potential. Studies using the lactoperoxidase catalyzed protein iodination technique in most cases have failed unfortunately to show any obvious differences in surface proteins between the lung-colonizing B16F1 and B16F10 variants (Nicolson *et al.*, 1977; Raz *et al.*, 1980). Rieber and Rieber (1981), on the other hand, used the lactoperoxidase method to show that B16LR6 cells with poor lung colonizing ability contained urea-releasable 18 kD and 75–80 kD components which were lacking in better lung colonizing variants such as the B16F10. As we have already seen, differences have also been found using this technique with brain colonizing and ovary colonizing variants (Brunson, Beattie and Nicolson, 1978; Brunson and Nicolson, 1979).

Glycoprotein labelling by the galactose oxidase-borohydride reduction technique (after removal of sialic acid) has shown B16F10 cells to have less of a major 78 kD glycoprotein (Raz *et al.*, 1980). This difference was more marked when cell surface sialoglycoproteins were labelled by periodate oxidation of sialic acid and subsequent reduction with borohydride, indicating that the 78 kD component is probably a sialoglycoprotein. In fact, expression of the 78 kD sialoglycoprotein correlated inversely with lung-colonizing ability. However, although differences in the surface chemistry of B16 variants may be found, it does not follow that these are responsible for selective organ colonization. According to Raz *et al.* (1980), differences in colonizing ability could not be ascribed to major qualitative differences in the range of cell surface properties they studied. These properties included cell surface glycoproteins, lectin binding affinities, ganglioside content, and acid phosphatase and 5′-nucleotidase activity.

Although the study above by Raz *et al.* (1980) did not report any qualitative differences between the B16 variants the authors did note some quantitative changes, namely in the reduction of the 78 kD sialoglycoprotein referred to above in B16F10 cells, a reduction in plasma membrane 5′-nucleotidase activity, and a reduction of WGA and SBA binding sites, the latter after sialic acid removal. Interestingly, B16LR6 cells had fewer con A receptors than either the B16F1 or the B16F10 variants, but whether this relates to their resistance to syngeneic cytotoxic lymphocytes is not known. Indeed, the relevance of the rest of these quantitative changes is unknown and was not even alluded to by the authors. More recently, and also more confusingly, Schroeder and Gardiner (1984) were not able to detect any differences in 5′-nucleotidase activity between B16F1 and B16F10 cells, emphasizing the variability between many of these reports. Furthermore, Warren, Zeidman and Buck (1975) were unable to show any differences between low and high lung-colonizing variants when they analyzed glycopeptides released by mild trypsin-pronase digestion.

Irimura, Gonzalez and Nicolson (1981) treated B16F1 and B16F10 cells with tunicamycin, an agent which blocks synthesis of the saccharide donor *N*-acetylglucosaminyl pyrophosphophorylpolyisoprenol, thereby interfering with protein glycosylation. Both B16 variants treated with tunicamycin showed minimal adhesion to endothelial cell monolayers and significantly reduced lung colony formation. Tunicamycin treatment induced changes in morphology (rounding) as the cells became less adherent, and it also caused a reduction in protein and DNA synthesis. Furthermore, complex carbohydrate synthesis was inhibited by about 90%, but it is difficult to ascribe the effects of the drug on colonization to any one event because of the multifactorial consequences of the treatment. Although the authors reported a reduction in cell surface sialogalactoprotein on treatment with tunicamycin, there were no differences between B16F1 and B16F10 cells and thus this could not explain the differences in organ-colonizing behaviour of these two variants. Interestingly, sialogalactoprotein modifications were not detectable with the lactoeroxidase iodination technique, but only by labelling with radiolabelled RCA_I (*Ricinus communis* agglutinin I) after removal of sialic acid groups which block binding of this lectin.

Other data implicating a role of carbohydrates in metastasis stem from the studies of Humphries *et al.* (1986) who treated B16 melanoma cells with swainsonine, a mannosidase I inhibitor which perturbs glycosylation yielding high mannose-type oligosaccharides.

These authors found that such treatment could reduce lung colonization by as much as 80%.

A wide range of biochemical parameters were studied by Bosmann *et al.* (1973) using low and high colonizing variants of the B16 melanoma. The low colonizing variant was actually line 26, which was derived from a subcutaneous B16 implant and used in the derivation of B16F1 and B16F10 by Fidler (1973a). The high colonizing variant was line 37, derived from line 26 after 11 cycles through the lung. With the parameters they studied, Bosmann *et al.* (1973) found little difference between the variants when near confluent cultures were examined biochemically. Using sparse cultures, however, these authors were able to show that relative to the low colonizing variant the high colonising variant contained less protein per cell, elevated levels of a number of glycosidases (β-galactosidase, α-fucosidase, *N*-acetyl-β-galactosaminidase, *N*-acetyl-β-glucosaminidase) and proteases (cathepsin-like and trypsin-like activities), increased total cell and cell surface-related glycosyltransferases, higher electrophoretic mobility, more neuraminidase susceptible sialic acid, and a higher partition ratio in a two-polymer aqueous-phase system (see below). This body of work was published within nine months of Fidler first describing the selection of colonizing variants (Fidler, 1973a), and the scene looked set for a rapid understanding of the biochemistry of metastasis. Sadly, this was not to prove the case, and it slowly emerged that the inherent variabilty of the cancer cell was likely to defeat any straightforward biochemical studies aimed at explaining metastasis.

More recently, Dobrossy *et al.* (1981) have compared aspects of the enzymatic biochemistry of B16F10 cells with that of B16LR6 cells, which are poor lung colonizers. They reported that B16F10 cells have elevated β-galactosidase, α-mannosidase, *N*-acetyl-β-galactosaminidase, and *N*-acetyl-β-glucosaminidase relative to B16LR6 cells, but decreased neuraminidase activity.

In another detailed biochemical study, Yogeeswaran, Stein and Sebastian (1978) showed that although B16F10 cells had 20–35% less total sialic acid than B16F1 cells, their cell surface-exposed sialylglycoproteins and gangliosides were moderately increased. Relative to B16F1 cells, B16F10 cells contained 80% more neuraminidase accessible sialic acid, the majority of which was contributed by the ganglioside G_{M3}. Overall, the B16 melanoma expresses a relatively simple ganglioside pattern, consisting almost entirely of G_{M3} with traces of G_{D1a}. According to Yogeeswaran *et al.* (1978) this pattern becomes more complex when cells are grown *in vivo* (although the contribution of host cells cannot be ruled out), and analysis also indicated an increase in G_{M3} at the expense of G_{D1a} in lung-colonizing variants. One particular sialylglycoprotein of M_r 66 kD was found to be present on B16F10 cells but not on B16F1 cells, although the significance of this observation requires further study. Finne, Tao and Burger (1980) have examined carbohydrate changes in the total cellular glycopeptides of a clone (Wa4b1) of the poorly metastasizing WGA resistant variant originally selected by Tao and Burger (1977). They found this variant to have a decreased sialic acid content and an increased fucose content of the complex carbohydrate chains, with both changes expressed at least partially at the cell surface. A decrease in cell surface sialic acid content is perhaps not surprising *in lieu* of the selection technique, since sialic acid serves as a binding site for the WGA lectin. In contrast to this work, Raz *et al.* (1980) could not find any difference in neuraminidase accessible sialic acid between B16F1 and B16F10 cells, and we are thus forced to question the generality of many of these biochemical studies.

The membrane lipid and enzyme contents of B16F1 and B16F10 cells were examined by Schroeder and Gardiner (1984). They found that relative to B16F1 cells, B16F10 cells had a lower cholesterol/phospholipid ratio, lower arachidonic acid and other polyunsaturated fatty acid content, but a higher phosphatidylcholine/phosphatidylethanolamine ratio. The full significance of these observations is uncertain, and although the lower amount of unsaturated fatty acids might decrease fluidity of B16F10 membranes, this could be offset to a degree by the lower cholesterol/phospholipid ratio.

Several lines of evidence point to the possible involvement in metastasis of tumour cell-associated fibrin (section 5.7.22). Gilbert and Gordon (1983) measured the cellular procoagulant activity of B16F1 and B16F10 cells and found that it decreased with culture density for both variants. Furthermore, lung colonization ability decreased in parallel from which they concluded that a relationship existed between procoagulant activity and metastatic potential. Parallel changes in activity, however, do not allow us to comment on cause and effect. These authors also noted that there was considerable overlap between the procoagulant activity of B16F1 and B16F10 cells, and thus their differences in lung-colonizing ability cannot be simply ascribed to differences in this behaviour. Interestingly, these authors reported differences in procoagulant activity between B16 variants from different sources, emphasizing a degree of interlaboratory variation for the B16 cell lines. Wang *et al.* (1980) examined the fibrinolytic activity of B16F1 and B16F10 cells and concluded that the former produced less plasminogen activator (PA) than the latter. Furthermore, lung colonies of metastatic B16 melanoma cells produced more PA than did cells in the primary, although in this case a possible contribution of host lung tissue cells could not be rigorously excluded. Nicolson, Winkelhake, and Nussey (1976) had previously failed to show any difference in PA activity between the cell lines, which Wang *et al.* (1980) attributed to the use of fewer cells in the assay system used by Nicolson and colleagues. Hart and Smith (1987) also found B16F10 cells to produce more PA than B16F1 cells, and these in turn produced considerably more PA than B16BL6 cells. All three sublines secreted a 33 kD PA-binding protein (presumably functioning like a PA-inhibitor), but interestingly no PA/PA-inhibitor complexes were found. It was postulated by the authors that the PA inhibitor might only bind to PAs of the urokinase type, though what advantage this might confer on the tumour cell is not clear (section 2.15).

Although tumour cell invasion is likely to be enhanced by the secretion of various enzymes, such products may not always be necessary for local invasion or metastasis (section 2.15). Sloane *et al.* (1982) found that cathepsin B activity was greater in solid tumours derived from B16F10 cells than in tumours derived from B16F1 cells. Of considerable interest was their observation that the activity of this lysosomal enzyme decreased with time in culture such that the observed difference in solid tumours was ultimately lost. The variability in cathepsin B activity with time in culture may explain why other studies have produced conflicting results suggesting, for example, that there were no differences in the activity of this enzyme between the variant lines (Nicolson, Brunson and Fidler, 1977). More recently, Sloane and her colleagues (1986) have shown that more cathepsin B activity (along with activity of the lysosomal glycosidase *N*-acetyl-β-glucosaminidase) is present in the plasma membrane of the metastatic amelanotic variant B16a than in the membrane of B16F1 cells. Thus increasing metastatic potential (following s.c. injection into the flank) appears to be associated with a shift in the location of these enzymes from a lysosomal distribution to a plasma membrane distribution. The localization of certain hydrolytic enzymes on the plasma membrane could result in focal dissolution of the ECM thereby promoting invasion and metastasis, although whether this scenario actually operates *in vivo* is another question.

The role of type IV collagenase in metastasis has been referred to elsewhere (section 2.15.1), but with respect to the B16 melanoma it is of interest that Murray, Garbisa and Liotta (1980) reported B16F10 cells to show relatively more activity of this enzyme than B16F1 cells. In comparing their data on metastasis with that of Weiss *et al.* (1982), it seems that both B16F1 and B16F10 cells can metastasize following i.m. injection, but the B16F10 cells may do so more rapidly. Such an observation is in keeping with the relatively higher type IV collagenase levels in the B16F10 variant.

Attention has also been focused elsewhere (section 2.15.2) on a correlation between the heparan sulphate (HS) degrading activity of certain tumours and their colonizing and metastasizing abilities. The HS degrading activity of B16F10 cell extracts, for example, is 1.5 × that of B16F1 extracts, while extracts from the more invasive B16BL6 cells are about 2.2 × more active (Nakajima *et al.*, 1988). This HS degrading activity is attributable to a specific endo-β-D-glucuronidase which can be inhibited by heparin or some chemically altered derivatives of it. When B16BL6 cells were pretreated with these inhibitors, a significant decrease in lung colonizing ability (not solely due to anti-coagulant effects) was noted.

Some interesting studies on the role of cell surface components have been performed by LeGrue and colleagues, who have used aqueous (single phase) 1-butanol to extract supposedly extrinsic cell surface components in a non-cytotoxic manner (LeGrue, 1982, 1985; LeGrue and Hearn, 1983). These authors found that immunization of mice with butanol extracts from B16F1 cells was immunoprotective when the experimental animals were challenged with either B16F1 or B16F10 cells. Conversely, immunization with B16F10 butanol extracts was found to promote lung colony formation on subsequent challenging with tumour cells. Furthermore, since butanol extracted cells may remain viable, they can be re-used in colonization assays. In such experiments, butanol extracted tumour cells were found to produce more lung colonies than non-extracted cells. The suggestion has been made that butanol extraction might selectively remove cell surface components which mask certain important membrane sites, such as binding sites for the endothelium, but much further research is required before we can interpret these results reliably. Almost amazingly, extracted tumour cells treated with the butanol extract show more normal lung colonizing potential on subsequent injection. This is presumed to be due to membrane reconstitution by material in the butanol extract, but the molecular nature of the membrane which results from this 'reconstitution' is not known. It is possible, for example, that the lung colonization results may be due to nothing more than adsorption of the material in the extract to the cell surface in a rather non-specific manner. More recent research from this group (Keren *et al.*, 1989) suggests that butanol extraction may remove an endogenous inhibitor of cell surface heparanase, an enzyme believed to play a key role in the haematogenic spread of at least some tumours (section 2.15.2). Activation of heparanase and possibly other degradative enzymes in this manner may promote lung-colonizing ability. Since the butanol extraction technique offers a unique system for studying many aspects of the role of cell surface determinants in metastasis, it would seem imperative that a more detailed analysis be undertaken.

The results from several research groups such as LeGrue and Hearn (1983) and Vollmers and Birchmeier (1983) clearly indicate that B16 cells express TATAs (section 1.7.5.1) on their surfaces, since it is possible to elicit an immune response against these cells in sygeneic mice. The extent of the responses are highly variable, however, and it appears to be particularly difficult to immunize against B16F10 challenge. According to LeGrue and Hearn (1983), this variant expresses a suppressor antigen which is thought to facilitate tumour growth by stimulating host suppressor cells.

A protein (M_r 65 kD) which is more or less specific to murine melanomas has been identified by Gersten, Hearing and Marchalonis (1981). This protein (also known as B700) is expressed on the surface of B16 cells, but it also represents the major protein species shed by these cells in serum free culture. Unfortunately, the role that this protein may play during immunization by B16 cells is not yet clear. Nevertheless, a rather interesting finding with respect to this melanoma protein is that it bears significant (but variable) homology to albumin (Marchalonis *et al.*, 1984). A panel of McAbs against the B700 antigen cross react with murine albumin but, somewhat surprisingly, they fail to cross react with the biochemically related bovine serum albumin. The function of B700 is obscure, although it is clear that it is not related to the human melanoma antigen p97, which shows a marked resemblance to transferrin (Brown *et al.*, 1982), and its relative molecular mass would indicate that it is also not related to the 375 kD mouse melanoma associated antigen isolated from B16 cells by Bhavanandan *et al.* (1980). According to Tomita, Montague and Hearing (1985), immunization with B700 significantly inhibits tumour growth and metastatic spread following challenge with the B16 melanoma.

Monoclonal antibody (McAb) techniques have recently been used by Falcioni *et al.* (1986) to examine a number of cell lines for the expression of a cell surface protein complex known as TSP-180. Analysis of immunoprecipitates from B16 cell lysates has shown that TSP-180 is composed of three main proteins of M_r 201 kD, 134 kD and 116 kD. The expression of TSP-180 was found to correlate with lung metastatic potential when B16F1 and B16F10 cells were compared, and also when some Lewis lung carcinoma variants were analyzed. Tumours which did not metastasize to the lungs (including the M5 reticulum cell sarcoma, sarcoma 180, and the CCL melanoma) failed to bind significant amounts of the anti-TSP-180 McAb. From a limited study of this sort, however, it is not possible to conclude that the expression of this antigen is always correlated with the ability of tumour cells to metastasize to the lungs. Nevertheless, the intriguing nature of the results gained so far suggests that further studies are warranted. A similar McAb-based approach was used by Kimura and Xiang (1986), who showed that the expression of a 72 kD cell surface glycoprotein (Met-72) on clones of B16F1 melanoma cells was quantitatively related to their lung colonizing ability. A later report confirmed the relationship between lung colonizing ability and Met-72 expression, and also showed that colonizing ability was independent of platelet aggregating activity (Kimura *et al.*, 1986). Analysis of Met-72 distribution in subcutaneous B16 tumours and their metastases by immunohistological techniques showed cells which displayed high levels of Met-72 expression to be located at the edges of the tumours or in perivascular locations, sites indicative of active invasion (Parratto and Kimura, 1988). These authors also reported using FACS selection to isolate Met-72 expressing cells from an ovary metastasis which had formed after the i.v. injection of B16F1 cells. Positive cells were cloned and recycled two more times with cloning, each time selecting for Met-72 expression and ovary colonization. The results showed that as Met-72 expression increased (28, 36, and 45 times background respectively) so did the percentage of mice with ovarian colonies (53%, 56%, 87%). Interestingly, by the third clonogenic selection the melanoma cells were colonizing the lungs relatively poorly, with less than half the experimental animals showing evidence of colonization of this organ. Parratto and Kimura (1988) conclude that Met-72 expression may be a generalized phenotype of B16 metastatic variants irrespective of their organ selectivity, but their results do not fully substantiate this conclusion.

Herd (1987) produced syngeneic McAbs against B16F1 and B16F10 cells and found that some of these could inhibit lung colonization while others actually promoted it. These results may be interpreted in the light of the immune stimulation-inhibition hypothesis of Prehn (1972) although since Herd (1987) did not use monovalent antibodies in her study, it is possible that the enhanced effects noted may have been due to the consequences of tumour aggregate formation. The technique used by Herd (1987) centred on the i.p. injection of crude McAb (1 ml) followed one day later by an i.v. injection of tumour cells ($25–100 \times 10^3$). Seven days later, experimental animals were injected i.p. with another sample of McAb, and the lungs were finally assayed for tumour growth on day 21. Interestingly, the antigen for one McAb (2G10) raised against B16F1 cells which was effective at inhibiting lung colonization by both B16F1 and B16F10 cells could not be immunoprecipitated from membrane extracts. Another McAb (3C10) raised against B16F1 cells inhibited lung colonization by B16F10 cells but promoted that of B16F1 cells. This McAb immunoprecipitated a number of antigens with relative molecular masses of 48 kD, 40 kD, 25 kD and 15 kD. A third McAb (3E9) raised against B16F10 cells behaved similar to 3C10 *in vivo*, but immunoprecipitated a 25 kD component only. All three McAbs reacted with both B16F1 and B16F10 cells, but whereas 2G10 and 3C10 reacted with 97–99% of cells, 3E9 reacted with only about 40%. Along with other studies, these investigations suggest that it is possible to modulate both artificial and spontaneous metastasis through passive antibody immunotherapy.

One of the major purposes of studying the biochemistry of tumour cells, however, is to identify cell surface determinants which correlate with metastatic potential in the hope that this may lead to the development of new therapeutic regimens. With respect to B16 cells, no single antigen has been consistently reported by different groups to correlate with either colonization or metastasis. Not surprisingly, for claims made by individual groups it remains to be explored as to whether there is any causal relationship between the expression of a certain antigen and metastasis or colonization.

Sheppard and his colleagues (1984) have reported a correlation between the ability of certain B16 clones to accumulate cAMP in response to treatment with either forskolin or melanocyte stimulating hormone (MSH) and lung colonizing potential. The responses of two specific clones to these activators of adenylate cyclase were found to change during culture, and interestingly enough they were paralleled by changes in lung colonizing capacity. These results are in general agreement with the earlier studies of Niles and Logue (1979) using different clones, although previously Niles and Makarski (1978) had shown the reverse relationship with uncloned B16F1 and B16F10 variants. In a note added in proof, Sheppard *et al.* (1984) report that although a similar correlation was detectable among B16BL6 clones, this was not the case when they analysed the responsiveness of clones of the K1735 murine melanoma. The general significance of this observation is thus left in some doubt, and there certainly is no evidence of any causal relationship between cAMP metabolism and lung colonizing ability. Subsequent work by Lester *et al.* (1989) showed that a particular B16F10 clone (F10C23) contained increased levels of a pertussis toxin (PT) sensitive G protein (G_{i2}) relative to a B16F1 clone (F1C29). G_i proteins, of which there are about four variants, act as transducing signals to inhibit adenylate cyclase activity following agonist binding to appropriate receptors (section 1.7.22). PT treatment of either B16 clone (which uncouples the G_2 protein from its receptor) did not, however, alter their cAMP levels. One interpretation of these results is that this particular G protein is not linked to cAMP production which, if anything, illustrates the problems of resolving any possible role for cAMP in metastasis. Notwithstanding the above, the observation by Lester *et al.* (1989) that a B16F10 clone with high lung colonizing ability contains significantly more G_{i2} protein than a B16F1 clone with low lung colonizing ability is of considerable interest. The possible significance of this G protein in metastasis is supported by experiments which show that its perturbation in F10C23 cells with PT is associated with decreased cell invasiveness through a synthetic basement membrane as well as decreased responsiveness to fibronectin, laminin and type IV collagen mediated directional motility in Boyden chambers. Interestingly, the effects of PT treatment on F10C23 cell behaviour could not be attributed to any alteration in their adhesiveness to matrix proteins and further examination of this system may shed some light on the haptotaxis/chemotaxis problem (section 3.4.8.4). Finally, it needs to be emphasized here again that the expression of high levels of PT sensitive G_{i2} in a B16F10 cell clone does not explain why the parent cells show enhanced colonization of the lungs relative to B16F1 cells.

3.4.7.3 *Biophysical correlates*

No significant differences in membrane fluidity, as determined by fluorescence polarization and photobleach recovery techniques, were found by Raz *et al.* (1980) for B16 variants with different colonizing ability. Fluorescence polarization is a technique for estimating membrane microviscosity based on measuring the distribution of an embedded fluorescent probe such as 1,6-diphenyl-1,3,5-hexatriene (DPH) in a population of cells. Photobleach recovery, on the other hand, estimates probe diffusion on individual cells by examining the return of a fluorescent compound (either embedded in the membrane or bound to the surface) to an area of the cell membrane which had previously undergone high energy bleaching.

Van Alstine *et al.* (1986) extended the studies of Bosmann *et al.* (1973) with respect to partitioning of B16F1 and B16F10 cells in dextran and poly(ethylene glycol) aqueous-phase systems. The partitioning of cells in to either the upper (PEG enriched) or the lower (dextran enriched) phase in such systems is influenced by both charge-related and non-charge-related cell surface properties. Certain ions which may be present in the aqueous buffer (notably phosphate) have different affinities for the two phases (giving rise to a potential difference), while other ions (notably sodium and choride) display minimal preference for either phase. By altering the ion content of the buffer it is possible to define two-polymer aqueous-phase systems of different potential differences, usually with the top phase positive. Van Alstine *et al.* (1986) found that the B16F10 variant showed a greater preference for the upper phase in a phosphate enriched system than did the B16F1 variant, but that this situation was reversed in the two NaCl enriched systems they studied. Partitioning into the upper phase in phosphate enriched systems is believed to correlate with DNA content, suggesting more active DNA synthesis in the B16F10 variant, although the cause of this difference is uncertain. The partition systems described above are hydrophilic in nature, but they may be rendered hydrophobic by fatty acid esterification of the PEG. In this system the hydrophobic acyl tails of the fatty acids (linked to PEG) are thought to intercalate into the lipid bilayer of the plasma membrane. Results from hydrophobic affinity partition studies with the B16 variants suggest that B16F1 cells are essentially more 'hydrophobic' than B16F10 cells. It is of interest in the light of homotypic adhesion studies with these cell lines (section 3.4.8.1), that increased hydrophobic affinity partitioning is thought to correlate with increased cell adhesion. Finally, it is worth noting that this study highlighted a considerable degree of heterogeneity in both hydrophilic and hydrophobic surface properties for B16F1 and B16F10 cells. Subtle changes in conditions from laboratory to laboratory could conceivably shift the proportions of cells underlying this heterogeneity, thereby drastically altering any averaged cell properties.

In a series of intriguing experiments, Baniyash, Netanel and Witz (1981) separated B16 melanoma cells on density gradients of colloidal silica. They showed that, on the average, cells prepared from solid B16F1 tumours were of higher density (1.073) than cells prepared from solid B16F10 tumours (1.067). Furthermore, low-density cells prepared from an unselected parental B16 tumour formed more lung colonies than did those of higher density. It seemed that this difference could not be attributable to host cell contamination, but the significance of these observations is difficult to put into the context of *in vivo* metastasis. Density changes could be related to cell cycle stages, which are known to affect both cell adhesion (Elvin and Evans, 1983) and lung colony formation (Suzuki *et al.*, 1977). A similar study to that of Baniyash and colleagues (1981), but with fibrosarcoma cells, also suggested an inverse relationship between cell density and lung colonization (Grdina *et al.*, 1977).

Further points relevant to selective metastasis will be discussed in some of the following sections, but generally speaking it seems that there is no universal explanation for this phenomenon. Indeed, because of the complexity of the metastatic process, it is not hard to envisage that any of a number of events (acting either alone or in association with others) could promote selectivity.

3.4.8 B16 adhesion, locomotion and metastasis

As will be explained in more detail in Chapter 5, cells can adhere to other cells of the same or different type (homotypic and heterotypic adhesion respectively), or they can adhere to substrates such as the basement membrane. During blood-borne metastasis, malignant cells must lodge in blood vessels before they can extravasate to form secondary lesions, but it must not be forgotten that lodgement alone cannot guarantee subsequent metastatic growth. Tumour cell lodgement may result from simple mechanical trapping or from some form of adhesive interaction between the circulating tumour cells and the endothelium. It might be supposed that mechanical trapping would be enhanced if tumour cells were bigger or if they formed aggregates either with each other or with other host cells. In general this is probably true (Fidler, 1973b; Liotta, Klienerman and Saidel, 1976), but the eventual outcome must be balanced against cell deformability and the possible destruction of larger cells or aggregates by shearing forces within the blood.

3.4.8.1 *Homotypic B16 adhesion*

The spread of tumours in the blood may involve the detachment of small homotypic clumps of cells following destructive *en masse* invasion of blood vessels or alternatively, single tumour cells may aggregate homotypically in the blood prior to entrapment. That neither of these processes may be that common is suggested by the monoclonal origin of many metastases. Nevertheless, it is clear that homotypic aggregates of tumour cells are more likely to form colonies after i.v. injection than are single cells, as has been shown for B16 cells by Fidler (1973b).

Much has been made of the observation that B16F10 cells display more homotypic aggregation than B16F1 cells and greater adhesion to homotypic monolayers (Nicolson and Winkelhake, 1975; Winkelhake and Nicolson, 1976). Furthermore, the selection of B16 cells with reduced homotypic aggregation potential resulted in cells with reduced lung colonizing potential (Lotan and Raz, 1983), while the selection of cells for homotypic aggregation ability resulted in cells with increased lung colonizing ability (Updyke and Nicolson, 1986) relative to their respective parental lines. Analyses based on well defined viscometric methods suggest, however, that under these particular conditions B16F1 cells can aggregate homotypically more extensively than B16F10 cells (Elvin and Evans, 1984). It should be noted that there are fundamental differences in the technical approaches used in the papers referred to above, since viscometric conditions result in near maximal aggregation within 10 min whereas other systems (such as rolling tubes) require as much as 1 h before near maximal aggregation is reached (Updyke and Nicolson, 1986). Presumably these differences in technique underly the reported differences in aggregation ability, although exactly how is not clear.

Interestingly, the early studies of Coman and colleagues (Coman, 1944; McCutcheon, Coman and Moore, 1948) using detachment techniques suggested that malignant cells may have decreased homotypic adhesiveness relative to normal cells. However, the generality of this conclusion has been questioned (Elvin and Evans, 1984, 1985) and there is now considerable evidence that different tumours have different homotypic adhesive capabilities and that this is not always related to metastatic ability. Elvin and Evans (1985), for example, isolated a number of B16F1 and B16F10 clones and showed that there was no relationship between lung colonizing ability and homotypic adhesiveness. Furthermore, in the study of Updyke and Nicolson (1986) in which cells with increased aggregating ability were selected from B16F1 cells, there was actually no relationship between the extent of homotypic aggregation and lung colonization. Thus all three selected cell lines used in lung colonizing experiments showed more homotypic adhesiveness than did B16F10 cells, but less lung colonizing ability.

The observations of Coman and others that tumours display decreased adhesiveness was interpreted as an explanation for how malignant cells might escape from the primary. Those groups that reported opposite results naturally found it convenient to interpret increased homotypic adhesiveness as being responsible for clump formation, which could promote tumour cell arrest and thereby increase metastasis. All this is very convenient for the scientists, who are continually searching for ways to interpret their results, but unfortunately it contributes little to our understanding of metastasis.

3.4.8.2 *Heterotypic B16 adhesion*

The ability to aggregate or adhere heterotypically might seem to be more important in the metastatic spread of tumours. Clumping with blood cells, or the attachment of tumour cells to the endothelium, for example, might promote lodgement and subsequent extravasation. Tao and Johnson (1982) examined the homotypic adhesiveness of B16 melanoma variants, as well as their heterotypic adhesion to aortic endothelial cells and the extracellular matrix laid down by them. They found that B16 variants with reduced lung colonizing and/or metastasizing ability showed reduced adhesion to endothelial cells and the ECM, whereas homotypic adhesion was either similar to or actually more than that for the controls. As mentioned previously, Fidler and his colleagues have repeatedly reported that in their hands B16F10 cells are more homotypically adhesive than B16F1 cells. If nothing else, the rather capricious behaviour of these cells highlights some inadequacies in its use as a model system. Furthermore, Poste and Fidler (1980) found B16F1 cells to adhere marginally but significantly more to intact human umbilical vein endothelium than did B16F10 cells. When studied for their attachment to subendothelial basement membrane, however, the B16F10 cells adhered much more significantly than did the B16F1 cells.

Fidler (1974 a and b) has suggested that in some cases (low lymphocyte to tumour cell ratios) immunized lymphocytes may bind to tumour cells *in vitro* without killing them, and the resultant clumps could be more easily trapped in small blood vessels following i.v. injection. He found that clumps also formed at higher lymphocyte to tumour cell ratios, but in these cases the numbers of lung colonies were reduced on subsequent i.v. injection, presumably because cytoxicity only becomes evident beyond a critical lymphocyte dose. B16LR6 cells selected for resistance to cytotoxic lymphocytes adhere less to lymphocytes and form fewer lung colonies than their parent B16F10 cells, indicating a possible correlation between these two phenomena in this experimental system (Fidler, Gersten and Budmen, 1976). These authors also suggested that B16F10 cells probably interact more with immunized lymphocytes than do B16F1 cells and that this interaction could contribute to enhanced lung colonization. However, lymphocyte–tumour cell clumps have not been quantified *in vivo* for the two cell lines, and furthermore, normal hosts are immunologically naive with respect to the B16 melanoma and thus intravenously inoculated tumour cells are unlikely to meet circulating, reactive lymphocytes. Even if heterotypic clumps with lymphocytes or other host cells could form in the circulation, it would not explain how B16F10 cells preferentially colonize both normal and ectopic lung tissue.

Platelets may also clump tumour cells, and such clumps may show a higher tendency to give rise to metastases. Gasic *et al.* (1973), however, have shown that initial entrapment of radiolabelled B16 cells in the lungs after i.v. injection is not significantly different in thrombocytopenic mice, and thus simple mechanical blocking by tumour cell–platelet aggregates would seem not to be of fundamental importance. Nevertheless, these authors also reported that the ability of B16 cells to colonize the lungs correlated with their ability to aggregate platelets *in vitro*. Furthermore, the i.v. injection of B16 cells was found to produce thrombocytopenia with accumulation of platelets in the lungs, but whether this correlated with enhanced lung colonizing potential was not explored. Tanaka, Tohgo and Ogawa (1986) used anti-platelet serum to reduce the number of circulating platelets, and found that the subsequent injection of B16 melanoma cells produced significantly fewer lung colonies. Menter and colleagues (1987) found that tumour cell association with platelets after i.v. injection of the amelanotic B16a variant was biphasic, peaking around 30 min and 24 h after injection. Most of the tumour cells were seen to be in direct contact with the endothelium, an observation in agreement with the earlier work of Roos and Dingemans (1979) which showed that although a thrombus containing fibrin, platelets and possibly leukocytes was associated with B16F10 cells arrested in the liver, the tumour cell projections which actually made contact with the endothelium were devoid of this material. The obvious conclusion here is that the melanoma cells adhere first to the endothelium and that platelet aggregation and fibrin deposition follow. It remains entirely conceivable, however, that as the malignant cells began to extravasate their motility relocated parts of the thrombus. In the study by Menter *et al.* (1987), evidence of intravascular platelet degranulation was found as early as 2 min after injection which could clearly influence tumour cell lodgement and extravasation. These authors suggested that the second peak in tumour cell–platelet association they detected (24 h) might coincide with tumour cell-induced endothelial cell retraction, but any possible relationship remains speculative.

The adhesion of tumour cells to platelets and/or the aggregation of platelets by tumour cell products are discussed more fully in section 5.7.2.2. With respect to the B16 melanoma, however, the study of Fitzpatrick and Stringfellow (1979) in which they compared the prostaglandin content of B16F1 and B16F10 cells is noteworthy. Somewhat surprisingly, in the light of studies on other cells, they found that both melanoma cell lines produced predominantly PGD_2, with lesser quantities of PGE_2 and PGF_{2a} and no PGI_2 (prostaglandin terminology is discussed in section 5.7.2.2). Perhaps of more interest, however, was their observation that B16F1 cells produced about 5 × more PGD_2 than did B16F10 cells. Although not as active as PGI_2 (which is produced predominantly by endothelial cells), PGD_2 can inhibit platelet aggregation in several species, including mice. In a subsequent report Stringfellow and Fitzpatrick (1979) showed that pretreatment of either B16F1 or B16F10 cells with the cyclooxygenase inhibitor indomethacin (which blocks prostaglandin production) resulted in an increase in lung colony formation. Taken together, the results from these two studies suggest that the inhibition of prostaglandin production in B16 cells might promote colonization by blocking production of the platelet inhibitory prostaglandin PGD_2, although there is no evidence of a causal relationship. When indomethacin treated B16 cells were injected along with exogenous PGD_2, the resulting number of lung colonies was reduced significantly for both tumour variants. It should be noted here, however, that the actual numbers of lung colonies involved were particularly small, with control B16F10 cells only producing an average of 5.1 lung colonies in one experiment and 2.4 in another. In the same experiments, involving i.v. injection of 2.5×10^4 cells, the B16F1 variant produced averages of only 1.7 and 0.6 lung colonies respectively. Such low levels of lung colonization might suggest that the injected tumour cells were not as viable as might have been liked, or that they had deviated from the original phenotypes selected by Fidler (1973a).

More recently, Honn, Cicone and Skoff (1981) have confirmed that PGD_2 can reduce colonization of the lungs, as well as of the liver and spleen, by using the amelanotic B16a variant. These authors have shown, however, that the intravenous injection of PGI_2 has a more marked effect, since they found that it reduced the number of lung colonies formed by as much as 70%. This effect was not due to the vasodilator properties of PGI_2, since PGE_2 which has similar vasodilator activity had no effect on lung colonization. The effect of PGI_2 (believed to be mediated in part by intracellular cAMP) could be potentiated by the i.p. injection of theophylline prior to the i.v. injection of tumour cells. Theophylline can induce the accumulation of cAMP by inhibiting its breakdown by the enzyme phosphodiesterase. Since the half-life of PGI_2 is relatively short (approx. 3–5 min) in physiological saline, there is some question as to exactly how it might operate *in vivo*. However, other evidence suggests that its half life may be promoted to around 60 min in plasma, possibly as a consequence of its interaction with albumin.

Since PGI_2 is a potent inhibitor of platelet aggregation, it was suggested that its anti-colonization effects are probably mediated through the blocking of tumour cell induced platelet aggregation. However, although it is known that B16 cells can induce the aggregation of human platelets (Gasic *et al.*, 1973), a more recent study found that cells from this tumour could not aggregate syngeneic mouse platelets (Stackpole, Fornabaio and Alterman, 1987). Although different B16 melanoma variants have been used in the studies reported above, the results of Stackpole and colleagues (1987) might lead us to question whether the anti-colonization effects of agents such as PGI_2 actually involve the inhibition of B16 induced platelet aggregation. There is a further complication, however, and that is that although most platelet aggregating agents are active in citrated plasma, some tumour cells appear unable to aggregate platelets under such circumstances (Gasic *et al.*, 1973). This clearly separates the platelet aggregating activity of some tumour cells from that due to ADP, for example, but it also brings the data of Stackpole and colleagues (1987) into question since these authors used citrated plasma. Other experiments have shown that B16 tumour cell induced platelet aggregation can occur in citrated plasma providing 2 mM $CaCl_2$ is added (Tohgo *et al.*, 1984). Since citrate is a calcium chelator, it would seem that its inhibitory effects are due to the removal of this divalent cation. The aggregates induced by B16 cells in $CaCl_2$ containing citrated plasma were found to be irreversible (unlike those induced by ADP), and they could be prevented by thrombin inhibitors such as MD-805 but not by ADP scavengers such as creatine phosphate/creatine phosphokinase. Tanaka *et al.* (1982), using heparinized plasma, have shown in fact that B16 cells can aggregate syngeneic platelets, although the malignant cells were from solid tumours and could have been contaminated with host cells. This is an important point since tumour-derived macrophages can release platelet aggregating factor (1-O-alkyl-2-O-acetyl-*sn*-glyceryl-3-phosphorylcholine or PAF). These authors also showed that B16 cells could induce significant fibrin formation which enmeshed the aggregates within minutes, and indeed by the end of such experiments (about 6 min) the preparation had clotted.

One possible view of these results is that the effects of the B16 tumour on platelet aggregation may be mediated via fibrin, and indeed the B16 tumour does contain significant procoagulant activity which involves direct activation of factor X (Tohgo *et al.*, 1984). Taken in conjunction with the work of others discussed above, however, it would seem most likely that both direct and indirect processes are involved in the interactions of tumour cells and platelets.

Other evidence supporting a role for interaction with platelets in the lung colonizing ability of B16 melanoma cells stems from the work of Agarwal and Parks (1983). These authors showed that the i.p. administration of forskolin (82 μg per mouse) prior to tail vein injection of B16F10 cells could reduce lung colonizing ability by more than 70%. Forskolin, a diterpene extracted from *Coleus forskohlii* plants, is a potent stimulator of adenylate cyclase and is also active in inhibiting B16F10-induced platelet aggregation *in vitro*. Of course, the *in vivo* effects of forskolin are likely to be markedly diverse and thus it is difficult to ascribe its anti-colonization effects solely to the inhibition of platelet aggregation.

Despite the suggestive results with PGD_2, PGI_2 and forskolin reported above, other agents which inhibit platelet aggregation (such as the dipyrimidol derivative RA233), have been shown to have mixed effects on B16 spontaneous metastasis and colonization (Maniglia *et al.*, 1982; Stackpole, Fornabaio and Alterman, 1987). Furthermore, Kimura *et al.* (1987) have failed to show any significant correlation between the lung-colonizing ability of a series of B16F1 clones and their platelet-aggregating activity *in vitro*, and thus we must conclude that the extent of platelet involvement in B16 colonization is far from clear.

Nicolson and Winkelhake (1975) and Elvin and Evans (1984) have shown that B16F10 cells aggregate more with lung cells than do B16F1 cells, and that this parallels their lung-colonizing ability. This is all the more surprising since the latter workers had shown that homotypically, B16F1 cells are more adhesive than B16F10 cells. Visual evidence that B1F10 cells aggregate more with lung cells than do B16F1 cells was obtained by Elvin and Evans (1984) through labelling the lung cells with the fluorescent lipid probe AFC_{16}. More mixed aggregates (fluorescent lung cells plus non-fluorescent B16 cells) were found when the lung cells were aggregated with B16F10 cells than with B16F1 cells. When tested against liver cells using the same approach, there was no visual evidence for pronounced heterotypic interaction using either the B16F1 or the B16F10 cell lines and thus the interaction with lung cells had a degree of selectivity. It should be noted here that this does not indicate exclusive specificity, merely that when tested against liver cells and lung cells, B16F10 cells show a tendency to aggregate more with the latter than do B16F1 cells. A demonstration of exclusive specificity (that B16F10 cells can *only* adhere to lung cells) would in fact be very surprising: B16F1 cells can colonize the lungs (it is merely a matter of degree) and both B16 cell types can form extra-pulmonary tumours. Nevertheless, the demonstration of a degree of adhesive selectivity may be important in tipping the balance of lung colonization in favour of B16F10 cells. Since the lung cells used in these experiments were not homogeneous, it was not possible to ascertain whether any particular cell type (such as the endothelial cells) was responsible for the differences in heterotypic adhesion displayed by B16F1 and B16F10 cells.

In a subsequent study, Nicolson *et al.* (1985) were able to show that more lung colonizing B16F10 cells attached to and infiltrated lung tissue than ovarian or heart tissue. Furthermore, they also found that more ovary colonizing B16-O10 cells attached to and infiltrated ovarian tissue than lung or heart tissue. Interestingly, Mareel *et al.* (1981) were unable to show profound differences in the invasive ability of B16 variants using the chick heart fragment invasion assay (section 4.8.4). To some degree, these different results probably reflect the nature of heart muscle which appears to be poorly invaded by malignant cells, but they also serve to focus our attention again on the limitations of model systems.

As a general observation, Nicolson *et al.* (1985) noted that B16 cells appeared to attach to exposed blood vessels on cut tissue surfaces, but this was not quantified. Netland and Zetter (1985) used cryostat sections of organ tissue to select for B16F1 cells which had enhanced heterotypic adhesiveness. Their model was based on that of Stamper and Woodruff (1976) in which lymphocyte adhesion to postcapillary venules was analysed on frozen sections of lymph node (section 2.12.2). There was no evidence, however, that Netland and Zetter (1985) were selecting for tumour cells that adhered preferentially to blood vessels. Nevertheless, these authors were able to obtain B16F1 cells with increased lung colonizing potential after only one *in vitro* selection step (from an average 19 colonies per mouse to an average 64 per mouse). After three such steps, there was a six-fold increase in colonizing ability. Variants subjected to four selection steps were found to have an altered morphology (fewer processes) and greater motility (particularly on fibronectin) with respect to B16F1 cells. However, these authors were not able to select variants for increased adhesion to brain tissue sections. Not surprisingly, such cells showed patterns of colonization that were not significantly different from their parent cells. Interestingly, brain-colonizing variants can be selected by other means (Brunson, Beattie and Nicolson, 1978) and these display enhanced adhesion to brain endothelium *in vitro*. Thus B16-B14B cells, which selectively colonize the brain meninges, adhere more to syngeneic brain endothelial cells than do B16F1 cells (Nicolson, 1982b). It is not known why it proved impossible in the study by Netland and Zetter (1985) to select for brain colonizing variants, but perhaps there were too few of such variants in the initial cell population.

Little is known of what molecular entities might be involved in the adhesion of B16 cells to endothelial cells, although Ramachandran *et al.* (1986) have shown that inhibition of cholesterol synthesis in endothelial cells facilitates the adhesion of B16F10 cells. One possible explanation for these results relates to the effects of cholesterol on membrane fluidity (section 1.7.9), since it is extremely unlikely that direct interaction between cholesterol molecules mediates adhesion.

Somewhat surprisingly, Ramachandran *et al.* (1986) failed to show any effects on tumour cell adhesion of treatment of the endothelium with glycosylation inhibitors such as tunicamycin and swainsonine, and this would suggest that oligosaccharide determinants on the endothelial cell surface might not be important in tumour cell adhesion. Previously, Irimura, Gonzalez and Nicolson (1981) and Irimura and Nicolson (1981) had shown that both lung colonization and adhesion to the endothelium by B16 cells are reversibly reduced following treatment of the melanoma cells (rather than the endothelium) with tunicamycin. Treatment with this agent resulted in a decrease of a particular high molecular mass cell surface sialogalactoprotein, and it was proposed that such a molecule probably served a crucial role in the adhesion of B16 cells to the endothelium. Raz and his colleagues (1980) have also provided evidence that sialogalactoproteins are important in lung colonization by B16 cells. Other studies by Humphries *et al.* (1986) have shown that pretreatment of B16 melanoma cells with swainsonine can reduce their lung colonizing ability by up to 80%.

More recently, Ferro *et al.* (1988) have prepared an extract from bovine aortic endothelial cells which is as active in promoting B16 cell adhesion as fibronectin and laminin. The principal component of this extract has an M_r >600 kD as determined by gel filtration. Since it remains at the origin, the major component runs differently in agarose gels when compared to laminin and fibronectin. The adhesive-promoting nature of this extract was confirmed by probing Western blots of such gels with B16 cells using a modification of the bioautographic method of Klebe *et al.* (1978).

Bioautography represents a rather ingenious approach to the problem of identifying adhesive proteins. In its simplest form, the technique involves chemical extraction of putative adhesive molecules, their separation by PAGE, and subsequent blotting onto nitrocellulose. The Western blots are then treated with a hopefully non-adhesive protein solution such as albumin, which acts as a blocking agent to cover parts of the filter not occupied by the separated proteins of the original extract. When the Western blots are then incubated with a suspension of cells, any active adhesive molecules can be recognized by the fact that they will bind the cells in question. Cells which do adhere to any specific molecules can be easily stained to enhance their contrast. Since appropriate standards can be run simultaneously with the protein extracts, it is a simple matter to determine the molecular mass of any separated molecule which is adhesive for the cells used in the test (reviewed in Ferro *et al.*, 1988).

Despite the reports referred to above, the exact mechanism by which B16 melanoma cells adhere to the endothelium remains uncertain.

Clearly, more research needs to be done in this direction, particularly since a major step forward in tumour biology will be achieved when the precise nature of the cell surface determinants involved in tumour cell–endothelial cell adhesion are identified. Further discussion on the molecular nature of tumour cell–endothelial cell interactions with reference to other experimental tumour systems can be found in section 5.7.2.1.

3.4.8.3 *Substrate adhesion of B16 cells*

The behaviour of tumour cells such as the B16 melanoma is influenced in several ways by the substrate. As will be discussed in more detail later (section 5.12) adhesion can govern the directional locomotion of tumour cells, thereby possibly contributing to the direction of invasion and to the process of extravasation. The extent of adhesion to the substrate also has profound effects on cell shape and growth. Raz and Ben-Ze'ev (1983) and Klein, Xiang and Kimura (1984) grew B16F1 melanoma cells on a poorly adhesive substrate made by coating dishes with 0.12% poly(2-hydroxyethyl methylacrylate) i.e. poly(HEMA). Both groups found that this caused a reduction in growth rate and the formation of cell aggregates. However, whereas the former group found that this process caused an increase in lung colonizing potential, these results could not be repeated by the latter. The results of Raz and Ben-Ze'ev (1983) are interesting since they indicate that decreased substrate adhesion and spreading correlate with lung-colonizing potential, whereas a number of reports have often claimed the opposite. In earlier studies, it had been shown that DNA synthesis (Raz and Ben-Ze'ev, 1982) but not protein synthesis (Ben Ze'ev and Raz, 1981) decreased when B16 cells were grown in spheroidal rather than spread configurations. The change in DNA synthesis, however, did not correlate with colonizing potential. Interestingly, Klein and his colleagues found that the growth of B16F1 cells on poly(HEMA) induced the formation of multinucleated cells, presumably by encouraging cell fusion. Although it has been suggested that cell fusion may contribute to tumour progression by increasing diversity, the full significance of this process is not clear. Multinucleate cells (being somewhat larger than normal) could conceivably display increased colonization behaviour as a result of enhanced mechanical trapping.

Briles and Kornfeld (1978) selected detachment variants by treating cultures of B16 cells with various concentrations of EDTA for different times. They found that the cells which detached more easily gave rise subsequently to fewer lung colonies that the parent cells, indicating a relationship between substrate adhesion and colonization. However, these authors were unable to demonstrate that B16F1 cells were more easily detachable than B16F10 cells despite the differences in their lung colonizing ability and thus as we have already seen, the relationship between adhesion and colonization is not generally applicable. Rieber and Castillo (1984) have reported that the pyrimidine analogue 5-bromodeoxyuridine (BrdU) causes a decrease in B16 cell detachment from the substrate and that it enhances cell flattening. Sequential solubilization with hydrophilic solvents and non-ionic detergent revealed that these behavioural changes were associated with an increase in a hydrophilic 140 kD glycoprotein in B16F1 cells. Different forms of the 140 kD glycoprotein were found in B16F1 and B16BL6 cells (being more hydrophilic in the former and hydrophobic in the latter), but precisely how these relate to changes in adhesion and colonizing or invasive ability are not clear.

Vollmers and Birchmeier (1983) prepared a range of monoclonal antibodies (McAbs) against B16 melanoma cells by syngeneic immunization with B16F1 cells. Although these cells are poor lung colonizers following i.v. injection, some of the McAbs the authors isolated nevertheless had marked effects on colonization. The McAbs were actually selected for their ability to inhibit the adhesion of B16 cells to tissue culture plastic: the inhibition was not absolute since with time (4 h) adhesion reached control levels. When B16 variants with high lung colonizing potential were pretreated with active McAbs (1 h), washed, and then injected into experimental animals, significantly reduced numbers of lung lesions were noted relative to controls. Similarly, i.p injection of active McAbs 4–5 h before i.v. injection of malignant cells significantly reduced the numbers of lung colonies. The effects of all of these McAbs required pretreatment in one form or another, which might suggest they operate early in the colonizing process (e.g. endothelial adhesion). Although the McAbs used in these experiments were not cytotoxic *in vitro* in association with rabbit complement, this does not mean that cell mediated cytotoxicity did not occur *in vivo*. Furthermore, although the McAbs interfered with adhesion to plastic surfaces it does not follow that they acted *in vivo* by inhibiting adhesion to the endothelium, or to any other structural component for that matter. It could also be that the McAbs block adhesion through steric hindrance and thus their precise mode of action is far from clear. Interestingly, the McAbs produced did not react with normal adult mouse tissues, but they did cross-react with transformed mouse cells and also with 6 out of 10 human melanomas.

It is worth noting here that anti-tumour assays involving pretreatment of either the malignant cells or the host would seem to be of little clinical relevance. A better type of assay, which more closely resembles the clinical situation, involves the treatment of mice with established tumours, although this too has its limitations. A popular model, for example, is based on the ascitic growth of leukaemic cells, with screening involving the i.p. injection of potential anti-tumour molecules. It is perhaps not surprising that many substances have potent anti-tumour activity in this type of assay since the peritoneal cavity essentially acts as an '*in vivo* test tube'. For many cytotoxic anti-tumour compounds, i.p. levels may be achieved that can destroy ascitically growing tumour cells without killing the host simply because of the fairly direct accessibility of the tumour cells in this situation. A more relevant model would seem to be one based on the eradication of an established tumour in the tissues of a host animal.

3.4.8.4 *Haptotaxis, motility and the invasive ability of B16 cells*

B16 melanoma cells are thought to respond haptotactically to substrate bound gradients of laminin and fibronectin, which are important molecules of the extracellular matrix (McCarthy and Furcht, 1984). It might be supposed that if a gradient of molecules like laminin and fibronectin existed from the luminal surface of the endothelium to the interstitial stroma, then haptotactic migration might direct the extravasation of malignant cells. Indeed, the endothelial luminal surface has a relatively small amount of fibronectin and laminin compared to the basement membrane, but unfortunately there is no evidence that this drives extravasation. Since tumour cells secrete proteolytic enzymes which digest fibronectin they may form their own gradient, but again there is no evidence that this is actually the case. Because cells in the blood are subjected to quite high concentrations of plasma fibronectin (300 $\mu g\ ml^{-1}$), it may well be asked how they could respond haptotactically to concentrations of substrate bound fibronectin which are a good deal less (peak 12.5–25 $\mu g\ ml^{-1}$). However, preincubation of melanoma cells with physiological concentrations of fibronectin has no effect on subsequent haptotaxis, presumably because soluble fibronectin interacts poorly with its receptor.

It is by no means clear as to whether quantitative differences in locomotion *in vitro* are of significance *in vivo*. With respect to the B16 melanoma system, the more invasive B16BL6 variant, for example, is actually consistently less motile (average speed about 6 $\mu m\ h^{-1}$) than the B16F10 variant (9 $\mu m\ h^{-1}$) when measured over three days using the agar droplet test. According to Raz and Ben-Ze'ev (1987), the lung-colonizing B16F10 variant is more motile than the non-colonizing B16F10-LR6 variant, but the difference from published drawings amounts to less than a cell diameter over 16 h. It remains an open question as to whether such minimal differences are of any significance *in vivo*. Using a Boyden-chamber based-system, Fligiel *et al.* (1986) found B16F10 cells to be significantly less motile than either B16F1 or B16BL6 cells, but this assay measures the *number* of cells which penetrate through a nitrocellulose filter rather than their actual speed of migration. All three B16 variants studied by Fligiel *et al.* (1986) were found to synthesize laminin, which was both secreted and bound to the cell surface. Interestingly, the supposedly less motile B16F10 cells expressed less cell surface laminin and also secreted less of this protein. The motility differences between the melanoma variants appeared not to be simply related to laminin synthesis, however, since the differences persisted even in the presence of considerable quantitites of exogenous laminin.

The invasive capabilities of B16 cells were examined by Nicolson and Brunson (1977) who employed the chick CAM assay described in section 4.8.6. These authors reported that B16F10 cells seeded onto the CAM could invade it within 12–24 h whereas B16F1 cells failed to do so. Despite this difference in invasive capabilities, according to Weiss *et al.* (1982) there is little difference between B16F10 cells and their parental type in ability to form spontaneous metastases after intramuscular injection.

3.4.9 B16 cell growth and metastasis

It is entirely conceivable that the metastatic ability of tumour cells relates to their growth potential: indeed, without growth tumours will not develop. However, is there a relationship between growth ability and metastatic potential? Stackpole, Alterman and Fornabaio (1985) explored this possibility using the 'null', colonizing and metastatic B16 cell variants referred to previously (section 3.4.6.1). Using this system, these authors found that the metastatic cells did not have the greatest growth potential and nor were they clonogenic in soft agar. Anchorage-independent growth seemed to correlate better with colonizing activity rather than metastatic activity as defined by their assay procedures. Interestingly, a number of groups have suggested that the drug sensitivity of human tumour cells growing in soft agar may have clinical relevance in the selection of particular chemotherapeutic regimens (Alberts *et al.*, 1980; Von Hoff *et al.*, 1981). In the light of the study by Stackpole, Alterman and Fornabaio (1985) it is worth questioning whether such assays actually test the chemosensitivity of cells with metastatic potential.

Vollmers *et al.* (1985) have prepared two monoclonal antibodies (McAbs) following injection of syngeneic C57Bl6 mice with B16F1 cells. These McAbs inhibit the growth of B16 cells in soft agar, cause cell rounding *in vitro* and inhibit their adhesion to plastic. Furthermore, B16 cells treated with them require higher than normal concentrations of serum for optimal growth. The authors consider these processes to be 'more normal' and the McAbs are thus referred to as 'normalizing antibodies' (NORM-1 and NORM-2). The effects of NORM-1 and NORM-2 on the *in vivo* growth of B16 cells was tested by first injecting the tumour cells i.v. (to allow lung colonies to become established) and then three days later injecting the McAbs either i.p. or i.v. (around 10-30 μg per mouse for NORM-2). Under these stringent conditions, NORM-1 and NORM-2 reduced lung colony formation by 70–90%. Furthermore, when B16 cells were injected subcutaneously and the McAbs were injected i.p. 3 days later, the weight of the primary tumour (after 15 or 30 days) was reduced by 50% with NORM-1 and by 80% with NORM-2 relative to controls. The original syngeneic immunization schedule which yielded these McAbs was chosen to maximize the generation of antibodies against tumour associated antigens, but the extent to which NORM-1 and NORM-2 antigens are present on normal cells has yet to be fully evaluated.

The full significance of these studies will become clear when the surface antigens against which the McAbs are directed are identified. Preliminary studies suggested that the NORM-1 antigen has an M_r of about 140 kD, while the NORM-2 antigen originally appeared to be a protein of M_r 59 kD. More recent work from the same laboratory, however, has suggested that the 59 kD protein is a contaminant and that the true antigen for NORM-2 is a glycoprotein with an M_r 83 kD on reduced SDS-PAGE (Wieland *et al.*, 1986). Under non-reducing conditions the molecule is slightly larger (M_r 105 kD) implying that it is probably a monomer. Its expression on B16 melanoma cells is markedly reduced when the cells are growth-arrested by either low serum or high culture density, or by treatment with TGF_{β}. When growth-arrested cells are restimulated, the glycoprotein reappears during S phase, but the full significance of these associations is not yet clear, and nor is it clear how the antibody exerts its effects *in vivo*.

3.4.10 The effects of differentiation

The tumorigenic potential of myeloid and erythroid leukaemic cells is decreased after the induction of differentiation by agents such as dimethylsulphoxide (Friend *et al.*, 1971). Furthermore, tumour-bearing mice live longer after *in vivo* treatment with differentiation inducing substances (Honma *et al.*, 1979). The differentiation of B16 melanoma cells, however, has been shown paradoxically to promote lung colonization, even though it was associated with the inhibition of cell proliferation (Bennett *et al.*, 1986). In this study, differentiation was assessed by increased pigmentation and increased tyrosinase activity after stimulation with the melanocyte stimulating hormone N-MSH (4-norleucine-7-*D*-phenylalanine) at high pH. Tyrosinase is a major enzyme involved in the synthesis of melanin, and the process of melanogenesis is influenced through MSH presumably via its effects on cAMP. Although differentiation was associated with an increase in cell size, Bennett and her colleagues (1986) were unable to show any differences in initial entrapment between differentiated and non-differentiated cells, although the induced cells apparently survived better in the lungs. Furthermore, there was no difference between the two cell types with respect to the cytotoxic effects of NK cells and macrophages, and thus non-specific immune mechanisms at least do not seem to be responsible for the underlying differences in colonizing potential. These results are perhaps not too surprising, since amelanotic B16 variants with colonizing ability already exist. Furthermore, the degree of pigmentation of human melanomas has little prognostic significance, and thus we must conclude that differentiation as assessed by melanization has little relevance to lung colonization, if not to metastasis and the malignancy of melanomas in general. To this we should add the possibility that melanization may actually not be a reliable indicator of differentiation in melanoma cells.

Nevertheless, since most melanomas are pigmented the pathway of melanogenesis offers scope for designing therapeutic regimens targeted to pigmented cells. Alpha-difluoromethylornithine (DFMO) has been shown to induce melanogenesis in B16F1 melanoma cells, although it is also remarkably toxic for them (Sunkara *et al.*, 1985). This agent is an irreversible inhibitor of ornithine decarboxylase, which is a rate-limiting enzyme in polyamine synthesis, and when added to B16 melanoma cells it causes a rapid depletion of putrescine and spermidine. Not surprisingly, the effects of DFMO on B16 melanoma differentiation can be reversed by the addition of exogenous putrescine. When DFMO is added to the drinking water of mice bearing the B16 melanoma it can induce a dose dependent inhibition of tumour growth. Polyamines are clearly important in the growth of cells (section 1.7.21), but their general role in differentiation is less certain. Interestingly, the inhibitory effects of interferon on the growth rates of certain cells are accompanied by the reduction of enzymes (including ornithine decarboxylase) involved in polyamine synthesis. When DFMO therapy (2% in the drinking water) was combined with s.c. interferon treatment (1000U per mouse), Sunkara *et al.* (1983) found a 96% reduction in B16F1 tumour growth (as measured by tumour weight after s.c. implantation). However, most mice were not without a detectable tumour burden, and thus although DFMO/interferon combination therapy reduces tumour growth, it is not curative. It should be noted that not all tumour cell types (e.g. L1210 leukaemic cells) respond to this combination therapy. Why B16 melanoma cells should be particularly sensitive to such treatment is not clear, although it is presumably related to the dual effects of this combination therapy on cell survival and melanogenesis. It must be emphasized, however, that the precise mechanism underlying the suppression of tumour growth by this form of treatment is not known. Although interferon may perturb polyamine synthesis, it also has other diverse effects on cellular mechanisms including interference with protein synthesis.

Xiang and Kimura (1986) have reported studies in which the differentiation state of B16 melanoma cells was shifted by simply changing growth medium conditions. They used a highly metastatic clone (BL6-10) of the B16BL6 variant and induced phenotypic changes by transferring the cells from Eagle's amino acid supplemented medium containing 10% newborn calf serum (EHAA) to Dulbecco's modified Eagle's medium containing 10% fetal calf serum (DMEM). Cells grown in EHAA had an epithelial morphology, were light grey in colour (amelanotic) and had considerable lung colonizing ability. When grown in DMEM, the cells became more spindle shaped, took on a black coloration (melanotic), and were less capable of forming lung colonies (mean colony numbers of 319 vs 44 respectively following injection of 3×10^5 cells). Cells grown in DMEM also had a slower growth rate and altered antigenic expression. These changes were reversible, but the media components involved have not been identified. The key antigen studied by Xiang and Kimura (1986) was a 72 kD cell surface determinant known as Met-72. Although the expression of this antigen correlates with lung colonizing potential, its precise relevance in metastasis remains to be determined (section 3.4.7.2).

Retinol (vitamin A) and its derivatives have been shown to have effects on cellular differentiation, and several groups have reported that pre-treatment of B16 cells with retinoids inhibits lung colonization (Lotan and Nicolson, 1981; Edward, Gold and MacKie, 1989), Unfortunately, like many other differentiaitng agents the retinoids exert extremely diverse effects on treated cells and it is difficult to relate the inhibition of colonization seen following pre-treatment of tumour cells with these agents to any specific cell characteristic. In any case, whether specific characteristics are correctly labelled as markers of increased differentiation or not is fraught with difficulties as we have already seen. Apart from inhibiting lung colonization, retinoids have been shown to reduce the adhesion of B16 melanoma cells to basement membrane components such as laminin and type IV collagen (Edward, Gold and MacKie, 1989), to reduce their ability to invade reconstituted basement membrane and to inhibit their type IV collagenolytic activity (Nakajima *et al.*, 1989), to lower their growth rate (Maniglia and Sartorelli, 1981), and to perturb their production of glycosaminoglycans (Maniglia and Sartorelli, 1981; Edward and Mackie, 1989).

3.5 *IN VIVO* MODELS FOR LYMPHATIC METASTASIS

As exemplified by the B16 malignant melanoma, most experimental studies of metastasis have involved haematogenous spread, even though a considerable number of human tumours metastasize via the lymphatic system. The experimental study of lymphatic spread was pioneered by Zeidman and Buss (1954) and a number of related model systems have now been developed for studying this phenomenon (Carr and Carr, 1982). The general approach in these model systems is to inject malignant cells into the footpad of an experimental animal and then to examine the draining popliteal lymph node. This is done in several animals over a range of time intervals and in this way a picture of the metastatic process may be constructed. Tumour cells may reach the subscapular sinus of the popliteal lymph node within 24 h of injection, but even earlier observations suggest how they enter the afferent vessels within the footpad. Some tumours (such as the dimethylbenzanthracene induced rat Rd/3 tumour) are thought to pass between the cells of the lymphatic endothelium with little tissue damage (possibly via patent inter-endothelial gaps), whereas the invasion of other tumours (such as the W256 carcinosarcoma) is associated with considerable endothelial damage. It is not clear whether tumour cells invade lymphatic vessels singularly or in clumps, since this actual step has not been seen *in vivo*, but some tumours (such as the rat 13762 mammary carcinoma) are found as cell clusters in the lumens of the lymphatic vessels. After reaching the subscapular sinus of the lymph node, the tumour cells proliferate before migrating down the radial sinusoids into the efferent lymphatic vessels and on to more proximal nodes. As explained in section 2.11, connections exist between the lymphatic and blood systems, and thus metastasis is seldom exclusively by lymphatic means. Indeed, the general approach of injection into the footpad has been used by many authors for the study of 'spontaneous' metastasis without recourse to the particular pathways involved.

Carr and Carr (1982) provide some circumstantial evidence in support of the possibility that tumour cells may be destroyed in the draining node. Thus the implantation of 5×10^6 Rd/3 tumour cells into the footpad of a syngeneic rat results in a few hundred cells being found in the draining popliteal node within 24 h, and a few thousand by 48 h. If the footpad is surgically removed at or before 24 h after implantation, lymph node metastases rarely occur. However, if the footpad is removed at a later stage then progressive metastases develop. These results suggest (but certainly do not prove) that the draining node may be able to destroy a few hundred tumour cells, thereby inhibiting metastasis if low numbers are involved. Certainly, if any cytotoxic mechanisms do operate at the level of the lymph node then they are effective in dealing only with small numbers of tumour cells.

3.6 THE NUDE MOUSE AS A MODEL SYSTEM FOR INVASION AND METASTASIS

The nude (*nu/nu*) mouse has played a limited but occasionally effective role in studies of the contribution of host effector cells to tumour destruction. Despite their lack of T cells, nude mice display an almost normal response to T cell independent antigens and they are able to respond to human tumours with elevated antibody titres (Adelman, Miller and Kaplan, 1980) although the significance of this is uncertain. A crucial observation relating to nude mice is that although they lack significant numbers of T lymphocytes, they do not have significantly more tumours than their normal littermates. This is now thought to be due to the presence of NK cells, although activated macrophages are also likely to contribute (section 1.8.2). Manipulation of the immune system of nude mice by treatment with anti-lymphocyte serum, X-irradiation or cyclophosphamide increases the likelihood of xenogeneic tumour acceptance and in some cases enhances colonizing capacity.

NK cells are not evenly distributed throughout the body, being particularly abundant (in decreasing order) in the peripheral blood, spleen, lymph nodes and bone marrow. If, as seems likely, NK cells play a role in the defence against tumours, then some sites in the body might be more effective in eradicating tumour cells than others. Indeed, the site of tumour implantation in nude mice seems to influence not only the tumour growth rate, but also invasive behaviour and the likelihood of metastasis. NK cells appear to display a degree of species selective activity (Hansson *et al.*, 1978) since such cells from the nude mouse can kill mouse YAC-1 lymphoma cells but not human K-562 leukaemia cells, even though the latter are highly sensitive to human NK cells (Naito *et al.*, 1987). This particular attribute of NK cells could be advantageous in the experimental establishment of xenogeneic tumours in nude mice.

The expression of NK cell activity in mice is genetically regulated and age dependent: the beige (*bg/bg*) mutant of the C57Bl/6 mouse strain, for example, is NK-cell-deficient, and young (3-wk) mice have fewer NK cells than do older mice (8–10-wk). It should be noted that beige mice apparently have essentially normal macrophage and T cell functions, whereas the NK cell activity of nude mice may be relatively enhanced. Recent developments in selective breeding have now yielded beige nude (*bg/bg* × *nu/nu*) mice which have both low NK cell activity and a T cell deficiency. Interestingly, the studies of Fodstad *et al.* (1984) and Naito *et al.* (1987) indicate that tumour cells grow as fast if not faster in adult nude mice as they do in beige nude mice, even though the latter are deficient in at least two types of defence cells. Whether these double mutants have increased macrophage activity to compensate for their lack of other defence cells or not is unknown.

The state of the tumour vascular system in experimental animals has important consequences if they are to be used as model systems for studying tumour therapy. When human tumours are grown in nude mice, for example, the vascular system and supporting stromal elements derive from the host while only the parenchymal tumour cells are of human origin (Warenius, Freedman and Bleehen, 1980). Few competent studies have been made, however, of vascular structure in experimental tumours. Generally speaking, although the vascular network of tumours is provided by the host, the nature of the tumour itself is not without an effect (Solesvik, Rofstad and Brustad, 1985). This is not surprising since angiogenic factors released by tumours, for example, are known to exert their effects in xenogeneic systems. According to Solesvik and colleagues (1985) the vascular structure of some human tumour xenografts is not unlike that in humans, even though the vessels are supplied by the host. This observation may explain in part why the responsiveness of some human tumour xenografts to both radiotherapy and chemotherapy may correlate with clinical effects. Of interest here is the report that human Kaposi's sarcoma cells implanted into nude mice give rise to typical Kaposi's sarcoma tumours, but they are of murine rather than human origin. It would seem that the human cells produce some factor which triggers growth of the mouse tumours, but as yet its nature is uncertain. The *hst* oncogene (originally found in human stomach cancers) has been isolated from Kaposi's cells, and sequence analysis of its product has shown it to be 45% homologous with basic FGF. Despite this relationship with a known growth factor and angiogenic substance, there is no evidence as yet that this protein has a causal effect in development of the sarcoma.

Despite the reports above on certain aspects of tumorigenesis, the nude mouse has not been particularly useful as a model system for studying the metastatic behaviour of human tumour cells. Indeed, malignant human tumour cells rarely metastasize when transplanted into nude mice, although this seems to be related to the nature of the transplanted tumour, the age and condition of the host, and the site of implantation. Prostatic carcinoma, for example, metastasizes poorly, although neuroblastomas have been reported to metastasize to lymph nodes, ovaries and the cerebrum with a distribution pattern similar to that shown in the donor (Hata *et al.*, 1978). It has been suggested that some human tumours may metastasize in nude mice only if they are implanted in their organ of origin, but results to date fail to support any generalizations in this direction.

Successful metastasis in nude mice has been observed for human renal carcinoma cells after their implantation into the kidney (Naito *et al.*, 1986) and for human colorectal carcinoma (HCC) cells after their implantation into the spleen (Giavazzi *et al.*, 1986). According to Jessup *et al.* (1989), there is a moderate correlation between the tumorigenic and metatstatic potential of HCC cells after injection into nude mice with the development of metastases in patients with colorectal carcinoma. Interestingly, these authors reported a more significant correlation between experimental tumorigenesis and time to recurrence in colorectal carcinoma patients following surgical treatment. They also showed that serum levels of carcinoembryonic antigen (CEA) in clinical patients correlated both with tumour growth and metastatic ability following the injection of HCC cells into nude mice. Although these results point towards the possible use of nude mice as prognostic indicators for malignant disease in humans, much more research is required before the validity of this system is esablished.

There is some indication that both tumour growth and metastasis may be greater in young nude mice than adult nude mice, presumably because NK cells are less abundant in the former. Unfortunately, efforts to study spontaneous metastasis in young (3-wk) nude mice are hampered by the fact that the time required for spread allows the host to develop competent NK cells which can attack tumour cells in the circulation thereby limiting secondary formation. Nevertheless, various studies indicate that NK cell effects are probably important in limiting tumour growth and spread (Hanna, 1980, 1982a). On the other hand, the results with beige nude mice reported above indicate an important role for other defence systems, probably macrophage related.

3.7 TRANSPARENT MODEL SYSTEMS FOR STUDYING CELL BEHAVIOUR DURING INVASION AND METASTASIS

The major drawbacks of most *in vivo* models for studying invasion and metastasis relate to technical features associated with tumour cell delivery, the number of tumour cells required, the physiological state of the host, and the difficulty of detecting small metastases in experimental animals. One other drawback of considerable significance concerns the inability to follow details of the invasive process *in situ*. Because of tissue opacity, model systems which are based on the formation of metastases *in vivo* usually fail to reveal information on how the metastases developed. Attempts to overcome this problem have resulted in the establishment of a number of *in vivo* models based on transparent tissues such as the hamster cheek pouch, the tadpole tailfin, gut mesentery and the rabbit ear chamber. Studies on these transparent systems, particularly the rabbit ear chamber, have provided data on how tumour cells behave *in situ* and this model system will now be discussed further.

3.7.1 The rabbit ear chamber

The use of the rabbit ear chamber for studying aspects of the motility of tumour cells *in vivo* was pioneered by Wood and his colleagues (Wood, 1958; Wood *et al.*, 1966; Wood, Baker and Marzocchi, 1967 a and b). The chamber basically consists of a porous floor separated by about 13 μm from an overlying coverslip via a plastic ring. The structure is inserted into the ear of an anaesthetized rabbit and left for a few weeks until blood vessels, lymphatics and connective tissue have grown within it. During emplacement the chamber fills with blood which clots, but this is soon replaced with normal tissue structures that grow upwards through the porous floor of the chamber. The host rabbit is anaesthetized only on emplacement of the chamber, and all subsequent manipulations can take place on the otherwise normal animal. Wood and his colleagues have reported some interesting observations using this model system. They found, for example, that PMNs and lymphocytes migrated within the chamber at similar speeds (typically around 5–7 μm min^{-1}) whereas macrophages and fibroblasts moved poorly (2 μm min^{-1}) showing prolonged stationary periods. They also reported that PMNs within small blood vessels could move on the endothelial cell surface with and against the blood flow, which in any case was prone to frequent reversal within the small vessels (Wood, Baker and Marzocchi, 1967b). The impression is gained that marginated PMNs probe the endothelial surface looking for intercellular gaps which might allow chemotactic factors to diffuse through. If this proves to be the case, then the relative accuracy of PMN extravasation in response to a tissue lesion can be understood.

When allogenic V_2 carcinoma cells (originally derived from a Shope virus papilloma) were introduced into the chamber they grew to fill it eventually. Tumour cells within the chamber were found to move at the same speed as PMNs and lymphocytes, but perhaps more interesting was the report by Wood, Baker and Marzocchi (1967b) that the tumour mass could undergo changes in differentiation. Indeed, over a period of time the same region of a tumour growing in an ear chamber was seen to express either undifferentiated or differentiated (acinar or squamous cell) histology. Lymphocytes were seen to invade or be phagocytosed by tumour cells in which they appeared to survive for some time. The full significance of this behaviour is not yet clear, although it may be related to the phenomenon of emperipolesis by which lymphocytes and neutrophils extravasate from blood vessels actually through the cytoplasm of the endothelial cells.

In other experiments, Wood (1958) labelled a suspension of V_2 carcinoma cells with trypan blue and introduced them into the rabbit ear chamber by injection into the central auricular artery just proximal to the chamber. This allowed him to record early events involving the marked cells, and he noted that the majority of injected cells were in fact destroyed within 10–15 min after injection. Surviving tumour cells had a marked tendency to stick to each other and to the endothelium. Within a few minues the adherent cells were surrounded by an observable thrombus which was often temporary, disappearing within 12–24 h. V_2 cells stimulate fibrin deposition, and the injection of heparin was found to decrease adherence to the endothelium. It would seem likely from these results that low levels of fibrin probably contribute to the adhesion of the tumour cells to the endothelium, but whether fibrin mediates adhesion directly (as a form of glue) or indirectly (more like a plug) is not clear. Adherence to the endothelium was most obvious in the capillaries, and although tumour cells could also adhere to arterioles and venules, tumour growth from single V_2 cells was only ever observed within the capillaries. Tumour cell adherence to the capillary endothelium was independent of leukocyte adhesion, capillary diameter, and the rate of blood flow suggesting that adhesion was not simply due to mechanical filtration.

Tumour cell division was first noted in 24 h, and invasion of the endothelium typically occurred after about 48 h. Occasionally, however, V_2 cells were seen to extravasate from blood vessels as early as 150 min after initial adherence. In virtually all cases, extravasation was noted to occur in the wake of emigrating leukocytes which may have created defects in the endothelium that were open to capitalization by nearby tumour cells as well as by other leukocytes. It is worth noting here that not all tumours extravasate in the pathway of emigrating leukocytes, and in many respects these observations of Wood and colleagues may now be considered to be unusual rather than representative.

V_2 carcinoma cells have been reported to move *in vitro* at about 0.7 μm min^{-1}. This rate of movement was not recorded *in vivo*, although the tumour cells did display active locomotion as they extravasated. About 72 h after introduction of the marked carcinoma cells, Wood was able to report the presence of a distinct metastasis (Wood, Baker and Marzocchi, 1967b).

The choice of trypan blue as a marker for the tumour cells in many of these experiments is somewhat surprising, given that this dye is used in many laboratories as a marker for dead cells since it is excluded by the intact plasma membranes of living ones. In retrospect, it would seem that a number of the tumour cells tracked by Wood and his colleagues using this technique may not have been fully viable, although some living cells may have actively phagocytosed the dye.

The rabbit ear chamber is a useful model system for observing cellular events which take place during invasion and metastasis. There are problems in identifying particular cell types from time lapse recordings, and some of the earlier work may have suffered from this difficulty. Although the identification of tumour cells *in vivo* is generally recognized as a technical problem, recent advances in cell labelling, video recording, and image analysis should encourage the repetition of these pioneering experiments. The main limitations of this system are first, that most studies have used allogeneic tumours, and secondly, that it provides no information on the molecular events involved in invasion and metastasis. This is best achieved by using other more appropriate model systems in concert.

3.8 THE EMBRYONIC CHICK AS A MODEL SYSTEM FOR INVASION AND METASTASIS

The embryonic chick was probably developed as a model system for studying invasion and metastasis for no reasons other than that eggs are cheap, plentiful, and easy to maintain in the laboratory. If the truth be known, they were probably also simply to hand at the time! Tumour cells can grow directly on the chorioallantoic membrane (CAM) of the embryonic chick when they are seeded through a window created in the shell. In essence, this is a test for tumorigenicity, but assays for both organ colonization and spontaneous metastasis are also possible with this system. Thus some tumours will colonize various organs of the embryonic chick after their injection into the major CAM vein and others will invade after seeding directly onto the CAM. Spontaneously invasive cells may form growths within the mesodermal layer of the CAM, and some may spread to distant organs in the developing chick. The CAM has also been used as a model system for assaying angiogenic activity, since many tumours will induce the directed growth of new blood vessels when they are seeded on to it (section 1.7.20).

Most normal cells other than macrophages do not invade the CAM, and these will not of course metastasize. However, invasion of the CAM mesoderm by tumour cells does not correlate well with their invasive behaviour in syngeneic hosts or in other model systems. Variants of the B16 melanoma selected for invasiveness of the CAM, for example, do not invade in other model systems based on the mouse bladder or dog femoral vein (Poste *et al.*, 1980). Cells selected for invasion through these latter tissues are invasive in the syngeneic host and will invade the CAM. Furthermore, i.v. injection into the chick CAM of B16F1 and B16F10 cells results in similar numbers of tumours in the lungs, whereas this clearly is not the case when they are injected i.v. into syngeneic mice (Chambers, Shafir and Ling, 1982). The lack of correlative significance shown by the CAM invasion model is a major detraction from employing this system. Nevertheless, Ossowski and Reich (1983 a and b) have made rather ingenious use of the CAM as a xenogeneic system. They studied metastasis to the embryonic chick lung of human HEP3 epidermal carcinoma cells following their implantation on the CAM. Spread could be inhibited by an antibody directed against human urokinase (plasminogen activator) which was not effective against chick PA. These results suggest that some human tumours utilize urokinase during invasion and this enzyme may therefore be of some benefit in metastasis. We have seen already, however, that the overall situation is somewhat more complex than this (section 2.15).

One other serious problem working with the CAM system is that metastases usually take a while to develop, during which time the chick becomes immunocompetent. The effects of this immunocompetency on tumour growth are not fully understood, but since most experiments usually employ xenogeneic combinations it is deemed prudent to restrict the time period over which tumour growth is studied to within day 18 of incubation. The problem this presents is one of recognition, since the CAM is not useful until around day 8 or later and during the remaining days metastases do not develop enough to be recognized naked eye or histologically. Chambers, Shafir and Ling (1982) have overcome this difficulty by making use of the differential sensitivity of rodent and chick cells to ouabain. Briefly, mouse cells are more resistant to ouabain than are chick cells, being able to survive concentrations two orders of magnitude greater than that capable of killing the latter. Mouse tumour cells which metastasize to the chick liver in the CAM system can be detected by a ouabain rescue technique in which the organ is dissociated and allowed to grow in medium containing levels of ouabain (2×10^{-5}M) toxic for chick cells but not mouse cells. The cell colonies which grow under these conditions must have come from tumour cells present in the liver (or within its blood vessels) even if secondary growths were not visible at the time of dissociation. Using this assay, Chambers and her colleagues (1982) were able to show that various tumours displayed similar degrees of trapping within the liver, and that differences in their abilities to form metastases were attributable more to differences in cell survival and growth rates. Although B16F1 and B16F10 cells appear not to spontaneously metastasize from the CAM when assayed by conventional procedures, these authors were able to show limited metastasis for both cell types using the oubain rescue technique.

There is another problem associated with the use of the embryonic chick CAM as a model sytem, both *in vivo* and *in vitro*, and that is that the results obtained are sensitive to the status of the CAM. Zwilling (1959) and Armstrong, Quigley and Sidebottom (1982) have reported better tumour growth when the CAM is slightly damaged, and Chambers, Shafir and Ling (1982) found that tumours formed more readily if cells were applied to the CAM immediately after opening the shell window. Opening of the shell may lead to injury of the CAM which is repaired over the next 48 h, and this sublethal damage could be crucial for subsequent metastasis. The possibility of CAM injury may explain why results with the CAM system do not always agree with other studies using different experimental approaches. Indeed, the CAM has been thought of as a less resistant system when compared to other assays for invasion and metastasis. The use of the CAM more as an *in vitro* model system is described in section 4.8.6.

Tickle and her colleagues have used the developing chick wing at stage 20–21 as a 3D model system for the study of invasion (Tickle, Crawley and Goodman, 1978 a and b; Tickle and Crawley, 1979). At this age of development, the chick wing is composed of loose mesenchyme bounded by a two layered ectoderm, and it consists of a bulge or bud about 1.5 mm across which protrudes about 1 mm. These authors grafted various normal and tumorigenic tissues to the wing bud and after 1 or 2 days they fixed and stained the tissue and examined it histologically for signs of invasion. Invasiveness was assessed qualitatively by the presence of single cells or small groups of cells away from the graft. They found that some normal cells (e.g. embryonic pigmented retina and heart ventricle cells) were not invasive, whereas others were (e.g. trophoblasts, lymphocytes, PMNs). Cells from some non-transformed cell lines (e.g. BHK and Nil 8) could also invade as could their virally transformed counterparts (e.g. Py-BHK and HSV-Nil 8). Only some invading cells (e.g. trophoblasts and HSV-Nil 8 cells) did so destructively. Interestingly, cells from various epithelial tumours could infiltrate the ectoderm but they seldom invaded the mesenchyme, even though carcinomas are seen clinically to penetrate underlying mesenchymal tissues. The carcinoma cells were found to form desmosomes with the ectodermal cells, from which it was suggested that adhesive interactions might limit invasion by carcinoma cells. Clinically, carcinomas are often seen to invade as cords or sheets of cells (Type 1 invasion) and although suggestive of high intercellular adhesiveness it is not proof of such. Within the chick limb, most carcinoma cells formed a smooth border with the the mesenchyme even in the absence of a basement membrane, so this structure alone cannot explain the failure of the cells to invade. An important comment made by Tickle *et al.* (1978b) was that a small minority of carcinoma cells could invade the mesenchyme, although in many cases these authors were uncertain as to the precise origin of these cells. That some cells (even if only a few) can invade the mesenchyme is likely to have significant impact on subsequent events since it is known that many tumours and their metastases arise from single cells.

Like other models, the developing chick wing bud is clearly an artificial system for studying tumour cell invasion. Nevertheless, it may prove useful for further studies since it offers a natural (although developing) 3D environment. The biggest problem facing this model system at present, however, is that it does not lend itself easily to quantification.

4 *In vitro model systems for the study of tumour cell invasion and metastasis*

4.1 INVASION AND METASTASIS *IN VITRO*

We have seen that metastasis is a complex process which involves a number of steps beginning with local invasion and detachment from the primary tumour. Because of its very nature, metastasis itself cannot be studied *in vitro*, but each of the steps involved may be subjected to investigation. It has been suggested that invasiveness is one of the more important phenotypic correlates of malignancy, and it is considered to play a pivotal role in the metastatic process. For this reason, many *in vitro* model systems have been designed to study the invasive behaviour of malignant cells. Considerable hope is held for these systems as proving grounds for potential anti-metastatic drugs, but in the first instance they are being used to further our currently fragmentary knowledge on the behavioural properties of malignant cells.

Until now, we have been using the term invasion rather loosely to mean the penetration of surrounding normal tissue by a malignant tumour. However, there are in fact two aspects to invasion, namely tissue penetration and tissue destruction. Normal cells can penetrate tissues: PMNs, for example, can penetrate through endothelial and epithelial barriers and also through collagenous connective tissues. Since PMNs contain a number of lytic enzymes they probably use these to break down certain tissue components (such as collagen) or to disrupt other barriers (such as cell junctions) in order to facilitate penetration. Normal cells such as PMNs are usually limited, however, in their extent of tissue destruction. Malignant cells, on the other hand, may penetrate normal tissues with considerable destruction and, generally speaking, both tissue penetration and destruction are progressive in time. Not surprisingly, the interpretation of many *in vitro* experiments on invasion is critically dependent on the criteria chosen in its definition.

4.2 GENERAL FEATURES OF *IN VITRO* MODEL SYSTEMS

It should be noted at the outset that invasion *in vitro* is almost certainly different from that seen *in vivo*, and furthermore that there is no ideal model system which allows the study of all the aspects of invasion. Nevertheless, the processes of invasion *in vitro* and *in vivo* may share enough in common for the limitations of the experimental systems not to act as a hindrance in the interpretation of experimental results.

What are the main advantages of *in vitro* systems for studying malignant cell invasion? As with models in general, the most significant benefits are associated with simplicity: the experiments are relatively easy to perform, they are open to experimental manipulation (e.g. by drug addition) without worrying about the effects of the host, and the results are easy to analyse. Paradoxically, it is also this simplicity which limits their usefulness: the contribution of host tissues in terms of cell type and architecture, hormones, growth factors, immunity, inflammatory reactions, and other physiological processes known to be important in tumour development and metastasis are missing. Under some circumstances, the host may enhance the possibility of metastases developing (by providing growth factors, for example), while it may also limit metastasis formation (through cytotoxic defence reactions, for example). Host variability is one reason why model systems have been developed, and yet it is the lack of this variability which usually limits the scope of the model. It should also be noted that *in vitro* studies require cells to be able to grow under experimental conditions: this obvious requirement is by no means trivial, since some primary cultures and even some established cell lines prepared from malignant tissues will not survive in certain *in vitro* invasive systems.

Invasion *in vitro* may be assessed in either two or three dimensional situations: both have their limitations, although the 3D system (in so far as it may preserve some tissue stucture) is considered to be more representative of the *in vivo* situation. In 3D systems, invasion is usually assessed following direct confrontation of normal, embryonic tissue fragments and aggregates of tumour cells. This clearly represents a highly artificial situation since clumps of adult tumour cells will not come into contact with the cut surfaces of embryonic tissues. However, as with other model systems, the important point is to understand and assess these limitations so that they may ultimately be taken into account.

To be of any relevance, *in vitro* studies of invasion must be correlated with parallel *in vivo* studies. Most *in vitro* research on invasion to date has used animal tumour cell lines, some of which have been passaged either *in vitro* or *in vivo* for many years, and their relevance to spontaneous human tumours is open to serious questioning. Furthermore, because of the heterogeneity of tumours it cannot be assumed that the invasive cells under study are representative of the entire tumour cell population. This problem may be overcome to some degree by cyclically selecting invasive cells, but because of phenotypic variability no assurance can be gained that the cells will be homogeneous in their behaviour. Regardless of which model system is employed in studying invasion, it needs to be recalled that histological studies have shown that tumours can invade either as single cells or as multicellular structures referred to variously as cords, columns or tongues. Some tumours show a preference for invading as solid cords and it may not be appropriate to use *in vitro* assays based on single cells in such circumstances.

4.2.1 Invasion and histocompatibility

In vivo studies of invasion and metastasis usually involve histocompatible systems in order to avoid the effects of the host's immune system. In special cases, particularly where human material is to be studied, the experiments are often modified through the utilization of immunocompromized xenogeneic hosts. There is also a strong temptation to use allogeneic or xenogeneic combinations in *in vitro* experiments, particularly where tissue supply is limited. In such situations we must ask ourselves if it is possible for histocompatibility relationships to effect biological processes (other than immune phenomena) which might influence the experimental outcome. Curtis and Rooney (1979) examined the degree of nuclear overlap between 2D confrontations of kidney tubule epithelial cells prepared from various congenic mouse strains. Such strains have different major histocompatibility complexes (MHCs) superimposed on the same highly inbred genetic background, and they are thus useful for studying the effects of varying the MHC. Curtis and Rooney (1979) found that the extent of nuclear overlap was less marked in confrontations between cells that differed at the K and D major histocompatibility loci (which are also involved in graft rejection), thus hinting at a possible relationship between invasion and histocompatibility type. Most studies using 3D model systems, however, have not pointed towards any significant effects of histocompatibility differences in invasion *in vitro*, but it must be admitted that the phenomenon has not been thoroughly explored.

4.3 TWO-DIMENSIONAL CONFRONTATION MODELS

This is perhaps the simplest model system for studying invasion *in vitro* and it was the method developed by Abercrombie and colleagues in their pioneering studies (Abercrombie and Heaysman, 1954, 1976; Abercrombie, Heaysman and Karthauser, 1957). Briefly, an explant of normal tissue (chick heart) is placed near that of a tumour so that they are separated by a distance of about 1 mm (see section 1.7.4.2). Cells emerge from the explants and migrate towards each other, ultimately coming into contact; it is the behavioural events which follow this contact which are of particular relevance to the invasive process. It is often thought that in such assays, the malignant cells invade the normal cells after contact. This general conclusion is misleading, however, since in many cases invasion is reciprocal, with malignant cells penetrating the normal cells and normal cells penetrating malignant cells. It may be argued that there could be a difference in the extent of invasion, with malignant cells invading normal cells more than that in the reverse situation. Unfortunately, this too turns out not to be so for every case. Furthermore, invasion may occasionally be seen in confrontations between two normal explants (Stephenson and Stephenson, 1978) and this raises problems with the validity of this assay system as a model of invasion *in vivo*.

Paddock and Dunn (1986) have recently made a detailed analysis of the contact interactions between FS9 mouse fibrosarcoma cells and embryonic chick heart ventricle fibroblasts (CHF cells) cultured together. Although these authors used one of the cell pairs that Abercrombie and his colleagues employed in their classic studies of contact inhibition, the experimental systems were not strictly identical. Nevertheless, using sparse mixed cultures, Paddock and Dunn (1986) were able to describe two aspects of FS9 cell behaviour which might be of significance in invasion:

(a) When FS9 cells contacted CHF cells they displayed a reverse contact inhibition response (i.e. contact promotion of locomotion).
(b) On colliding with CHF cells, FS9 cells tended to turn towards the point of initial contact.

Thus not only do FS9 cells fail to show heterotypic contact inhibition of locomotion (unlike the CHF cells), but their responses after contact might be thought to promote invasion. There is no evidence, however, that the invasion of explant outgrowths of CHF cells by outgrowths of FS9 cells is actually promoted. This apparent paradox illustrates the point that two different systems are being studied: although contact in sparse cultures may lead to promotion of locomotion, this is not necessarily the case when the invasion of populations of cells from explants is considered. The observations of Barski and Belehradek (1965) are of interest in this respect, since they found that the invasion of malignant murine fibroblasts was inhibited by coherent sheets of embryonic mouse heart cells but not by more loosely arranged heart cells. Not surprisingly, we should be very guarded on how we apply these results to the phenomenon of invasion *in vivo*.

Recently, Parish *et al.* (1987) have prepared a McAb against a 37 kD protein (p37) on FS9 cells which inhibits their invasion of CHF cells at concentrations as low as 1 μg ml^{-1}. This protein is found on a number of cell lines (including some from human tumours), but it has not been located on normal mouse tissues using immunohistochemical techniques. It is relatively poorly expressed on mouse L929 cells which have been shown to be non-invasive in 2D confrontation systems with CHFs. The studies of this research group indicate that the presence of p37 correlates strongly with 2D invasion, but its relevance to metastasis *per se* remains to be explored. Indeed, later work has shown that p37 is not a direct product of FS9 cells but is in fact a constituent of contaminating *Mycoplasma hyorhinis* (Dudler *et al.*, 1988). Despite the foreign origin of p37, there is some evidence that transformed cells which are not contaminated with mycoplasma and which do not express p37 nevertheless may still be sensitive to the invasive inhibitory effects of the specific McAb. One possible interpretation of these results is that some transformed cells express a cell surface determinant immunologically cross-reactive with p37, but the details of this possibility remain to be worked out.

Interestingly, invasion by normal cells characterizes many *in vitro* model systems, particularly those using mesenchymal cells (such as embryonic fibroblasts) as normal controls. This suggests that it may be worth exploring the use of other cell types (such as endothelial and epithelial cells) in assays of this sort. In unpublished studies from my laboratory of the invasion of bovine aortic endothelial (BAE) monolayers by B16 melanoma cells, we have noted that the tumour cells almost invariably invade by migrating under the BAE cells. Presumably as a consequence of this underlapping, the BAE cells are lifted off the substrate and they roll back on themselves allowing for the seemingly inexorable progression of the melanoma cells. At least two explanations indirectly related to adhesive changes may be put forward to explain the ability of tumour cells to underlap normal cells. One possibility is that highly invasive tumour cells spontaneously release proteolytic enzymes which disrupt the substrate attachment mechanisms of normal cells more than that of the tumour cells themselves. Alternatively, tumour cells might release toxic products which damage normal cells leading to a weakening of their substrate adhesiveness and consequent underlapping. When confrontations between B16 melanoma cells and BAE cells are treated with trypan blue to detect dead cells, it is not unusual to see a zone of dead BAE cells in the rolled back fringe. Whether this death is the result of toxins produced by the melanoma or whether it occurs because of the loss of substrate contact on rolling back of the BAE cells, cannot unfortunately be determined.

Although the ability of B16 cells to invade BAE cell sheets when they come into contact is visually striking, tumour spread *in vitro* is actually more marked when a non-locomotory mode is employed. The basis for this type of spread is the ability of B16 melanoma cells in culture to spontaneously detach and relocate themselves elsewhere in the dish. Invasive islands can thus be established amongst an endothelial sheet quite remote from the position of first contact between the normal and malignant cell fronts. The processes involved in this phenomenon are not entirely clear. However, the basis for detachment may be changes in adhesiveness associated with the cell cycle (Elvin and Evans, 1983), while cell relocation may arise as a consequence of fluid movement due to mechanical vibrations while in culture.

The situation with respect to invasion by carcinomas is not clear. Stoker, Piggott and Riddle (1978), for example, failed to find invasion of fibroblast monolayers by human mammary carcinoma cells. As will be described in more detail later, carcinoma cells also fail to invade collagen gels (Schor, 1980). Furthermore, Tickle, Crawley and Goodman (1978a) have reported that carcinomas invaded chick limb bud mesenchyme only poorly. HeLa cells, on the other hand, are a carcinoma cell type which have been reported to invade monolayers of CHFs.

Despite a considerable number of limitations in both our understanding of contact inhibition and in its general applicability, it has been suggested that the lack of contact inhibitory behaviour shown by some tumour cells *in vitro* might be expected to provide a plausible explanation of their invasive nature *in vivo*. Given the variable results of studies on contact inhibitory behaviour between cellular outgrowths of normal and tumorigenic explants, however, it would seem not to be of considerable predictive significance in the study of invasion (section

1.7.4.2). Abercrombie and his colleagues were clearly aware of the limitations of their model system since if contact inhibition was only exhibited in 2D confrontation cultures then the phenomenon would be of little relevance in understanding the mechanisms of invasion *in vivo*. Normal cells can invade when tissue fragments are fused (Armstrong and Armstrong, 1973), although this may not always be the case (Weston and Abercrombie, 1967), and normal cells move within heterotypic aggregates during sorting out (Steinberg and Wiseman, 1972). Interestingly, Wiseman (1977a) found that single chick heart or liver cells invaded aggregates of limb bud cells whereas Weston and Abercrombie (1967) using the same cell types, but as tissue fragments, found no invasion. Inert materials such as glass beads can also invade cell aggregates thus questioning the need for motility at all in the invading cell (Wiseman, 1977b). Using an *in vivo* model system in which cell pellets or tissue fragments were implanted in the developing chick bud, Tickle, Crawley and Goodman (1978 a and b) found that both normal and polyoma transformed BHK cells invaded wing mesenchyme. Since both these cell types display contact inhibition *in vitro* (Erickson, 1978 a and b) it would seem that any interpretation of 3D invasiveness in terms of contact inhibitory behaviour as determined from 2D studies may have little applicability. Thus, although our knowledge of the invasive process is incomplete, the evidence suggests that there may be considerable differences in the behaviour of cells in 2D and 3D environments and therefore we should interpret results from 2D confrontation experiments with some caution.

4.4 INVASION OF CELL MONOLAYERS

Various types of cell monolayers have been chosen to study tumour invasion using different experimental systems. Endothelial or epithelial cell monolayers, for example, have been grown on plastic, filters, collagen, and natural and artificial extracellular matrices. Invasion has been quantified in a number of ways, usually visually but also electrophysiologically, and reference to many of these systems will be made in the following sections. Invading tumour cells are often faced with single layered cellular barriers of endothelial, epithelial or mesothelial nature and thus there is some relevance for using a model system incorporating these cell types. However, cells are known to behave differently and to display different morphologies on different substrates and there is no assurance that *in vitro* substrates (even when modified by the cell monolayers) are identical to their *in vivo* counterparts. Although in its original form the test cells used in the monolayer invasion model were not subjected to flow conditions, this can be readily achieved either by simply shaking the cultures or by incorporating them into a viscometer (in which laminar flow can be induced) as suggested by Elvin, Drake and Evans (1983).

One important feature of all studies of this type is that the monolayers employed must be intact. This is often not the case, and the supposedly invasive behaviour of some cell types may actually reflect the presence of an incompletely formed monolayer. Furthermore, cells in monolayers take on a polarized morphology with the upper surface usually being less adhesive than the basolateral cell surface (DiPasquale and Bell, 1974). Invasive tumour cells spreading via the blood system translocate across the endothelial barrier in both directions, from the basal surface during intravasation and from the inner surface during extravasation. This is not always the case for tumours invading epithelial or mesothelial barriers and every effort should be made to check the general validity of the model system. Invading tumour cells are not likely to meet a monolayer of fibroblasts *in vivo* and the validity of studies of this sort are questionable.

4.4.1 Invasion of endothelial cell monolayers

With an estimated 2 kg of endothelium in the human body, there is obviously plenty of opportunity for interaction with tumour cells. A wide range of normal and malignant cells were studied by Kramer and Nicolson (1979) for their abilty to adhere to bovine aortic endothelial (BAE) cell monolayers, to induce endothelial retraction, and subsequently to invade the monolayer. It is noteworthy that adherent cells did not spread on the endothelium. Such behaviour required contact with the subendothelial ECM. Some normal cells such as fibroblasts could adhere to monolayers of endothelial cells and induce their retraction, but they failed to invade (as indicated by a failure to underlap the endothelial cells). Such cells became incorporated into the structure of the monolayer. Most tumour cells, on the other hand, were capable of all three processes, but it is noteworthy that two lymphomas (RAW117-P and RAW117-H10) adhered poorly, failed to induce retraction, and did not invade. Later research with the RAW117 variants has shown that these cells do adhere to endothelial cells, but that their adhesion is more or less selective for syngeneic endothelia derived from the particular organs to which they metastasize (section 5.7.2.1). Typically invasive normal cells such as monocytes and neutrophils displayed similar behaviour to invasive tumour cells, yet peritoneal macrophages failed to invade. The reasons for these differences remain obscure. Interestingly, cells which attached to endothelial monolayers did so at or near junctional regions. There are at least three possibilities which might explain this observation:

(a) Adhesive receptors on cells of the monolayer may be restricted to the junctional region, possibly as a consequence of their redistribution during monolayer formation.
(b) Contact of seeded test cells with underlying ECM components may occur through incomplete endothelial junctions. It is known that many tumour cells are more adherent to the ECM than to the endothelium (Nicolson, 1982a).
(c) Diffusion of factors may occur through the junctional region. Small molecular mass chemotactic factors, for example, may diffuse through junctional complexes and stimulate the accumulation of responsive cells at the junctional site.

Which if any of the above possibilities underlies the phenomenon of non-random adherence is not clear.

Finally, three important points need to be borne in mind when interpreting the results of *in vitro* studies employing endothelial cell monolayers:

(a) Most studies have used xenogeneic combinations of tumour cells and endothelial cells.
(b) Most studies have used endothelial cells derived from major vessels, such as the aorta, which are invaded only rarely. There may be considerable differences between endothelial cells derived from major arteries or veins, small vessels and capillaries. Both capillary (Folkman and Haudenschild, 1980) and aortic endothelial cells (Feder, Marasa and Olander, 1983) can form capillary-like structures *in vitro*, but their similarities have not been fully explored. Interestingly, these *in vitro* vessels may represent 'inside-out' capillaries since they contain basement membrane-like material in their lumens.
(c) It seems likely that endothelial cells from different organs may have recognizably different cell surface determinants (Auerbach *et al.*, 1985). The consequences of these differences, which could play an important role in metastasis, have not been considered in most studies.

4.4.2 Invasion of mesothelial cell monolayers

The invasion of monolayers of mesothelial cells from the peritoneum by ovarian tumour cells has been studied by Niedbala, Crickard and Bernacki (1985). Epithelial tumours of the ovary spread by direct extension into the peritoneal cavity followed by tumour implantation onto the peritoneal surfaces. Niedbala and his colleagues (1985) grew human peritonel mesothelial cells on BAE extracellular matrix and seeded them with human ovarian tumour cells. They found that the tumour cells adhered to the mesothelial cells, induced their retraction, and subsequently invaded in a manner similar to that described above with endothelial cells. Interestingly, adhesion was markedly reduced by flow which agrees with clinical observations that implantation occurs more frequently in areas where the flow of peritoneal fluid is minimal. The tumour cells were also found to adhere more to the ECM than to the mesothelial cells, which again is in broad agreement with clinical observations suggesting that tumour cells localize at sites where the mesothelium is defective. In contrast to studies with B16 cells, ovarian carcinoma cells failed to show significant signs of matrix degradation. However, this may not be totally unexpected in the light of clinical observations which suggest considerable localized growth of ovarian cancer within the confines of the peritoneal cavity.

Shinkai and colleagues (1988) have recently reported on an extract of normal rat liver cells which can inhibit the invasion of rat AH 130 ascites hepatoma cells through a rat mesothelial cell monolayer *in vitro*. About 60% of the inhibitory activity localizes with a 3–4 kD fraction, while the rest is found in a fraction of M_r >25 kD. Co-injection of this anti-invasive material with AH 130 cells into the rat peritoneal cavity can prevent the development of invasive and metastatic lesions (3/8 animals showed invasive lesions when injected with the crude extract, 0/3 after injection with a <10 kD filtrate, and 21/24 controls). These results are obviously very preliminary in nature, lacking in a significant number of repeats, and revealing nothing of any substance about the biochemistry of the extract. However, the authors claim to be able to extract similarly active factors from rat lung and thymus, and it is clearly of interest to resolve the nature of the active components as soon as possible.

4.4.3 Invasion of epithelial cell monolayers

Tumour cells can invade through epithelial monolayers *in vitro*. Elvin, Wong and Evans (1985), for example, have shown that W256 carcinosarcoma cells can induce retraction of MDCK epithelial cells and that the gaps can be colonized by the tumour cells. In many ways this is similar to the behaviour described by Kramer and Nicolson (1979) for the invasion of endothelial cell monolayers and to that described by Niedbala *et al.* (1985) for the invasion of mesothelial cells. Indeed, it is likely that similar processes are involved in the three systems, although at present these are poorly characterized. The system described by Elvin and colleagues (1985) is based on an electrically tight epithelial monolayer which, aside from confirming monolayer integrity, is amenable to the electrophysiological recording of invasion as described below (section 4.6).

4.5 INVASION OF FILTERS

Most filter-based assay systems utilize modifications of the chamber described originally by Boyden (1962) and subsequently used widely in the study of the chemotactic behaviour of leukocytes (Wilkinson, 1982). At its simplest, the Boyden chamber is composed of two compartments separated by a porous filter. This filter is either of the Millipore-type (approx. 150 μm thick with tortuous pores) or of the Nuclepore-type (approx. 10 μm thick with regular holes). The filters are made of different materials with different adhesive and adsorptive properties for different cells and proteins. Using the thicker Millipore-type filters, an experiment is usually stopped while the cells are still within them, and migration is quantified by measuring the depth of penetration from the surface of the filter to the leading front of cells (usually defined as the two furtherest cells in the same plane of a single field as viewed through a light microscope). With the thinner Nuclepore type filters, an estimate of migration is gained from counting the number of cells per high powered field of view on the bottom of the filter. It is important to appreciate that these two measurements assess different aspects of the migratory response: one measures how far the fastest subset of cells has moved within a given time, while the other measures how many cells have moved a set distance in a given time. One other feature of these two assay systems warrants discussion and that is the nature of the pores themselves. The tortuous, cave-like pores of Millipore-type filters offer a 3D surface for migrating cells which may confer some advantage to them when compared to the almost 2D, regular pores of Nuclepore-type filters. In a study of the invasive ability of Walker 256 carcinosarcoma, Elvin, Wong and Evans (1985) selected two sub-populations which differed markedly in their adhesiveness but which responded chemotactically to FMLP. Admittedly, the non-adherent, suspension growing W256S cells did not migrate as far as the substrate attached W256A cells, but it is worth pondering on how essentially non-adhesive cells can migrate. Indeed, if a cell cannot gain friction through purchase on a substrate then it will not be able to move upon it. One possible explanation is that the ability of non-adherent W256S cells to migrate through Millipore filters might arise because of changes in cell adhesiveness due to the presence of the chemoattractant. Since FMLP was shown not to be required for locomotion of these non-adherent cells on Millipore filters, however, the authors concluded that this explanation was unlikely to be appropriate. An alternative explanation is that the physical nature of the filter substrate was somehow responsible for allowing the non-adherent cells to move. In fact, Elvin, Wong and Evans (1985) suggested that the W256S tumour cells were able to gain purchase within the pores of the Millipore filter, perhaps by extending and then expanding pseudopodia to give traction in a manner analagous to that suggested for the movement of leukocytes through 3D matrices (Brown, 1982; Haston *et al.*, 1982; Lackie and Wilkinson, 1984). Thus, given a suitable substrate, even non-adherent cells may be able to move and invade, and this clearly has consequences on the interpretation of data from *in vitro* models of invasion.

The main limitation to the use of porous filters as substrates for studying the process of invasion is that 8–12 μm pores are not found within *in vivo* connective tissue barriers such as the basement membrane. Thus invasion *in vivo* is not a simple matter of motility, whether in response to a chemotactic gradient or not, since invasion usually involves the disruption of physical barriers. Recognition of the physical limitations of the standard Boyden chamber led to modifications in which connective tissue material such as collagen was incorporated within the pores of the filter (Evans *et al.*, 1983) and to the growth of endothelial or epithelial cells (Cramer, Milks and Ojakian, 1980; Evans *et al.*, 1983; Elvin, Wong and Evans, 1985) on its surface. The Boyden chamber has also been adapted to provide laminar flow conditions (Elvin, Drake and Evans, 1983) in an attempt to model the conditions which single cells might be subjected to within the blood vessels. The motility and chemotactic behaviour of rat leukocytes was found to be dependent on flow with stimulation of movement at relatively low shear rates (about 22.5 sec^{-1}) when compared to conditions in which flow was absent. These experiments, however, have not yet been repeated with tumour cells. It should also be recalled that flow within small blood vessels may temporarily reverse and, furthermore, that observations of blood vessels within rabbit ear chambers have shown that cells may move with, across and even against the flow of blood (Wood, Baker and Marzocchi, 1967b). It is not yet clear how cells move under experimental laminar flow conditions in comparison to physiological conditions, and thus it remains uncertain as to the relevancy of results from such studies. Further developments to this type of model system have actually involved the replacement of the filter or cellulose support altogether with tissues such as the amnion, CAM or the bladder wall as described in later sections.

4.6 ELECTROPHYSIOLOGICAL ASSESSMENT OF INVASION

Since normally motile cells will move into the filter of a standard Boyden chamber (providing its pore size is appropriate), this assay system actually models motility rather than invasion *per se*. In order to model invasion rather than motility, cellular barriers (endothelial or epithelial in nature) have been grown on one surface of the filter and extracellular matrix components have been incorporated into the pores within the filter itself. Now one of the main technical limitations of the standard Boyden chamber assay system is that assessment is not continuous: to show that cells have moved it is necessary to fix and stain the filters first and then to measure the depth of penetration relative to appropriate random locomotion controls. Recently, however, the fact that electrically tight epithelial or endothelial monolayers can be grown on filters has been utilized to monitor invasion continuously.

The procedure is remarkably simple: a standard Boyden chamber is constructed and the upper chamber is seeded with epithelial or endothelial cells which are able to form electrically tight monolayers *in vitro*. The cells establish a confluent monolayer on the surface of the filter (lining the bottom of the upper chamber) and their resistance is monitored electrophysiologically (Cramer, Milks and Ojakian, 1980; Evans *et al.*, 1983). The resistance measurements determined are thought to result from the patency of the tight junctions which form occluding belts around the apical region of each cell. When the tight junctions are disrupted by cation chelation with EDTA, the electrical resistance of the monolayer drops. The process is entirely reversible as shown by continuous recordings: removal of EDTA allows the junctions to reform and electrical resistance returns (Evans *et al.*, 1983). Initial studies used this electrophysiological system to model the invasiveness of PMNs through a barrier of Madin-Darby canine kidney (MDCK) epithelial cells. Whereas PMNs will spontaneously migrate into the filter of a standard Boyden chamber, they were found to penetrate an epithelial barrier only under chemotactic conditions (Evans *et al.*, 1983). This makes sense intuitively, since PMNs should not be able to penetrate spontaneously either epithelial or endothelial barriers, yet they should be able to do so under inflammatory conditions. Continuous monitoring of MDCK monolayer resistance showed that the integrity of the monolayer could be regained after removal of the chemotactic stimulus and excess PMNs.

In a subsequent study, Elvin, Wong and Evans (1985) examined the ability of W256 carcinosarcoma cells to invade a barrier of MDCK epithelial cells. These authors originally chose W256 cells because they reacted to the presence of a chemotactic signal (FMLP). However, this feature of their behaviour (in which they paralleled PMNs) appeared to be unnecessary for invasion through the cell monolayer. Thus, at least as far as W256 cells are concerned, there is a major difference in how they penetrate an epithelial barrier when compared to that of PMNs. Whereas PMNs invaded only in response to a chemotactic signal, the W256 cells were clearly able to invade in the absence of any exogenous signals. Invasion by W256 cells was associated with MDCK cell retraction, leaving holes in the monolayer which were colonized by the tumour cells. What aspect of the W256 cells induces epithelial retraction allowing invasion of the monolayer to occur is not clear, but the possibility of enzyme involvement warrants further study since both *in vitro* and *in vivo* evidence point to the excessively erosive nature of this tumour cell line.

Although B16F10 cells have been selected for lung colonizing ability, results from my laboratory have consistently shown them to be unable to penetrate either low- (c. 300 ohm cm^2) or high- (c. 1 kohm cm^2) resistant MDCK cell monolayers. It might be thought that B16F10 cells would have high invasive ability through tight intercellular junctions. That this need not be the case, however, is indicated by estimates of lung junctional resistance of <35 ohm cm^2 (Crone and Christensen, 1981). In other words, during the original selection process (Fidler, 1973a) virtually no selection pressure may have been exerted for the ability to penetrate high resistance junctions. Interestingly, Conn and Knudsen (1989) have recently reported that B16F10 cells penetrate LLC-PK_1 pig kidney cell monolayers only when their junctional resistance is induced to drop from around 300 ohm cm^2 to about 60 ohm cm^2 by treatment with the phorbol ester TPA. Such results indicate a relatively poor invasive character for B16F10 cells, although they still may be superior in this respect to B16F1 cells (section 3.4.8.4).

The model system described above, in which tumour cells invade an epithelial barrier in an apical to basal direction, is applicable to a number of clinical situations, particularly those involving spread by implantation. Bladder metastases, for example, are often derived from tumour fragments carried in the urine from primary growths in the renal pelvis and ureter. Development of metastases in this example is associated with tumour cell invasion of the bladder wall from the apical epithelial surface (Willis, 1973). Tumours which spread via the blood, on the other hand, must penetrate an endothelial barrier in two directions: first to enter the vessel, and secondly to leave the blood system. Although endothelial monolayers which are electrically tight have been maintained under *in vitro* conditions, they have not yet been employed in electrophysiological studies of tumour cell invasion.

4.7 INVASION OF COLLAGEN GELS

Leighton and his colleagues pioneered the use of three dimensional model systems for the study of tumour cell behaviour *in vitro*. Their early work involved the establishment of an artifical matrix composed of cellulose sponge and clotted plasma (Leighton, 1951; Leighton *et al.*, 1959). Normal cells or tissues were grown in the matrix which was then inoculated with tumour cells and subsequently examined histologically for the extent of invasion. Invasion was usually associated with necrosis, which Leighton *et al.* (1967) attributed to a smothering effect of the tumour cells by which they restricted the supply of nutrients to the normal cells they encompassed.

One of the main problems with this early version of a 3D model for invasion was that tumour cells with fibrinolytic activity could lyse the plasma clot and disrupt the matrix. It should be noted here that the cellulose framework employed was a poor surface for tumour cell adhesion and without the clot the tumour cells would not invade. In order to improve upon the stability of the matrix, Leighton *et al.* (1967) infiltrated the cellulose sponge with collagen. Results from these studies showed that the degree of tissue invasion by tumour cells was related to the orientation of connective tissue cells and fibres within the matrix, since penetration was greater when the cells and fibres were arranged radially to the growing tumour front. Such observations support the notion that tumours often spread locally along preferential planes, but fall far short of offering any proof.

More recent work has since shown the cellulose sponge support to be unneccesary for the study of cell behaviour in this type of model system (Elsdale and Bard, 1972; Schor, 1980; Schor, Allen and Harrison, 1980). Indeed, Ehrmann and Gey (1956) had suggested earlier that rat tail collagen (type I) gels could provide a 3D environment for cell culture, and Cuprak and Lever (1974) had observed that hamster fibrohaemangiosarcoma cells could migrate into a collagen gel matrix. One of the major disadvantages of the collagen coated cellulose sponge system used by Leighton's group was its lack of transparency, and thus invasion could only be determined after rather laborious histological sectioning. This has proven not to be a problem in collagen gels. One other important advance of the more recent developments was the establishment of procedures for quantitating cell behaviour in 3D gels (Schor, 1980).

Schor and his colleagues (1985a) have recently reported observations on the behaviour of RPMI 3460 Syrian hamster melanoma cells seeded either directly on to type I (rat tail tendon) collagen gels, or on to gels covered with monolayers of either human foreskin fibroblasts or BAE cells, or on to gels covered with the extracellular matrices of these cells (obtained after cell removal with 0.2 M ammonia solution). It was found that melanoma cells adhered more to both ECMs and proliferated more upon them than they did to plain collagen gels. Both ECMs also acted as barriers to invasion, since although a substantial number (30%) of cells could invade plain collagen gels after 6 h incubation, only 10% and 0.3% of added cells could penetrate into the collagen gel below the fibroblast and endothelial ECMs respectively. This is somewhat misleading since a significant number of melanoma cells could in fact breach both ECM barriers, but because they remained adherent to the underside of the ECMs they were excluded from the invasion assay. Treatment of plain gels or exposed ECMs with a combination of trypsin and elastase, or chondroitinase ABC, had no effect on melanoma cell invasion even though considerable amounts of protein and sulphated proteoglycans were believed to be solubilized from the ECMs. Apparently, all detectable fibronectin is removed from the gel after treatment with trypsin and elastase, but even this had no effect on invasion. This suggests some form of direct, fibronectin independent adhesion of the melanoma cells to collagen, possibly via a collagen receptor. Alternatively, the melanoma cells may have migrated through the gel by gaining purchase on the scaffolding via cell extensions which may curl arround collagen fibres and not actually adhere to them via specific receptors. We have seen that such a form of binding has been suggested for the locomotion of lymphocytes in collagen gels (Haston, Shields and Wilkinson, 1982) and for the invasion of a non-adherent variant of the Walker 256 carcinosarcoma (Elvin, Wong and Evans, 1985). The extensive penetration of plain collagen gels by RPMI 3460 melanoma cells (Schor *et al.*, 1985a), was not found to be associated with detectable gel degradation. Furthermore, the degradation of ECMs by melanoma cells did not facilitate invasion, bringing into question the need for enzymes in invasion. That tumour cells could invade collagen gels without significant collagenolytic activity was suggested earlier from the studies of Cuprak and Lever (1974).

It should be noted here that cells of mesenchymal origin whether normal or tumorigenic are able to invade collagen gels, whereas most normal and tumorigenic cells of epithelial origin cannot. Pigmented retinal epithelial cells from the chick embryo represent an exception to this observation since these cells have been reported by Docherty, Forrester and Lackie (1987) to invade type I collagen gels (1 mg ml^{-1}). Interestingly, the invasive behaviour of these cells is severely curtailed if the type I collagen gels are coated with type IV collagen (0.6 mg ml^{-1}) plus the adhesive molecule laminin (0.1 mg ml^{-1}). Pigmented retinal epithelial cells might represent an unusual epithelial cell type since (unlike other epithelial cells) they contain the intermediate filament protein vimentin, they lack desmosomes, and their proliferation seems not to involve a stem cell population. Generally speaking, however, the observations suggest a fundamental difference in the 3D behaviour of cells of mesenchymal and epithelial origin, even if only in an *in vitro* environment. That normal or tumorigenic epithelial cells cannot invade collagen gels whereas both normal and tumorigenic mesenchymal cells can, places restrictions on the use of this model system for studying the invasive ability of all types of malignant cells. It may be of some significance, however, that the invasion of collagen gels *in vitro* is predominantly by cells which display Type 2 invasive behaviour (section 2.1), and thus this model system may only be appropriate for studying this type of invasion (based on the motility of single cells rather than penetration *en masse*).

There are important differences in the ways normal and tumorigenic cells which are capable of invading collagen gels actually do so. According to Schor *et al.* (1982), the migration of both normal human foreskin fibroblasts and hamster melanoma cells into collagen gels is critically dependent on stiffness of the gel and cell density. Maximum invasion for both cell types is seen around 2 mg ml^{-1} of collagen, but whereas the melanoma cells invade in proportion to cell density, invasion by normal fibroblasts is inversely related to cell density. The different effects of density on the migration of normal and transformed cells into collagen was confirmed in a later and more extensive study (Schor *et al.*, 1985c) and a cell density migration index (CDMI) describing the effects in a single numerical value has been proposed.

Briefly, the CDMI relates the migratory behaviour of cells at two different densities as follows:

$$\text{CDMI} = \log\,[\%(\text{low density})/\%(\text{high density})]$$

where the terms '%(low density)' and '%(high densiity)' represent the percentages of cells which have migrated into collagen gels when dishes are seeded with 10^3 and 10^4 cells cm^{-2} respectively. Cells whose migration is inversely proportional to cell density thus have positive CDMI values, whereas negative values are displayed by cells whose migration increases proportionally with cell density. Cells whose migration into collagen is largely unaffected by cell density have CDMI values approaching zero. It should be noted that the CDMI value reflects the average behaviour of cells within a population, and does not describe the heterogeneity of migratory phenotypes expressed by individual cells.

Using CDMI data from a wide range of normal, fetal and

transformed fibroblastic cell types, Schor *et al.* (1985c) were able to define empirically certain ranges of CDMI values which categorized the cell types. Thus most transformed cells had CDMI values <-0.4, some transformed and fetal cells ranged between -0.4 and 0, some fetal and normal cells ranged between 0 and $+0.4$, while most normal cells had CDMI values $>+0.4$. The difference in CDMI values between fetal and normal fibroblasts suggests that this cell type undergoes an 'isoformic' transition at some point in its developmental history, and an aspect of this change can be recorded *in vitro* by monitoring an increase in the CDMI of fetal cells as a function of passage number. A similar isoformic change, this time involving a decrease in the ability to display anchorage independent growth, has been suggested to occur during the development of Syrian hamster embryo fibroblasts (Nakano and Ts'o, 1981).

When ostensibly normal fibroblasts were isolated from a number of malignant tumours of diverse origins (including breast cancer), the CDMI values obtained were found to fall within the fetal range rather than the normal range (Schor *et al.*, 1985b). Fibroblasts obtained from benign lesions, however, displayed normal CDMI values. Interestingly, skin fibroblasts obtained from 10 out of 15 apparently unaffected first degree male and female relatives of patients with a strong family history of certain cancers showed a similar defect in that they too expressed CDMI values largely within the fetal range. Schor *et al.* (1985b) suggest from these findings that the expression of fetal CDMI values may precede the development of clinically recognizable malignancy, but whether studies of this sort will identify individuals at risk from developing cancer remains to be evaluated fully.

At this stage we might ask as to how changes in fibroblast cell behaviour could affect the development of cancers, including those of epithelial origin? Schor *et al.* (1985b) discuss two hypotheses:

(a) *The genetic hypothesis*

This conventional interpretation suggests that the genetic abnormalities responsible for the expression of fetal behaviour in fibroblasts from patients with cancer co-existed in the cells (e.g. epithelium) which were subject to malignant change, and that somehow these genetic changes contributed to the induction of malignancy. Some evidence in support of this hypothesis stems from the observation that cancers with a genetic basis develop progressively in individuals whose cells inherit a particular mutation (section 1.7.2). According to Knudson (1971), susceptibility to familial cancer is inherited through a germline mutation, and the cancer actually develops when a second somatic mutation occurs in the homologous allele. The study by Schor *et al.* (1985b) included ostensibly normal fibroblast cell lines from forearm skin biopsies of patients with familial adenomatous polyposis (also known as familial polyposis coli). This hereditary disorder is characterized by the development of hundreds of adenomatous polyps in the large intestine. Since a tendency towards malignancy is thought to be associated with each of the numerous polyps, the eventual onset of carcinoma of the colon appears almost inevitable. Skin fibroblasts from patients with familial adenomatous polyposis (FAP) express a number of altered phenotypic characteristics apart from perturbed migration into collagen (Kopelovich, 1982; Schor *et al.*, 1985b) which, since the disease has a distinct genetic basis, may reflect particular chromosomal aberrations. Recently, Bodmer *et al.* (1987) have localized the gene for FAP to chromosome 5, but it must be admitted that how a germ line mutation in this chromosome (resulting in heterozygosity) might affect cell behaviour as shown by Schor *et al.* (1985b) is unknown. Karyotypic aspects of FAP are discussed further in section 1.7.2.

(b) *The mesenchymal–epithelial dysfunction hyothesis*

This more radical hypothesis suggests that mesenchymal–epithelial interactions play a significant role in tissue differentiation and maintenance and that changes in the mesenchymal cell which perturb these interactions may predispose epithelial tissues to malignant change. Certainly, fetal fibroblasts transplanted into the adult mammary gland can promote hyperblastic change in the epithelial cells and render them more sensitive to transformation by carcinogenic agents (Sakakura, 1983). This hypothesis clearly has considerable implications in our understanding of cancer development and warrants further study.

More recently, Schor and Schor (1987) have provided evidence that both fetal fibroblasts and the aberrant fetal-like fibroblasts of cancer patients release a migration stimulating factor. This factor (approx. M_r 55 kD) stimulates the migration of normal adult fibroblasts in high density (i.e. confluent) culture, and its effects are inhibited by hyaluronidase (10 U ml^{-1}). One interpretation of this result is that the factor may function by stimulating adult fibroblasts to synthesize and release hyaluronic acid, which in turn would stimulate migration.

In an effort to model certain aspects of extravasation from blood vessels, Schor and his colleagues also examined the behaviour of melanoma cells after seeding onto BAE cells which had been allowed to grow to confluence on the collagen gel. Under these circumstances the melanoma cells proliferated over 4–5 days to form foci which pushed back the endothelial cells as they expanded. Most melanoma cells were adherent to the upper surface of the underlying ECM, but a few penetrated into the collagen gel. Bogenmann and Sordat (1979) have reported that mouse sarcoma cells can rapidly penetrate an endothelial barrier growing on a collagen gel with minimal damage, whereas human carcinoma cells tend to destroy the monolayer by overgrowth and they do not simply invade the space between the gel and the endothelium. Kramer and Nicolson (1979) studied the effects of both non-tumorigenic and tumorigenic cells on endothelial cells grown in culture. They found that cells of either type could adhere to the endothelial monolayer and cause its retraction, but only invasive cells whether normal (e.g. neutrophils) or tumorigenic in nature could actually migrate under the endothelial barrier. Invasive cells appeared to adhere in the vicinity of cell junctions and migration was between the endothelial cells rather than through them.

In *in vitro* studies employing endothelial cells it is worth enquiring as to how similar the cultured cells are to their *in vivo* counterparts. Endothelial cells have been successfully grown from both capillaries and major vessels from a wide range of experimental animals as well as from human tissues. In most circumstances, the cells synthesize an extracellular matrix which contains the major components of true basement membranes (fibronectin, laminin, type IV collagen, nidogen, and proteoglycans), they develop a junctional network containing both tight and gap junctions, they express factor VIII antigen, and their apical surface is non-adhesive for platelets (reviewed in Jones, 1982). These similarities between endothelial cells *in vitro* and *in vivo* give some credence to their use in model systems, although clearly many of the features of blood vessels are missing.

4.8 INVASION OF ORGANS, TISSUE FRAGMENTS AND AGGREGATES

The use of organs rather than reconstituted tissues in the study of tumour cell invasion was pioneered by Wolff and colleagues, who studied the invasion of cultured embryonic organs by various types of tumour cells (Wolff and Schneider, 1956; Wolff, 1967). Since that time, numerous organs and tissues have been employed in invasion systems including heart, lung, bladder, kidney, skin, omentum, calvaria, scrotal sac and the chorioallantoic membrane from experimental animals, as well as prostate and amnion from human sources. A number of these approaches will be discussed further below.

4.8.1 Invasion of aggregates

Many vertebrate cell types, although often of embryonic origin, will adhere to form aggregates under appropriate conditions (Moscona, 1952, Moscona and Moscona, 1952; Steinberg, 1963, 1970; Trinkaus and Lentz, 1964). Moscona (1957) co-aggregated chick embryo cells with mouse melanoma cells in order to study aspects of cell–cell interaction between normal and malignant cells. An essentially similar approach was adapted by Schleich (1973). He found that when normal human tissue was aggregated with HeLa cells, the normal tissue aggregated first and then the HeLa cells attached to the aggregates. Within 48 h the aggregates consisted of external HeLa cells with a necrotic centre of normal cells. In order to study tumour cell invasion, Schleich (1973) added tumour cells to preformed aggregates of normal cells. Using this approach, he concluded that invasion of human endometrial aggregates by HeLa cells required some sort of lesion in the aggregate (perhaps only one cell diameter in depth) before the tumour cells could penetrate. This would be in keeping at least in part with the poorly invasive nature of this tumour cell type as seen, for example, on collagen gels (Cuprak and Lever, 1974).

An interesting technical variation for studying invasion using aggregates was reported by Umbreit and Erbe (1979). These authors separated aggregates of BHK cells and hamster melanoma cells with a nylon mesh which allowed only single cells to pass through. They found that radiolabelled melanoma cells could detach from their aggregates, pass through the mesh and bind onto the BHK aggregates at about twice the rate of the reverse process. This experimental system seems to be appropriate for studying the detachment behaviour of malignant cells which may be related to their ability to escape from the primary tumour *in vivo*.

There are three major problems associated with the use of aggregates to study invasion:

(a) Although the cells are arranged three dimensionally, they do not always closely resemble normal tissues.
(b) Both normal and malignant cells can invade under these experimental conditions.
(c) Cells within aggregates may be actively motile and they can sort themselves out in a manner which may have little to do with malignant invasion.

4.8.2 Sorting out within aggregates

When mixtures of cells from different organs are co-aggregated, the cells often sort out into a concentric 'sphere-within-sphere' arrangement. This arrangement is an equilibrium configuration since it can be reached either from aggregating single cells of two tissue types or from fusing two aggregates, each of a different cell type (Steinberg, 1964). Cells from different tissues of the embryonic chick sort out according to a particular hierarchy. Thus, in the order liver cells, neural tube cells, heart ventricle cells, pigmented retina cells, limb bud cells and back epidermal cells, each cell type sorts out externally to those that are listed after it. Liver cells, for example, would envelop any of the other cell types in a mixed aggregate. According to Steinberg's differential adhesion hypothesis, the cells sort out in this hierarchical manner because of differences in their strengths of cell–cell adhesion (Steinberg, 1964, 1970). In other words, cells in mixed aggregates sort out in order to minimize their total free energy of adhesion (i.e. with the more adherent cells sorting out internally to the less adherent cells). This can perhaps best be illustrated by recourse to a simple model. Imagine an initially random mixture of two cell types (A and B). Now suppose cell type A has a greater strength of homotypic adhesion (W_{aa}) than cell type B (W_{bb}). According to the differential adhesion hypothesis, A cells will sort internally from B cells providing the heterotypic adhesions (W_{ab}) are of intermediate strength (i.e. $W_{aa}>W_{ab}>W_{bb}$. If $W_{aa}>W_{bb}>W_{ab}$, then separation of the cells into independent aggregates (i.e specific sorting out) would be expected. Specific sorting out is seen in some experimental systems such as sponge cell aggregation, for example, where cells from two species may sort out according to kind. Interestingly, in this and perhaps all other examples of specific sorting out, the cells first pass through a non-specific phase where intermingling occurs. Wiseman and Strickler (1981) have reported a correlation between sorting out and the presence of desmosomes (section 5.8.1.4). In a limited study they found that externally segregating cells tended to have fewer desmosomes than internally segregating ones, and they offered this observation as evidence in support of differential adhesion as a factor in sorting out. The conflicting studies of Overton (1977, 1979), however, suggest this correlation is not universal.

The differential adhesion hypothesis has important implications for the invasive process using model systems based on the aggregation technique, since it implies that cells will sort out according to their interaction energies to reach an equilibrium situation in which one cell type aggregates internally to the other. Thus to all intents and purposes, the differential adhesion hypothesis predicts that when aggregates of two cell types of different adhesion potential are apposed *in vitro*, one will apear to invade the other. It is not clear, however, how the differential adhesion hypothesis applies to malignant adult cells rather than normal embryonic cells. It may be that differential adhesion expresses itself in aggregate sorting out only under conditions where the cells are highly mobile. This is probably not the case in normal, adult tissues *in vivo*, although the cells of malignant tissues may have more mobility. The extent to which it is desirable to use the differential adhesion hypothesis as a basis for explaining sorting out and invasion *in vitro* is clearly open to question.

One interesting observation with respect to sorting out *in vitro* is that different arrangements exist for 2D cultures and 3D aggregates. Limb bud cells, for example, sort out internally to most cell types in aggregate form, but in 2D cultures the limb bud cells surround islands of the other cell types (Garrod and Steinberg, 1973; Steinberg and Garrod, 1975). Hypotheses can be put forward to explain these observations, but more than anything else they tell us to be wary when comparing results from different experimental systems.

Lackie and Armstrong (1975) have examined the possibility that invasiveness might reflect differences in adhesion between confronting cell populations. Rabbit PMNs, for example, invade chick heart embryo fibroblasts and, in a direct analogy with the differential adhesion hypothesis, it might be postulated that this is favoured by heterotypic adhesive interactions being intermediate in strength between the homotypic ones. Unfortunately, Lackie and Armstrong (1978), using aggregation techniques to assess adhesiveness, were unable to provide any support for this possibility. Thus, in this system at least, homotypic interactions seemed to be favoured over heterotypic ones, and the differential adhesion hypothesis fails to explain fully the observed results. The possible involvement of the cadherins (Ca^{2+}-dependent adhesion molecules) in sorting out is discussed in section 5.3.1.2.

4.8.3 Multicellular tumour spheroids

Multicellular tumour spheroids are, in fact, nothing other than aggregates of tumour cells. They are primarily of interest as model systems for studying aspects of therapy at the avascular stage of tumour growth (reviewed in Sutherland *et al.*, 1981; Mueller-Klieser, 1987). Both normal cells and tumorigenic cells will form spheroids (i.e. aggregates), the prime difference being that many tumour spheroids will grow in culture, albeit to a limited size. Small aggregates of tumour cells show an initial phase of exponential growth until diameters of 50–200 μm are reached, when the growth rate begins to decline. Non-proliferating cells (believed to be arrested in G_0) begin to accumulate in the central region while proliferating cells are restricted to the more peripheral zones (the outer 50–100 μm of larger spheroids). Under appropriate conditions, spheroids eventually reach a maximum size of about 1–4 mm diameter and they fail to show any further increases in size even if the culture medium is regularly replaced. Continued growth can only be obtained if the spheroids are implanted *in vivo* and vascularization (angiogenesis) takes place. It is thought that some *in vivo* tumours may undergo an avascular growth phase akin to that illustrated by spheroids. The presence of restricted growth zones in spheroids mimics that seen *in vivo*, where tumour cells proliferate preferentially in areas adjacent to microvessels. Presumably this reflects the supply of oxygen and nutrients as well as the ability to remove waste products. Cells within spheroids are more resistant to ionizing radiation than cells in monolayer culture and the effects of radiosensitizing drugs as well as antiproliferative agents have been studied extensively in the multicellular spheroid system. Spheroids allow the effects of these drugs to be tested in the absence of vascularity, as is thought to occur in the early stages of tumour growth, and their 3D structure also allows for tests to be made of drug penetration ability. Tests may be made exclusively under *in vitro* conditions or the spheroids may be implanted in the peritoneal cavity of experimental animals which are then treated with the appropriate therapy. It should be noted, however, that the spheroid microenvironment does not replicate that of tumours *in vivo* since many stromal elements due to the presence of normal host cells are missing. What limitations this may have on the use of spheroids as a model system is unfortunately often overlooked.

Multicellular tumour spheroids have been used in studies of tumour cell invasive capacity. Zamora, Danielson and Hosick (1980) have reported on the invasion of endothelial monolayers growing on collagen gels by cells from murine mammary tumour spheroids (25–500 μm in diameter). They found that spheroids could attach to endothelial monolayers within 2 h and that this was followed by local retraction of the monolayer and migration of tumour cells from the spheroid onto the underlying collagen gel. Some tumour cells were then able to invade laterally, moving underneath the endothelial cells, while others penetrated into the collagen gel. Except for the occasional invasion of the collagen gel by carcinoma cells, this behaviour does not appear to be significantly different from that shown by the invasion of endothelial monolayers growing on collagen gels by single tumour cells as described previously (section 4.7).

4.8.4 Invasion of tissue fragments

Mareel and his colleagues combined organ culture and aggregation techniques through the development of a confrontation assay in which embryonic tissue fragments are incubated with tumour cell aggregates (Mareel, Kint, and Meyvisch, 1979; Mareel, 1982; Mareel and De Mets, 1984). A typical experiment involves the gyratory coculture over several days of a precultured tissue fragment (usually embryonic chick heart) with an aggregate of tumour cells, although in some experiments the tumour cells may be single in order to accelerate invasion. The experiment is usually performed in gyratory suspension in order to avoid the spreading of cells on the substrate, and precultured fragments are preferred because thay offer a well-defined border for subsequent analysis of invasion. Embryonic tissues are used in preference to adult tissues because they are thought to maintain their normal structure, growth and differentiation patterns better *in vitro*. The confronted pairs are examined histologically and invasiveness is assessed by determining the extent of degeneration of the tissue fragment and the amount of its occupation by tumour cells. These criteria are somewhat restrictive, since although malignant invasion *in vivo* is often associated with tissue destruction, the extent to which this occurs is variable. Various studies using different organs suggest that malignant cells first invade the connective tissue framework and only later do they replace epithelial structures. Furthermore, the pattern of invasion is also affected by tissue organization since organs with considerable connective tissue are invaded by fine strands or even solitary cells, whereas organs with minimal connective tissue are often subject to invasion in bulk. The pattern of invasion is also likely to vary with the malignant cell type under study, but the extent of contribution of the malignant cell and of the tissue being invaded to the overall pattern of invasion is not clear.

The invasion of embryo chick heart fragments by Kirsten murine sarcoma virus-transformed MO_4 fibrosarcoma cells (known to metastasize after implantation into syngeneic C3H mice) was found to be inhibited reversibly by microtubule inhibitors such as podophyllotoxin (0.3 $\mu g\ ml^{-1}$) but not by antiproliferative agents such as cisplatin (1 $\mu g\ ml^{-1}$). Although it is known that microtubule inhibitors can influence cells in a manner not directly related to their effects on microtubule kinetics, results such as these suggest that the effects of microtubule inhibitors on invasion are due not to their cytotoxic properties or to their antiproliferative nature, but to their perturbing effects on microtubular organization. Interestingly, murine neuroblastoma cells selected for vincristine resistance were reported to invade heart tissue fragments at concentrations of vincristine which would normally inhibit invasion of MO_4 cells, thus indicating that the prime anti-invasive effect of microtubule inhibitors arises from direct action on the tumour cells even though such substances interact with the heart cells as well.

Microtubule inhibitors do not interfere with locomotion *per se* since treated cells can still move. Thus, invasion would not seem to be a consequence of active (random) movement alone. Evidence from studies on a wide variety of cells suggests that microtubule inhibitors interfere with directed locomotion, presumably through disruption of the normal cytoplasmic microtubular arrangement (Mareel and De Mets, 1984). Directed locomotion, a form of persistence of direction, should not be confused with chemotaxis which is directed locomotion in response to a gradient of some chemoattractant (Wilkinson, 1982). Although evidence from studies with microtubule inhibitors is taken to indicate that directed locomotion is important in invasion, at least under *in vitro* conditions, there is no proof that this is the case. Invasion is obviously not a passive process and although firm evidence is lacking, it may be hypothesized that active probing by cell processes contributes to invasive ability. If this is so, cells lacking a normal microtubular framework may not meet the physical requirements in order to succeed in invading. Keller and colleagues (1985) have shown that microtubule inhibitors elicit polarization and stimulate the motility of both PMNs and W256 carcinosarcoma cells, but the projections which characterize the polarization response are more bleb-shaped than normal. Such rounded blebs may not serve the same functions or act as efficiently as normal cellular processes. Thus, although cellular probes may be extended in cells treated with microtubule inhibitors, whether such cells succeed in invading or not may be a function of the physical resistance of their 3D environment and not actually related to directed locomotion *per se*. Furthermore, when cell aggregates are placed on plastic substrates, the cells emigrate outwards in a directional manner. It might be supposed that this pattern of locomotion arises as a consequence of contact inhibition of locomotion in all directions but the external one, although there is no convincing evidence that this is the case. Regardless of the underlying mechanism, cells may always escape from aggregates in a directional manner and thus the results with microtubule inhibitors may be a function of the experimental system. It is also possible that negative chemotactic factors may exist, stimulating directional migration outwards from the aggregate, but there is as yet no evidence for these in mammalian systems, and we must admit that the cue for the postulated directional migration from aggregates remains uncertain.

Growth pressure arising from tumour cell proliferation has been suggested as one mechanism by which tumour cells may invade (Easty, 1975) and the growth of MO_4 cells in this confrontation system quantitatively influences invasion as judged by replacement of heart tissue (Mareel, Kint and Meyvisch, 1979). However, since growth inhibitors which do not interfere with the microtubule complex permit invasion in the confrontation system (Mareel and De Mets, 1984) it would seem that invasion involves more than growth alone. Thus growth influences the rate of invasion (possibly through increasing the number of invasive cells) but it is not causally related. This conclusion is in accord with clinical observations which suggest that tumour growth is not necessarily predictive of invasive behaviour (section 2.13.1).

When MO_4 cells were seeded onto heart monolayers rather than tissue fragments, the tumour cells adhered to the heart cells and eventually migrated between them, but there was no evidence of tissue degeneration as seen when fragments were invaded. It was presumed that heart cells in 2D form were resistant to destruction (Mareel, Kint and Meyvisch, 1979) although it may be that this form of culture simply allows dilution of the agents responsible for tissue damage. Tissue necrosis is not an absolute pre-requisite for invasion and indeed invading cells by-pass necrotic areas (Latner, Longstaff and Lunn, 1971). This might suggest that the criteria chosen to describe invasiveness by Mareel and colleagues are unneccesarily restrictive, although it is interesting to note that in the earlier work of Wolff (1967), both normal cells and tumour cells invaded embryonic organs, the main difference being the extent of tissue destruction by invading tumour cells.

More recently, Mareel and his colleagues have used this model system to study the role of cell surface carbohydrates in invasion. According to the results of Bolscher *et al.* (1986), only cells with certain (unknown) carbohydrate alterations show invasive capacity. Furthermore, the processes of immortalization and morphological transformation do not lead to invasion unless they are coupled to such alterations. Carbohydrate changes were detected by an increase in the size of some unknown glycopeptides following gel filtration of protease digested surface glycoproteins extracted from transformed cells. Some of these alterations may have been due in part to additional sialic acid residues, but direct evidence for this is lacking. HSU and R1C cell lines, both derived from baby rat kidney cells, were among the cell types studied by Bolscher *et al.* (1986). The non-tumorigenic (but morphologically transformed) HSU cells were derived by transfection with a small fragment (the left hand 0–7.2%) of adenovirus type 12 DNA. R1C cells, on the other hand, are tumorigenic and these were derived by transfection with a larger part (0–16%) of the adenovirus type 12 genome. In agreement with the general conclusion above, glycopeptides derived from R1C cells were found to be of greater apparent molecular weight than those derived from HSU cells. This difference is presumably due to the effects of the larger Ad12 genome fragment used in the transformation of the R1C cells.

The HSU and R1C cell lines were also used in experiments in which cell surface carbohydrate content was altered by culture with ET-18-OCH_3, the ether analogue of the alkyl-lysophospholipid 2-lysophosphatidylcholine. This agent, and other lysophospholipid analogues, has been shown to inhibit metastasis of the Lewis lung carcinoma in mice (Berdel *et al.*, 1981). It is incorporated into cell membranes where it induces an increase in fluidity as detected by fluorescence polarization techniques (Storme *et al.*, 1985). Presumably, membrane incorporation also effects other cellular processes, but these have not been analysed in detail. Pretreatment of non-tumorigenic HSU cells by culturing with ET-18-OCH_3 was found to increase the size of glycopeptide fragments obtained following protease digestion of cell surface glycoproteins. The carbohydrate alteration of HSU cells induced by pretreatment with the lysophospholipid (20 $\mu g\ ml^{-1}$) was found to promote invasion of chick heart fragments following their incubation with single, suspended HSU cells (rather than aggregates). That at least part of the carbohydrate alteration involved sialic acid was suggested by experiments in which treatment with neuraminidase abolished the effects of the lysophospholipid on invasion. When the R1C cells were treated with the lysophospholipid, no further increase in glycopeptide size was noted. Essentially similar results were found with NIH 3T3 cells and their transformed counterparts transfected with human T24 bladder carcinoma DNA, but untreated NIH 3T3 cells showed some spontaneous invasion of the chick heart fragments. If invasiveness is related to altered carbohydrate content, then control NIH 3T3 cell populations presumably contain some variants which have the necessary carbohydrate changes.

Three technical points are of interest with respect to this study:

(a) The effects of the lysophospholipid on invasion were transient, disappearing about 64 h after removal. Invasion was more marked after the effects of the agent had worn off.

(b) Cells were pretreated with the agent which was thus not present during the invasion assay.
(c) Because the agent is toxic, it was used at a level which killed no more than 25% of the cells.

In a related study, Storme *et al.* (1985) found that ET-18-OCH_3 (10 μg ml^{-1}) actually inhibited the invasion of chick heart fragments by malignant MO_4 cells. This group has now attributed these results to the fact that in these earlier experiments the lysophospholipid was present continuously in the invasion assay, thus offering it the possibility of interfering with the carbohydrate content of both the MO_4 and the chick heart cells. Whatever the explanation for these disparate results, we must admit that we know remarkably little about how cell surface chemistry affects invasive behaviour. Interestingly, the inhibition of invasion seen when the malignant cells were cultured with heart fragments in the continuous presence of ET-18-OCH_3 was not associated with any noticeable effects on microtubule assembly or directional migration, indicating that these processes alone do not guarantee invasion (Storme *et al.*, 1985).

In an effort to mimic *in vivo* the characteristic features of brain tumour invasion, Steinsvag, Laerum and Bjerkvig (1985) developed a model system in which fetal rat brain fragments were confronted with spheroids of syngeneic BT_5C glioma cells. The brain fragments were pre-cultured for 20 days at which time they consisted of three layers: an outer layer of astrocytes and oligodendrocytes, a middle layer of circularly arranged astrocytic and neuronal processes, and a predominantly neuronal centre. These fragments were then confronted with the spheroids on semi-solid agar and co-cultured for up to 30 days. Histotypic changes occurred in the brain fragments during this time even without the spheroids present. Furthermore, as the tumour spheroids increased in size their central regions became necrotic, beginning at a diameter of about 1500 μm. Dividing tumour cells within the spheroids were usually confined to the outer 200 μm. Invasion in this system began with adherence between a brain fragment and a tumour spheroid, usually within 24 h. Within 3 days, tumour cells had reached the middle layer of the fragment and gradual replacement of the middle and outer parts of the brain fragment with tumour cells was seen to occur, although total replacement was not evident over the course of the experiment. The invading cells remained contiguous with the tumour mass and there was no indication of invasion by single cells. These results are fundamentally different from those of Wang and Nicolson (1983) who noted invasion of neonatal murine cerebral and cerebellar tissue by single B16-B15b cells within

4 h of culture, and essentially complete degeneration of the brain tissue by 5 days. It should be noted, however, that Steinsvag and colleagues (1985) studied invasion from tumour spheroids whereas Wang and Nicolson (1983) used a single cell suspension of tumour cells. B16-B15b cells are a derivative of the B16 melanoma cell line selected 15 times for brain colonization (Miner *et al.*, 1982), but in the absence of appropriate control studies the relevance of the selection procedure on *in vitro* invasion is difficult to put into perspective.

In later work, Bjerkvig, Laerum and Mella (1986) used the aggregation system to study the invasive behaviour of two rat glioma cell lines, BT_4Cn and BT_5C. In this work normal fetal rat brains (18 d gestational age) were dissociated enzymatically and the single cells were allowed to aggregate spontaneously. This usually took about 48 h, and was followed by morphological cell differentiation over the next 18 d. At this age, the brain aggregates had a well developed neurophil with neurons (including myelinated nerve axons), synapses, astrocytes and oligodendrocytes. When such aggregates were confronted with BT_5C glioma cells in stationary culture, the tumour cells adhered to the aggregate and began to destroy the normal tissue. By 12 d in culture the tumour cells had replaced most of the normal tissue. The tissue destructive activity of BT_5C cells was detectable within 72 h when aggregates were cultured with conditioned medium from these tumour cells, suggesting that they secrete lytic factors. The BT_4Cn cells also adhered to and progressively invaded the normal brain aggregates, but they did not cause rapid tissue destruction. In this study the authors attempted to correlate the *in vitro* pattern of behaviour with that seen *in vivo*, and in fact there appeared to be good correlation between the two. Thus the BT_5C cell line was characterized by solid invasion with high lytic activity, whereas the BT_4Cn cells invaded singly with reduced lysis of normal brain cells.

The type of model system in which invasiveness is studied *in vitro* using tissue fragments and aggregates of tumour cells has a number of limitations:

(a) There are differences in the extent and time course of invasion when using single tumor cells or aggregates of such cells. Whether one system is more realistic than the other is probably not too important since both differ considerably from the *in vivo* situation.
(b) During prolonged culture the centres of tissue fragments may become necrotic and this could influence the extent of invasive behaviour, particularly if tissue destruction is used as an indicator of invasion. Experiments with polyoma transformed BHK21 cells

have shown that these cells do not invade necrotic adult mouse kidney fragments, and that in partially necrotic material they bypass the necrotic zones (Latner, Longstaff and Lunn, 1971). Experimentally, some attempts have been made to reduce spurious necrotic effects by restricting the diameter of tissue fragments to less than 0.4 mm, thereby allowing diffusion to maintain the viability of the cells. Experiments have shown that malignant cells can invade tissue fragments in the absence of central necrosis and that the tissue destruction which accompanies invasion is not due to hypoxic conditions.

(c) As we have seen in section 4.8.2, aggregates of different normal, embryonic cell types can sort themselves out in an arrangement where one is inside the other. Furthermore, normal tissue fragments will associate with other normal tissue fragments and their cells can intermingle. Although Weston and Abercrombie (1967) demonstrated only minimal mixing of cells between fused fragments of embryonic chick heart ventricle, many other experiments based on fusion of homotypic tissues have shown considerable intermixing (Armstrong and Armstrong, 1973, 1978). Interestingly, inert particles of metal, glass and plastic can be translocated to the interior of aggregates of liver and heart cells (Wiseman, 1977b). The ability of both inert particles and normal cells to translocate inside aggregates and/or embryonic tissue fragments imposes severe restrictions on the validity of invasion systems based on this methodolgy.

(d) Embryonic tissues, fragments and aggregates appear to be composed of particularly motile cells and this behaviour might be associated with the ability of normal cells and even inert particles to translocate internally. Because normally non-invasive cells or particles can appear to invade in this model system, associated tissue destruction might appear to become a key criterion in discriminating between invasion by normal and malignant cells. However, as discussed elsewhere, the extent of destruction shown by invading tumour cells is variable.

(e) In systems using heart muscle it should be remembered that this tissue is rarely invaded *in vivo*. Apart from being an embryonic tissue, the heart mucle used in these experiments is also undergoing reorganization following fragmentation. The surfaces of tissue fragments become lined with tangentially arranged cells (up to a few layers deep) soon after they are cut and placed in culture. Although fibroblastic in appearance, the true origin of these cells is uncertain and they apparently disappear when the fragments are cultured with aggregates of either normal or malignant cells. The importance of the lining cells in the invasive process is uncertain, but since they represent the first cellular barrier they may play a significant role in early events such as adhesion between the tissue fragment and the aggregate. This emphasizes another limitation of the system, since embryonic tissue fragments and maligant cell aggregates are unlikely to come into contact *in vivo*.

What then are the advantages for using assays of this type? The main advantage, as mentioned previously, is one of experimental simplicity. A more important question we should ask is to what extent does the model mimic invasion *in vivo*? There are several lines of evidence which offer some confidence in using this type of model system:

(a) The same criteria for invasion *in vivo* may be applied *in vitro*. Thus invasive malignant cells are seen to occupy host tissue and induce destruction, both of which progress with time. However, not all tumour cells destroy normal tissue as they invade and these results may reflect nothing more than a judicious choice of tumour cell types.
(b) The histopathology of invasion *in vitro* may show a remarkable resemblance to that *in vivo*.
(c) There appears to be at least some correlation between invasion in this *in vitro* system and invasive and metastatic capacity *in vivo*. Again, however, the point needs to be made that these correlations do not hold for all types of tumour cells. Although the predictive significance of this model has not been fully tested, the spontaneous acquisition of tumorigenicity by ST-L mouse lung cells corresponded with the development of invasiveness as assessed *in vitro* by this technique (Kieler *et al.*, 1979).

4.8.5 Amnion invasion

Several reports have described an amnion invasion system for studying tumour cell behaviour *in vitro* (Liotta, Lee and Morakis, 1980; Russo *et al.*, 1981; Russo, Thorgeirsson and Liotta, 1982; Thorgeirsson *et al.*, 1982.). In this model, fresh human amnion (about 10 cm^2) supported on a Millipore filter is clamped into a chemotactic chamber and tumour cells are seeded onto the upper surface. Tumour cells which succeed in migrating through the amnion may be collected within the filter, correctly identified, and counted to provide a quantitative assessment of invasion. There is some variation in thickness between different amnions and between different portions of the same amnion, and this needs to be taken into account in studies where attempts are made to quantify invasion. It has been observed that tumour cell migration in this system is decreased if the culture medium contains serum, so the medium is usually supplemented with 0.1% fetuin (Russo, Thorgeirsson and Liotta, 1982). The reason for this migration decrease in the presence of serum is unknown and further research on this point is required. Varani, Orr and Ward (1978a) have suggested a correlation between the low levels of protease inhibitor activity in human serum and the decreased adhesivness of tumour cells which possess protease activity, but fetuin has no known protease activity and thus it is uncertain how the observation of Varani *et al.* (1978a) might relate to studies using the amnion model system. The use of serum-free conditions has its advantages since serum components might interact with the reagents under study using this model system.

The typical human amnion is a transparent membrane about 0.5 mm thick with an area of about 1000 cm^2. It is composed of a single layer of epithelium overlying a continuous basement membrane, which is in itself attached to a non-vascular, collagenous stroma. Tight junctions and desmosomes are found between the epthelial cells and their surfaces display microvilli and a relatively prominent glycocalyx. The epithelial cells are bound to the basement membrane by hemi-desmosomes. The basement membrane (which is thus of epithelial rather than endothelial origin) contains the typical components of such structures including Type IV collagen, laminin, fibronectin and various proteoglycans. The underlying stroma is known to contain collagen types I, III, and V and fibronectin (Alitalo *et al.*, 1980). The amnion is impermeable to labelled high molecular weight glycoproteins (Liotta, Lee and Morakis, 1980) and the stromal layer alone is impermeable to colloidal carbon particles (Russo, Thorgeirsson and Liotta, 1982). Although the amnion is not invaded by human endothelial cells or fibroblasts, other cells such as macrophages, PMNs and different types of tumour cells have been shown to do so (Russo *et al.*, 1981; Russo, Thorgeirsson and Liotta, 1982). Human PMNs require a chemotactic stimulus to migrate through the amnion whereas tumour cells such as human breast carcinoma do not, which is in line with the results of Elvin, Wong and Evans (1985) using the electrophysiological assay described earlier. Interestingly, treatment of PMNs with chemoattractants such as FMLP induces secretion of type IV collagenase (Mainardi, Dixit and Kang, 1980), which is supportive of the concept that enzyme release might be important in tissue invasion, if only for some cell types.

Invasion of the amnion is thought to follow cell adhesion to its surface. Since the upper surface of epithelial cells presents a poor surface for tumour cell attachment (but not for PMN attachment), the amnion is usually used in this assay system after it has been stripped of its epithelium by treatment with 0.1 M ammonium hydroxide. The assay system as it is employed for tumour cell migration is thus a model of basement membrane and connective tissue invasion and not one of migration through a contiguous cellular barrier.

Adhesion of A-431 human squamous epidermal carcinoma cells to the amnion basement membrane is inhibited but not abolished by anti-laminin antibodies indicating that tumour cells probably use more than one adhesive mechanism (Russo, 1985). The invasion of amnion by M5076 murine reticulum sarcoma cells is responsive to but not absolutely dependent upon the chemoattractant FMLP, and is reduced by metalloproteinase inhibitors such as TIMP (Murphy, Cawston and Reynolds, 1981) but not soybean trypsin inhibitor (Thorgeirsson *et al.*, 1982). These results would suggest that invasion through this type of connective tissue matrix is crucially dependent upon metalloproteinases, which include collagenases capable of degrading the major collagenous components of the amnion (collagen types I, III, IV and V). It should be borne in mind, however, that the spectrum of active enzymes produced by tumour cells may be modulated according to their environment or substrate (Heisel, Jones and Lang, 1981) and thus other enzymes may be important at other steps in the metastatic cascade.

One significant advantage of this model system is that the basement membrane of the amnion can be radiolabelled in culture with ^{14}C-proline. When the amnion is subsequently denuded and used in invasion studies, the amount of radioactivity released into the medium serves as an indicator of degradative activity during the invasive process. The release of radioactive material into the culture medium increases with time during the invasion of denuded amnion by murine M5076 cells, but no such increase is seen with non-invasive fibroblastic cells (Russo, Thorgeirsson and Liotta, 1982). However, there appears to be no simple relationship between basement membrane degradation and invasive capacity, since MCF-7 human breast carcinoma cells are highly invasive in this system but display only limited basement membrane degradation. It may be that only a certain amount of degradation is required (the rest being surplus to that actually required) and/or some cells may be more deformable or forceful than others thus requiring less basement membrane degradation. MCF-7 cells which migrate through the denuded amnion and penetrate the underlying Millipore filter can be recovered and recycled through the invasion system. A five-fold increase in the rate of invasion through the amnion has been reported for MCF-7 cells on retesting after a single passage (Russo, Thorgeirsson and Liotta, 1982). It would be useful to know if this increased invasive behaviour correlates with any changes in proteolytic activity. These results with MCF-7 cells should be compared with those of Mareel and his colleagues discussed above, in which the MCF-7 cell line failed to invade in the tissue/aggregate confrontation sytem. The only explanation available at the moment for these results is that cells behave differently in different systems: this serves to remind us yet again of a major limitation of model systems.

In order to test the invasive properties of tumour cells through an endothelial barrier, the model system described above has been modified by growing endothelial cells on the amnion after having first stripped off the epithelium. The presence of an endothelial layer decreased tumour cell invasion from two- to four-fold indicating that a single layer of cells serves as a barrier of some sort to invasion. Although tumour cells adhere to the upper surface of endothelial cells, their rate of adhesion is less than that to exposed basement membrane (Poste and Fidler, 1980) and this difference might underlie the observed decrease in invasion.

4.8.6 *In vitro* invasion of the embryonic chick chorioallantoic membrane

The embryonic chick chorioallantoic membrane (CAM) has been used as a model system to study invasion under both *in vitro* and *in vivo* conditions. The use of the CAM as an *in vitro* model system for invasion was pioneered by Easty and Easty (1973, 1974) who reported preliminary studies with a number of tissues including the CAM, chick amnion, hamster, rat and mouse omentum, and rat scrotal sac. Each of the membranous tissues was spread over an expanded metal grid previously covered with a film of agar, placed in a Petri dish, and seeded with a known number of tumour cells. After various time intervals, tumour cell invasion was evaluated by histology using direct staining (if the tumour cells had distinct characteristics) or autoradiography (for which the tumour cells were prelabelled with ^{3}H thymidine). These authors found that the CAM was the most suitable tissue of those that they studied for analysing invasive behaviour *in vitro*. Non-tumorigenic C3H mouse L cells invaded the CAM poorly (13% at 48 h), as did mouse peritoneal macrophages (17%) in these studies. The authors attributed this small amount of movement (up to 50 μm for occasional cells) to ectodermal damage caused during preparation, which of course detracts from the use of the CAM as an *in vitro* model system. A poorly malignant, rodent lung cell line (P4) was found to invade the CAM moderately (38%), whereas the P4T line (its highly malignant variant) invaded extensively (79%). Colcemid

(5 μg ml^{-1}) which affects microtubule formation and reduces directional cell movement had little effect on the extent of CAM invasion contrary to the results of Mareel and De Mets (1984) reported earlier, although admittedly in a different system. Invasion of the CAM seemed not to correlate with cell adhesiveness to the ectodermal surface, and these authors quoted Walker 256 carcinosarcoma cells as an example of a tumour which was poorly adhesive but nevertheless invasive, usually in a destructive manner. Similar results with this cell type were obtained by Elvin, Wong and Evans (1985) who studied the invasion of monolayers of MDCK epithelial cells by electrophysiological and ultrastructural means. Easty and Easty (1974) estimated rather crudely that the tumours they studied invaded at an average rate of at least 2-3 μm h^{-1}.

Various improvements have been made to the original system for studying invasion of the CAM *in vitro*. Poste *et al.* (1980), modified an approach suggested by Hart and Fidler (1978) by culturing the CAM as a sheet on a teflon supporting ring. Radiolabelled tumour cells were then added to the relatively non-traumatized endodermal epithelium, and invasion was quantified by measuring the amount of radioactivity in the CAM after various time intervals. Some of the tumour cells migrated through the entire depth of the CAM and these could be collected from the basal chamber. When nutrient agar was included in the basal chamber, these invasive cells were able to divide to form colonies which were recoverable in due course.

4.8.7 Invasion of blood vessels *in vitro*

Several laboratories have studied the invasion of tumour cells through portions of excized blood vessels. Poste *et al.* (1980), for example, isolated segments of vein into which they inserted a supporting porous cylinder of a pore size (<50 μm) which enabled single cells to pass through. Each vein segment could be placed on the supporting cylinder either with the endothelium facing inwards or with it facing outwards after eversion of the blood vessel. The cylinder with attached vein segment was then placed in a flow-through system in which medium was recycled over its inner surface to maintain viability. The outer surface of the endothelium was bathed in an external medium, which only communicated with the inner recycling medium via the thickness of the vessel wall and the porous tube combined. Tumour cells could be added to the system via injection ports into the external medium so that they came into contact with the outward facing surface of the endothelium. Cells which invaded through the wall of the vessel and passed through the pores of the supporting cylinder could be collected in the recycled medium which bathed the inner aspect of the cylinder. Although an ingenious system, the technique uses blood vessels of much greater diameter than the microvessels which are usually invaded *in vivo*. The isolated blood vessels also presumably suffer some damage during isolation and preparation which is difficult to account for, and which may be associated with the release of bioactive molecules that could affect processes such as adhesion to the endothelium. Finally, the tumour cells are added to the external medium which is essentially static, and thus they are not able to flow down the lumen of the vessel as would occur *in vivo*. Such criticisms do not detract from other features of the model, but they need to be taken into account when assessing the validity of the assay as a model system for invasion *in vivo*.

Jones and his colleagues have recently developed a model system for studying invasion based on an 'artificial blood vessel wall' (Jones *et al.*, 1981). In fact, the model consists of multilayers of rat smooth muscle cells (strain R22CID) on which a layer of bovine endothelial cells is grown. The two cell types in the artificial wall become separated by an irregular basement membrane-like structure, much as may be seen *in vivo*. These authors found that the ability of HT1080 human fibrosarcoma cells to rapidly hydrolyze the smooth muscle multilayer was considerably retarded by the presence of the endothelial cell layer. At present, the mechanism by which endothelial cells inhibit the proteolytic activity of tumour cells is not known, but supernatants from endothelial cells have been shown to inhibit the plasminogen activator activity of HT1080 cells (Heisel, Jones and Lang, 1981). Although the lytic activity of HT1080 cells is reduced by the presence of endothelial cells it is not totally blocked, and the tumour cells invade the endothelium, often proliferating between the two cell layers of the artificial blood vessel wall. Ultimately, the tumour cells may completely destroy the artificial structure. It should be noted that the inhibition of tumour cell lytic activity is not restricted to endothelial cells since other cell types (including human fibroblasts) are capable of similar activity. Interestingly, the presence of tumour cells was found to decrease the synthesis of smooth muscle extracellular matrix components such as elastin and collagen (as determined by incorporation of ^{3}H proline). This raises the possibilty of a two-pronged attack by invading tumour cells, destroying matrix components on the one hand and lowering their rate of synthesis on the other, but how these results relate to *in vivo* studies remains to be evaluated.

Intuitively, a positive correlation between the ability to invade blood vessel walls and metastatic capacity would seem a strong possibility. However, we are beginning to appreciate that tumour cell behaviour is riddled with the unexpected and thus it is no surprize that Aulenbacher *et al.* (1984) have reported that a non-metastasizing rat tumour (BSp73AS) is more active at invading excized blood vessels than its metastasizing variant (BSp73ASML). Despite this, the metastasizing variant has higher cathepsin B-like proteinase activity than the non-metastasizing variant. Two points to note here, however, are that the metastasizing variant spreads via the lymphatics rather than the blood, and that the enzymes present, although potentially capable of assisting invasion, were recorded from inside rather than outside the cells.

4.9 A COMPARISON OF *IN VITRO* AND *IN VIVO* MODELS

Model systems have distinct advantages for the experimental analysis of invasion and metastasis, but results from studies based on such systems cannot be simply extrapolated to the natural *in vivo* situation. We have now examined a number of model systems for studying invasion and metastasis, and it should be clear that each has its own advantages as well as its own shortcomings. All *in vitro* systems lack various cellular and humoral components which may reflect upon the ability of malignant cells to establish metastases. There is evidence, for example, that organ-associated NK cell activity may influence the metastatic process (Wiltrout *et al.*, 1985). Indeed, the enhancement of metastatic capacity in young nude mice is thought to be due to their lower levels of NK cell activity (Hanna, 1980). Furthermore, Horak, Darling and Tarin (1986) have shown that some organs secrete soluble factors which decrease the survival of tumour cells *in vitro* while others promote their attachment and survival and that this is in a manner consistent with the distribution of metastases *in vivo*. Although we have seen that experiments can be done with animal models, these too have their limitations when results are extrapolated to the human situation. The experimental use of animals is also subject to ethical considerations. The nude mouse is gaining support as a model system for studying the invasive and metastatic behaviour of human tumours *in vivo*, although most human tumours metastasize poorly in adult nude mice (Sharkey and Fogh, 1979). More aggressive tumour variants may be selected by grafting lung fragments which have been previously invaded in culture by human tumour cells (Kerbel, Man and Dexter, 1984). One particular advantage of some animal models is that invasiveness *in vivo* may be studied by video and photographic techniques. Early work in this area was performed by Wood (1958), who used the rabbit ear chamber to study blood vessel invasion, and more recently by Haemmerli and Strauli (1978) and Haemmerli, Arnold and Strauli (1982).

Generally speaking, it is useful to employ a spectrum of possible model systems since in this way any limitations specific to one particular assay may be overcome. This approach implies the use of one tumour in a series of studies, and for many laboratories the B16 malignant melanoma has been the system of choice. The reason for this, namely the availability of variant cell lines of different lung colonizing potential, is readily apparent. We have seen, however, that the characterization of fundamental differences between the B16F1 parental cell line and its various derivatives has not always been an easy task. Indeed, despite the enormous amount of research activity spent on this and other similar experimental systems, we are forced to admit that we are still some distance from resolving the crucial differences between metastatic and non-metastatic cells. Nevertheless, multiple studies using a range of model systems do provide information on certain tumour cell characteristics which correlate reasonably well with metastatic behaviour.

5 The adhesive and locomotory behaviour of tumour cells

5.1 GENERAL ASPECTS OF CELL ADHESION

There are two important types of cell adhesion: (a) cell–cell adhesion, and (b) cell–substrate adhesion. It is likely that several different mechanisms of adhesion are used within each of these categories. Some may be used in both cell–cell and cell–substrate adhesion, but most evidence so far suggests that cells employ generally different mechanisms for the two types of adhesion.

Cell adhesions of either type may be categorized by whether they are homophilic or heterophilic in nature, although in reality this has not been done with all suggested adhesive molecules (Fig. 5.1). In homophilic adhesions, as exemplified by L-CAM (the adhesion protein of liver intermediate junctions), the adhesion develops from binding between pairs of L-CAM molecules, one on each cell (Edelman, 1983). In heterophilic adhesions, on the other hand, the bonds form between a cell surface molecule on one cell and a complementary molecule on the surface of another cell or on the substrate, akin to receptor–ligand interactions. This type of adhesion is characterized by bonding of cell surface receptors to molecules such as laminin, fibronectin, vitronectin and collagen. It is not restricted to cell–substrate interactions since Ng-CAM, which is involved in neuron–glial cell adhesion, probably binds in a heterophilic manner (Edelman, 1984). In this latter example, however, the complementary molecule has yet to be conclusively identified. In another variant of heterophilic binding, a cell adhesion ligand may link identical receptors on different cells. This is the type of interaction which is thought to mediate adhesion between sponge cells (Weinbaum and Burger, 1973). Note that when used in the adhesive context the abbreviation 'CAM' refers to a cell adhesion molecule rather than the chick chorioallantoic membrane. Throughout this text

1. Homophilic adhesion, e.g., N-CAM

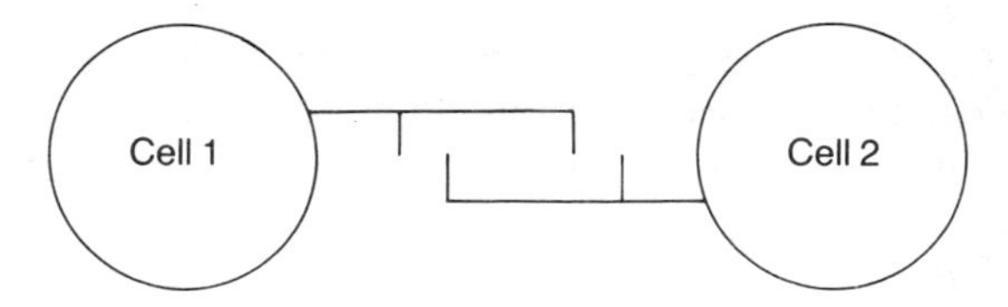

2. Heterophilic adhesion, e.g., Ng-CAM

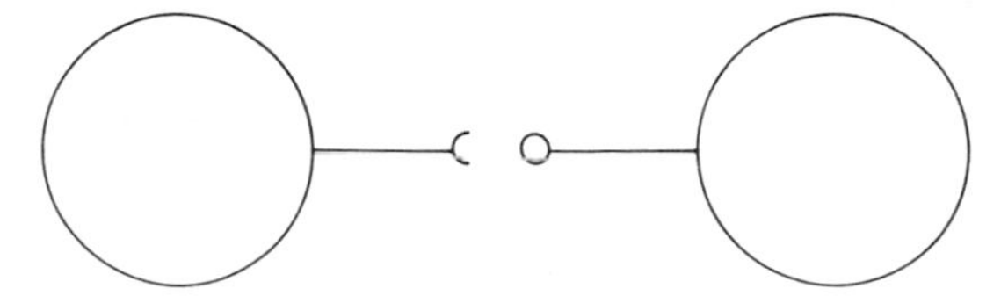

3. Heterophilic substrate attachment

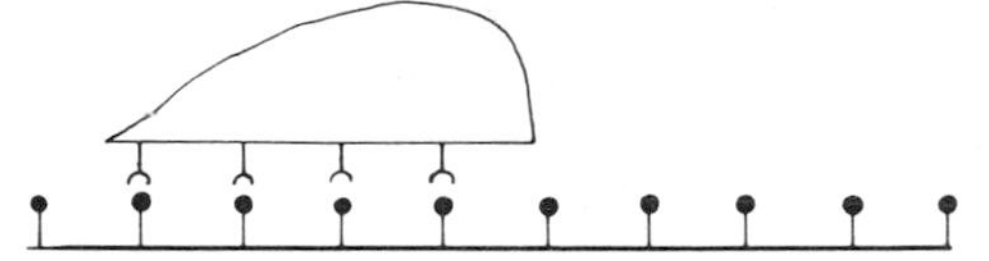

4. Heterophilic adhesion, e.g., sponge aggregation factor

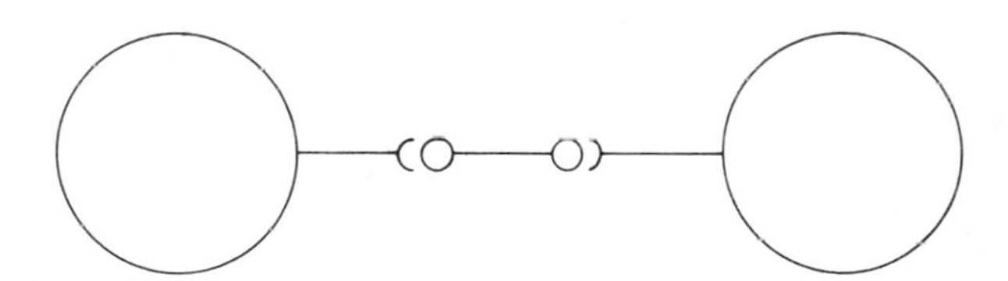

Figure 5.1 Classification of cell adhesions. The two principal types of cell adhesion are illustrated. Homophilic cell adhesion is mediated by identical molecules on each cell (1), whereas heterophilic adhesion involves different molecules on each cell and is more akin to receptor–ligand type interactions (2). Two other variants of heterophilic adhesion may be identified. One is typified by the adhesion of cells to substrate-bound molecules such as fibronectin (3), whereas the other involves a cross-linking molecule as illustrated by sponge aggregation factor. Such factors are thought to aggregate cells by cross-linking cell surface receptors referred to as baseplates (4). N-CAM: neural cell adhesion molecule; Ng-CAM: neuron–glia cell adhesion molecule.

cell adhesion molecules are always preceded by a distinguishing qualifier e.g. Ng-CAM, L-CAM etc.

Cell adhesions may also be divided into two general categories on the basis of ultrastructure. Within this scheme, one category is represented by adhesions which have a recognizable ultrastructure (i.e. the cell junctions), while the other involves interactions between molecules which are currently unresolvable at the ultrastructural level. This is not to say that the second type of interaction is lacking in some microanatomical order, only that the molecules involved are not arranged on or within the cell membrane so as to form a distinct structure.

There is some indication that adhesions develop through a two stage process. In the first stage a rapid, relatively weak bond is formed, while in the second stage the adhesions develop to become more stable. This is analogous to the situation described for standard contact adhesives where an initial, relatively weak bond is formed on first contact and chemical bonding then develops to form a stronger, more permanent attachment. In cellular terms, initial binding might involve non-specific, physical interactions (perhaps based on general charge distribution), whereas stable binding might involve interaction between specific chemical groups (as between ligand and receptor) and the development of cell junctions.

5.1.1 Ionic requirements for adhesion

The function of divalent cations such as Ca^{2+} and Mg^{2+} in adhesion has had a long and chequered history. An early idea (reviewed in Curtis, 1973) was that Ca^{2+} might act as a cationic bridge linking negatively charged groups on the surfaces of apposed cells. Grinnell (1978) later suggested that it might act more through its effects on the cytoskeleton, and for a large number of cells this may indeed be the case. Chong, Parish and Coombe (1987), for example, studied cell–cell and cell–substrate adhesion in a variety of systems and provided evidence for a general involvement for active metabolism, Ca^{2+} and a fully functional cytoskeleton.

In at least some systems, Mg^{2+} and other divalent cations appear to be able to substitute for Ca^{2+} in promoting adhesion, although their effectiveness is generally reduced. It should be recalled here that divalent cations such as Ca^{2+} and Mg^{2+} serve important roles in metabolism, intracellular signalling, and numerous other cytoplasmic processes, and thus it is difficult to distinguish between their direct and indirect effects on cell adhesion. Note also that some adhesions such as that between cytotoxic T cells and their tumour cell targets appear to be Mg^{2+} rather than Ca^{2+}-dependent (section 5.7.2.4).

Other lines of evidence suggest that some adhesions may be classified according to the effects of Ca^{2+} on the enzymatic separation of adherent cells. The so called Ca^{2+}-dependent adhesions are highly sensitive to proteolysis by enzymes such as trypsin, but they are protected from enzymatic activity by the presence of Ca^{2+}. Ca^{2+}-independent systems, on the other hand, require high concentrations of trypsin before they are degraded, and this action is not protected by Ca^{2+}. Examples of Ca^{2+}-dependent adhesion molecules include the cadherins such as L-CAM (section 5.3.1.2) while examples of Ca^{2+}-independent adhesion molecules include N-CAM and Ng-CAM. More recent data, based on molecular sequence analysis, has shown that receptors for adhesive ligands such as fibronectin contain multiple Ca^{2+} binding sites and this probably explains their sensitivity to Ca^{2+} depletion by chelating agents.

Several research groups have implicated Mn^{2+} in adhesion, principal among them being Rabinovitch and De Stefano (1973), Edwards *et al.* (1975) and Grinnell (1984). According to Grinnell (1984), who studied the substrate attachment of suspension adapted BHK cells, Mn^{2+}-dependent adhesion is highly specific, requires the continuous presence of the cation, occurs on a variety of substrata, involves protease-dependent cell surface determinants, and can be inhibited by antibodies against the receptor for WGA (wheat germ agglutinin). The adhesion assay employed by Grinnell (1984) had rather specific protein and ionic requirements, however, which leads us to query the physiological relevance of Mn^{2+}-dependent adhesion. Interestingly, Sonnenberg, Modderman and Hogervorst (1988) have shown that platelet adhesion to laminin mediated by the integrin VLA6 requires the presence of Mn^{2+}, Co^{2+} or Mg^{2+}, but not Ca^{2+}, Zn^{2+} or Cu^{2+}. At a concentration of 2mM, both Mn^{2+} and Co^{2+} appeared to increase the affinity of VLA6 for laminin to such an extent that they could not be displaced by competitive antibody.

5.1.2 Adhesion and the cell cycle

A variety of cellular changes occur during the division cycle, and many of these reflect alterations in determinants present on the cell surface. It would be surprising in the light of these changes if cell surface mediated events such as adhesion were not affected. In fact, there is evidence that both cell–cell and cell–substrate adhesion are altered during progression through the cell cycle. With respect to cell–substrate adhesion, for example, Ohnishi (1981) has observed by scanning electronmicroscopy (SEM) that mouse L cells flatten and apparently adhere more to the substrate during S phase. It is also commonly observed that adherent cells tend to round up as they enter into mitosis and this is usually taken as an indication of reduced cell–substrate adhesion. Many cell types show a reduction in cell surface fibronectin during M phase (Stenman, Wartiovaara and Vaheri, 1977), which may be responsible at least in part for the rounding up observed. Using a collecting lawn assay system to study cell–cell adhesion, Hellerqvist (1979) showed that CHO-K1 cells arrested in G_1 were more adherent than those in S phase. Elvin and Evans (1983) used aggregation kinetics to show that Balb/c 3T3 cells were relatively poorly adherent in S phase, and that they displayed an increase in adhesion as they proceeded into and through M phase. More recently, Cross and ap Gwynn (1987) employed rotary aggregation to show that the adhesiveness of L929 and CHO-K1 cells was maximal during S and late G_2 phases, and minimal in mitosis. In this study, adhesiveness was assessed by counting the numbers and types of contacts between cells after 30 min of aggregation. Discrepancies between the studies referred to above probably reflect a combination of different cell types, different techniques for assessing adhesion, and different synchronization procedures. Cross and ap Gwynn (1987) in fact reported similar results to Hellerqvist (1979) and Elvin and Evans (1983) when they used synchronized cell populations, but opposite results when they used non-synchronized cell populations. Exactly how synchronization affects cell adhesion is not clear, since both chemical and non-chemical synchronization procedures can yield similar results (Elvin and Evans, 1983).

The data above indicate that both cell–substrate and cell–cell adhesions change through the cell cycle, although the matter is far from being fully resolved. In the light of these results it is worth exploring the possible involvement of the cell cycle in tumour spread, particularly in blood borne metastasis where adhesion to the endothelium is likely to play a prominent role. Using the lung colonization technique in which FSA 1233 fibrosarcoma cells were injected intravenously, Suzuki *et al.* (1977) reported that lung tumour formation was highest for cells in S phase, decreased slightly for those in G_2 and M, and was lowest for cells in G_1 phase. These results clearly indicate a cell cycle dependency for lung colonization, but whether it reflects cell cycle dependent increased tumour cell–endothelial cell adhesion or some other process remains to be evaluated. Athough cell cycle effects may well influence colonization at the lodgement stage, they may also influence metastasis at other stages by affecting detachment from the primary tumour and aggregation within the blood. The role of cell adhesion in metastasis is considered in more detail in section 5.7.

5.1.3 Short- and long-range adhesions

Derjaguin and Landau (1941) and Verwey and Overbeek (1948) independently formulated a theory (now known as the DLVO theory) which attempted to describe the interactions between colloid particles which could lead to colloid stability or adhesion. Briefly, DLVO theory proposes that adhesion between colloid particles is determined by the interaction of their repulsion energies (resulting from the fact that their surface charges are of the same sign) and their attractive energies (resulting mainly from long-range Van der Waals interactions). According to London (1937), there are three principal groupings of Van der Waals interactions:

(a) *Orientation interactions*

These are due to dipole–dipole effects.

(b) *Induction interactions*

These are due to dipole–induced dipole effects.

(c) *Dispersion interactions*

Now known more popularly as London forces, these are due to the effects of induced dipole–induced dipole interactions.

Clearly, the first two types of interactions are absent unless the interacting molecules contain permanent dipoles, but the third (induced) type is always present. London dispersion forces appear to dominate in biological systems, although orientation interactions may not be without effect.

DLVO theory has been placed in a biological context by replacing the colloidal particles with cells (a step unfortunately not without considerable assumptions). As illustrated in Fig. 5.2, which represents a theoretical interaction energy curve for two cells, adhesion can occur at two separation distances known as the 1° and 2° minima. At some intermediate distance, the interaction energy becomes strongly repulsive and this represents a potential energy barrier which prevents the two surfaces coming close enough together to form adhesive bonds in the 1° minimum. It is generally thought that adhesion in the 2° minimum is too weak to be of relevance in biological systems, but it may be that such weak adhesions form prior to the developemnt of stronger bonds. This possibility has not been fully explored, but for strong adhesions to form it would seem clear that the potential energy

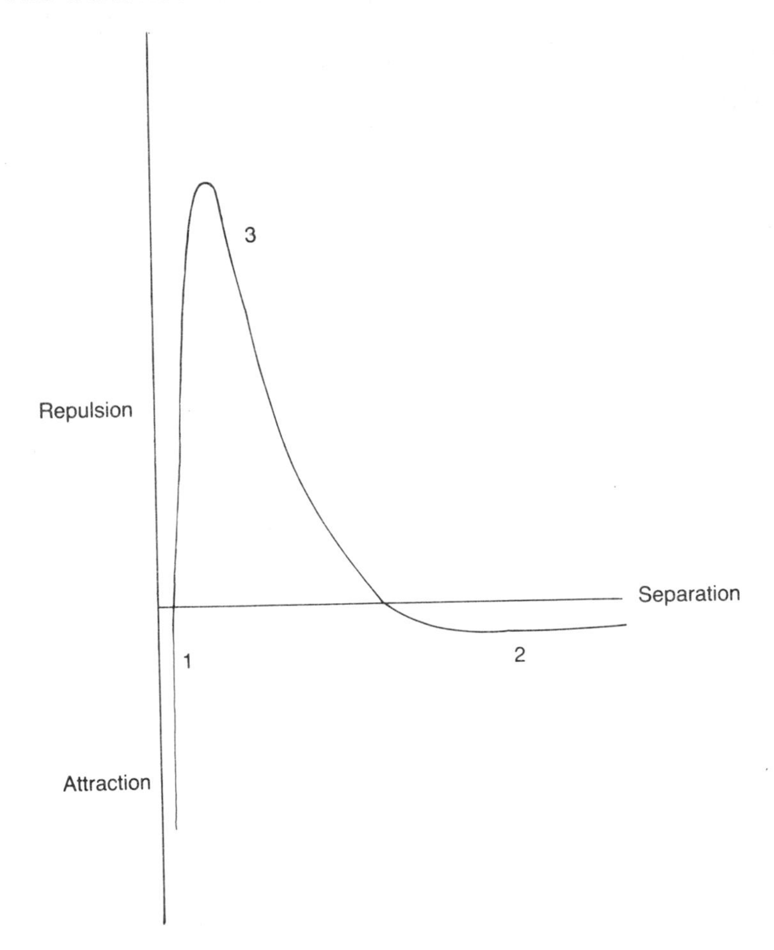

Figure 5.2 Theoretical relationship of the energies of interaction between two cells (after Curtis, 1967). The figure depicts theoretical interaction energies between two cells as a function of the distance between them. According to DLVO theory, the total interaction energy between two cells or particles is represented by the summation of their repulsion and attraction energies. Calculations suggest that adhesion is likely at two separating distances: (1) when the cells are very close (the 1° minimum) and (2) when they are some distance apart (the 2° minimum). The actual energy values and separation distances involved vary with conditions, although it should be obvious that adhesion in the 2° minimum is relatively weak. Typical estimates suggest that such weak adhesions might exist when cells are separated by 50 nm or more. At separating distances between the 1° and 2° minima (3), calculations suggest the presence of a strong repulsive force which presents a potential energy barrier to close approach. One way by which this potential energy may be reduced is for cells to extend surface probes of low radius of curvature (Bangham and Pethica, 1960).

barrier must first be overcome. Biophysically then, we are presented with something of a paradox since there seems to be plenty to keep cells apart but little to get them together, as indeed they can and do. According to Bangham and Pethica (1960), one way by which this energy barrier can be significantly reduced is if cells approach each other via narrow projections. Considerable ultrastructural evidence suggests that cells may indeed behave in this way, but there is some doubt as to whether the cell projections seen in most cases are actually small enough in diameter (around 60 nm) to allow the energy barrier to be overcome. One way round this difficulty is to assume that adhesion occurs through bridging molecules which could extend some distance beyond the lipid bilayer. Such entities are essentially projections with molecular dimensions. An alternative way round the energy barrier might involve interaction between microdomains of different net charge. Based on a study of Ehrlich ascites tumour cells treated with positively-charged colloidal iron hydroxide, Weiss and Subjeck (1974 a and b) have proposed that adhesions might develop between cell regions of different charge densities. For technical reasons, however, it was impossible to actually prove that adhesions involved such sites.

There is some evidence that long-range interaction can confer a degree of specificity on resulting adhesions, but there is little to suggest that this plays a major role in the recognition interactions of animal cells. Both recognition and adhesion probably involve primarily short-range interactions and, as will become clear, there is now abundant evidence for the existence of specific molecules involved in these two phenomena. For the interested reader, Curtis (1973) provides a review of the possible actions of long-range forces in biological systems, and Nir *et al.* (1983) discuss the application of DLVO theory to the aggregation of phospholipid vesicles.

5.2 MEASURING ADHESION

It is not technically possible to measure the strength (i.e. force) of cell adhesion *per se*. It is possible to estimate the force of de-adhesion, however, and this is most simply done by measuring the force required to pull two cells apart or to prise a cell off its substrate. Close examination of the process, however, shows many pitfalls to be present (section 5.2.3).

In a simple, ideal system the strength of adhesion would be constant over the whole area of contact, but in biological systems this is not the case. First, cells probably have multiple mechanisms of adhesion, each one of which may be of a different strength and present in different combinations at different times. Secondly, the precise area of contact in cellular systems is unknown. Cells in close apposition need not be adherent and thus the force of adhesion per unit contact area cannot be reliably deduced. Thirdly, many types of cell adhesions are likely to be dynamic stuctures, breaking down, reforming, and changing their affinity, distribution and surface density.

5.2.1 Aggregation systems

Estimates of the relative adhesiveness of cells can be made from studies in which single cells in suspension are induced to collide and form multicellular aggregates. Several methods have been proposed for the quantitation of relative cell–cell adhesion using this system:

(a) *Rate of aggregation*

(1) Reduction in the number of single cells per unit time; (2) reduction in total particle number per unit time (note that the term 'particle' includes both single cells and aggregates); (3) rate of build up of aggregates. The basic assumption underlying these methods is that the rate at which cells aggregate is a function of their adhesiveness. This is commonly thought to be the case, but there is no *a priori* reason for accepting such a relationship. Some cells, for example, might require a good deal of time to develop strong adhesions, whereas other cells might aggregate quickly but loosely, and never form strong adhesions. Edwards *et al.* (1979), for example, reported that variously transformed BHK21 cells could be either more or less adherent than their parental line when assessed in short-term aggregation assays, but all were less adherent than their parent in long-term (24 h) assays.

(b) *Maximum aggregate size*

Proponents of this approach argue that aggregate size reflects adhesive strength. Aggregates are thought to build up until a maximum size (governed by the shearing forces of aggregation) is reached. Since aggregates can build up at different rates, this approach could yield misleading information unless size is measured under stabilized conditions i.e. in the plateau phase of aggregation (Evans and Proctor, 1978).

(c) *Particle distribution profiles*

An estimate of the adhesiveness of a cell suspension may be gained by analyzing the frequency distribution of different particle size classes.

More adherent cell suspensions are presumed to yield a higher frequency of larger aggregate size classes than poorly adherent cell suspensions.

Although not one of the above methods measures the strength of adhesion directly, this does not limit their practical application. In experimental terms, major problems arise only when comparisons are attempted between different studies where different methods of quantification have been employed.

In an effort to derive an absolute method for measuring adhesiveness based on the rate of aggregation, Curtis (1969) introduced the viscometer into cell aggregation studies. Briefly, in any aggregating system the degree of aggregation is proportional to the number of collisions induced by shear in the medium. It is thus related to the period of aggregation, the magnititude of the shearing forces, the conditions of flow within the system, the initial particle concentration, and the volume fraction of aggregating particles (Evans and Proctor, 1978). If comparisons between different studies are to be made, then these parameters must be taken into account. Viscometers provide an environment for doing just this, since they allow the establishment of laminar flow conditions at known and selectable shear rates. Since few research groups have used similar techniques, however, considerable limitations still exist in making comparative studies.

5.2.1.1 *Problems with aggregation systems*

Aggregation systems are clearly artificial in nature and there is abundant evidence to suggest that results gained from these systems are sensitive to experimental conditions. Among the conditions which are thought to affect aggregation are the following: (a) temperature; (b) pH; (c) flow conditions e.g. turbulent or laminar; (d) viscosity of the aggregating medium; (e) ion composition; (f) serum; (g) different culture media; (h) cell type; (i) cell concentration; (j) cell products; (k) cell preparation methodology; (l) cell cycle.

It is worth noting that many studies have considered cell suspensions to be homogeneous in the sense that they are composed of cells of equal adhesiveness. There is absolutely no evidence to justify this assumption, and indeed it would seeem that a suspension of cells is in fact heterogeneous in respect to this phenotypic trait (Evans and Proctor, 1978).

5.2.1.2 *Tumour cell aggregation*

Numerous studies of tumour cell adhesion have been made using the aggregation system, but the limitations referred to above need to be borne in mind when discussing their application. There is a general opinion in the literature that tumour cells aggregate less than their normal counterparts, but such an interpretation appears to be overly simplistic. For example, as mentioned above, Edwards *et al.* (1979) have reported that virally transformed BHK21 cells always aggregate less than their non-transformed parental cells in long-term, rotation-mediated aggregatory systems. Cassiman and Bernfield (1975) and Wright *et al.* (1977), on the other hand, have reported that a variety of transformed cells actually aggregate faster than their non-transformed counterparts. In yet other studies, Dorsey and Roth (1973) reported little difference in the aggregation performances of 3T3 and SV40-3T3 cells (as judged by the rate of disappearance of single cells), and yet the virally transformed cells were considerably more adherent when assessed by the rate of collection of single cells on preformed homotypic aggregates. Another study of 3T3 and SV40-3T3 cells, this time based on viscometric techniques, has shown that the transformed cells are consistently less adhesive than their non-transformed counterparts, providing the cells are obtained from cultures of comparable cell densities and aggregation occurs at shear rates less than 45 sec^{-1}. At higher shear rates, the aggregation profile reverts with the 3T3 cells becoming less adherent than the SV40 transformed variants (Elvin and Evans, 1982). The obvious conclusion from these and many other studies is that generalizations concerning tumour cell adhesiveness cannot really be susbstantiated.

Aggregate formation is likely to influence metastasis in several ways. The resultant cell clusters are likely to increase the possibility of lodgement, and they also may offer a protective environment with the internal cells shielded by those located more externally. On the other hand, if aggregates develop with host defence cells then the outcome might be more detrimental to the tumour cell. The role of cell adhesion in metastasis is discussed further in section 5.7.

5.2.2 Attachment systems

The measurement of adhesion by attachment is not significantly different from the use of methods based on aggregation rate, the only variation really being in the nature of the substrate. Most studies have measured the rate of attachment of cells to plastic, usually that of tissue culture grade since this has been treated to promote the adhesion and spreading of most tissue cells. Attachment is clearly a function of the physical nature of the substrate, but it is important to understand that this can be modified by the adsorption of molecules from the suspending medium, such as vitronectin and fibronectin (both of which are present in serum and may promote adhesion) or albumin (which may inhibit adhesion). The substrate may also be modified by secretions from the cells themselves, and these secretions may also either promote or inhibit cell attachment.

It should be obvious that the rate of cell settlement from suspension is crucial in attachment assays, and thus physical conditions of the assay system (fluid depth, local eddies, viscosity and so on) will severely affect measured rates of attachment. Unfortunately, numerous authors have overlooked this point, and many studies may well reflect settling times rather than attachment rates. Settling times may be estimated by the application of Stoke's Law or they may be determined by direct microscopic observation.

Cell attachment is clearly different from cell spreading, although the latter is dependent on the former. The criteria for assessing attachment are best kept independent of those for spreading, but many authors have used the presence of spread cells to assess attachment. One simple way of measuring attachment is to allow cells to settle onto an appropriate substrate which forms the lower surface of a small incubation chamber. Following due consideration of the physical conditions involved, adherence may be assessed by counting the number of cells which remain in suspension per unit time following inversion of the chamber (Kellie, Evans and Kemp, 1980). In order to adhere in this assay, cells must bind to the substrate with a force which is able to resist gravity.

The collecting lawn technique of Walther *et al.* (1973) is an interesting variant of the attachment approach in which the substrate is first coated with cells, so what is really measured is a form of cell–cell attachment. This technique is obviously one of considerable relevance when modelling the adhesion of suspended cells to a polarized monolayer. This is clearly the case, for example, when tumour cells circulating in the blood interact adhesively with the lining endothelium.

5.2.3 Detachment systems

In detachment systems, an estimate of adhesion is made from the force required to detach two cells or to remove a cell from its substrate. This approach was first used by Coman and colleagues (Coman, 1944, 1953, 1961; McCutcheon, Coman and Moore, 1948) who employed miniature crowbars (connected to a mechanical sensing system) to prise cells apart or off their substrates. Their results tended to focus attention on the relatively poor adhesiveness of tumour cells when compared to that of normal cells, but we have seen already that the universality of this conclusion is questionable. There are three major problems with detachment systems:

(a) Detachment is not the reverse of attachment. Indeed, detachment and attachment are fundamentally different processes and results from the two systems are not strictly comparable.
(b) When detachment occurs it is often never clear if the cleavage zone passes between the cells or within the cells. In the latter case, any measurements made reflect the internal strength of the cell membrane rather than the strength of the adhesive bond.
(c) The force required to detach cells also depends on the mode of separation (Evans and Proctor, 1978). A peeling mode requires much less force to separate two bodies than a tensile mode.

The nature of these problems is perhaps made more clear by analogy with peeling adhesive tape off paper. Clearly, the process of pulling the tape off is fundmentally different from that involved in putting it on. When the tape is pulled off it often tears the paper so that separation of the two is not clean. Furthermore, it is easier to peel the tape off starting at one corner than to pull it all off in one movement.

5.3 MECHANISMS OF CELL–CELL ADHESION

Possible mechanisms of cell–cell adhesion will be described in terms of recognizable junctions, and in terms of molecular interactions which apparently result in cell–cell adhesion in the absence of overt structural rearrangements.

5.3.1 Intercellular junctions

There are four major types of intercellular junctions in mammalian cells, excluding direct cytoplasmic bridges (Woodruff and Telfer, 1980) and junctions such as the synapse associated with nerve cells. Since epithelial cells display the major junctions of interest in malignancy they will be used as a model system to describe the general properties of mammalian intercellular junctions, with reference being made to other cell types only when required. The reader should be aware that many of the details of junctional structure are not common to different cell types. Cells of the choroid plexus, for example, have their tight junctions located basally rather than apically, and the ridges of endothelial tight junctions partition to the E face rather than the P face on freeze-fracture (see below). General reviews on the structure and function of intercellular junctions are provided by Farquhar and Palade

(1963), McNutt and Weinstein (1973), Staehelin (1974) and Colaco and Evans (1983) among others.

5.3.1.1 *Tight junctions (zonulae occludentes)*

Tight junctions run in a continuous belt, usually around the apical end of each epithelial cell, connecting it to its immediate neighbours (Farquhar and Palade, 1963; Schneeberger and Lynch, 1984). They serve as permeability barriers maintaining the composition of the internal milieu by delimiting luminal and abluminal compartments. The barrier function of the tight junction can be demonstrated vividly under the electron microscope by the failure of electron-dense colloidal lanthanum hydroxide to diffuse through the junction.

Tight junctions also make a major contribution to maintaining epithelial polarity by restricting the diffusion of membrane-located ion pumps and various receptors to either the apical or the basolateral cell surfaces. The apical region often contains enzymes important in the uptake of substances from the external medium whereas the basolateral surface contains the Na^+/K^+ ATPase, hormone receptors and histocompatibility antigens. It should be noted that cell–substrate interactions are also of importance in determining epithelial polarity, since polarity is reversed if collagen is plated over a monolayer of isolated thyroid epithelial cells (Chambard, Gabrion and Mauchamp, 1981). Furthermore, tight junctions appear not to be necessary for the polarized distribution of all surface proteins. When MDCK cells are plated in <5 uM Ca^{2+} they form monolayers, but no tight junctions can be detected. Under such conditions, some proteins can still become restricted to the apical surface (Vega-Salas *et al.*, 1987). Presumably cytoskeletal components are involved in this polarization, but the actual events are as yet unknown. Dragsten and colleagues have provided evidence that the tight junction is also important in restricting the diffusion of membrane lipids (Dragsten, Blumenthal and Handler, 1981; Dragsten, Handler and Blumenthal, 1982). These authors first used fluorescently labelled lectins to show that glycoproteins could not move beyond the tight junctions. In fact, the labelled glycoproteins did not display lateral diffusion within the plane of the membrane at all, possibly because of anchoring via cytoskeletal components. Dragsten and colleagues (1981, 1982) then used fluorescent lipid probes to study the diffusion behaviour of membrane lipids. They showed that some lipids were free to pass the tight junction whereas others were not and that this depended on whether the lipid probes partitioned into the inner or the outer membrane lipid leaflet. Probes which entered the inner leaflet were free to diffuse past the tight junction, whereas those in the outer leaflet were not. Other experiments showed that the lipid probes were not transferred between neighbouring cells, and that the ability of the lipophilic probes to diffuse through the tight junction did not correlate with electrical resistance i.e. lipid diffusion did not perturb the sealing function of the tight junction.

As viewed by conventional transmission electronmicroscopy (TEM), tight junctions in transverse sections appear as somewhat punctated regions between cells where the outer lipid leaflets of the apposed membranes appear to fuse, or to kiss as it has been more evocatively described. In freeze-fracture preparations, where the fracture plane passes along the hydrophobic region between the inner and outer leaflets of the cell membrane, the tight junction is visualized as a linear array of closely packed particles. This array forms an anastomosing network of 6–8 nm ridges on one fracture face of the membrane (predominantly the protoplasmic or P face, but this may depend on fixation). The other fracture face (the exoplasmic or E face) has a network of grooves which corresponds more or less to the ridges on the P face as determined from the examination of complementary replicas (Chalcroft and Bullivant, 1970), and the net result is a quilt-like structure where the ridges and grooves represent the lines of sewing (Bullivant, 1981).

The ridges seen on freeze-fracture preparations are presumed to represent the key protein components of the tight junction. The simplest explanation of such images is that the tight junction proteins of two cells abut end on to form the seal or kiss. Bullivant (1981), however, has subjected complementary freeze-fracture preparations to close scrutiny and he has concluded that the protein particles in fact make contact in an offset manner, with one side of each ridge coming into contact with a side of the other. Such an offset form of contact may improve the sealing capacity of the tight junction. A full explanation of the terminology used in describing freeze-fracture and freeze-etch preparations is given in Branton *et al.* (1975) to which the reader is referred.

The number of ridges (i.e. tight junction sealing strands) seen on analysis of freeze-fracture replicas of various epithelia roughly correlates with their degree of electrical resistance. Low resistance epithelia such as the renal proximal convoluted tubule (which has a resistance of around 6 ohm cm^2) have only 1–2 junctional strands, whereas electrically tight eipthelia such as the urinary bladder (1–2 kilohms cm^2) typically have about 4–8 sealing strands (Schneeberger and Lynch, 1984). The relationship, however, may not always be so clear-cut. One reason for this is that the tight junction is a dynamic structure. Alternatively, the ridges may contain regulatable pores, but firm evidence for this is still lacking (Madara, 1988).

Little is currently known about the molecular nature of the tight junction. That proteins might be involved in tight junction structure was suggested by the studies of Griepp *et al.* (1983), who reported that the development of tight junctions in monolayers of MDCK cells (as measured by transepithelial resistance) was sensitive to the presence of the protein synthesis inhibitor cyclohexamide, providing it was added to the cells within about 8 h of passaging. The lack of any effect thereafter was probably due to synthesis of the necessary proteins having already taken place, even if the cells were kept in suspension. Stevenson (1987) employed a rat monoclonal antibody raised against partially solubilized liver plasma membranes (thought to be relatively enriched in tight junctions) in a study of tight junction biochemistry. The McAb reacts with a phosphoprotein (ZO-1) of M_r 210 kD on Western immunoblots of MDCK cell extracts, and a similar phosphoprotein of M_r 225 kD on mouse tissue extracts (Anderson *et al.*, 1988). Immunofluorescent labelling indicates that ZO-1 is located near tight junctional regions, but because of problems of resolution its precise location is unclear. Because of its ease of solubility during extraction procedures, ZO-1 is presumed to be a peripheral membrane protein lying in a cytoplasmic zone immediately subjacent to the tight junction kiss. More recently, Citi *et al.* (1988) have characterized a molecule associated with tight junctions which they refer to as cingulin, because of its ring-like distribution around avian brush border cells. Like ZO-1, cingulin is a peripheral rather than an integral membrane component, but it has the interesting property of being heat stable. It also differs from ZO-1 in terms of molecular mass, being present in immunoblots as two bands of 140 kD and 108 kD. Immuno-electron microscopical studies suggest that its location is on the endofacial surfaces of the tight junction. Since neither ZO-1 nor cingulin appears to be located in the kissing zone itself, the precise nature of the molecule(s) which actually span between cells linked via tight junctions remains to be determined.

5.3.1.2 *Zonulae adherentes (intermediate junctions or belt desmosomes)*

Zonulae adherentes are usually found in epithelial cells just basal to their tight junctions. Each zonula adherens may strengthen cell adhesion in the vicinity of the tight junction and, since they circumscribe the cell, they may also have an occluding function like the tight junction (Gumbiner and Simons, 1986). Gumbiner and Simons (1987) suggest further that the zonula adherens may influence the assembly and/or positioning of the tight junction in the apical region of the epithelial cell, while Atsumi and Takeichi (1980) and Kanno *et al.* (1984) have suggested that this type of junction might also influence gap junction assembly and maintenance, thereby affecting intercellular communication. Viewed by TEM, the zonula adherens has an intercellular space (about 15–25 nm wide) occupied by amorphous, low-density material. It also has a cytoplasmic plaque which acts as an anchor for the meshwork of actin filaments referred to as the terminal web. The cytoplasmic plaque is rich in vinculin, a 130 kD molecule first described by Geiger (1979). Geiger *et al.* (1985) have used the presence of vinculin to classify adherens junctions into two types, one involving cell–cell adhesion (the intermediate junction) and the other involving cell–substrate adhesion (focal contacts). There are differences between these two types of junctions, however, the prime one being that the 215 kD protein talin (Burridge and Connell, 1983 a and b) is found only in the latter. The role of talin in cell–substrate adhesion is described more fully in section 5.4.2, and the structure of focal contacts is described separately in section 5.4.7. Apart from vinculin, the plaque of intermediate junctions also contains actin, α-actinin and the 83 kD protein plakoglobin, which was first found in spot desmosomes (maculae adherentes). Plakoglobin is thus a molecule common to the two main types of adherens junctions (Cowin *et al.*, 1986). In freeze-fracture preparations, zonulae adherentes display membrane particles which are irregular in size and arrangement.

The major Ca^{2+}-dependent adhesion molecule of the zonula adherens has been isolated by different research groups from different tissues. It is known by a variety of names including L-CAM (Gallin *et al.*, 1983), cell-CAM 120/80 (Damsky *et al.*, 1983), E-cadherin (Ogou, Yoshida-Noro, and Takeichi, 1983), uvomorulin (Peyrieras *et al.*, 1983), Arc-1 (Behrens *et al.*, 1985), and A-CAM (Volk and Geiger, 1984; 1986 a and b). Generally, these adhesive proteins have been classified as cadherins since they are all Ca^{2+}-dependent (reviewed in Gumbiner, 1988; Takeichi, 1988). Most native cadherin adhesion molecules appear to be intrinsic membrane glycoproteins with an M_r of about 124 kD, giving rise to a major trypsin cleavage product of about 81 kD (although considerable variation has been reported). Although the cadherins contain *N*-linked carbohydrates these do not seem to be necessary for proper functioning since tunicamycin has been shown not to affect E-cadherin activity (Shirayoshi *et al.*, 1986). The cadherin family is composed of at least three subclasses of similar molecular composition (723–748 amino acids), but they have different spatio-temporal distribution profiles in the tissues and they have different immunological specificities:

(a) *Epithelial cadherin*

This subclass includes the original E-cadherin discovered on F9 teratocarcinoma cells, mouse blastocyst uvomorulin, human mammary carcinoma cell-CAM 120/80, dog kidney epithelium Arc-1, and chick hepatocyte L-CAM. All of these molecules are found primarily on epithelial tissues.

(b) *Neural cadherin*

This subclass is found primarily on non-epithelial tissues such as nerve tissue and muscle, and it includes the original N-cadherin (Hatta and Takeichi, 1986), chick cardiac intercalated disc A-CAM (Volk and Geiger, 1984, 1986 a and b) and chick neural retina N-Cal-CAM (Bixby *et al.*, 1987). Note that the Ca^{2+}-independent neural cell adhesion molecule N-CAM is not a cadherin, but is a member of the immunoglobulin superfamily (Cunningham *et al.*, 1987).

(c) *Placental cadherin*

This third subclass is detected on both epithelial and non-epithelial tissues, but it occurs predominantly in the placenta (Nose and Takeichi, 1986).

In most epithelial cells, the cadherins are believed to be more or less restricted to the region of the zonula adherens junction itself. Geiger and his colleagues have been able to show that a major part of A-CAM is exposed in the junctional cleft (reviewed in Geiger *et al.*, 1987). In some cell lines, such as MDCK strains, the junctional cadherins have been shown to have a more diffuse distribution soon after plating, but they are subsequently recruited to the lateral surfaces where junctions develop. It is thought that this reflects junctional assembly, with the molecules moving from a cytoplasmic location after synthesis to the cell surface, where they are subsequently incorporated into developing zonulae adherentes.

Antibodies to the zonula adherens adhesive protein dissociate monolayers of MDCK cells when added soon after plating of the cells (Behrens *et al.*, 1985). In established cultures of high-resistance strains of the MDCK cell line, however, such antibodies are ineffective, possibly because thay cannot gain access to the protein which becomes localized to the intermediate junction and therefore basal to the restrictive tight junction (Gumbiner and Simons, 1987). This interpretation could be easily tested by growing the cells on a porous filter support and applying the antibodies from the basolateral aspect. Other evidence that the cadherins promote cell–cell adhesion has been provided by the studies of Nagafuchi *et al.* (1987). These authors co-transfected L cells, which express little cadherin, with plasmids selected for G418 resistance and E-cadherin cDNA expression. Stable transformants were selected in G418 and screened for E-cadherin expression by immunofluorescent techniques. Transfectants expressing E-cadherin were found to acquire Ca^{2+}-dependent aggregating activity, and this correlated with the amount of E-cadherin present.

It has been known since the classic studies of Townes and Holtfretter (1955) that mixed pairs of embryonic cell types can sort out from each other according to their tissue origins (section 4.8.2). Many of these examples of selective cell adherence correlate with the presence of different cadherins on each cell type, although the cadherins themselves are not strictly tissue specific. In fact, it appears that cadherins have two binding specificities. One allows them to bind uniquely with cadherins of the same subclass, while the other allows them to bind more generally to other subclasses. This interpretation is in accord with the molecular structure of the cadherins, which shows that they have both unique and public (shared) amino acid sequences, and it also explains aggregation behaviour *in vitro* which usually proceeds through an initial non-specific stage. Geiger *et al.* (1987) have shown that chick hepatocytes (which express L-CAM) can form heterotypic zonulae adherentes with lens epithelial cells (which express A-CAM). Although nothing is known of the adhesive forces involved, if heterotypic junctions are less energetically favourable than homotypic ones it might be expected that cells will move from heterotypic to homotypic involvement. Such a progression would be in accord with the differential adhesion hypotheis of Steinberg (1964). However, cells are probably capable of utilizing a number of adhwsive mechanisms, and it would be overly optimistic to explain all aggregation and sorting out phenomenon in terms of the cardherins alone.

Changes in the level of expression of the cadherins are seen during development, and in some cases these correlate with major changes in cell behaviour. The loss of E-cadherin by mesodermal cells occurs as they migrate in development, and neural plate infolding also correlates with the loss of E-cadherin from the ectodermal cells involved. The major question to be answered here, however, relates to the possible relevance of the cadherins in invasive and metastatic behaviour. Unfortunately, at this time little is known on whether aspects of the malignant behaviour of tumour cells (such as changes in motility) correlate with changes in cadherin expression.

5.3.1.3 *Gap junctions*

Gap junctions are found between epithelial cells at various points along their lateral membranes (Revel and Karnovsky, 1967; Colaco and Evans, 1983). They are believed to represent communicating structures through which ions, nucleotides and other small molecules can pass between cells. The gap junction is characterized by conventional TEM as having a reasonably regular intercellular space of about 2–4 nm. Oblique sections show the intercellular space to contain structural units, often with a central dot suggestive of a channel or pore. Examination of freeze fracture replicas from cells with gap junctions shows the P fracture face to contain structural units similar to those seen in oblique sections. Each structural unit (or **connexon**) of the gap junction is made up of six particles (or **connexins**) which enclose a small central channel (1.6–2 nm wide in mammalian cells). The channels of apposing connexons on different cells align axially during gap junction formation, thereby effectively joining the cytoplasm of the contributing cells. There is a restriction on the movement of cytoplasmic components between the cells, however, since the diameter of the central channel enables it to act as a molecular sieve, allowing only molecules of particular size to pass through. Studies using synthetic, fluorescent linear probes (about 1.4–1.6 nm wide) suggest an upper size limit for transfer through junctions in human cells equivalent to an M_r of about 800 D (Flagg-Newton, Simpson and Loewenstein, 1979). Larger molecules seem to be able to pass through certain insect cells (upper size equivalent to an M_r of about 1800 D), but as with human cells this is probably dependent on molecular charge as well as structural conformation. Although the amount of charge on a molecule may make a contribution to the efficiency of transfer, gap junction communication is not particularly ion selective. Thus both positively and negatively charged species can pass through the junction, with it being slightly more permeable to cations than anions.

Each of the six subunits which make up a connexon in the cells of a given tissue is thought to be composed of the same junctional protein. Two different proteins have been put forward as the gap junctional molecule of mammalian liver. Most researchers support a 28 kD molecule, but one group argues in favour of a 16 kD protein. The two molecules are not structurally related, and the latter is claimed to be present in gap junctions in organs other than just the liver (Finbow *et al.*, 1987). According to Paul (1986), the size of a cloned rat liver cDNA is consistent with a gap junction protein of M_r 28 kD. Furthermore, the 28 kD protein can form functional channels when reconstituted into an artificial phospholipid bilayer (Young, Cohn, and Gilula, 1987). Antibodies directed against this putative junctional protein significantly reduce conductance both between cells *in vitro* and across reconstituted synthetic bilayers. Experiments with reconstituted bilayers need to be interpreted with some caution, however, since there is evidence that some non-junctional proteins can allow conductance to occur when they are incorporated into artificial lipid bilayers. The introduction into frog oocytes of mRNA prepared from the rat liver cDNA clone isolated by Paul (1986) yields communicating channels on translation which are qualitatively different from the endogenous frog oocyte channels (Dahl *et al.*, 1987). This result, however, does not specifically exclude other proteins (such as the 16 kD form) from playing some role in junctional formation. According to Nicholson *et al.* (1987), the liver gap junction is composed of two closely related components, the 28 kD form described above and another 21 kD molecule. The precise role of the smaller component remains to be defined, and the topographical distribution of both is not clear.

The cDNA obtained from a rat liver library by Paul (1986) recognizes a liver cell mRNA that is not found in heart and lens cells, although cells of these organs have gap junctions. This raises the possibility that there may be more than one gap junction protein, and that the different types may have some tissue specificity. Kistler, Christie and Bullivant (1988) have since shown that homologies exist between the putative gap junction proteins of the liver (21 and 28 kD), heart (47 kD) and lens (70 kD), and that in effect they compose a superfamily of gap junctional molecules or connexons. A noted exception is the 26 kD protein MIP described from the lens (Revel *et al.*, 1986), which thus may not be a gap junctional molecule. The 16 kD protein of Pitts' group was not analysed in the study by Kistler and colleagues (1988), but other workers have shown it to be unrelated to the proposed superfamily.

Other studies have now made it clear that the three principal gap junction proteins are not restricted in their distribution to the tissues in which they were first described. Beyer, Goodenough and Paul (1988) have thus suggested a new terminology which is based on the molecular mass of each protein as predicted from its cDNA. In this scheme the liver gap junction protein is referred to as connexin 32, that of the heart as connexin 46, and that of the lens as connexin 43.

Communication between cells can be studied electrically by measuring conductance changes and visually by either observing the transfer of fluorescent molecules or by studying autoradiographs developed after the administration of radiolabelled metabolites. The relationship between gap junctions, electrotonic coupling and metabolic coupling was examined in an elegant study by Gilula, Reeves and Steinbach (1972). These authors used three cell lines in their research: Don cells (a Chinese hamster fibroblastic cell type), DA cells (an 8-azaguanine-resistant clonal derivative of Don cells), and A9 cells (an 8-azaguanine-resistant derivative of mouse L cells). Don cells are able to incorporate exogenous hypoxanthine into their nucleic acids since they contain the enzyme inosinic guanylic pyrophosphorylase (IPP^+). DA cells and A9 cells, however, are IPP^- since they lack this particular enzyme and cannot incorporate the exogenous purine into their DNA. When Don cells were mixed with DA cells, Don–Don groups were able to incorporate exogenous ^3H-hypoxanthine (as shown by autoradiography), whereas DA–DA groups failed to do so. In heterotypic (Don–DA) cell clusters, however, both cell types incorporated the isotope. This is in agreement with the earlier studies on metabolic coupling of Burk, Pitts and Subak-Sharpe (1968) and Subak-Sharpe, Burk and Pitts (1969), who showed that mutant IPP^- cells can incorporate hypoxanthine if they are grown in contact with wild type IPP^+ cells. Gilula, Reeves and Steinbach (1972) extended this study, however, by co-culturing Don cells (IPP^+) with A9 cells (IPP^-). In this experiment, A9 cells never incorporated the isotope, whether they were in contact with Don cells or not. When examined ultrastructurally, both the Don and DA cells were found to be able to form gap junctions, whereas the A9 cells were deficient in this ability. Their observations that all coupled cells formed gap junctions and that all cells which were not coupled lacked gap junctions provide strong correlative support for the involvement of gap junctions in cell communication. More definitive evidence for such a role has come from the establishment of communicating channels in artifical membranes into which the gap junctional protein connexin has been incorporated (see below).

Qualitative changes in junctional conductance of a quantal nature have been reported and these are thought to be due to the opening and closing of single junctional channels (Loewenstein, Kanno and Socolar, 1978). Recent quantitative recordings from pairs of rat lacrimal gland cells suggest that single channels have an average conductance of about 120 pS when fully open (Neyton and Trautmann, 1985). A picosiemen (pS) is a unit of conductance equivalent to 10^{-12} ohm^{-1}. Much larger conductance changes may result from the synchronous opening and closing of a number of channels. The control of channel opening is not fully understood, but it is thought to be brought about by altering the arrangement of the component subunits of the connexon, a process which can be induced (at least experimentally) by altering the concentration of intracellular Ca^{2+}. In low concentrations of Ca2+

(about 10^{-6} M), the subunits appear to be tilted within the membrane maintaining a patent central channel. When cytoplasmic Ca^{2+} reaches a concentration more typical of extracellular levels (about 10^{-2} to 10^{-3} M), however, the subunits become arranged more parallel to the junctional axis, sliding radially to effectively close the channel in a shutter-like fashion (Unwin and Ennis, 1984). The activation of channel closure by Ca^{2+} offers a mechanism whereby cells which die, or otherwise become leaky to the extracellular environment, can be isolated from the communicating network. The precise role of Ca^{2+} in modulating channel conductance is unclear, but there is mounting evidence that it may in fact act indirectly by perturbing cytoskeletal components of the cell. Loewenstein and his colleagues have since suggested that channel opening may be regulated by phosphorylation. Thus a cAMP-dependent protein kinase (with serine and threonine phosphorylating activity) has been shown to raise junctional permeability, while the tyrosine specific kinase of c-*src* has been shown to reduce it (Wiener and Loewenstein, 1983; Azarnia and Loewenstein, 1987). This association with c-AMP and cellular oncogene related processes links intercellular communication with cell responses to hormones and growth factors via the intracellular signalling network, but exactly how the link-up is made is not yet resolved (see sections 1.7.1 and 1.7.22). Like $pp60^{c-src}$, the product of v-*src* also inhibits junctional communication. The recent studies of Rose (1988) highlight two possible modes of action of $pp60^{src}$:

(a) Intermediate phosphorylation. In this model the oncogene product stimulates polyphosphoinositide breakdown yielding DAG which activates PKC in the presence of cytoplasmic Ca^{2+} (section 1.7.22). Activated PKC is then presumed to phosphorylate certain as yet unknown intermediates which ultimately induce channel closure.
(b) Direct Ca^{2+} involvement. Here the action of the *src* gene product is thought to lead to elevated intracellular Ca^{2+} as before, but this cation then directly induces channel closure.

Other studies have shown gap junctional conductance to be sensitive to the concentration of CO_2 in the bathing medium (Turin and Warner, 1977). This effect is most likely manifested via cytoplasmic acidification, and indeed junctional conductance is inhibited when the intracellular pH (pH_i) is lowered (Spray, Harris and Bennett, 1981). Conductance between isolated pairs of rat hepatocytes is completely (but reversibly) inhibited when the pH_i is artificially lowered from around 7.2 to 6.1 (Spray *et al.*, 1986). Changes in pH_i contribute to the intracellular signalling pathway, but exactly how these effects tie in with the ideas of Loewenstein and colleagues on phosphorylation-mediated control of gap junctional conductance is not clear. Indeed, because the intracellular signalling systems are so intimately interwoven with each other it is extremely difficult to resolve between the effects of changes in intracellular Ca^{2+} and pH.

Studies with filipin, which forms complexes with cholesterol that are recognizable ultrastructurally, suggest that the latter is deficient in areas of the plasma membrane where tight and gap junctions exist. This may serve to stabilize the membrane in the junctional region and could conceivably serve to facilitate adhesion (Robenek, Jung and Gebhardt, 1982). Other studies on regenerating liver (where there is considerable cell proliferation) show that both the number and the size of gap junctions are considerably reduced under these conditions (Meyer, Yancey and Revel, 1981). However, the extent to which similar changes occur in other proliferating tissues such as tumours is uncertain.

5.3.1.4 *Desmosomes (maculae adherentes)*

The junctional structures referred to as spot desmosomes (or more simply, just desmosomes) are local adhesion sites about 0.1 to 1.5 μm in diameter that are found between most epithelial cells, but not in the pigmented retina or in the lens. Cardiac muscle, in which they are a component of the intercalated discs, is the major non-epithelial tissue where they are found. As visualized under conventional TEM, the intercellular zone (25–35 nm wide) of the desomosome consists of an electron-dense midline which may be connected to the plasma membrane by cross-bridges. The structures each side of the midline are thought to be contributed by different cells of each linked pair. It would seem that the 'glue' of desmosomes is likely to be in the intercellular zone, and recent work has suggested that there are at least three major glycoproteins in this area. The desmosomal plaque is seen under TEM as a 15–20 nm wide electron dense zone close below the plasma membrane of each cell contributing to the desmosome. Conventionally, the cytoplasmic plaque is thought to be the anchorage structure for tonofilaments, which are actually intermediate filaments of cytokeratin, desmin (in the heart) or vimentin (in the arachnoid). However, the tonofilaments appear not to reach into the recognizable dense plaque region but instead are associated with the adjacent cytoplasm. There is some evidence that 3–5 nm diameter protofilaments may attach the tonofilaments to the plaques. In itself, the desmosomal plaque may not be an homogeneous structure since Steinberg and colleagues (1987) have identified two component parts, referred to as the inner and outer dense plaques, in bovine muzzle desmosomes. According to Steinberg's group, the outer dense plaque represents the plaque generally seen on TEM and it is located closer to the plasma membrane than the inner (less) dense plaque. In freeze-fracture preparations, the desmosomal plasma membrane contains clusters of 12–20 nm elongated particles on the P fracture face.

It should be noted that although desmosomes are almost ubiquitous in all vertebrate epithelia, desmosomes from different tissues appear to differ. Furthermore, not all desmosomes studied so far contain the same constituent molecules. Despite molecular differences, desmosomes may form under *in vitro* conditions between cells from vertebrates as diverse as humans and the frog, suggesting that there has been some evolutionary conservation (Mattey *et al.*, 1987).

The molecular architecture of the desmosome is still to be fully described, and there are some differences between research groups on the precise location of certain of the constituent molecules. In part this reflects a combination of different tissues, different isolation procedures, and different approaches. The procedure for isolation of desmosomes makes use of their resistance to solubilization by non-ionic detergents in different strength buffers (Skerrow and Matoltsy, 1974 a and b). The preferred technique for localization of desmosomal constituents involves immunolabelling and subsequent TEM of isolated junctions, the results of which are subject to some variability because of problems with antibody penetration. There is also some confusion with respect to the terminology of the molecules which contribute to the structure of the desmosome. In this text, two general classes of desmosomal molecules will be referred to: (a) the membrane glycoproteins, and (b) the plaque proteins. Their distribution as described here largely follows the work of Steinberg's laboratory (Gorbsky and Steinberg, 1981; Steinberg *et al.*, 1987), with contributions also from the groups of Franke and Garrod (Franke *et al.*, 1987; Mattey *et al.*, 1987). Because of current difficulties with terminology, an alternative classification of consituent molecules based on molecular mass is to be preferred (Table 5.1), but the reader should note that these too vary considerably between different studies.

(a) The membrane glycoproteins. This category contains all the significantly glycosylated desmosomal proteins. It includes the material which either extends across the membrane or is exclusively extracellular in nature.

1. **Desmoglein I.** This protein has an M_r of about 150–165 kD but there is some variation (as seen also with other desmosomal proteins) depending on conditions of isolation

Table 5.1 Desmosomal constituent molecules

(a) The membrane glycoproteins

M_r (kD)	*Name*	*Location*
150	desmoglein I	transmembrane
110	desmocollin I	transmembrane?
97	desmocollin II	transmembrane?
140		stratified squamous epithelia
125		bovine muzzle and tongue; uvomorulin-like
22		midline?

(b) The plaque proteins

M_r (kD)	*Name*	*Location*
250	desmoplakin I	all desmosomes
215	desmoplakin II	stratified epithelia
240	desmocalmin	all desmosomes?
200	D1 antigen	minor component
83	plakoglobin	all adherens-type junctions
75	basic protein	restricted distribution

and electrophoresis. Desmoglein I in fact represents a family of about three glycoproteins, all of which are thought to be transmembrane molecules, extending from within the plaque to the intercellular space.

2. **Desmocollin I and II.** There appear to be at least two members of this family of glycoproteins (about 110 and 97 kD respectively). Like desmoglein I, these molecules are also thought to extend from within the plaque into the intercellular zone. Both desmoglein I and the desmocollins are Ca^{2+}-binding proteins and, since desmosome formation is Ca^{2+}-dependent, it has been suggested that these molecules may be important in desmosome assembly. Mouse polyclonal antisera raised against the desmocollins stain suprabasal but not basal cells in both bovine and human epidermis, a difference which may be pertinent to epidermal differentiation. These antisera also react with the arachnoid layer of the meninges (which possess desmosomes associated with vimentin tonofilaments) and they were found to stain 11 out of 12 meningiomas by Parrish *et al.* (1986). Since there is no reliable marker for meningiomas, it has been suggested that these antisera may be

useful in the diagnosis of tumours of arachnoid origin.

3. **The 22 kD component.** The precise location of this small glycoprotein is uncertain, but it appears to be associated with the desmosomal midline.

Other desmosomal glycoproteins recently described include a 140 kD component from bovine tongue epithelium (Jones, Yokoo and Goldman, 1986), and a 125 kD uvomorulin-like molecule from bovine muzzle and tongue epithelium.

(b) The plaque proteins. These proteins are poorly glycosylated, if at all, and they are thus thought not to be located in the desmosomal intercellular zone.

1. **Desmoplakin I and II.** The two main members of this family of plaque proteins (about 250 kD and 215 kD respectively) are thought to extend from just below the plasma membrane to deep within the cytoplasm where they may be involved with the tonofilament network. Desmoplakin I occurs in all desmosomes so far examined, whereas desmoplakin II is more characteristic of stratified epithelia.
2. **Desmocalmin.** This molecule (240 kD) has calmodulin-binding properties which suggests an important role in the assembly and maintenance of desmosomal structure (Tsukita and Tsukita, 1985).
3. **The D1 antigen.** This 200 kD protein is a relatively rare component of some desmosomes (Franke *et al.*, 1987).
4. **Plakoglobin.** This protein (about 83 kD) is probably more or less restricted to the plaque. According to the research of Franke and colleagues (Cowin *et al.*, 1986; Franke *et al.*, 1987), plakoglobin is localized in the plaques of both intermediate junctions (zonulae adherentes) and spot desmosomes, but is not to be found on hemidesmosomes. Cowin *et al.* (1986) have shown that plakoglobin is present in various cells as a soluble 7S cytoplasmic form. This soluble form may represent a precursor, or alternatively it may accumulate in cells of altered physiological state as junctions are broken down. Studies on the presence of the soluble form of this molecule in malignant cells remain to be done.
5. **The basic protein.** Also known as band 6 protein, this molecule of about 75 kD is thought to be restricted to the desmosomal plaque. It is not found in all desmosomes, however, and thus it may not serve a critical function (Kapprell, Owaribe and Franke, 1988).

Occasionally structures representing half-desmosomes (not to be confused with hemidesmosomes) are visualized by TEM. These structures, also known as maculae adherentes imperfectae, have the dense desmosomal plaque in one cell only, and usually some intercellular material. The typical cytoplasmic components of the other cell involved in forming a true desmosome are missing. Such half-desmosomes are often seen in malignant tissue, but their relevance is uncertain (Weinstein, Merk and Alroy, 1976). They may represent an aberrant form of the true desmosome, or they simply may reflect desmosomes in the process of formation or in the process of detaching, thereby lowering overall intercellular adhesion.

It would seem that not all of the constituent molecules of the mammalian desmosome have been described, and ongoing research should reveal more relevant data both on the nature and the distribution of these molecules. Available evidence suggests that the desmosomal glycoproteins may show considerable variability depending on tissue source, and that this may be due to carbohydrate variation. Since hybrid desmosomes can form between cells from diverse vertebrate species, it is suggested that the recognition and adhesive properties of the desmosomal glycoproteins may predominantly involve conserved protein domains.

Desmosomes seem to be more prevalent in tissues subject to mechanical stress than in tissues which are not, and this may reflect one of their major roles *in vivo*. The distribution of stress forces may occur via the intermediate filament network which is effectively linked from cell to cell via the desmosomes. Like all junctions, desmosomes also contribute to cell adhesion. In at least some *in vitro* systems, cells which form numerous desmosomes will sort out internally from cells which form fewer (Overton, 1977; Wiseman and Strickler, 1981), but this is not always the case. In terms of the differential adhesion hypothesis (Steinberg, 1964), cells which sort out internally may be more adhesive than cells which assume a peripheral position in mixed cell aggregates (section 4.8.2).

It would seem premature to comment on the precise role of junctions in overall cell adhesion until other adhesive processes can be taken into account, and thus no firm conclusions can be reached as yet on the adhesive role of junctions in malignancy. It is not without interest, however, that Gail and Boone (1971) found that 3T3 cells formed adhesions with each other that lasted about three times longer than did adhesions between their SV40 transformed counterparts, and that McNutt, Culp and Black (1973) reported finding more adherens-type junctions in the former than the latter. Other aspects of cell junctions in relation to the transformed phenotype are discussed further in section 1.7.12.

5.3.2 Molecular mechanisms of cell–cell adhesion

Not all intercellular adhesive mechanisms involve cell junctions. Among the best currently described non-junctional adhesive mechanisms are the neuronal cell adhesion molecule N-CAM, and the leukocyte adherence-related proteins (section 5.7.2.4). N-CAM is not a single entity; in fact in chicken neural tissue it is represented by a family of three related polypeptides (N-CAM_{160}, N-CAM_{130}, N-CAM_{120}) which primarily mediate the adhesiveness of neuronal cells to each other and to muscle (reviewed in Rutishauser, 1983). The two larger of these polypeptides are transmembrane molecules, whereas the smallest is a surface determinant linked to the membrane by a phosphatidylinositol containing anchor (section 1.7.5). All three polypeptides are derived from a single gene by alternative RNA splicing. N-CAM adhesion is Ca^{2+}-independent and homophilic in nature, involving *N*-terminal immunoglobulin-like domains (Cunningham *et al.*, 1987). This partial similarity with immunoglobulin structure has led to its classification within this superfamily. Indeed, it has been suggested that the specialized functions of the immunoglobulin superfamily may have evolved from a precursor initially involved in mediating cell–cell adhesive interactions. Interestingly, N-CAM displays some similarity with other adhesive molecules including the leukocyte adherence-related proteins mentioned above. Although there is also some limited similarity between fibronectin and N-CAM, the latter does not contain either RGD or REDV (see section 5.4.2).

Direct evidence for a role of N-CAM in adhesion was gained by showing that liposomes reconstituted with this molecule were capable of aggregating to each other as well as adhering to neuronal cells (Rutishauser *et al.*, 1982). These results are of fundamental importance since many adhesive molecules have been identified indirectly. The general approach in identifying a possible adhesion molecule has involved selection of antibodies which inhibit cell attachment. When polyclonal antisera are used, the adhesion molecule may be indentified by competitive inhibition of the anti-adhesion effects of the antibody with purified membrane determinants. Antibodies which inhibit attachment, however, need not bind directly to adhesion molecules: they may be effective indirectly through steric hindrance or they may operate through secondary effects such as interference with cytoskeletal components. One way of overcoming at least some of these problems is to purify the molecule and test its adhesive effects in a cell-free system such as one based on liposomes.

Gel electrophoresis of N-CAM indicates a considerable degree of heterogeneity which is attributable largely to variations in the content of an unusual oligosaccharide, polysialic acid. When this is removed by neuraminidase the heterogeneity disappears, and the three polypeptides are resolved. In the embryo, N-CAM is richly polysialated whereas in the adult it is much less so. The extent of polysialation correlates inversely with adhesiveness, presumably as a consequence of steric hindrance and charge repulsion, and this developmental modulation of N-CAM may have important consequences in histogenesis. Cells expressing the embryonic form of N-CAM may be more motile allowing cell movements to occur early in development, whereas conversion to the adult form later in development may tend to stabilize cell interactions within tissues. Cells which undergo relatively marked relocation during embryogenesis may undergo a shift in N-CAM expression. Thus, although N-CAM is expressed in cells of the neural crest this is lost during their time of migration. Furthermore, the loss of N-CAM from neural crest cells correlates with the appearance of fibronectin in the pathway used by the cells as they migrate to form the dorsal root ganglia, and it has been suggested that these changes might play a major role in this aspect of development (Thiery *et al.*, 1982). Since these changes in adhesion and motility are associated with changes in N-CAM expression, various groups have been encouraged to look at its expression in malignant tumours. RSV transformation of both cerebellar cell lines (Greenberg *et al.*, 1984) and retinal cells (Brackenbury, Greenberg and Edelman, 1984) leads to a decrease in N-CAM expression but the full significance of these observations is not yet clear. N-CAM is expressed on both B16F1 and B16F10 melanoma cells, although it appears that only the two larger polypeptides are produced. Studies of N-CAM biosynthesis indicate that there are no qualitative differences in N-CAM expression in the two melanoma variants despite their different lung colonizing abilities (Linnemann and Bock, 1986). The possibility exists, of course, that some subtle variation of the cell surface distribution of N-CAM may influence the colonizing abilities of the two melanoma variants, but further evidence is required before any conclusions can be reached.

N-CAM not only mediates cell–cell adhesion, but it also influences other cellular processes such as synapse formation, gap junctional communication, and the effects of other Ca^{2+}-dependent cell adhesion molecules. These pleiotropic effects are probably mediated by the close apposition of cells resulting from homophilic N-CAM binding, and by the presence of polysialic acid which, through either charge or steric effects, is likely to influence other cell surface related phenomena.

Extensive homology (90%) between human and murine N-CAMs has allowed mouse N-CAM cDNA to be used in localizing the gene for human N-CAM. *In situ* hybridization experiments by Nguyen *et al.*

(1986) have shown the gene to be localized to the distal portion of the long arm of chromosome 11 (11q23). Interestingly, Ewing's tumour of the bone is characterized by the translocation t(11;22)(q23–24;q12) which might involve the N-CAM gene. Cell lines from this tumour express a relatively small amount of N-CAM relative to a neuroblastoma cell line, but this is unlikely to be attributable to a chromosomal rearrangement of the N-CAM gene since none has been detected using available methodology (Lipinski *et al.*, 1987).

Many of the molecules involved in cell adhesion are glycoproteins, which poses the problem as to whether the protein or the attached carbohydrate is more important. At least three properties of cell surface carbohydrates support a general role for them in cell adhesion:

(a) *Surface location*

The external coat of a cell (the glycocalyx) is rich in carbohydrates making them ideally placed for mediating adhesion.

(b) *Structural diversity*

The carbohydrates display considerable structural diversity which is appropriate for a role in recognition and selective or specific interaction, as shown by many adhesive mechanisms.

(c) *Presence of specific lectins on the cell surface.*

Lectins with recognized cell adhesion promoting activity are present on many cells as specific carbohydrate receptors.

Roseman (1970) and Shur and Roth (1975) proposed a novel mechanism for cell recognition and adhesion based on the mechanism of carbohydrate chain elongation. In essence, it was suggested that cell–cell adhesion could result when a growing carbohydrate side chain of a glycoprotein or glycolipid on the surface of one cell interacted with an appropriate glycosyltransferase on the surface of another. Shur (1983), for example, has provided evidence that the Ca^{2+}-dependent adhesion of embryonal carcinoma cells is mediated via a cell surface galactosyltransferase interacting with cell surface lactosaminoglycan, presumably via *N*-acetylglucosamine residues. It is not impossible to envisage a mechanism founded on differences in either time or space which would confer some degree of specificity on this type of interaction, but its *in vivo* significance is not clear.

The use of carbohydrates as adhesion molecules is phylogenetically widespread, with viruses, bacteria, yeasts, slime moulds and sponges all utilizing such systems (reviewed in Sharon and Lis, 1989). Cell surface carbohydrates have also been implicated in the binding of spermatazoa to eggs in animal groups as diverse as echinonderms and mammals. One of the major vertebrate cell types which adheres to carbohydrates is the hepatocyte. Chick liver hepatocytes, for example, have been shown to bind to surfaces derivatized with GlcNAc, while rat hepatocytes bind to Gal derivatized surfaces. Carbohydrate-specific receptors on mouse hepatocytes are thought to mediate interaction with tumour cells which metastasize to the liver (Cheingsong-Popov *et al.*, 1983). For most tumours spreading via the blood the endothelium will represent the first contact site during extravasation. The importance of carbohydrates in mediating adhesive contact between tumour cells and the endothelium is illustrated by the interaction of rat hepatocarcinoma cells with endothelial cell monolayers which is inhibitable by the monosaccharides α-D-mannopyranoside and GalNAc (Stanford, Starkey and Magnuson, 1986). Studies with the glycosylation inhibitor tunicamycin have also suggested that the interaction of metastatic cells with host tissues and cells may be carbohydrate-mediated (Irimura, Gonzalez and Nicolson, 1981), but it should be borne in mind that tunicamycin also has effects on protein synthesis. Raz and colleagues (1981, 1984) have described the presence of galactoside-specific endogenous lectins on the surfaces of human and murine tumour cell lines, including the B16 melanoma. These endogenous lectins have been postulated to mediate cell–cell adhesion through binding to complementary carbohydrates on the surfaces of adjacent cells. Fetuin, the major glycoprotein of fetal calf serum, was found to bind to tumour cells and promote their intercellular adhesiveness, but this is unlikely to be the only mechanism by which homotypic aggregation is mediated.

A mutant cell line of 3T3 cells known as AD6 has a defect in glycoprotein synthesis due to a decreased ability to acetylate glucosamine. AD6 cells thus lack GlcNac and they exhibit poor substrate adhesiveness. When exogenous GlcNac is added to AD6 cells they can adhere and spread like normal 3T3 cells (Willingham *et al.*, 1977). The behaviour of this cell line indicates the importance of carbohydrates in adhesion, but it does not prove that these molecules are used directly in adhesion. Since AD6 cells are non-tumorigenic in syngeneic mice, it is clear that poor substrate adhesiveness is not necessarily characteristic of the transformed state.

The data implicating a role for carbohydrates in cell adhesion must be contrasted with other results suggesting direct protein involvement. Adhesion of cells to fibronectin, for example, can be inhibited by small peptides but not by exogenous sugars (section 5.4.2). Interestingly, although Discoidin I (one of the key molecules involved in slug formation in the slime mould *Dictyostelium discoideum*) has been characterized as a galactose-specific lectin, its role in slug formation is dependent on the presence of the tripeptide arg-gly-asp and not on its carbohydrate-binding ability.

Overall, the obvious conclusion is that both protein and oligosaccharide sequences are involved in cell adhesion, and that some cell types may use either or both of these mechanisms. Additionally, one type of sequence may be more important in cell–substrate adhesion while the other could act primarily in cell–cell adhesion. The observation reported above for Discoidin I, for example, fits in more with the suggestion that this molecule is involved in cell–substrate adhesion rather than cell–cell adhesion, but the general validity of this proposal seems doubtful. We must add to these comments the potential inhibitory effects of oligosaccharide side chains, as seen in the probable steric hindrance and/or charge effects of the polysialic acid of N-CAM.

5.4 MECHANISMS OF CELL–SUBSTRATE ADHESION

A wide variety of substrates exist, both artificial and natural, for which it might be supposed that an equally wide variety of adhesive molecules would be required. This looks unlikely to be the case, however, and it seems that a small handful of molecules may well suffice. Some controversy surrounds the adhesion of cells to tissue culture surfaces such as glass or plastic, even though the biological relevance of such adhesions is questionable. Numerous reports indicate that the majority of cell types require serum proteins for adhesion and spreading on such surfaces, although certain cells appear to be able to adhere and spread in serum-free medium. Adherence and spreading under such conditions is to be distinguished from passive, non-physiological adsorption of cells onto a relatively clean surface. Cells which do adhere and spread under serum-free conditions probably secrete or have attached to their surfaces adhesion promoting molecules such as fibronectin, vitronectin or laminin (Grinnell and Feld, 1980). Some cells, such as HepG2 human hepatoma cells, synthesize all three of these molecules, but they do not spread equally well to them in attachment assays. In fact, HepG2 cells adhere and spread on laminin and vitronectin, but not on fibronectin (Barnes and Reing, 1985).

There is considerable evidence that the mechanisms by which cells adhere to substrates are different from those involved in cell–cell adhesion. Nevertheless, as will become clear in section 5.8, considerable homologies exist between various molecules involved in both cell–cell and cell–substrate adhesion. Furthermore, at the structural level it is clear that the substrate adhesion sites known as hemidesmosomes share considerable features with cell–cell desmosomes, although their molecular components are not identical.

5.4.1 Hemidesmosomes

Hemidesmosomes are thought to be involved in the attachment of some epithelial cells to their underlying basement membrane. They are macular or spot-weld like structures with a close structural resemblance to half a desmosome, although it appears that they are composed largely (but perhaps not exclusively) of a different set of proteins. Both spot desmosomes and hemidesmosomes serve as anchoring sites for tonofilaments (i.e intermediate filaments) and in this they differ from zonulae adherentes (also known as belt desmosomes) which are associated with actin microfilaments. As visualized by convential TEM, hemidesmosomes in epidermal cells of adult amphibian skin can apparently extend fine filaments (about 12 nm in diam.) through the underlying basement membrane, where they may unite to form anchoring fibrils which then mesh with dermal collagen fibres (Ellison and Garrod, 1984). In this way a structural link exists between the cytoplasmic intermediate filament system of epidermal cells and the collagen of the dermis. The intermediate filament network is further linked from cell to cell via spot desmosomes thus providing an extensive system which may have considerable involvement in the distribution of shear forces and tissue stabilization. A similar structural network may exist in the human since anchoring fibrils have been shown to extend from below the basement membrane of the oral mucosa to the overlying epithelial cells (Susi, Belt and Kelly, 1967).

5.4.2 Fibronectin

Fibronectin (native M_r about 440 kD) has been identified as a key molecule in the attachment of a variety of cell types to both natural and artificial substrates. There are two major types of fibronectin, one (pFN) being found in the plasma while the other (csFN) is found on the surface of many cells as a striking fibrillar array (reviewed in Yamada and Olden, 1978; Ruoslahti, 1988). Although convenient, this general categorization based on distribution is not strictly true since both types have been found in the blood and both may be incorporated into basement membranes and the extracellular matrix. Furthermore, recent evidence suggests that there may be as many as 12 fibronectin variants, arising as a consequence of alternative RNA splicing. All the variants consist of two similar polypeptide chains (about 62 nm long)

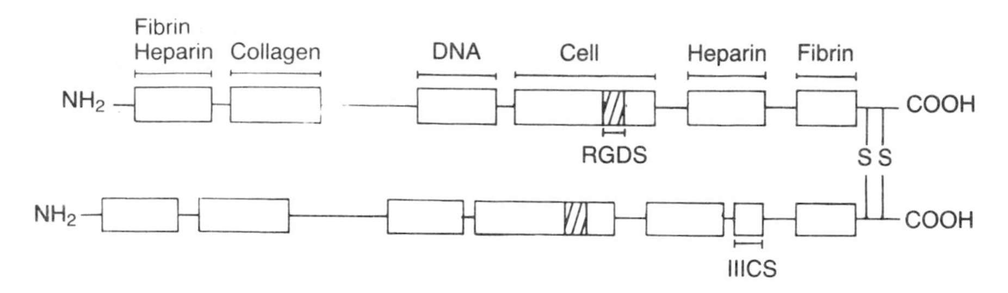

Figure 5.3 The structure of fibronectin (after Humphries *et al.*, 1987). The figure illustrates the arrangement of the principal domains of fibronectin which is a disulphide-linked dimer. The major cell-binding domain contains the sequence RGDS which has been shown by competitive-inhibition experiments to be important in the adhesion of many cell types to fibronectin-coated substrates. Some cells, however, can adhere to different regions of the fibronectin molecule. B16F10 melanoma cells, for example, have been shown to adhere to sequences contained within the type III connecting segment (IIICS) of the larger fibronectin polypeptide. The smaller polypeptide results from alternative splicing of type IIICS during fibronectin mRNA processing.

connected via disulphide bonds near their C terminal ends (Fig. 5.3). The entire sequence of human fibronectin has been determined from recombinant DNA studies (Kornblihtt *et al.*, 1985). Each fibronectin chain displays three different types of homologous sequences (types I, II and III) which are arranged to form a number of functional domains along the molecule. The type I repeat is about 45 amino acids long, the type II repeat about 60 amino acids long, and the type III repeat about 90 amino acids long, with each polypeptide chain containing a little less than 2500 amino acids. It seems that the major difference between soluble pFN and insoluble, polymerized csFN is the lack of specific type III repeats in the former. Overall, the fibronectin molecule contains about 5% *N*-linked carbohydrate, although this varies between the fibronectin types. It appears that there may also be some *O*-linked sugars attached to to a specific type III segment (type IIICS). The carbohydrate is not essential for the biological activities of fibronectin, and indeed there is some evidence that it actually may interfere with receptor binding since carbohydrate depleted fibronectin (produced in the presence of tunicamycin which inhibits *N*-glycosylation) is more adhesive for human skin fibroblasts than normal fibronectin (Jones, Arumugham, and Tanzer, 1986). Other evidence suggests that the carbohydrate may serve a protective role, possibly by increasing fibronectin resistance to proteolytic attack (Olden, Pratt and Yamada, 1979).

One of the key biochemical differences between pFN and csFN that explains their different distributions is their solubility: pFN is soluble at neutral pH whereas csFN is soluble around pH 11. Fibronectin is present in the plasma at about 300 μg ml^{-1}, but since it binds to fibrin much less is found in serum. Plasma transglutaminase (activated Factor XIII) can covalently cross-link fibrinogen to fibronectin, and it was presumably the formation of fibrinogen–fibronectin complexes (which can form in the cold) that led to its initial identification as 'cold insoluble globulin' by Morrison, Edsall and Miller (1948). Factor XIIIA can also covalently link fibronectin to collagen as well as fibronectin molecules to themselves, and fibronectin polymerization is enhanced by heparin. The amount of polymerized, fibrillar fibronectin found on the surface of normal cells appears to vary with the cell cycle, being lowest during M phase (Stenman, Wartiovaara, and Vaheri, 1977). Although tumours usually have relatively decreased levels of csFN (five- to ten-fold or more), this is not entirely attributable to their often greater mitotic activity (section 1.7.5.1).

The cell surface receptor for fibronectin has proven elusive to characterize for a number of reasons, the prime one being that plasma fibronectin binds extremely poorly to cells unless it is first bound to a solid substrate. Nevertheless, considerable progress has been made in this direction relatively recently using both avian and mammalian cells. Generally speaking, fibronectin binds to the surface of cells with only moderate affinity, as suggested by a Kd of about 10^{-7}M for BHK cells (Akiyama and Yamada, 1985). Quantitative binding studies with the same cell type indicate that there may be as many as 10^5 free fibronectin receptor sites per cell. Steps towards identifying the molecular nature of the receptor have made use both of affinity chromatographic techniques (primarily with mammalian cells) and immunolabelling techniques (primarily with chick cells).

In mammalian cells, the major fibronectin receptor appears to be a glycoprotein complex which can be resolved into two distinct bands (Pytela *et al.*, 1985), now known to be of about 150 kD and 130 kD as determined by PAGE analysis. The bands represent α and β subunits, the former of which is processed into two polypeptides which are linked together by disulphide bonds (Fig. 5.4). Molecular sequencing of the human fibronectin receptor has now been accomplished, from which its homology with other adhesive receptors is made quite clear. This homology suggests that many adhesive receptors may be members of a molecular superfamily, the details of which are described more fully in section 5.8. Interestingly, the α subunit of the human fibronectin receptor contains five sequences with homology to known Ca^{2+}-binding proteins such as calmodulin. It is thought that Ca^{2+}

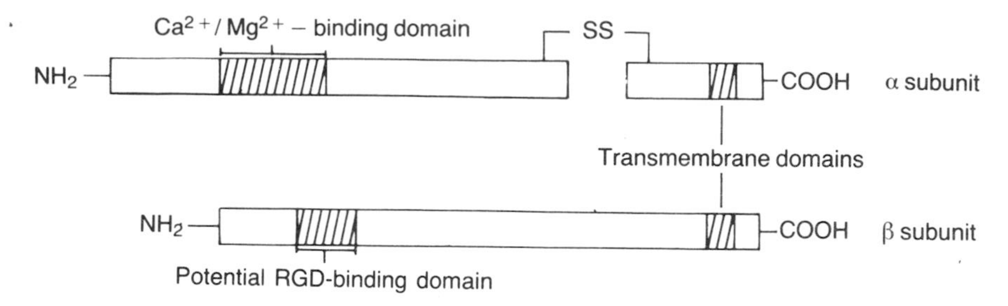

Figure 5.4 Proposed structure of the human fibronectin receptor (after Argraves *et al.*, 1987). The fibronectin receptor (also known as VLA-5) is a member of the integrin superfamily of adhesive receptor molecules (section 5.8). It is composed of non-covalently associated α and β subunits, with the former being processed into two polypeptides which are linked to each other by disulphide bonds. Interaction between fibronectin and its receptor is Ca^{2+}-dependent, which is presumably accounted for by the presence of a Ca^{2+}/Mg^{2+}-binding domain on the α subunit. The β subunits of known integrins share extensive homology, whereas the α subunits vary more between different members of the superfamily. The potential RGD-binding domain shown on the β subunit is that region in gpIIIa (residues 109–171) thought to be involved in the binding of platelet to RGD-containing molecules (S'Souza *et al.*, 1988). SS: disulphide bond.

stabilizes the structure of the fibronectin receptor. The major human receptor for fibronectin is now thought to be equivalent to VLA5, but other cell surface determinants including VLA3 and gpIIb/IIIa can also bind this adhesive protein.

In the chick system, a monoclonal antibody (JG22, or its subclone JG22E), which inhibits the adhesion and spreading of CEFs on fibronectin, has also been shown to recognize a non-covalently complexed cell surface antigen. This antigen, however, resolves into three glycoproteins of 155 kD, 135 kD and 120 kD (Hasegawa *et al.*, 1985). Immunostaining of adherent cells has shown that the complex co-distributes with fibronectin fibrils (Chen *et al.*, 1985). Another monoclonal antibody known as CSAT has also been used to identify the fibronectin receptor on chick cells (Knudsen, Horwitz and Buck, 1985). Like the JG22E antigen above, with which it is presumably identical, the CSAT antigen can be resolved into three components. Despite this behaviour on PAGE gels, the chick fibronectin receptor is considered to be a heterodimer. The simplest interpretation of this paradox is that two heterodimers exist, with one subunit (the 120 kD β chain) being common to both dimers (Buck and Horwitz, 1987). Unlike the fibronectin receptor described on at least some mammalian cells, the CSAT antigen is somewhat promiscuous, binding to both vitronectin and laminin, as well as to fibronectin. Note that the CSAT

Table 5.2 The superfamily of adhesion receptor molecules

Family component		*Subunit M_r (approx. kD)*	*Ligand*
β1	chick integrin	155/120*	FN,VN,LM,Co?
		135/120*	
	mammalian FNR	150/130	FN
	VLA1	165/140	Co,LM?
	VLA2 (gpIa/IIa)	165/130	Co(I–IV,VI),LM
	VLA3	135/130	Co(I,IV),FN,LM
	VLA4⁺	150/130	FN
	VLA5 (FNR;gpIc/IIa)	150/130	FN
	VLA6	150/110	LM
β2	LFA–1	180/95	ICAM–1
	Mac–1	170/95	C3bi
	p150/95	150/95	C3bi?
β3	gpIIb/IIIa	150/105⁑	FN,VN,vWF,FB,TSP?,Co?
	mammalian VNR	125/115⁑	VN,FB,vWF,TSP

* Approximate M_r assuming two heterodimers rather than a single heterotrimer (see text).
⁺ The α chain of VLA4 is almost identical with that of LPAM–1, a murine lymphocyte cell surface molecule involved in their traffic to Peyer's patches (section 2.12.2).
⁑ The β chains of these molecules are believed to be identical.
After Hynes (1987), Buck and Horwitz (1987), Takada *et al.* (1988) and others. Molecular masses are estimates after electrophoresis under non-reducing conditions: they vary in different cell types and in different reports. Note that many other molecules involved in adhesion do not belong to this superfamily.
Abbreviations: Co: collagen; FB: fibrinogen; FN: fibronectin; FNR: fibronectin receptor; TSP: thrombospondin; VN: vitronectin; VNR: vitronectin receptor; vWF: von Willebrand factor; C3bi: inactivated complement component.

antigen was renamed integrin, and that this name has since been applied to the superfamily of related adhesion receptor molecules (Table 5.2).

The use of various double immunolabelling combinations has shown that integrin distribution co-localizes with fibronectin outside the cell, and actin, talin and vinculin inside the cell. Competitive binding experiments have suggested that fibronectin binds to integrin which crosses the membrane and binds in turn to talin, which then binds to vinculin (reviewed in Buck and Horwitz, 1987; Burridge, Molony and Kelly, 1987). Although actin co-localizes with these molecules, exactly how it is linked into the system is uncertain. Alpha actinin (α-actinin) was once thought to be a likely candidate for linking, but it now seems more likely that this molecule functions to bind actin filaments together (Burridge and McCullough, 1980). Another possibility is the group of molecules known collectively as HA1, but although they may cap the ends of actin filaments there is no evidence as yet to implicate a direct

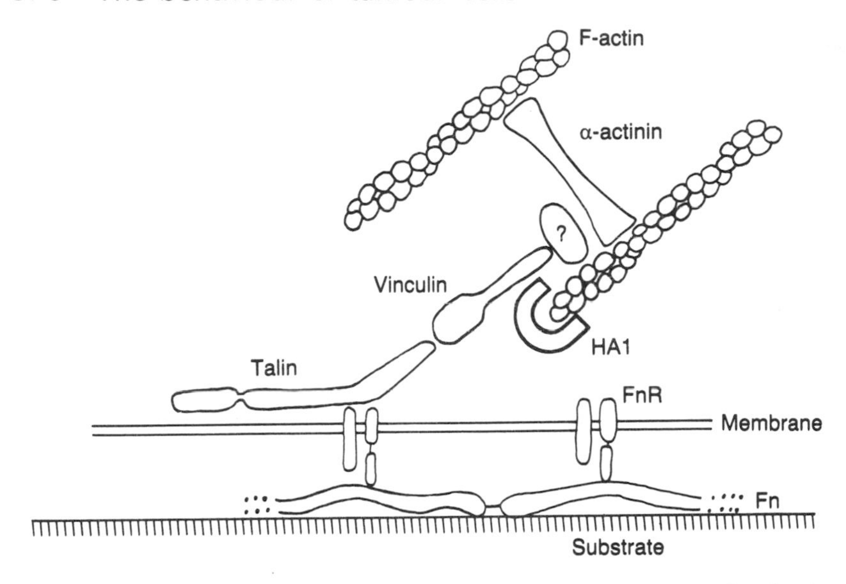

Figure 5.5 The postulated topography of molecules implicated in cell adhesion to substrate-bound fibronectin (after Burridge *et al.*, 1987). The figure shows part of a fibronectin dimer (Fn) adsorbed to a solid substrate. The transmembrane fibronectin receptor (FnR) binds through an extracellular domain to fibronectin and through an intracellular domain to talin, which in turn binds to vinculin. How the fibronectin receptor links to the actin cytoskeleton is not yet known. Shown in the figure is a hypothetical molecule linking vinculin to F-actin threads (stress fibres). Vinculin may also be linked to the plasma membrane more directly via another protein, as yet unidentified. Alpha-actinin (α-actinin is shown cross-linking actin threads, and the HA1 proteins are depicted as terminal caps although firm evidence is lacking for this arrangement.

role for them in linking actin to the system of receptor-related molecules (Wilkins and Lin, 1986). A schematic model for adhesion to substrate bound fibronectin is shown in Fig. 5.5, although it should be clear that other components of this system may as yet await discovery.

Different lines of research have indicated that other classes of fibronectin receptors may exist. On platelets, for example, the receptor for fibronectin appears to be the membrane glycoprotein known as gpIIb/IIIa, which actually binds a number of proteins including fibrinogen, fibronectin, vitronectin and von Willebrand factor, and possibly also collagen and thrombospondin. All of these proteins contain the 3 (or 4) amino acid sequence arginine-glycine-aspartic acid (-serine), whose importance in adhesion will become more clear below. Here we should note that because amino acid sequences can be rather long, it is convenient to use three-letter or, even better, single-letter abbreviations. The peptide sequence arginine-glycine-aspartic acid (-serine) can thus be represented as arg-gly-asp (-ser) or RGD(S).

Another possible cell surface receptor for fibronectin centres around a 47 kD glycoprotein which was first identified by chemically cross-linking membrane determinants to substrate adsorbed fibronectin (Aplin *et al.*, 1981). The nature of this experimentally induced linking was such that only cell determinants closely associated with fibronectin (within about 1.6 nm) would be bound, and these could easily be identified on subsequent purification. Unfortunately, antibodies raised against the 47 kD glycoprotein were found to be inactive in inhibiting adhesion of BHK cells to fibronectin (Hughes, Butters and Aplin, 1981). One likely interpretation of these results is that the 47 kD glycoprotein is topographicaly related to fibronectin, but it is not itself the cell receptor.

Gangliosides rich in sialic acid may also act as cell surface receptors for fibronectin since they can competitively inhibit fibronectin-mediated adhesion (Kleinman, Martin and Fishman, 1979). Furthermore, radiolabelled fibronectin has been shown to bind directly to adsorbed gangliosides (Perkins *et al.*, 1982), and experiments with fluorescent gangliosides have shown that there is a direct association between these molecules and fibronectin at the cell surface (Spiegel *et al.*, 1985). The function of charged gangliosides as receptors for fibronectin seems to lie with the nature of their oligosaccharide groups, which is at variance to the body of data implicating amino acid sequences in fibronectin-mediated adhesion. This might suggest that gangliosides could act non-specifically, rather than as direct receptors for fibronectin. Indeed, gangliosides seem to inhibit cell adhesion to a number of substrates, which would be in accord with their generalized effects on cell adhesion. However, sialylated gangliosides bind to fibronectin with some domain specificity, and the site has been tentatively assigned to the *N*-terminal region.

Since gangliosides are too short to span the lipid bilayer of the plasma membrane, it is difficult to see how they can manifest the cytoskeletal rearrangements which accompany cell adherence and spreading on fibronectin. Indeed, Perkins *et al.* (1982) suggested that fibronectin binding via gangliosides probably does not lead to such reorganization. Mugnai and Culp (1987), have proposed an indirect form of interaction, possibly via an intermediate glycoprotein. These authors have shown that human neuroblastoma cells adhere and extend neurites on pFN via two cell binding sites, one containing RGDS (see below) while the other involves interaction with cell surface gangliosides. Ganglioside-mediated neurite formation is apparently considerably slower (16–18 h) than RGDS-mediated neurite formation

(2–4 h). Studies with other cell types have suggested that cell surface gangliosides such as G_{M1} might act as receptors for mediating fibronectin fibrillogenesis (Spiegel *et al.*, 1986). Thus NCTC 2071A cells, which are a ganglioside-deficient derivative of the mouse L929 clone, are unable to organize fibronectin on their surfaces, even though they actively synthesize the molecule. This deficiency in organizing cell surface fibronectin can be overcome when the cells are exposed to exogenous gangliosides. Since NCTC 2071A cells can adhere and spread on fibronectin coated substrates, it would seem that adhesion and fibrillogenesis are mediated by different receptors, at least for this cell type. If fibrillogenesis is independent of cytoskeletal reorganization, then this can take into account the problem of transmembrane linkage via a ganglioside receptor rather nicely.

Other research, primarily on M21 human melanoma cells, has shown that gangliosides such as G_{D2} and G_{D3} can inhibit their adhesion to fibronectin-coated substrates (Cheresh *et al.*, 1986). Double immunolabelling experiments have shown that these gangliosides co-localize with a promiscuous gpIIb/IIIa-like receptor which recognizes vitronectin, fibronectin, fibrinogen and von Willebrand factor, and which may be found in focal contact sites (Cheresh *et al.*, 1987). The gangliosides do not appear to bind the adhesive molecules, however, and instead they may act by modulating the activity and/or specificity of the actual receptor in a manner known to be Ca^{2+}-dependent. On balance, it seems that neither the gangliosides nor the 47 kD glycoprotein mentioned above will prove to be true fibronectin receptors, although it must be admitted that the position is still not particularly clear.

Fibronectin can be digested experimentally by enzymes and the fragments produced may be recovered for testing of cell binding affinity. Using this approach, the cell binding site has been found to lie in a small fragment (about 108 amino acids long) near the middle of the molecule. In more recent work, the precise amino acid sequence involved in cell binding was determined by testing various synthetic peptides for this activity using normal rat kidney (NRK) cells. The smallest peptide found with reasonable adhesion-promoting activity consisted of four amino acids arranged in the sequence arginine, glycine, aspartic acid, serine, with the last in the series being less crucial for activity than the others (Pierschbacher and Ruoslahti, 1984 a and b). Since serine in the sequence RGDS can be replaced by various amino acids without significant loss of activity, a large number of RGD-containing molecules (including type I collagen and thrombin) will at least have the potential to act as adhesive ligands. Interestingly, rat fibronectin (unlike other fibronectins studies so far) expresses two

RGD sequences, the expression of which is controlled by RNA splicing, and both of these could be involved in cell adhesion.

Although the RGD sequence has been found in various molecules this does not necessarily provide them with cell adhesion promoting activity (Pierschbacher and Ruoslahti, 1984b). Fibrinogen, which contains the sequence RGD, is thus not adhesive for NRK cells. Somewhat contrary to the point made above specifically for fibronectin, this would suggest that amino acids neighbouring the RGD adhesive sequence (and presumably also the location of carbohydrate groups in glycoproteins) are likely to influence cellular adhesiveness. There are also likely to be differences in the receptors on the surfaces of specific cells which might influence their binding and spreading on RGD-containing molecules. The RGD sequence is widespread phylogenetically, being found in viruses and bacteria (reviewed in Ruoslahti and Pierschbacher, 1986). This raises the possibility that viral and bacterial adhesion to mammalian host tissues could be mediated via receptors for this sequence.

In contrast to the above, Yamada and Kennedy (1985) have used BHK cells rather than NRK cells to study adhesion to fibronectin. According to their experiments, the pentapeptide GRGDS is substantially more active at promoting adhesion and spreading than RGDS. This suggests, as alluded to above, that different cell types might be influenced to different degrees by the amino acids which flank the minimum adhesion determinants. Interestingly, the reverse sequence SDGRG has similar activity to the forward sequence: the fact that this sequence is found on type II histocompatibility antigens may be of significance in immune surveillance, but as yet there is no clear evidence that this is the case. The reverse sequence DGR occurs twice in fibroblast growth factor, but whether this is of any significance in FGF binding to its receptor is uncertain.

The peptides RGDS and GRGDSP inhibit the spreading of cells to fibronectin and to other adhesive molecules such as vitronectin and laminin (see below), but their effects are dependent to a degree on cell type (Silnutzer and Barnes, 1985). The spreading of HeLa human cervical carcinoma cells to laminin, for example, is inhibited by both peptides whereas the spreading of C6 rat glioma cells is unaffected. Interestingly, the spreading of MRC5 human diploid fibroblasts on con A is also inhibited by these peptides. Since there is no evidence that fibronectin is directly involved in con-A-mediated cell spreading, these results indicate the need for caution in interpreting peptide competition effects.

When purified fibronectin is adsorbed onto an otherwise poorly adhesive surface, cells added subsequently may adhere, spread, become polarized and move. In fact, for a wide variety of cells, fibronectin, laminin and vitronectin seem to exert similar behavioural patterns (see below). In general, for many cell types this behaviour involves interaction of the substrate bound molecules with free cell surface receptors. In so far as fibronectin is concerned, its binding to NHK and BHK cell types appears to be mediated primarily by RGDS. Thus not only does soluble fibronectin at high concentration inhibit the adhesion and spreading of these cells on substrate-adsorbed fibronectin, but RGDS containing peptides do likewise (Yamada and Kennedy, 1985). High fibronectin concentrations (up to 10 mg ml^{-1}) are necessary because this molecule binds poorly to cells when it is in the soluble state. It should be emphasized here, however, that the binding of some cell types to fibronectin is not mediated exclusively by the RGDS sequence.

One of the fibronectin domains located near the C-terminal end has a relatively strong affinity for heparin and also for less sulphated glycosaminoglycans such as heparan sulphate, the chondroitin sulphates, and dermatan sulphate. It seems that sulphated cell-surface proteoglycans can interact with the heparin binding domain of fibronectin and that the adhesions which develop are morphologically different from those which form following cellular interaction with both the heparin and cell-binding domains, or the cell-binding domain alone (section 5.4.6).

With respect to invasion and metastasis, it has been proposed that malignant cells might be able to locomote haptotactically in response to substrate bound gradients of fibronectin (section 5.12). Experimentally, McCarthy, Hagen and Furcht (1986) have shown that B16F10 cells migrate in a dose-dependent manner up a gradient of fibronectin, with a peak response in their system of around 12.5 μg ml^{-1}. When various fragments of fibronectin were tested similarly, haptotactic activity was seen to lie in a 75 kD trypsin-derived fragment containing the RGDS sequence. However, a smaller 11.5 kD pepsin-derived fragment (domain IV) also containing RGDS was unable to promote such activity, and in fact it was only poorly adhesive for B16F10 cells. These conflicting results suggest that more than RGDS might be involved in the adhesion of B16F10 cells to fibronectin. This possibility is supported by the observation that although RGDS (up to 10^{-3}M) can detach B16F10 cells from surfaces coated with fibronectin at 1 μg ml^{-1}, it is much less effective when the fibronectin concentration is increased to 5 μg ml^{-1}. In pursuing this observation, Humphries and his colleagues (1986) have compared the adhesion and spreading of B16F10 cells to fibronectin and its fragments with that of BHK cells by measuring the degree of competitive inhibition elicited by various peptides. For BHK cells, spreading to fibronectin was found to be inhibited by GRGDS, RGDS and SDGR in that order of decreasing effectiveness. Similar results were found for Balb/c 3T3 cells and primary chick fibroblasts, but for B16F10 melanoma cells the order was RGDS, SDGR and then GRGDS. Furthermore, the peptides GRGD and GRGES (with a conservative asp to glu conversion) were substantially active in inhibiting spreading of B16F10 cells, whereas they were relatively inactive in inhibiting the spreading of BHK cells. The full significance of the inhibition of spreading of B16F10 cells on fibronectin by GRGES is not yet clear (see below), although this is the only cell type tested so far which responds in this way. Taken together, the data above suggest that B16F10 cells might adhere to a peptide sequence other than RGDS. In fact Humphries *et al.* (1986) were able to show, using a series of trypsin-derived fibronectin fragments, that a site in the type III connecting segment is adhesive for B16F10 cells. According to their observations, the 75 kD fragment containing the cell binding sequence RGDS is not adhesive for B16F10 cells. The type III connnecting segment (type IIICS) is a sequence of 120 amino acids which is found only in the larger of the two fibronectin polypeptides, and it is apparently spliced out of the smaller partner of the dimer. Sequence analysis has shown that type IIICS contains a hydrophilic, four-amino-acid sequence, arg-glu-asp-val (REDV), which is not unlike RGDS. When this sequence was tested *in vitro*, it was found to inhibit the adhesion and spreading of B16F10 cells on fibronectin but was without effect on BHK cells. The observation that B16F10 cells cannot adhere and spread effectively on the RGDS-containing 75 kD fragment and yet the RGDS peptide can inhibit spreading on fibronectin was attributable by Humphries *et al.* (1986) to cross-reactivity between RGDS and REDV. Later studies identified an even more active fragment of type IIICS, but its sequence strucure has yet to be determined (Humphries *et al.*, 1987). Clearly, the results suggest that B16F10 cells adhere to fibronectin using an amino acid sequence different from that shown by BHK cells. Interestingly, chick fibronectin does not contain the REDV sequence and thus it would be interesting to test the extent to which B16F10 cells adhere and spread on this molecule. The current hypothesis is that fibronectin contains at least two cell binding sites and that various cell types may show some degree of specificity in their binding. Whether additional binding sequences exist remains to be seen, but it is worth noting that these may well require different receptors. So far, there is no evidence for cell surface fibronectin receptors of different site specificity.

Although GRGDS is not the most effective peptide for inhibiting B16F10 adhesion to fibronectin, when both peptide and melanoma cells are coinjected into host mice there is a dramatic inhibition of lung colonization relative to controls (Humphries, Olden and Yamada, 1986). The effectiveness of the peptide in inhibiting lung colonization was a function of the concentration of peptide used and the number of cells injected: 3 mg of peptide per mouse almost completely inhibited colonization (97.3% inhibition) when 5×10^4 cells were injected, but was much less effective when 15×10^4 cells were injected (68.2% inhibition). Additional experiments by this group showed that GRGDS had no effect on B16F10 tumorigenicity and nor was it directly cytotoxic for these cells, so its inhibitory effects are unlikely to be mediated by these processes. *In vivo* experiments with radiolabelled tumour cells, however, showed that GRGDS accelerated the loss of cells from the lungs and this points the way to a possible explanation of the observed effects of the peptide on lung colonization. Interestingly, GRGDS did not show any significant inhibition of lung colonization even though it is now known to influence B16F10 melanoma adhesion and spreading on fibronectin (see above). Taken together, the results suggest that although GRGDS may affect melanoma cell retention in the lungs, it is not necessarily through its effect on inhibiting the adhesion of the tumour cells to fibronectin. What alternative scenarios might be possible? One worthy of consideration is that some of the peptides interfere with platelet aggregation and that this indirectly influences tumour cell retention in the lungs. Whether this might be the case or not, however, remains to be explored fully (section 5.7.2.3).

Although fibronectin is clearly implicated in the substrate adhesion of numerous cell types, its role in cell–cell adhesion is more controversial. Since fibronectin is found on the cell surface along with free fibronectin receptors, it might be thought a simple matter for the two to match up on apposed cells and for adhesion to result. This does not appear to be the case, however, suggesting perhaps that csFN has no free (or accessible) cell binding sites. For many cell types, exogenous pFN appears to have little if any effect on cell–cell adhesion, but purified csFN has been reported to promote the aggregation of chick embryo and BHK cells (Yamada, Olden and Pastan, 1978). It has also been claimed that csFN is active in promoting the agglutination of red blood cells, a function at which it appears to be about 200 × more active than pFN. More recently, however, the agglutinating activity of csFN has been attributed to contamination with the extracellular matrix protein tenascin (Chiquet-Ehrismann *et al.*, 1986). Tenascin has a unit substructure very similar in size (240 kD) to that of fibronectin, although it tends to form oligomers rather than dimers. It seems likely that the six armed oligomers (hexabrachions) isolated by Erickson and

Inglesias (1984) from csFN preparations and originally thought to be a precursor or separate form of fibronectin are in fact contaminants of tenascin. Although tenascin has good haemagglutinating activity, it is not very effective at promoting cell–substrate adhesion and in fact it actually induces the rounding of primary CEFs (Chiquet-Ehrismann *et al.*, 1988). It seems that tenascin binds to immobilized pFN, possibly masking the GRGDS binding site. Cytotactin, an extra cellular matrix protein composed of subunits of around 200 kD (Grumet *et al.*, 1985), is thought to be a tenascin-like molecule (Chiquet-Ehrismann *et al.*, 1988), although how its role in mediating glia–neuron adhesion can be equated with the anti-adhesive effects of tenascin on CEFs is not clear. Interestingly, tenascin may turn out to be a valuable marker for oncogenesis. Normal adult mammary gland, for example, does not contain tenascin but the molecule is re-expressed in the mammary glands of mice bearing a chemically induced mammary adenocarcinoma. Similar re-expression of tenascin in the adult is apparently not seen in benign mammary tumours.

Finally, it should be noted that not all cells have an absolute requirement for fibronectin in order to adhere and spread. Virtanen *et al.* (1982), for example, have shown that human embryonal fibroblasts do not require either endogenous or exogenous fibronectin for spreading, although its absence prevents formation of focal contacts. According to Curtis *et al.* (1983), fibroblasts can adhere to plastic (polystyrene) in the absence of fibronectin providing the substrate bears a relatively high density of hydroxyl groups. Furthermore, according to these authors the images detected by interference reflection microscopy (which represent focal and close adhesive contacts) are essentially the same for fibroblasts grown on adsorbed fibronectin or on clean hydroxylated plastic. Curtis (1987) has raised the possibility that fibronectin might not be an adhesive molecule *per se*, but rather an activator of adhesion. Although superficially not supported by most available evidence, this intriguing possibility has yet to be fully explored. Cells such as fibroblasts clearly have the ability to interact with a number of cell adhesion molecules other than fibronectin such as laminin, vitronectin and collagen, and interactions with these molecules will be discussed further below.

5.4.3 Laminin

Laminin (Fig. 5.6) is a large (900 kD) protein found in virtually all basement membranes where it forms non-covalently linked complexes with type IV collagen, heparan sulphate proteoglycan, and the 150 kD protein nidogen (Dziadek and Timpl, 1985; Paulsson *et al.*, 1986).

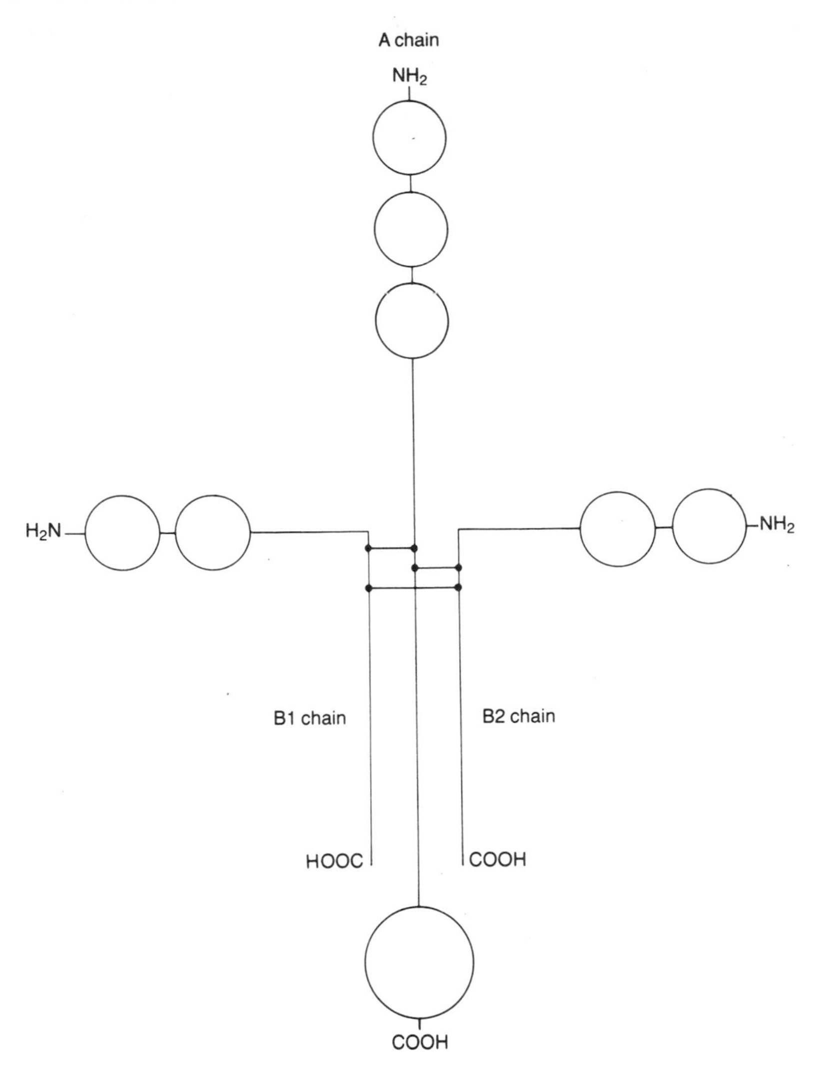

Figure 5.6 The structure of laminin (after Iwamoto *et al.*, 1987). Laminin is composed of three chains, A (450 kD), B1 (225 kD) and B2 (205 kD), which are arranged in a cross-like shape and are held together by disulphide bonds. Two globular domains (thought to be involved in binding to collagen) are present on each of the short arms of the cross, and a larger single domain (with affinity for heparin) is present at the end of the long arm. Note that the long arm is composed of all three chains, whereas the short arms are not. The sequence GDPGYIGSR, which promotes B16F10 melanoma adhesion to laminin and inhibits lung colonization *in vivo*, is found on the short arm of the B1 chain near the intersection site.

Nidogen, prepared from the EHS tumour, actually shares many similarities with the 158 kD entactin, prepared from Reichert's membrane (Carlin *et al.*, 1981), but the relationship between these two molecules is not yet completely clear. A cell surface receptor (67 kD) for laminin with high affinity (2×10^{-9}M) has been isolated by affinity chromatography from various cell types including carcinoma and sarcoma cells, and the B16 melanoma. Generally speaking there are about $5–10 \times 10^4$ unbound laminin receptors per cell although of course this depends on the cell type.

More recent studies by Elices and Hemler (1989) and others have since shown multiple receptors for laminin to be present on a range of mammalian cells. At least four of these are members of the integrin superfamily of adhesion receptor molecules (Table 5.2), namely VLA1, VLA2, VLA3 and VLA6. In this respect laminin appears to parallel fibronectin which itself is recognized by at least five different receptors. Also like fibronectin, the amount of cell-surface-bound laminin appears to decrease on transformation, at least for some epithelial cell types. It seems that laminin may play an important role in regulating the formation of the basement membrane (section 2.10.1) by initiating the deposition of insoluble supramolecular complexes. This role is supported by studies which show that the addition of laminin to thyroid cells can induce basement membrane formation (Garbi and Wollman, 1982).

Rotary-shadowed samples of purified laminin examined under the electron microscope show an asymmetric, cross-shaped structure with three short arms and one long arm. The dimensions of the molecule are roughly 115 nm × 75 nm along the arms. Two small globular domains are located near the ends of the short arms, while a larger domain (possibly comprising three smaller domains) is found at the end of the long arm. Electrophoretic separation of the components of laminin shows a single A chain (about 450 kD), a B1 chain (225 kD) and a B2 chain (205 kD) to be present. Part of each chain forms a short arm while the rest projects down the long arm, and the three are linked together by interchain disulphide bonds as shown in Fig. 5.6. Laminin contains about 13% predominantly *N*-linked carbohydrate on a weight for weight basis with protein, although recent evidence suggests that this may extend up to 27%.

The type IV collagen-binding activity of laminin seems to located primarily within the distal globular domains of each short arm (Rao *et al.*, 1982). With respect to cell-binding, however, it seems that several sites may exist. One such site seems to be located near the intersection of the arms of the laminin molecule (Edgar, Timpl and Thoenen, 1984). Using information gained following the determination of the primary structure of the B1 chain (Sasaki *et al.*, 1987), an active cell binding site for laminin was located by Graf *et al.* (1987) to the nine amino acid sequence Gly-Asp-Pro-Gly-Tyr-Ile-Gly-Ser-Arg (GDPGYIGSR). This nonapeptide promoted HT-1080 (human fibrosarcoma) and CHO cell adhesion when it was coated onto a plastic substrate, and it also blocked cell adhesion to laminin when it was added to the cell incubation medium. On a molar basis, however, the nonapeptide was less than 1% as active as laminin in promoting cell adhesion, suggesting that other sites on laminin may also have cell binding affinities. The extent to which the effectiveness of the peptide in promoting adhesion might be influenced through the vagaries of binding a small peptide to plastic, however, is not clear. When the nonapeptide was bound to albumin its effectiveness at promoting cell adhesion was increased by about an order of magnitude, suggesting that in the nonapeptide the active sequence may not be optimally displayed to cells. This possibility is supported by the observation that the nonapeptide is derived from a cysteine rich domain in which correct positioning of intrachain disulphide bonds could be of functional significance. The nonapeptide was also found by Graf *et al.* (1987) to be chemotactic for B16F10 melanoma cells, although it was only about 30% as effective in this regard as was laminin. Again, it could be argued from this result that either the active site is not correctly displayed or sequences other than GDPGYIGSR might be involved in eliciting chemotactic behaviour in B16F10 cells. Certainly, the latter interpretation would be in keeping with results suggesting that some of the biological effects of laminin are mediated by parts of the molecule other than those near the intersection of the arms. Neurite extension, for example, has been associated with a distal portion of the long arm which has heparin-binding activity (Edgar, Timpl and Thoenen, 1984). Furthermore, since the GDPGYIGSR sequence appears only to block the adhesion of cells to laminin mediated via the 67 kD receptor and not others, then it would indeed seem likely that additional sequences are involved.

Other cell-binding sites on laminin have in fact been located to different regions by Terranova *et al.* (1983), Timpl *et al.* (1983), Goodman *et al.* (1987), and Charonis *et al.* (1988). The emerging consensus from these studies is that laminin not only has multiple cell adhesion sites, but also that these sites may function with a degree of cell selectivity. Thus HT-1080 and CHO cells adhere to the GDPGYIGSR sequence identified by Graf *et al.* (1987), whereas a range of other cell types including the K1735 murine melanoma and BAE cells adhere to RYVVLPRPVCFEKGMNYTVR, a sequence identified in the inner globular domain of the short arm of the B1 chain by Charonis *et al.* (1988). Both neuronal cell adhesion and outgrowth are promoted by the sequence SRARKQAASIKVAVSADR on the laminin A chain (Sephel *et al.*, 1989), whereas only their attachment is promoted by GDPGYIGSR.

Laminin affects the adhesion, morphology, growth and migration of a number of cell types, particularly those of epithelial origin. Some fibroblasts appear to secrete and respond to laminin (Couchman *et al.*, 1982), but most cells of mesenchymal origin appear unable to use it as an adhesive protein. Laminin influences the metastatic and colonizing abilities of tumour cells as suggested by its enhancing effects when injected into mice along with melanoma cells (Barsky *et al.*, 1984). Co-injection of melanoma cells with antibodies to laminin or with a fragment of laminin which binds to cells but not to type IV collagen inhibits lung-colonizing ability. The general concept is that surface-bound laminin might facilitate adhesion of malignant cells to basement membrane collagen thereby enhancing their extravasation and the possibility of metastatic growth. Furthermore, cell lines which express laminin on their surfaces are more adherent and more motile than laminin-deficient lines, and this is likely to influence their metastatic ability. The general importance of laminin in the metastatic process is shown by the studies of Terranova *et al.* (1984) in which malignant cells exposed to laminin for 8 days displayed increased affinity for it and also show increased metastatic ability. Furthermore, laminin-deficient variants selected from murine 3-methylcholanthrene-induced fibrosarcomas have low malignant potential and do not metastasize, whereas variants which express laminin are highly malignant and spontaneously metastasize in most animals. When the laminin-deficient cells are allowed to bind laminin, the *in vivo* differences between the two variants becomes less marked (Malinoff *et al.*, 1984). Interestingly, in this study the laminin deficient fibrosarcoma variants were selected by incubation with human serum (containing antibodies against blood group B antigen) and active complement, the idea being to lyse cells which expressed α-D-galactopyranosyl groups on their surface. Since this particular group terminates the carbohydrate structure of laminin, then only laminin deficient cells should survive the complement-mediated, lysis selection regime. Since it was noted previously that some transformed cells actually have less surface laminin than their normal counterparts, it may be that metastatic cells rich in laminin comprise only a minority of the total population. This would certainly be in keeping with the general view of phenotypic diversity in tumour populations.

Iwamoto *et al.* (1987) have recently shown that both GPDGYISGR and the pentapeptide YIGSR can block the invasion of a reconstituted basement membrane matrix by B16F10 cells. Both peptides also reduced the lung-colonizing ability of B16F10 cells following tail vein injection of pretreated cells, with pronounced reduction at doses in the region of 1 mg per mouse. The amide forms of the peptides appeared to be more active, possibly because this moiety neutralizes the negative charge on the arginine. The effects of the peptides appeared not to be due to direct cytotoxicity or to alteration of the tumorigenicity of the melanoma cells. Similar inhibitory effects were seen when the nonapeptide and the tumour cells were injected into separate tail veins indicating that co-incubation prior to inoculation was not essential. Interestingly, a sequence very similar to GDPGYIGSR is found in both EGF and TGF_{α}, but the nonapeptide appears not to exert its effects through binding to the EGF receptor. Iwamoto *et al.* (1987) proposed that the peptides worked by reacting with the laminin receptor on B16F10 cells, blocking their adhesion to the basment membrane and thereby reducing lung colonizing ability. This scenario is likely to be relevant only if binding to the basement membrane is a crucial event in extravasation. We have seen, however, that other processes may be involved in lodgement and thus alternative scenarios may be just as viable, if not more so. In this light, it would be worthwhile exploring the effects of the peptides on platelet interactions since these could markedly influence trapping and subsequent extravasation (section 5.7.2.2). Since the fibronectin peptide GRGDS can also inhibit lung colonization by melanoma cells, it would seem that a number of different types of cellular interactions might be involved in the metastatic process.

5.4.4 Vitronectin

Vitronectin and fibronectin represent the two most important cell adhesion molecules found in plasma. Both are also found at the cell surface and in the tissues, and indeed immunofluorescent techniques have shown considerable overlap in their distribution profiles. For many years fibronectin was thought to be the only significant plasma molecule which mediated cell–substrate adhesion, although both Grinnell, Hays and Minter (1977) and Knox and Griffiths (1979) had shown that serum contained two distinct proteins which promoted adhesion and spreading. Later, Hayman *et al.* (1982) were able to make use of a novel method to confirm that at least one other molecule had been overlooked. The technique they employed was based on the 'bioautographic' method of Klebe *et al.* (1978). In essence, plasma proteins were separated by SDS-PAGE, transferred to nitrocellulose, and then probed with NRK cells (reviewed in Ferro *et al.*, 1988). Using this procedure, NRK cells were found to adhere to two plasma proteins, one of about 220 kD (fibronectin) and the other of about 70 kD (vitronectin). Although a useful method for identifying adhesive proteins, the bioautographic technique is not without limitations. As originally used by Hayman and colleagues (1982) the adhesive molecule must remain active under reducing conditions (SDS plus 2-mercaptoethanol), and it must also be present in sufficient quantity to be detected after diffusion blotting. Indeed, Ferro *et al.* (1988) reported better success using electroblotting of molecules separated under non-denaturing conditions. They were able to show that fibronectin and laminin were adhesive for B16F10 melanoma cells, and they also identified a large (>600 kD) molecule or molecular complex present in extracts from endothelial cell cultures which promoted melanoma adhesion.

In retrospect, it appears that vitronectin or serum spreading factor was first described over 20 years ago by Holmes (1967). Unfortunately, it remained poorly characterized for a number of years, possibly because it was thought to be a fragment of fibronectin. It can be isolated from fresh citrated plasma (in which it is present at about 200–300 μg ml^{-1}) by chromatography on glass bead columns, and its name reflects this attribute as well as its ability to promote cell adhesiveness (Hayman *et al.*, 1983). Human plasma vitronectin runs on SDS-PAGE as two polypeptides with molecular masses of about 65 and 75 kD. Vitronectin extracted from various types of fetal bovine serum runs on SDS-PAGE predominantly as polypeptides of 65 and 80 kD, with the latter dominating (Hayman and Pierschbacher, 1985). Interestingly, vitronectin appears to be depleted from fetal bovine serum intended for *in vitro* use, presumably during its commercial preparation, but it is nevertheless considered to be the principal adhesive protein in the *in vitro* culture of many cell types. According to Underwood and Bennett (1989), vitronectin in serum is more effective than fibronectin in promoting cell attachment because the latter molecule coats substrates poorly in the presence of other serum proteins.

When Ill and Ruoslahti (1985) prepared vitronectin from human serum rather than plasma, they found that it co-isolated with an 82 kD component which proved to be a complex of thrombin-antithrombin III. Vitronectin appears to bind to this complex through a cryptic site on antithrombin III which is exposed only when the latter binds to thrombin. This behaviour is very similar to that of another serum protein called S-protein, which is involved in the coagulation and complement pathways. S-protein binds to thrombin–antithrombin III complexes and it is thought that this may actually prevent thrombin

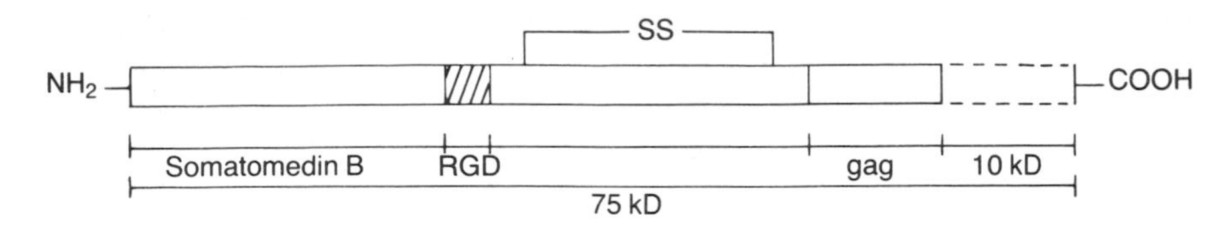

Figure 5.7 The structure of vitronectin (after Suzuki *et al.*, 1984). Vitronectin prepared from human plasma exists in two forms. The larger form, a polypeptide of 75 kD, is processed to the 65 kD form by cleavage of a C-terminal 10 kD fragment (dashed lines). The N-terminal amino acid sequence of vitronectin is identical to that of somatomedin B. An RGD-containing cell attachment segment is found adjacent to the somatomedin B sequence, while a glycosaminoglycan-binding region (gag) is located towards the C-terminal end of what corresponds to the shorter 65 kD polypeptide. SS: disulphide bond.

from antithrombin III inactivation. S-protein also inhibits the formation of the cytolytic terminal complex of the complement cascade by binding to soluble C5b–7 complexes. Since it reduces complement complex attachment to cell surfaces, S-protein may also protect bystander cells against complement mediated lysis.

Sequence analysis of both vitronectin and S-protein has shown them to be identical (Jenne and Stanley, 1985). Both molecules show homology at their amino terminals with somatomedin B (Suzuki *et al.*, 1984; Jenne and Stanley, 1985), a molecule once thought to have mitogenic activity but this has now been ascribed to contamination with EGF. Interestingly, there is some concentration of vitronectin in Cohn fraction IV of human blood, the fraction enriched in the somatomedins as well as in α- and β-globulins and albumin.

The cell attachment promoting activity of vitronectin is located next to the somatomedin B domain (Fig. 5.7), the key sequence being arg-gly-asp-val (RGDV) which is obviously closely related to the RGDS of fibronectin (Suzuki *et al.*, 1985). This suggests that both the fibronectin and vitronectin receptors bind RGD related sequences. Pytela *et al.* (1985), however, have isolated the receptor for vitronectin (125/115 kD) on human MG-63 osteosarcoma cells and rat fibroblasts and shown that it is different from that for fibronectin on the same cell types. There is also some evidence that the vitronectin receptor has greater affinity for small RGD-containing peptides than does that for fibronectin, possibly as a consequence of the subtleties of molecular conformation favouring vitronectin receptor binding over the other. When incorporated into phosphatidylcholine liposomes, the 125/115 kD receptor mediates their specific binding to vitronectin but not to fibronectin. As we have seen, the situation appears somewhat different in birds, since the fibronectin receptor integrin binds both fibronectin and vitronectin, as well as laminin.

The vitronectin domain next to that with cell binding affinity has three possible sites for *N*-linked glycosylation, and this is followed by a heparin-binding domain and finally a C-terminal domain. This lattermost domain contains a site susceptible to proteolytic cleavage, which probably explains the existence of the two molecular mass forms seen on electrophoresis. Unlike fibronectin, vitronectin does not bind to gelatin, but it does bind to a limited extent to type IV collagen. The physiological relevance of this binding, however, is not yet clear.

Vitronectin is immunologically related to epibolin, a 65 kD glycosylated plasma protein so named because it promotes epiboly, a term describing the coordinated movement of epithelial cell sheets (Stenn, 1981 a and b). Molecular sequencing is required, however, before the two can be definitively equated. The epithelial cell movement-promoting activity of epibolin is apparently enhanced by an unknown cofactor, but as yet no such enhancing cofactor has been identified for vitronectin.

5.4.5 Collagen

As a widely dispersed component of the interstitial stroma and basement membranes, collagen in its various forms (Table 2.2) is appropriately located to function as an adhesive substrate for a wide variety of cells. A glycoprotein known as colligin (47 kD), which acts as a receptor for type IV collagen, has been isolated from cultures of mouse parietal endoderm cells, embryonal carcinoma cells, and hepatoma cells (Kurkinen *et al.*, 1984) but its cell surface location is suspect (Saga *et al.*, 1987). Although some cells may bind directly to type IV collagen via a collagen receptor, others could conceivably take advantage of the binding of laminin to collagen so that adherence between cells and type IV collagen may be via a laminin bridge.

Anchorin CII is a 31 kD cell membrane protein on chick chondrocytes which acts as a receptor for type II collagen and probably others (Mollenhauer and von der Mark, 1983). Its mammalian counterpart (34 kD) has been isolated from sheep fibroblasts (Mauch *et al.*, 1988). Recently, Dedhar, Ruoslahti and Pierschbacher (1987) have isolated three polypeptides (250 kD, 70 kD, and 30 kD) from MG-63 human osteosarcoma cells which probably serve as a cell surface receptor complex for type I collagen. When incorporated into liposomes the polypeptides promote binding to type I collagen, and their effects are inhibited by RGD-containing peptides. The RGD sequence appears twice on the $\alpha1$ chain and four times on the $\alpha2$ chain of type 1 collagen. Interestingly, MG-63 osteosarcoma cells have now yielded three RGD-dependent receptors: one for type I collagen, one for fibronectin, and one for vitronectin.

On-going research indicates that there may turn out to be a relatively large number of collagen receptors, at least four of which are known to be of the integrin type (Table 5.2). VLA2, which binds to a range of collagens, is identical to the platelet gpIa/IIa complex which functions in a Ca^{2+}-independent manner on non-activated platelets. At least one collagen receptor, the 90 kD CRIII protein of human fibroblasts, is known to occupy a transmembrane location (Carter and Wayner, 1988). In cell labelling studies CRIII co-localizes with vimentin suggesting that it may act as a bridge between extracellular collagen and the cytoskeleton.

5.4.6 Glycosaminoglycans and proteoglycans

When certain substrate-adherent cells are treated with chelating agents such as EDTA or EGTA, they may detach leaving behind foci containing molecules such as fibronectin and various proteoglycans (Culp, 1976). This substrate-attached material (SAM) also contains various intracellular proteins, particularly actin, suggesting that the foci of SAM may represent adhesion sites in association with stress fibre components. It has been estimated that cells detached by EGTA may leave behind 1% of their protein and phospholipid and as much as 15% of their surface carbohydrate. As explained below, it seems that SAM contains the remnants of two types of cell contacts: **close contacts** with a separating distance between a cell and its substrate of about 25–30 nm, and **focal contacts** with a separating distance of about 10–15 nm (Laterra *et al.*, 1983a). Close contacts are thought to develop as a consequence of the binding of cell surface heparan sulphate (HS) proteoglycan (present within SAM) to the heparin-binding domain of substrate attached fibronectin. In order for the tighter focal contacts to develop, it seems that substrate-bound fibronectin must bind to both cell surface HS proteoglycan and to the fibronectin receptor. Although cells will adhere and spread on fibronectin fragments containing the cell-binding domain, they do not form tightly adherent focal contacts unless the heparin-binding domain is also present (Woods *et al.*, 1986). Interestingly, mixtures of purified cell-binding and heparin-binding domains from fibronectin do not fully replace the native molecule implicating a precise steric relationship between the two in generating maximum adhesion. In other studies with heparanase, removal of >80% of cell surface HS proteoglycan from Balb/c 3T3 cells following treatment with this enzyme inhibited their spreading on fibronectin coated substrates (Laterra *et al.*, 1983b). Furthermore, 3T3 cells can adhere and spread on platelet factor 4 (PF4) coated substrates in a manner which appears to be almost entirely HS proteoglycan mediated, although as would be expected they form only close contacts and not focal contacts (Lark *et al.*, 1985). Taken together these results clearly imply an important role for cell surface HS proteoglycan in cell adhesion.

Analysis of SAM obtained from Balb/c 3T3 cells and their SV40-transformed variants has highlighted further important differences betweeen ostensibly normal cells and virally transformed cells. Thus SV40-3T3 SAM contains predominantly HS proteoglycan with a smaller amount of chondroitin sulphate (CS) proteoglycan, whereas the proteoglycans of 3T3 SAM are 90% CS with the remainder HS. Interestingly, SV40-transformed Balb/c 3T3 cells form mainly close contacts, whereas their non-transformed parent cells form more focal contacts. The precise relationship between proteoglycan content and cell adhesion is difficult to define, however, other than that focal contact development requires cell surface HS proteoglycan to be present. The principal reason for this is that the proteoglycans undergo changes in their binding affinities for fibronectin. Newly synthesized SV40-3T3 SAM has a high affinity for fibronectin which soon decreases, whereas even long-term SAM from 3T3 cells retains high affinity for this molecule. These changes in affinity probably arise as a consequence of proteoglycan processing, and indeed it has been shown that SV40-transformed 3T3 cells display more proteolytic and endoglycosidic catabolism of HS proteoglycan than do their non-transformed counterparts (Lark and Culp, 1984 a and b; Wightman, Weltman and Culp, 1986). Whereas HS proteoglycan seems to promote cell adhesion to fibronectin, there is some evidence that CS proteoglycan and/or hyaluronic acid (HA) can inhibit it. The appearance of CS proteoglycan and HA in the long-term SAM of some cell types thus might represent a mechanism for modulating focal contact mediated adhesion.

With respect to B16 melanoma cells, Maniglia and his colleagues (1985) have shown that there is a partial correlation between lung-colonizing efficiency and the glycosaminoglycan (gag) content of the extracellular coat. Correlation was not absolute since two clones which differed in their colonizing ability displayed similar gag levels. These authors also reported in this study that the more efficient lung-colonizing variants were more resistant to detachment from Petri dishes by EDTA, but whether this was a direct result of higher gag content in the extracellular coat or not is unknown. It is of interest to note that Kramer, Vogel and Nicolson (1982) have reported more

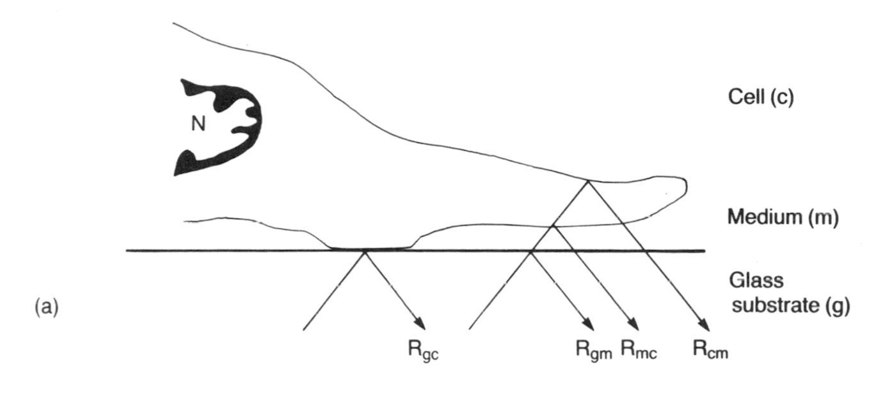

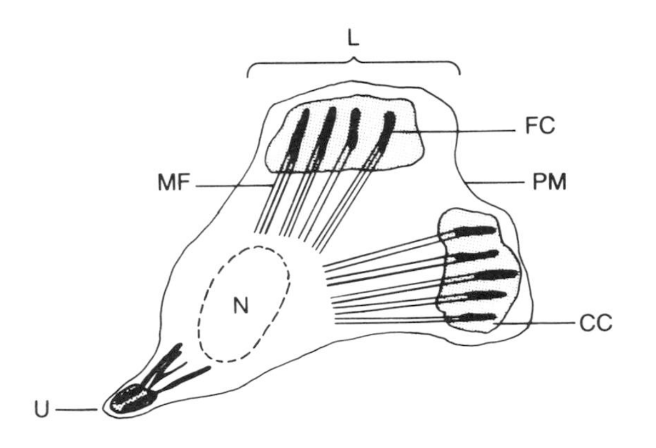

Figure 5.8 Interference reflection microscopy and the visualization of focal and close contacts. (a) As explained in the text, interference reflection microscopy (IRM) is based on the reflection of an incident light source, conveniently (but not necessarily) of monochromatic nature. When the plasma membrane (PM) of a cell is in direct contact with the glass substrate, reflection (R_{gc}) arises at the substrate–cell interface because of differences in their refractive indices. If a cell is separated from its substrate by some distance, three reflection fronts are likely (R_{gm}, R_{mc} and R_{cm}). Under appropriate illuminating conditions, reflection from the interface of the upper cell surface and the surrounding medium (R_{cm}) can be excluded. The remaining two reflection fronts can interfere resulting in bright and dark zones compared to the background. The intensity of the resulting image indicates the distance of the lower surface of the cell from the substrate, with the brighter images indicating further separation. Focal contacts typically appear black with a monochromatic light source, whereas close contacts appear grey (after Verschueren, 1985).

active HS proteoglycan metabolism in more efficient lung-colonizing melanoma variants.

Chick neural retina cells release extracellular complexes of proteins and glycosaminoglycans (known as **adherons**) which have been implicated in cell–substrate adhesion. The principal components of these complexes are a glycoprotein of 170 kD (the C_1H_3 antigen) and heparan sulphate (HS) proteoglycan. Cole, Schubert and Glaser (1985) have proposed that the C_1H_3 antigen and HS proteoglycan together represent a cell surface receptor for secreted, substrate-bound C_1H_3 antigen. In their model, cell surface HS proteoglycan (associated with cell surface C_1H_3 antigen) binds to substrate-bound C_1H_3 antigen and induces a conformational change that allows homophilic interaction between cell surface- and substrate-bound C_1H_3 antigens thereby forming the adhesion. Although implicating proteoglycans in cell adhesion, the relevance of this particular type of interaction in tumour biology is uncertain.

5.4.7 Focal contacts

Several different ways exist for observing the nature of the gap between adherent cells and their substrates. Living cells may be looked at from the side, for example, by employing specially modified microscopes, while the gap associated with fixed and embedded cells may be observed in either transverse sections or tangential sections which skim the lower surfaces of adherent cells. Another method involves interference reflection microscopy (IRM), which was introduced into biology by Curtis (1964). In this technique, phase changes due to reflection of incident light between the substrate and the medium, and between the medium and the lower surface of a cell, yield an interference pattern which corresponds to the closeness of contact between the cell and the substrate. If a cell is in direct (molecular) contact with its substrate, reflection arises from the substrate–cell membrane interface (Fig. 5.8). Because of the relatively high refractive

---

(b) A schematic representation showing the generalized relationship between close and focal contacts. The cell shown in the figure is envisaged to be shifting from stationary to motile behaviour. Broad areas of grey intensity which represent close contacts (CC) are seen within two lamellipodia (L) extending from the cell. Dark focal contacts (FC), associated with cytoplasmic microfilaments (MF), lie within the close contact zones. FCs are also seen in the tail or uropod (U). The nucleus (N) is barely visible because of the optical conditions, and other cytoplasmic organelles are not apparent. (PM), plasma membrane.

index of the membrane, reflection in this situation is low and the contact zone appears darker when imaged with a monochromatic light source (reviewed in Verschueren, 1985).

Studies of various cell types in culture with IRM has revealed discrete areas underneath the cells which, because of their relatively close apposition with the substrate compared to the rest of the surface, are thought to serve a role in cell adhesion. As mentioned previously, these areas of relatively close substrate apposition may be classified as being either focal contacts or close contacts (Izzard and Lochner, 1976). Focal contacts, typically 0.25–0.5 μm wide × 2–10 μm long in chick heart fibroblasts, appear black in IRM using a monochromatic light source. They tend to have a radial distribution, and are concentrated towards the periphery of the cell (particularly its processes). Extensive focal contacts are not found in highly motile cells. As indicated by the studies of Shure *et al.* (1979) and others, focal contacts are in fact associated more with tightly adherent, stationary cells. It seems that focal contacts occur in such cells at sites where stress fibres (prominent bundles of actin microfilaments) approach the cell membrane (Izzard and Lochner, 1976; Heath and Dunn, 1978). According to the studies of Geiger *et al.* (1980) and Singer (1982), focal contacts are also associated with α-actinin and vinculin. Extracellular fibronectin is known to be associated with actin filaments and several studies have shown that this molecule contributes to focal contact sites (Rees *et al.*, 1978; Singer, 1982). This may not be the case for all cell types, however, since HeLa cells appear to be able to form focal contacts in fibronectin-depleted medium, in the presence of antibodies against fibronectin, and in the presence of the monovalent ionophore monensin (1 μM) which inhibits the secretion of fibronectin and other proteins (Morgan and Garrod, 1984). One interpretation of these conflicting results is that focal contacts might undergo a maturation process, passing from a fibronectin-negative form in motile cells to a fibronectin-positive form in stationary cells (Singer, 1982). It should be noted here that focal contacts are not absolutely essential for cell–substrate adhesion and spreading, although as we have seen their presence appears to correlate with increased cell attachment for many cell types.

Using both IRM and fluorescence microscopy to look at the distribution of fibronectin within focal contacts of WI-38 human fibroblasts, Birchmeier *et al.* (1980) have suggested that actin filaments within the cell might be offset in their association with substrate-bound fibronectin. In this model, the plasma membrane in the zone of focal contact is envisaged as a corrugated sheet with fibronectin and actin fibres in alternating corrugations. Other studies, however, imply more direct overlapping between these two components (Hynes and Destree, 1978; Heggeness, Ash and Singer, 1978). As mentioned above, however, there is some debate as to whether fibronectin is found in focal contacts at all. In fact, Chen and Singer (1980) have claimed that fibronectin is found within close contacts but only around the edges of focal contacts. Similarly, Damsky *et al.* (1985) have shown that the fibronectin receptor integrin is closely associated only with the peripheral regions of focal contacts. Clearly, differences in technique and the use of different cell types contribute to these conflicting observations on the presence of fibronectin in focal contacts.

Grinnell (1986) has shown recently that certain cells growing in low-serum medium concentrate fibronectin in the region of focal contacts, whereas this molecule is removed from these sites in the presence of serum and concentrated in patches on the cell surface. This process is actually dependent on bulk phase protein (usually supplied in culture by serum), and it is likely to be promoted by the shearing forces generated through cell contractile processes. This would suggest that focal contacts are probably labile structures, and indeed Streeter and Rees (1987) have reported that the lifetimes of focal contacts can be quite short, some lasting only 10–15 min. Dynamic focal contacts would be expected if they were to be utilized in the locomotory activity of cells. This might seem surprising since increased numbers of focal contacts generally tend to correlate with restricted cell motility. However, the number and extent of development of focal contacts in cells grown *in vitro* might represent an over-response to growth on an inflexible, artificial substrate. According to this idea, focal contacts *in vivo* are less developed and more dynamic structures, but for obvious practical reasons this possibility cannot be substantiated.

Using the fluorescence photobleach recovery technique in association with microinjection of fluorescently labelled vinculin, α-actinin and actin, Kreis *et al.* (1984) have been able to show that these three proteins exist in two pools. One pool is structural in nature and represents adhesive sites such as the focal contacts, while the other pool is represented by free (cytoplasmic) molecules. Continuous interchange seems to be possible between these pools, confirming the dynamic nature of adhesive sites. Interestingly, urokinase-type plasminogen activator has recently been localized to focal contact sites in normal human fibroblasts and in HT-1080 sarcoma cells (Pollanen *et al.*, 1988). The presence of this enzyme at focal adhesive sites may affect modulation of the latter, thereby influencing locomotory ability.

In their recent study, Streeter and Rees (1987) compared the morphology and behaviour of NRK cells spreading on either plasma fibronectin or the peptide Gly-Arg-Gly-Asp-Ser-Cys (GRGDSC). Using IRM, they found that NRK cells formed mainly point contacts on GRGDSC but focal contacts on fibronectin. The point contacts (90–200 nm diam.) remained after detergent extraction of the cells, stained poorly or not at all for vinculin, were not associated with microfilament bundles, and were further (145–160 nm) from the substrate than focal contacts (30–80 nm for these cells). Some normal focal contacts were formed by cells spreading on the peptide, but Streeter and Rees (1987) attributed these to the possible presence of endogenous fibronectin. Generally speaking, spreading of NRK cells on the peptide was slower and less extensive than on fibronectin, and the quantities required differed by up to five orders of magnitude. These results are consistent with the observations of Laterra *et al.* (1983a), who concluded that focal contact formation required adhesion to both the cell-binding (RGDS) and heparin-binding domains of fibronectin (section 5.4.6).

Maupin and Pollard (1983) have shown that areas of adhesion of HeLa cells to the substrate which appear to be morphologically similar to focal contacts are associated with a network of clathrin (the molecule involved in the formation of coated endocytic pits). More recently, Nicol and Nermut (1987) have shown patches of clathrin on the inner surface of substrate-attached NRK cell membranes at sites which might correspond to the point contacts described by Streeter and Rees (1987). The structures detected by Nicol and Nermut (1987) appeared not to be associated with significant microfilament bundles or with vinculin, and with IRM they appeared as dot, U or sigmoidal shaped structures slightly less dark than focal contacts. The presence of clathrin in these sites would be consistent at first sight with a dynamic nature for such contacts, but Nicol and Nermut (1987) concluded from time-lapse studies that they could persist for more than two hours. These authors concluded that the clathrin-coated structures represented sites of frustrated endocytosis, in which an attempt was being made by the cell to endocytose substrate-bound material.

Many transformed cells typically lack both focal contacts and stress fibre bundles, and this correlates with reduced levels of fibronectin and poor attachment to substrates. Indeed, when fibronectin is added back to transformed cells not only do they become more spread, but they also develop focal contacts and microfilament bundles (Ali *et al.*, 1977; Willingham *et al.*, 1977). Fibroblasts transformed with RSV have fewer focal contacts than their non-transformed counterparts (David-Pfeuty and Singer, 1980), and Marchisio *et al.* (1984) have shown that focal contacts reappear when chick embryo chondrocytes infected with temperature-sensitive RSV mutants are shifted from permissive to restrictive temperature. Reports from other research groups, however, suggest that there is no simple relationship between decreased focal contact expression and malignancy. Thus a range of human carcinoma cell lines have been shown to form extensive focal contacts under *in vitro* conditions (Haemmerli and Strauli, 1981). Despite these conflicting reports, a number of other studies have indicated a relationship between the lack of focal contacts and metastatic potential as assessed by lung colonization. High-colonizing variants of the K1735 melanoma system, for example, are poorly attached to the substrate and have fewer and smaller focal contacts than low-colonizing variants (Raz and Geiger, 1982). Interestingly, the low-colonizing variant forms compact tumours with adherens-type junctions after subcutaneous injection, whereas the other variant produces tumours in which the cells are more loosely bound. Similarly, Cottler-Fox *et al.* (1980) have reported that a non-metastasizing murine carcinoma variant forms both close and focal contacts, whereas a metastasizing variant forms less extensive close contacts and generally fails to form focal contacts. These results may not seem too surprizing if the lack of focal contacts is indicative of increased motility. However, this comment should be evaluated in the light of other studies indicating that some highly metastatic cells are more tightly adherent to the substrate than their poorly metastatic variants, and that enhanced locomotory ability *in vitro* does not always correlate with metastatic behaviour.

Finally, it should be noted that the dark band seen around the periphery of some cells and thought to be a more or less continuous focal contact is actually generated by a thin (<100 nm) cytoplasmic veil. This was confirmed by total internal reflection fluorescence (TIRF) microscopy (Gingell, Todd and Bailey, 1985). In TIRF, the angle of incidence of a pencil of laser light is adjusted so that it is totally internally reflected at the interface between a substrate and the medium (Axelrod, 1981). Although no light is transmitted in this situation, an evanescent wave which travels parallel to the interface is generated and this can excite fluorescent dyes which are located close (<200 nm) to the interface. Gingell and colleagues (1985) treated cell cultures with fluorescently labelled beads and examined these under appropriate conditions. Dark images resulted from where the beads were excluded because of the close apposition of the cell and its substrate, and these images were then be used to map the topography of the apposition zone. The location of the peripheral dark band seen by IRM did not correlate with sites of close interaction between the cell and its substrate.

5.4.8 Close contacts

Close contacts, despite their name, are characterized by a gap slightly wider than that for focal contacts. Unlike focal contacts, close contacts appear not to contain vinculin and it has been postulated that they probably represent a more labile adhesive site. Using monochromatic IRM, close contacts in chick heart fibroblasts appear in peripheral regions of the leading lamellae as broad, relatively uniform grey areas in which the focal contacts are usually found. According to Izzard and Lochner (1976), the focal contact represents an area of the cell separated from its substrate by about 10–15 nm, while the close contact is separated by about 30 nm, and these results are in broad agreement with many ultrastructural studies. This correlation may be fortuitous, however, since Gingell (1981) has pointed out that the calculations used by Izzard and Lochner (1976) are in error. In contrast to focal contact sites, close contacts are found predominantly under motile cells or moving cell extensions. Because their contours are continually changing, their shape and dimensions are not defined. As mentioned above, close contacts rather than focal contacts predominate in many transformed cell types, but this correlation does not have universal applicability.

5.4.9 The fibronexus

Whereas both focal and close contacts tend to be found more towards the periphery of adherent cells, Singer (1979) has identified another contact site (prominent in stationary fibroblasts) which is located more under the central regions of well spread cells. The transmembrane association between fibronectin and actin in these sites is colinear, and he has coined the term **fibronexus** to describe the resulting association. Thus in the fibronexus contact between fibronectin and actin is along the microfilament bundles rather than at their termini. One interpretation of these observations is that the fibronexus represents a relatively stable extracellular matrix contact site. An alternative, but related possibility is that the fibronexus is involved in fibronectin fibrillogenesis (section 5.4.2).

Interestingly, Singer (1979) observed fibronexuses with small fibronectin tufts emerging from them at points of cell–cell contact in SV40-transformed hamster embryo fibroblasts. This is not without interest given the earlier report that fibronectin can promote the aggregation of chick embryo fibroblasts and BHK cells (Yamada, Olden and Pastan, 1978).

5.5 THE MODULATION OF ADHESION

Many cells within the body, such as epithelial cells, seem to be firmly anchored in place, and indeed it would seem impossible for most organs to operate optimally if this were not the case. The functioning of other cells of the body such as the white blood cells, however, requires active cell motility and this almost certainly requires some degree of modulation of cell adhesion. It is possible to envisage a cell which may be relatively fixed at one moment yet which can become free to move when required. This possibility would seem to demand that cell adhesions should be reversible. Indeed, locomotion itself probably requires reversible adhesions so that the back of a cell can flow forward to establish a new projection with new adhesions at the front of the cell.

How might cell–substrate adhesions be modulated? As our understanding of the molecular nature of adhesive interactions unfolds, various ways by which adhesion may be modulated become evident. The most simple way is to digest adhesive molecules such as those in focal contacts by various secreted enzymes, although optimal functioning will require a degree of substrate specificity and localized activity. Another possibility might involve modulation of the interactions between the various molecules implicated in adhesion at focal contacts, such as fibronectin, HS proteoglycan, talin, and vinculin. A calcium-dependent protease (CDPII) which digests talin has been identified in focal contact plaques, but it is not known to what extent this might influence adhesive processes (reviewed in Burridge *et al.*, 1987). A possible role for HS proteoglycan catabolism in modulating adhesion was discussed earlier (section 5.4.6) as was the possible involvement of CS proteoglycan and/or HA. Other studies have suggested that phosphorylation of adhesion-related molecules might control their interactions and thereby modulate adhesion. Most of the molecules involved in these phosphorylations are the products of viral oncogenes, and their activities will be discussed further below.

It is commonly thought that the strength of an adhesion results solely from the stickiness of the 'glue'. However, it is well known from standard adhesion technology that stickiness is also a function of abhesives (which decrease adhesion) and tackifiers (which can increase adhesion). The roles of such modifiers in biological adhesion is uncertain. We have already seen that some gangliosides may act essentially as tackifiers in fibronectin-mediated adhesion, and that oligosaccharide side chains may act like abhesives (as seen, for example, in the polysialic acid of N-CAM). As yet, however, these relationships remain to be fully explored.

5.6 ONCOGENES AND CELL ADHESION

The products of at least four viral oncogenes, all with tyrosine kinase activity, have been found closely associated with focal contacts (reviewed in Burridge *et al.*, 1987). The product of the *src* oncogene ($pp60^{src}$), for example, is a tyrosine kinase that appears to be concentrated in the region of focal contacts before they disrupt soon after transformation (Rohrschneider, 1980; Nigg *et al.*, 1982; David-Pfeuty and Singer, 1980). According to the studies of Chen and colleagues (1985), newly expressed viral $pp60^{src}$ is localized at the cytoplasmic surface of the cell membrane in sites which correspond to a degree with areas of degradation of extracellular fibronectin. They argued that phosphorylation of membrane or cytoplasmic components by the viral kinase might induce local protease activity which could lead to fibronectin degradation and reduced adhesion. One of the proteins $pp60^{src}$ phosphorylates is the 130 kD protein vinculin, which is found in association with focal contacts. As discussed above, vinculin has been put forward as one of the proteins which may anchor microfilaments into focal contact regions, although the validity of this viewpoint remains to be firmly established. It has been proposed that changes in the phosphorylation pattern of vinculin following transformation by RSV could lead to the loss of focal contacts and a corresponding reduction in adhesion, but more recent studies have not been in general support of this possibility. Kellie *et al.* (1986), for example, failed to confirm a relationship between vinculin phosphorylation, $pp60^{src}$ concentration in adhesion plaques, and decreased adhesion. This conclusion was reached using chick embryo fibroblasts transformed with the temperature sensitive RSV mutant LA32. At their restrictive temperature (41°C) such cells appear morphologically normal with well-formed adhesion plaques and prominent microfilament bundles. However, they also contain a fully active $pp60^{src}$ kinase, have $pp60^{src}$ concentrated in their adhesive plaques, and contain vinculin which is extensively phosphorylated on its tyrosine residues. LA32 infected cells at their permissive temperature also contain $pp60^{src}$ in their adhesion plaques and they have a fibronectin matrix indistinguishable from that of normal cells. This last observation suggests that $pp60^{src}$ may not be responsible for reduced fibronectin deposition as discussed above, although LA32 infected chick cells may lack an intermediate protease whose activities are modulated by $pp60^{src}$.

It may be that the principal substrate for phosphorylation by $pp60^{src}$ which leads to reduced cell–substrate binding is not vinculin, but some other molecule involved in adhesion. Evidence exists to indicate that both talin and integrin can be phosphorylated on tyrosine residues, and the region of the talin–integrin binding site appears to be in the same location as the site for phosphorylation (Buck and Horwitz, 1987). In fact, phosphorylation of integrin appears to reduce binding to both talin and fibronectin. These results are highly suggestive that changes in phosphorylation levels acompanying transformation might result in decreased cell–substrate adhesion, but further studies supporting this possibility and analysing the role of $pp60^{src}$ are required. Interestingly, treatment of 3T3 cells with PDGF activates its receptor tyrosine kinase activity and is associated with a transient disruption of focal contacts and a redistribution of vinculin to the perinuclear region (Herman and Pledger, 1985). Under these conditions, vinculin is lost from focal contacts within 2.5 min, and this is followed within 5-10 min by actin stress fibre disruption. Presumably these events are involved in the induction of competence by PDGF, but exactly what role they play is uncertain.

In other studies, Maher *et al.* (1985) have used a high affinity polyclonal antibody against phosphotyrosine to study the distribution of tyrosine phosphorylated proteins in normal and transformed cells. It is worth noting here that in normal cells, tyrosine phosphorylation accounts for about 0.03% of the total pool of phosphoamino acids, but this may be increased about 10-fold in some transformed cells (Hunter and Sefton, 1980). Maher and colleagues (1985) found that 3T3 cells infected with Abelson murine leukaemia virus (A-MuLV) contained elevated levels of phosphotyrosine which was prominent at cell–cell contacts where it correlated with F-actin distribution. Since the 160 kD product of v-*abl* can phosphorylate its own tyrosine residues, this is probably the source of much of the labelled phosphotyrosine in A-MuLV transformed cells. Normal 3T3 cells, however, are also labelled with antibody against phosphotyrosine and the staining is restricted to discrete sites thought to represent focal contacts at the cell periphery. In the epithelial cell lines MDCK and PtK2 (from canine and potoroo kidneys respectively), phosphotyrosine labelling is localized to the termini of stress fibres in focal contacts and to sites of cellular interaction. Primary chick embryo fibroblasts (rather than cell lines) also showed labelling of phosphotyrosine, but this decreased on passaging. Interestingly, this decrease in the presence of phosphorylated tyrosine residues correlated with an increase in stress fibre formation. Unfortunately it is still too early to reflect fully on the *in vivo* relevance of these changes to cells in culture.

5.7 CELL ADHESION AND METASTASIS

If we consider metastasis via the blood to be composed of a series of discrete steps (Fig. 2.1) then we can envisage adhesion to be of possible importance in the following events:

(a) *Detachment from the primary*

This may be facilitated by the poor homotypic adhesiveness of cells with metatstatic potential, although reduced heterotypic adhesion may also facilitate detachment since tumours are commonly composed of several cell types (including macrophages and fibroblasts). Tumours may also contain varying amounts of fibrous connective tissue components and thus, as an alternative possibility, detachment from the primary may require decreased cell–substratum adhesiveness.

(b) *Formation of embolic aggregates*

Experimental evidence has been provided which supports the notion that both homotypic and heterotypic aggregate formation may promote the metastatic process. Increased homotypic adhesion within blood vessels, for example, may facilitate metastasis by enhancing tumour cell lodgement. Increased heterotypic adhesion of tumour cells to blood components such as platelets, leukocytes, and monocytes or to elements of the clotting cascade such as fibrin, could facilitate metastasis in a similar manner.

(c) *Extravasation*

This step in the metastatic cascade would be facilitated by the heterotypic adhesion of tumour cells to endothelial cells or to basement membrane components such as laminin and fibronectin.

With these points in mind, it would seem futile to study one aspect of cell adhesion in isolation and then attempt to give it some general applicability, although this is a common approach. Furthermore, tumour cell adhesion (like cell adhesion in general) is probably a variable trait. In cell locomotion, for example, adhesions must be made and broken as the cell translocates. Too strong an adhesion might result in immobility and too weak an adhesion might result in slip. Tumour cell adhesion to the endothelium probably needs to be variable since the tumour cells must first adhere from the bloodstream and then de-adhere in order to migrate out into the tissues. In fact, tumour cells often remain adherent to the undersurface of the basement membrane rather than emigrating deeply (Schor *et al.*, 1985a), and this could be due to their strong adherence to this structure. It is clear from this outline that adhesion is a crucial aspect of the metastatic process: cells almost certainly must adhere at some stage to metastasize but as we shall see, adhesion to host defence cells can also serve to limit the metastatic process.

Both homotypic and heterotypic adhesions within the blood might influence metastatic outcome. The question of immediate interest here concerns whether such adhesions can form at all. Using an aggregation system similar to that described by Evans and Proctor (1978), it is possible to determine average collision times under laminar flow conditions of around $\pi/2G$ (where G is the shear rate). Since aggregates can develop at shear rates of 100 sec^{-1} and higher, it is clear that adhesions can form within the millisecond range. Shear rates of this magnitude are in the physiological range for small blood vessels, and thus these theoretical considerations imply that aggregates may develop within at least some parts of the blood stream. In other studies, Weiss and Harlos (1972) determined the adherence time for Ehrlich ascites tumour cells which had settled onto glass coverslips to be around 1630 sec. Although different cell types were used in the two studies, it maybe that cell–substrate adhesion is a good deal slower than cell–cell adhesion, where any receptors involved could be mobile on both contact surfaces. Indeed, relatively widely spaced substrate adhesive molecules may require activation of cytoskeletal components and cytopodial extension in order to affect a reasonable degree of adhesion, and thus an assessment of the rate of adhesion can become confused with the rate of spreading. Alternatively, cells settling onto artificial substrates such as glass coverslips may have to process the surface to some degree before significant adhesion is possible.

Using a different mathematical approach, Weiss and Dimitrov (1984) have calculated adherence times for sarcoma cells of 0.005–50 sec and typical transit times in capillaries 50–100 μm long of around 0.0007–2 sec. Although there are many assumptions in these theoretical calculations, the results for adhesion and passage times at least overlap in orders of magnitude. Thus the development of cell adhesions while flowing through a small blood vessel would seem entirely possible. In any case, we have seen that in some situations of turbulent flow vortices may be established where cells could become trapped for considerable lengths of time (section 2.9.1). There is also an indication that stasis may develop from vessel occlusion through the blocking effects of the tumour cell, with or without platelet aggregation and thrombus development. Interestingly, Weiss and Dimitrov (1984) have also shown that passage through small blood vessels is probably the source of a large amount of malignant cell death, and this probably contributes significantly to metastatic inefficiency. They proposed that cell death arises as a consequence of the generation of tension at the malignant cell surface as it passes through a narrow capillary. When this tension, generated by the friction of passage, exceeds some critical value the plasma membrane ruptures and cell death results.

5.7.1 Homotypic adhesion

Although much has been claimed of the original observations of Ludford (1932), Cowdry (1940) and Coman (1944, 1953, 1961) that malignant cells are often less adherent than their normal counterparts, we have seen that there is little reason for accepting this as being universally true. With respect to metastasis, a number of workers have suggested that cells capable of a high degree of lung colonization following tail vein injection adhere homotypically to a greater extent than cells which colonize these organs less (Nicolson and Winkelhake, 1975; Winkelhake and Nicolson, 1976; Fidler and Nicolson, 1978; Raz *et al.*, 1980). This also appears not to be universally applicable since other workers have either found no such relationship or actually found the reverse to be the case. Elvin and Evans (1984), for example, found B16F1 cells to be more homotypically adhesive than B16F10 cells at all conditions except those where cell contact was minimal. These results, based on aggregation rate, were confirmed by analysis of aggregate distribution profiles. Thus parental B16F1 cells were found to form more aggregates of larger size classes than high lung colonizing B16F10 cells aggregated under identical conditions. Hausman (1983) examined the homotypic aggregation ability of three Morris hepatoma cell lines which differed in their metastatic abilities. He found that a line which rapidly metastasized to lung in the buffalo rat failed to aggregate, one that metastasized less rapidly showed poor aggregation, and one which did not metastasize at all aggregated considerably. Fusion of plasma membranes from the tumour cells which aggregated well to those which failed to aggregate conferred aggregating potential on the recipients, but whether this affected metastasis or not was unfortunately not explored. Further study by Elvin and Evans (1985) led to the isolation of a number of clones of both B16F1 and B16F10 melanoma variants. These authors tested the clones for adhesiveness as assessed by their aggregation performance under controlled physical conditions, and then measured their lung colonizing ability after tail vein injection into syngeneic mice. No correlation between homotypic adhesiveness and lung colonizing ability emerged from this study. We are thus obliged to conclude that although some reports may show a relationship between these two processes, such a correlation is without general applicability.

A rat metastatic model system based on the BSp73AS tumour and its variant BSp73ASML has been studied by Werling *et al.* (1985). Both of these tumour lines form lung colonies after tail vein injection, but only the ASML variant can metatasize to the lungs (via the lymphatics) after subcutaneous injection. Analysis of homotypic aggregation kinetics has shown that the ASML metastatic variant can aggregate faster than the non-metastasizing AS form, though after about 3 h of aggregation there was a decline in the percentage of aggregated ASML cells from about 75% to 40%. Aggregation of the AS type peaked at about the same level after 5 h of aggregation, and the percentage of aggregated cells remained high over the next 45 h. SEM observations of aggregated cells showed the AS type to form tight clusters whereas the ASML metastatic variant formed looser clusters of rounded cells. Werling and colleagues (1985) concluded from their observations that the metastatic variant was less homotypically adhesive than the non-metastasizing variant, although with complex aggregation kinetics it is difficult to interpret how the differences in adhesive behaviour between the two cell lines might influence their metastatic capabilities.

5.7.2 Heterotypic adhesion

For tumour cells spreading via the blood, it might be supposed that heterotypic interaction with the endothelium is likely to be of considerable importance. However, tumour cells may also interact while within the circulation with various host defence cells, including NK cells, lymphocytes and monocytes. A number of studies have shown that circulating tumour cells are often trapped in heterotypic emboli along with cells of the immune system (Liotta, Kleinerman and Saidel, 1976; Glaves, 1983). According to Glaves (1983), as many as 75% of circulating Lewis lung emboli contain attached leukocytes, the majority of which are lymphocytes, although monocytes and rare PMNs may also be involved. Such multicellular emboli are generally fairly small, with most distributed in the 5–15 cell range. Existing evidence would suggest that these heterotypic emboli can either contribute positively or negatively to the metastatic process. Thus Schirrmacher and Appelhans (1985), for example, have shown that peritoneal macrophages bind more to low-colonizing Eb lymphoma cells than to high-colonizing Esb cells. The macrophages exert a cytostatic effect on Eb but not Esb cells, and it is postulated that this might contribute to the differences in organ colonizing ability of the two lympoma variants. Orr and Mokashi (1985) have shown that tumour cells can be incorporated into PMN aggregates induced by the chemoattractants FMLP or C5a. Unlike the study of Schirrmacher and Appelhans (1985), however, aggregation with PMNs appeared not to have any adverse effects on the tumour cells, even though these cells can yield products cytotoxic for tumour cells when activated (Clark and Klebanoff, 1979; Glaves, 1983). Indeed, the authors hypothesized instead that such aggregation might actually increase metastatic spread by enhancing the trapping of cells and by causing vascular damage. Starkey and her colleagues (1984 a and b) have provided evidence from *in vitro* studies that the presence of peritoneal exudate PMNs or activated macrophages (but not lymphocytes or resident peritoneal macrophages) can lead to an increase in the number of RT7-4bs rat hepatocarcinoma cells which bind to bovine aortic endothelial monolayers. Furthermore, heterotypic aggregates of tumour cells with PMNs were found to be retained more in the lungs of rats after i.v. injection compared to that for homotypic tumour aggregates. Noteworthy, the heterotypic aggregates also formed more lung colonies after injection compared to homotypic aggregates. Although peritoneal exudate PMNs and activated macrophages enhanced tumour cell adhesion to the endothelium, a degree of adhesion was also noted in the complete absence of leukocytes, and thus there would seem to be no obligatory role for leukocytes in tumour cell adhesion to the endothelium.

Several experimental approaches have examined the adhesion of tumour cells to organ cell suspensions, or to organ fragments. In many of these examples, the precise nature of the host cell involved cannot be identified. Nevertheless, there is some indication that results from these types of assays reflect patterns of *in vivo* metastasis. Schirrmacher *et al.* (1980), for example, have shown that the ability of liver-colonizing lymphoma variants to adhere to hepatocytes is a reflection of their ability to colonize that organ. Similarly, Phondke and colleagues (1981) have shown that leukaemic cells which colonize the spleen in preference to the lungs adhere to the former and not to the latter. Netland and Zetter (1984) made use of the cryostat section technique of Stamper and Woodruff (1976) to show that lung-colonizing B16F10 melanoma cells and liver-colonizing reticulum cell sarcoma (M5076) cells adhered preferentially to syngeneic lung and liver respectively. The binding of B16F10 cells to cryostat sections of lung was markedly inhibited by treatment with pronase or tunicamycin, suggesting either direct or indirect glycoprotein involvement in adhesion. Interestingly, treatment with trypsin or neuraminidase showed only mild inhibitory effects.

Although it is difficult to conceive how heterotypic adhesion to stromal cells might influence lodgement, they may nevertheless effect metastatic outcome through contributing to cell retention and survival. The two murine leukaemic cell lines L1210 and P388 were shown by Kamenov and Longenecker (1985) to colonize the bone marrow of recipient animals after tail vein injection. These authors reported that cells from both lines adhered to normal bone marrow cells *in vitro* and *in vivo*, and they also adhered to stromal bone marrow cells prepared from Dexter (long-term bone marrow) cultures. The argument put forward was that this adhesive interaction might underly leukaemic cell retention in the bone marrow as is seen, for example, in acute lymphocytic leukaemia. There is some question, however, as to whether the L1220 and P388 murine cell lines are fully appropriate for studying this type of leukaemia.

5.7.2.1 *Endothelial cell–tumour cell interaction*

Although malignant tumour cells which spread via the blood have to become lodged in a vessel before they can extravasate, we have seen that this lodgement need not necessarily involve adhesive interaction between the tumour cells and the endothelium. Nevertheless, there is considerable evidence that some tumours at least show preferential adhesion to endothelial cells derived from particular target organs. Furthermore, a number of studies have shown that enhanced adhesion to endothelial cells correlates wih increased metastatic ability as assessed by organ colonization techniques. Takenaga and Takahashi (1986), for example, treated P-29 cells (derived from a low metastatic clone of the Lewis lung carcinoma) with the tumour promoter 12-O-tetradecanoyl-phorbol-13-acetate (TPA) and measured the effects of treatment on adhesion and lung colonization. They found that treatment with TPA enhanced P-29 cell adhesion to both Petri dish surfaces and to bovine pulmonary artery endothelial cells, and it also significantly promoted the number of lung colonies which developed after i.v. injection. These changes in adhesive and colonizing behaviour were coincident with an increase in phosphorylation of a 54 kD cellular protein. Like other phorbol esters, TPA binds to protein kinase C and many of its effects are probably mediated through the activities of this enzyme. The synthetic diacylglycerol 1-oleoyl-2-acetylglycerol (OAG) is known to activate protein kinase C directly, and when added to P-29 cells this too caused an increase in phosphorylation of the 54 kD protein. Exogenous phospholipase C (PLC), which can generate diacylglycerol from hydrolysis of phosphatidylcholine (section 1.7.22), had similar effects to treatment with OAG. Both these agents, however, were without effect on adhesion and metastasis, so the significance of phophorylation of the 54 kD protein is not clear. This illustrates an important point central to much of the confusion which surrounds a good deal of experimental cancer research, namely that simple correlations are not indicative of cause or effect. In most experimental studies an association of one sort or another may be identified, but it is usually the case that more extensive research is required before definitive evidence is obtained.

It is generally recognized on structural and physiological grounds that endothelial cells of different organs can subserve different functions. This suggests that the endothelial cells of the body are not all alike, although clearly they share many aspects since they have a common derivation. Auerbach and his colleagues (1985) have examined a small panel of capillary endothelial cells for antigen expresssion using enzyme linked immunosorbent assay (ELISA) techniques, and they have found that such cells can express organ-specific antigens. The ELISA technique is simply an immunological assay for antigens in which enzymes such as peroxidase or phosphatase are coupled to antibodies which bind the antigen. The presence of bound antibody is then detected by using enzyme substrates which yield a coloured reaction product.

In many cases, specific endothelial antigens have not yet been properly identified, but in others the nature of the antigen is clear. Brain endothelial cells from the mouse, for example, express brain-related Thy-1. The mouse Thy-1 (or theta) antigen is actually a marker for certain T lymphocytes, but it is known to be expressed on brain cells. That only brain-derived capillary endothelial cells express this marker suggests that environmental cues (perhaps during development) lead to the expression of particular cell surface markers. Whatever the reasons, brain capillary endothelial cells are immunologically different from the capillary cells of other organs, such as the ovary. Previously, Alby and Auerbach (1984) were able to show that glioma cells and ovary-derived teratoma cells had different patterns of adhesiveness to these two capillary types. What antigens might be responsible for these adhesive differences, however, is not known although it would appear not to be MHC-related since adhesive combinations involved cells from various mouse strains.

The ovarian teratoma line used by Alby and Auerbach (1984) had a preference for metastasizing to the ovaries after either s.c., i.p. or intracardiac (left ventricle) injection. In adhesion studies, cells from this line bound preferentially to ovary-derived rather than brain-derived capillary endothelial cells. Similarly, cells from the glioma bound preferentially to brain-derived rather than ovary-derived capillary endothelial cells. Although the brain is the normal site of glioma growth, the authors had no data on the metastatic capabilities of this tumour. These results suggest that tumour cells can recognize and adhere to particular organ-related antigens on endothelial cells. Furthermore, the fact that in one case this correlates with *in vivo* metastasizing behaviour provides circumstantial evidence that adhesive interactions between certain malignant cells and the endothelium of particular organs may contribute to preferential patterns of tumour dissemination. In a more recent and more extensive study, Auerbach and colleagues (1987) examined the adhesiveness of a panel of mouse tumour cells to brain, liver, lung and thoracic duct endothelium of mouse origin, and to bovine aortic endothelium. These results confirmed that tumours show differences in their adhesive preferences to endothelia, but the authors did not extend their previous observations on a possible correlation between metastasis and selective endothelial adhesion. Interestingly, C755 mammary adenocarcinoma cells adhered more to thoracic duct endothelial cells than did sarcoma 180 cells, but what significance this might have to the lymphatic distribution of carcinomas compared to that for sarcomas is uncertain (section 2.5). In fact, neither the adenocarcinoma or the sarcoma cell types showed any preference for adhesion to thoracic duct endothelial cells relative to that towards bovine aortic endothelial cells and, furthermore, the metastatic profiles of the tumour lines used are not clear.

As Alby and Auerbach (1984) point out, the adhesion assay they used in their study has considerable limitations since the results are influenced by a number of parameters, some of which are difficult to keep constant. Nevertheless, the consistency they achieved in their study suggests that the differences in tumour cell adhesion to endothelial cells that they reported may be of some relevance to the metastatic process. Unfortunately, the experimental control obtainable with *in vitro* experiments is not obtainable *in vivo*, and thus we can only speculate on the true significance of these results. It is worth noting that many *in vitro* models of tumour cell adhesion to the endothelium use endothelial cells derived either from the bovine aorta or the human umbilical vein. The organ specificity suggested for endothelial cells by Auerbach *et al.* (1985) questions the relevance of these studies. However, it is apparent that all of these studies relate to a model system of some sort, each of which is not without its own set of limitations (section 3.1). Regardless of the nature of the endothelial cells used, information of value is still likely to be gained.

In vitro experiments by Kramer and Nicolson (1979) suggest that the initial arrest of tumour cells (regardless of how it occurs) may lead to endothelial retraction thereby exposing subendothelial components, which in turn could trap more tumour cells and activate inflammatory cascade reactions. Kramer, Gonzalez and Nicolson (1980) showed that B16F1 and Hs939 melanoma cells (of mouse and human origin respectively) adhered relatively slowly to BAE monolayers, and much more rapidly to their underlying matrix. They suggested that fibronectin, which is largely absent from the endothelial cell surface *in vitro* (Birdwell, Gospodarowicz, and Nicolson, 1978) and *in vivo* (Linder *et al.*, 1978), was probably responsible for the rapid attachment to the matrix. It would seem, however, that any one of several molecules including laminin, collagen, elastin, thrombospondin and fibronectin may be involved in cell adhesion to the matrix. Furthermore, the prime site for dictating selectivity of metastasis is likely to involve interaction between the surfaces of the endothelium and the circulating tumour cells, rather than between the tumour cell surface and the ECM or basement membrane components.

Little is known of the molecular species which might mediate selective binding of tumour cells to the endothelium. Starkey *et al.* (1984 a and b) have shown that simple sugars such as 60 mM α-methyl mannoside and 60 mM *N*-acetylglucosamine can partially inhibit the adhesion of rat hepatocarcinoma cells to bovine aortic endothelial cells. This inhibitory activity was found in serum-free, low-glucose conditions and thus its physiological relevance is questionable. Nevertheless, as indicated earlier, these authors provided evidence for at least two processes by which tumour cells may adhere to the endothelium, one directly and the other indirectly via peritoneal exudate PMNs or activated macrophages. Since PMN and macrophage adhesion to the endothelium was not affected by simple sugars, it would seem that different molecular mechanisms are probably responsible for the two processes.

A different approach, based on bioautography (section 3.4.8.2), has been used by Nicolson and colleagues in an attempt to identify endothelial determinants involved in tumour cell adhesion (reviewed in Nicolson, 1988). These authors studied the adhesive and invasive behaviour of variants of the RAW117 murine lymphoma cell line. The parental cells of this line (RAW117-P) metastasize relatively poorly, whereas RAW117-H10 cells metastasize preferentially to the liver, and RAW117-L17 do so to both the liver and the lungs. When tested for selective adhesion, the variants behaved more or less according to their metastatic abilities. Thus RAW117-L17 cells displayed a high rate of adhesion to lung endothelial cells, and both these cells and RAW117-H10 cells adhered at a high rate to liver sinusoidal endothelial cells.

The parental RAW117-P cells, on the other hand, adhered relatively poorly to both murine lung and liver endothelial cells. When the RAW117 variants were tested for their ability to invade syngeneic lung, liver and heart fragments, the parental line invaded all three only moderately after 48 h, the RAW117-H10 variant invaded the liver preferentially, and the RAW117-L17 variant invaded both the lung and the liver tissues extensively, and the heart only moderately. Again, these results are more or less in accord with the preferential metastasizing ability of these lymphoma variants.

This distinctive behaviour of the RAW117 variants makes them ideal candidates for detailed studies of the molecular basis of tumour cell–endothelial cell interaction. Nicolson and his colleagues first subjected three types of endothelial cells (BAE, murine lung and liver) to detergent extraction, separated the extracted components on PAGE gels, and then Western blotted them onto nitrocellulose. The blots were then probed with parental RAW117-P cells and the liver metastasizing RAW117-H10 variant. In general, the results were in accord with the metastasizing abilities and preferences of the two variants. Thus the RAW117-H10 cells bound more to the separated extracts than did their parental cells, and they also bound to the various endothelia in their order of metastatic preference (i.e. liver>>lung>>BAE). Five glycoproteins (of M_r 20 kD, 25 kD, 30 kD, 32 kD, and 48 kD) were identified as the major endothelial determinants responsible for tumour cell binding.

In order to identify possible tumour cell determinants involved in the preferential binding to lung and liver endothelial cells, Nicolson and colleagues next made use of their observation that the RAW117 variants bound poorly to separated BAE components. Thus detergent solubilized RAW117 cell extracts were pre-adsorbed on fixed BAE cells to remove non-specific determinants, and then they were allowed to bind to fixed liver endothelial cells. Using this simple selective strategy, Nicolson's group have identified eight glycoproteins (of M_r 14 kD, 18 kD, 20 kD, 49 kD, 60 kD, 90 kD, 180 kD, and 390 kD) on RAW117 cells which bind selectively to murine liver endothelial cells.

At the present time, little more is known of the identities and precise roles of the various determinants. It is not clear, for example, whether the RAW117 determinants bind selectively to the identified endothelial determinants, and nor is it clear that any of these molecules actually play a major role in selective adhesion. Recent observations indicate that RAW 117-endothelial cell adhesion is not inhibitable by GRGDS, so that a prominent role for many of the major integrin superfamily receptors can be ruled out (section 5.8).

These results contrast somewhat to those of Ferro *et al.* (1988) who have shown that B16F10 melanoma cells, selected originally for their ability to colonize the lungs, nevertheless adhere rapidly and extensively to BAE cells. Using a similar bioautographic approach to that of Nicolson's group, the former authors identified a BAE component of extremely high molecular mass to which the melanoma cells adhered (section 3.4.8.2). Since different experimental systems have been used by the two groups it is not appropriate to make comparisons, but the twin technical problems of complex formation and protein degradation should be borne in mind. Independent studies by Rice and Bevilacqua (1989) have shown that activation of human umbilical vein endothelial cells by various cytokines such as tumour necrosis factor (TNF), interleukin-1 (IL-1), or endotoxin induces the expression of at least two cell surface receptors which are adhesive for particular tumour cell lines. One receptor, described as an inducible cell adhesion molecule of M_r 110 kD (INCAM-110), mediates the adhesion of both human and murine (B16F10) melanoma cell lines to activated endothelium. The other induced molecule studied by these authors promotes the adhesion to activated endothelium of HT-29 human carcinoma cells but not that of melanoma cells. This molecule is in fact identical with ELAM-1, known from other studies to be involved in the adhesion of neutrophils to activated endothelium (section 2.12.1). That tumour cells can utilize molecules normally employed in the extravasation of non-transformed cells probably comes as no surprise given what we now know of their flexibility. Since an antibody against INCAM-110 does not completely inhibit B16F10 adhesion to activated endothelium, other endothelial determinants of similar function may remain to be discovered. Nevertheless, because tumour cells can themselves secrete cytokines the possibility exists that they may induce adhesive receptors on endothelial cells as they circulate in the blood or after they become trapped mechanically in small blood vessels (section 2.13.5).

5.7.2.2 *Platelet–fibrin–tumour cell interaction*

The endothelial lining of blood vessels normally represents a non-thrombogenic surface. This is achieved by the secretion of prostacyclin (which inhibits platelet interactions) and plasminogen activator (which yields the fibrinolytic enzyme plasmin), and by synthesis of the cell surface glycoprotein thrombomodulin (which binds thrombin and alters its substrate specificity). When the endothelium is injured and sub-endothelial components are exposed, however, the rapid adhesion and aggregation of platelets ensues. The terms **adhesion** and **aggregation** have precise meanings, particularly with respect to platelet interactions: thus platelet adhesion refers to their interaction with a solid substrate, whereas platelet aggregation refers usually to platelet–platelet (but also platelet–tumour cell) interaction. At least experimentally, platelet aggregation appears to have two phases:

(a) The primary phase. This is the reversible phase of aggregation which follows shape change and pseudopodia formation after stimulation with agents such as ADP, thrombin or collagen.
(b) The secondary phase. This is the irreversible aggregation phase which occurs following α-granule secretion induced by stronger stimulation. The molecular events involved in this phase are not clear, but they are probably mediated by the four major adhesive proteins secreted by platelets: fibrinogen, fibronectin, von Willebrand factor and thrombospondin (see below).

The interactions of platelets are markedly affected by certain products of arachidonic acid metabolism, particularly thromboxane A_2 (TXA_2) and the prostaglandins (reviewed in Gorman, 1979). An outline of some of the pathways involved in arachidonic acid metabolism is provided in Fig. 5.9. Under the influence of a cyclo-oxygenase, arachidonic acid may be metabolized to endoperoxide intermediates (PGG_2 and PGH_2). These in turn may be metabolized either to the prostaglandins PGD_2, PGE_2, or $PGF_{2\alpha}$ of a cyclic prostaglandin (PGI_2) and its metabolites, or to TXA_2 and its metabolite thromboxane B_2 (TXB_2). Alternatively, under the influence of 5-lipoxygenase, arachidonic acid may be metabolised to 5-hydroperoxy-6,8,11,14-eicosatetraenoic acid (5-HPETE) which is then converted to its epoxy acid, the leukotriene LTA_4, and subsequently to LTC_4 by the addition of glutathione, to LTD_4 by the loss of glutamic acid, and to LTE_4 by the loss of glycine. LTB_4 is formed separately from LTA_4 by the addition of water (reviewed in Marx, 1982). LTB_4 is noteworthy in that it is an extremely potent chemoattractant for PMNs, and it also plays a major role in immunoregulation (Rola-Pleszczynski, 1985). LTB_4 as well as lipoxin A and LTA_4 have been shown to enhance NK cell cytotoxic activity, although the latter two molecules are less active in this respect. Another lipoxygenase can yield 12-hydroperoxy-eicosatetraenoic acid (12-HPETE) and its hydroxy metabolite (12-HETE), but these need not concern us here except in so far as to note that 12-HETE is a leukocyte chemoattractant.

Briefly, in prostaglandin nomenclature, the letters A to I indicate the ring structure and the subscript represents the number of unsaturated double bonds in the side chains. Prostaglandins with the subscript 1 are derived from the unsaturated fatty acid dihomo-γ-linolenic acid, whereas those with the subscript 2 are derived from arachidonic acid.

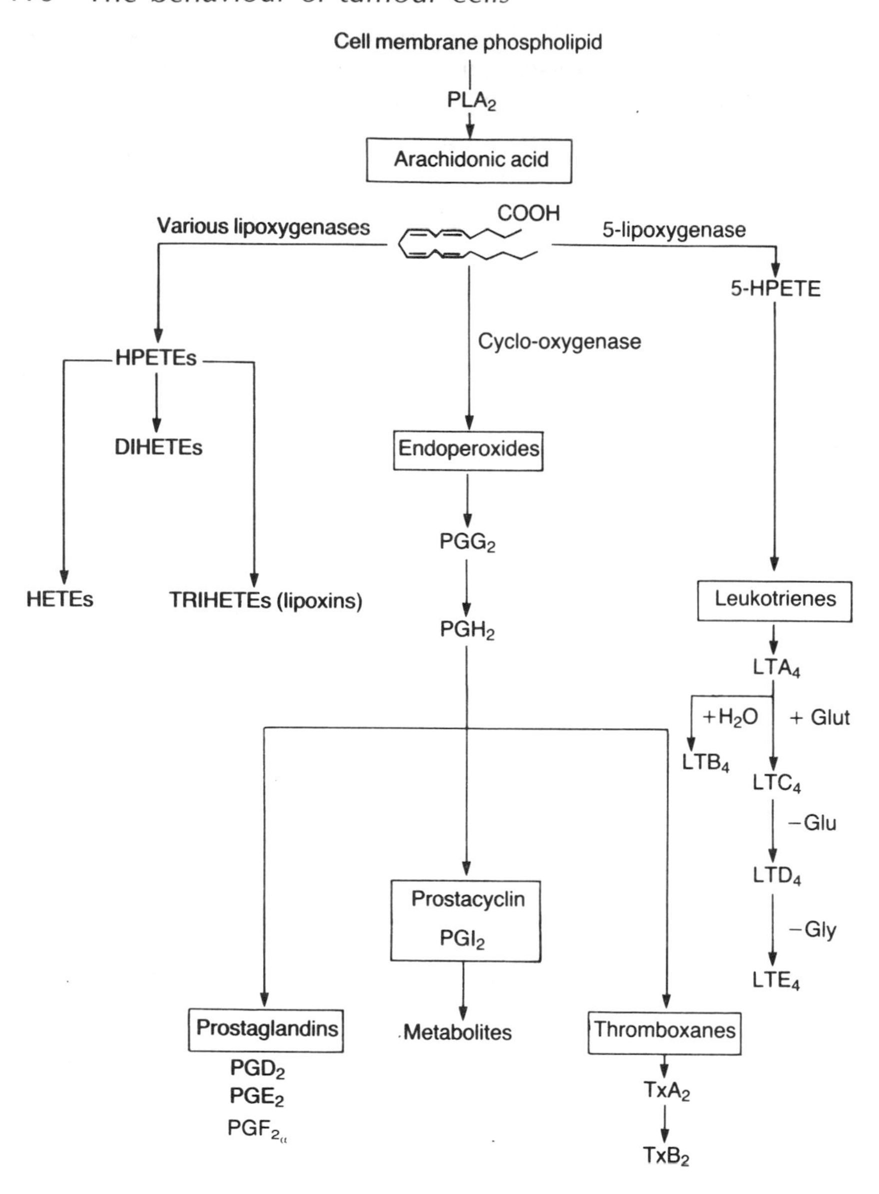

Figure 5.9 The metabolism of arachidonic acid. Arachidonic acid, obtained from plasma membrane phospholipids, is metabolized to yield a number of products with potent physiological effects. Its metabolites include the leukotrienes, the endoperoxides, the prostaglandins, the cyclic prostaglandin PGI_2, and the thromboxanes. Refer to the text for abbreviations and an explanation of the pathways involved (section 5.7.2.2). Note that the sequence of events depicted in this diagram can be incorporated into Fig. 1.4b summarizing the phosphatidylinositol-

Note that the different derivatives have different functions, and attention must be paid to the subscripts! The α and β subscripts in the F series relate to stereospecificity. Prostaglandin activity was first identified in semen, and it was assumed that the active principle originated from the prostate gland. In fact, the seminal vesicles are probably the main source of the prostaglandins present in semen, but by the time this was known it was too late to change the name! The leukotrienes (LTs) are so-called because they were originally shown to be made by leukocytes and they all have three conjugated double bonds: here the subscript refers to the total number of double bonds present in each molecule.

TXA_2 has potent platelet-aggregating activity and it is also a vasoconstrictor, even though it has a half-life of about 30 sec. Conversely, prostacyclin (PGI_2) is a potent inhibitor of platelet aggregation and is a vasodilator. Although other molecules are involved, it seems that TXA_2 and PGI_2 between them markedly influence platelet interactions. There is, however, an interesting twist in the story, for it seems that despite utilizing the same precursor molecule, TXA_2 is produced more by platelets than by the endothelium, and PGI_2 is produced more by endothelial cells than by platelets. Thus platelet activation could yield a molecule which induces aggregation, but its effects could be modulated by an anti-aggregatory product of the endothelium.

Both tumour cells and their shed vesicles contain substances which have procoagulant activity (Dvorak *et al.*, 1981) and can induce platelet aggregation (Gasic *et al.*, 1978). The adhesive interactions between tumour cells and platelets (and/or fibrin) are likely to influence metastasis in a number of ways (reviewed in Gasic, Tuzynski and Gorelik, 1986). These may be summarized as follows:

(a) The tumour cells may interact with platelets and/or fibrin while in the circulation and the resulting heterotypic aggregates could enhance the possibility of mechanical entrapment. It has been reported that thromobogenic tumours injected into thrombocytopenic mice form fewer lung colonies than they do when injected into normal mice (Gasic, Gasic and Stewart, 1968).

based intracellular signalling pathway, and that different pathways of arachidonic acid metabolism are favoured in different cells. Macrophages and PMNs produce predominantly 5-lipoxygenase metabolites, platelets yield mainly 12-lipoxygenase metabolites, while lymphocytes on the other hand yield minimal lipoxygenase metabolites. In relative terms, thromboxane production is favoured in platelets whereas prostacyclin production is favoured in the endothelium. The so-called 'slow-reacting substance of anaphylaxis' is a mixture of LTC4, LTD4 and LTE4.

Furthermore, anti-platelet serum has been show to reduce the extent of lung colonization and spontaneous lung metastasis of some experimental tumours (Pearlstein *et al.*, 1984). Skolnik *et al.* (1984) reported that lodgement of rat fibrosarcoma cells in the livers of experimental animals after intraportal injection was reduced to 53% following treatment with anti-platelet serum. Not surprisingly, thrombocytopenic rats showed a corresponding decrease in levels of the platelet product 5-hyroxytryptamine (5-HT). When rats were treated with ketanserin (a 5-HT antagonist with selective action on 5-HT_2 receptors on platelets and the endothelium) tumour cell lodgement was reduced to 72%. The authors concluded from these results that 5-HT (released primarily from activated platelets) probably played a role in tumour cell lodgement within the vascular system of the liver, and that this subsequently influenced metastatic outcome. In support of this conclusion, the authors also reported that intraportal injection of tumour cells resulted in a two-fold increase in 5-HT levels in the liver when compared to the injection of saline as a control. The precise role of 5-HT in metastasis remains uncertain, however, since it is not a particularly powerful platelet aggregating substance. It may operate more indirectly by stimulating leukocyte adhesion to the endothelium thereby inducing endothelial damage which could promote the trapping of circulating tumour cells. It is worth noting that whereas inflammatory reactions in rats and mice are particularly sensitive to 5-HT this is not the case in humans, and thus its clinical relevance requires further evaluation.

Mechanical entrapment cannot explain the selectivity of metastasis displayed by some tumour types (section 2.14.6). Of interest in this respect is the observation that tumours which aggregate platelets tend to produce more metastases in the first capillary bed they meet than do tumours which do not aggregate platelets. Studies with i.v. injected radiolabelled B16 melanoma cells, however, have shown that thrombocytopenia does not affect initial (5 min) cell entrapment (Gasic *et al.*, 1973). Furthermore, according to Chopra *et al.* (1988), W256 tumour-induced platelet aggregation actually occurs after malignant cell arrest in this particular experimental system. Despite these observations with W256 cells, in other experimental models tumour cell–platelet–fibrin complexes can be seen within minutes of the intravascular injection of tumour cells, and thus the short term involvement of these components in the mechanical trapping of some tumours cannot be rigorously excluded. Given these conflicting results, it is obviously difficult to generalize on the role of platelets in the trapping of tumour cells.

Of interest in the study of Chopra *et al.* (1988) was their report that W256 cells display a membrane protein immunologically related to gpIIb/IIIa, which is thought to be involved in platelet aggregation (sections 5.7.2.3 and 5.8). Since W256-induced platelet aggregation appears to operate only after tumour cell lodgement, the gpIIb/IIIa-like receptor may be functionally cryptic until tumour cell adherence to the endothelium occurs. A number of other tumours of both human and rodent origin (including a B16 amelanotic variant) have been shown to contain mRNAs for the gpIIb/IIIa related protein.

(b) The platelets and/or fibrin may act as a bridging structure between circulating tumour cells and the endothelium or basement membrane thereby facilitating their arrest. Available evidence suggests that most circulating tumour cells have only a fleeting involvement with platelets and fibrin, and generally speaking there is litle evidence that these components act as bridges between the tumour cells and the endothelium or basement membrane (Crissman *et al.*, 1985). Nevertheless, it should be pointed out that platelets can adhere to the three principal components of the sub-endothelial matrix (laminin, collagen and fibronectin) via a variety of receptors including VLA6 (laminin), gpIa/IIa (collagen), and gpIIb/IIIa and VLA5 (fibronectin). Note that gpIa/IIa (VLA2) is a laminin receptor on some cells but a collagen receptor on platelets (Elices and Hemler, 1989). Although some platelet adhesion interactions (such as that involving gpIIb/IIIa) are dependent on platelet activation, other interactions such as that to laminin mediated by VLA6 are not.

(c) The platelets could assist in breaching of the endothelium and/or the basement membrane by extravasating tumour cells. This possibility is based on tumour cell activation of the platelet release response. Platelets can release an array of cell products, some of which (e.g. histamine) could lead to alterations in the patency of the endothelium thereby enhancing tumour cell extravasation. Other products (e.g. fibronectin, thrombospondin) may activate the adhesive and/or locomotory machinery of the tumour cells. Platelets also release proteases (including collagenase and cathepsin B) and glycosidases (including a heparin sulphate endoglycosidase), roles for all of which have been implicated in tumour cell invasion.

(d) Platelets and fibrin may be deposited as a cocoon about the tumour cells and this could be protective against host defence systems. For some tumours at least, such an effect may last less than 24 h since within this time tumour-associated platelet/fibrin masses tend to disappear. Tumours with more stable platelet/fibrin cocoons are obviously likely to benefit more if this scenario is relevant. Evidence that fibrin may protect tumour cells from host-mediated cytotoxic events is provided by recent experiments reviewed in Gasic, Tuszynski and Gorelik (1986). It seems that the anticolonization effects of anticoagulants are dependent on the activity of NK cells. Thus the effects of heparin or warfarin on lung colonization by B16F10 cells are negated if the host mice are pretreated with agents, such as anti-asialo-G_{M1}, which depress NK cell activity. Furthermore, although both anticoagulants and NK cell activators, such as poly(I:C), can cause a significant reduction in lung colonization, the two classes of agents can act synergistically, thus reducing lung colonization even further.

(e) As an alternative to the process described immediately above, a platelet-fibrin coccoon may trap tumour cells within it thereby preventing their release and the establishment of secondaries unless they have some means of escape. The balance between these opposing possibilities is likely to be different for different tumours, and thus the outcome in terms of enhanced or restricted metastatsis is difficult to predict. Fibrin may be digested under the influence of enzymes either secreted or activated by tumour cells. Fibrinolysis is probably continually active *in vivo*, and indeed we may view the coagulation and fibrinolytic systems to be normally in dynamic equilibrium. We have seen that tumour cells can produce the fibrinolytic enzyme plasminogen activator in either the urokinase- or the tissue activator-forms and this may help during invasion. According to Dang *et al.* (1985), fibrinogen fragment D can disrupt cultured endothelial cell monolayers and the production of this lytic product via plasminogen activator produced from tumour cells might help extravasation. Weimar and Delvos (1986) unfortunately were not able to repeat these observations, and they suggested instead that endothelial disorganization occurs when a suitably adhesive material (perhaps a tumour–platelet thrombus) forms on the surface of endothelial cells *in vitro* or within a vessel lumen *in vivo*. Local endothelial disorganization (coupled with the phagocytic and invasive behaviour of endothelial cells) might represent an early step in the dissolution of an occluding thrombus, and this situation could be capitalized on by the malignant tumour cells incorporated into the thrombus.

(f) Both fibrin and platelet products may be angiogenic leading to tumour vascularization and subsequent growth stimulation.

(g) Platelets adherent to tumour cells may release PDGF which could enhance the growth of the tumour.

Lerner *et al.* (1983) attempted to classify the processes by which tumour cells might induce platelet aggregation. The precise molecular mechanisms unfortunately are not yet clear and the system requires further study. We may summarize the situation as follows:

(a) Tumour cells may either release products such as ADP or have on their surfaces certain determinants which might act by:
    1. Directly aggregating platelets.
    2. Inducing the secretion of aggregatory agents from within platelets.

(b) Tumour cells may release or have on their surfaces products which activate the clotting system leading to platelet aggregation. There are several ways by which tumours might achieve this:
    1. Indirect activation. Tumour cell-induced damage to the endothelium could lead to the exposure of thrombogenic surfaces such as the basement membrane.
    2. Direct activation.
    (i) Tissue factor activity. Tumours have been shown to contain tissue factor-like activity (probably a phospolipoprotein) which activates factor VII in the coagulation cascade. These in turn activate factor X (Fig. 5.10). Note, however, that Kadish and colleagues (1983), found that both normal and neoplastic cells had tissue factor activity, and they were unable to find any correlation between its activity and transformation.
    (ii) Direct activation of factor X. Some tumour cells may directly activate factor X, thereby short-circuiting factor VII involvement. Falanga and Gordon (1985) have isolated a cysteine protease (M_r 68 kD) from tumour cells capable of this activity.
    (iii) Prothrombinase assembly. Some tumour cells (and their shed vesicles) appear to present phospholipid surfaces suitable for the assembly of the prothrombinase complex (composed of factors Xa, V, prothrombin and Ca^{2+}) essential for the conversion of prothrombin to thrombin (Dvorak *et al.*, 1981).

The studies of Tohgo *et al.* (1984) confirm that tumours have at least two ways of aggregating platelets. Thus B16 and 3LL tumour cells induce platelet aggregation essentially via the clotting cascade, whereas cells from other tumours such as MH134 (a hepatoma of C3H mice)

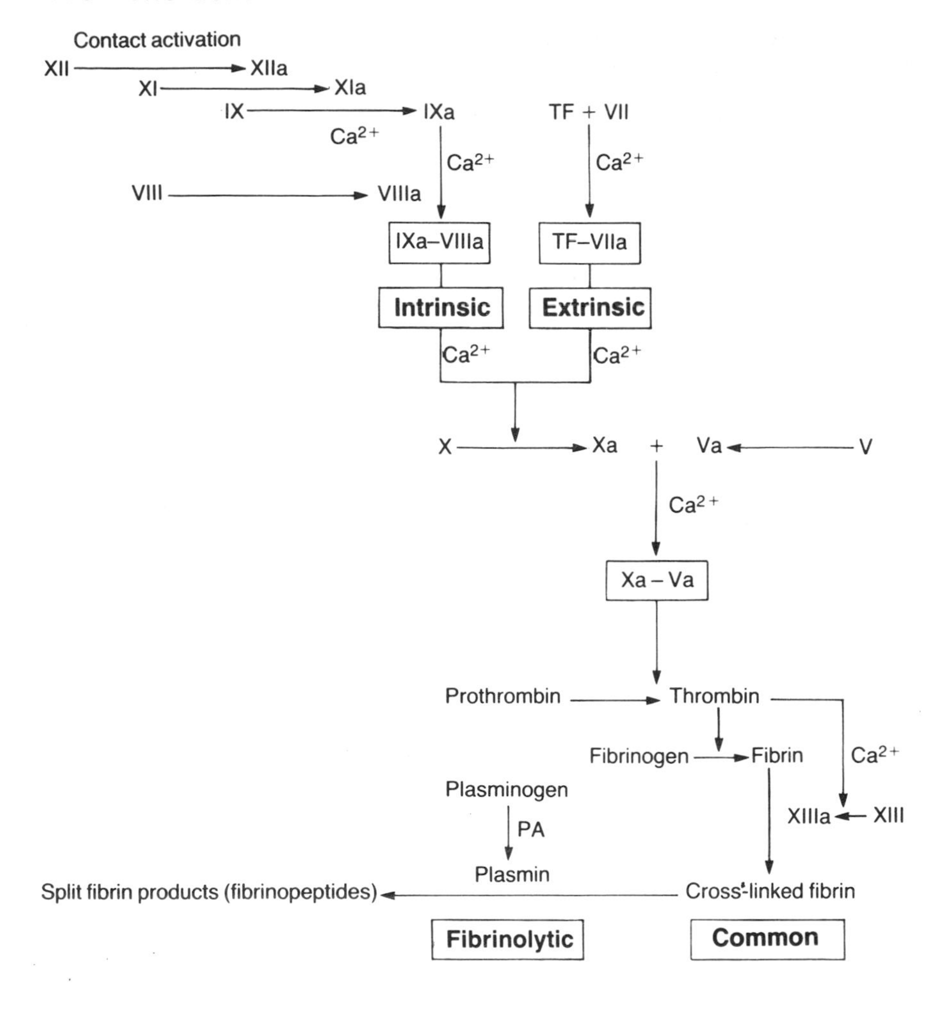

Figure 5.10 The coagulation cascade (after Dvorak, Senger and Dvorak, 1983). The coagulation cascade is composed of intrinsic and extrinsic pathways which unite in a common pathway leading to fibrin deposition. The intrinsic pathway begins with the activation of factor XII which is brought about through contact with surfaces such as glass or collagen. Factor XIIa is then involved in the activation of factor XI, whose activated product (in association with Ca^{2+}) subsequently activates factor IX. Factor IXa forms a complex with activated factor VIII on membrane surfaces and this then activates the common pathway by first catalyzing the conversion of factor X to its active form. Factor Xa forms a complex on membrane surfaces with activated factor V. This complex converts prothrombin to thrombin which then cleaves fibrinogen to yield fibrin. Fibrin can then be cross-linked by factor XIIIa. In the fibrinolytic pathway, fibrin is enzymatically split into short fibrinopeptides through the activity of plasmin, which is derived from plasminogen under the influence of plasminogen activator (PA). The intrinsic pathway is not thought to be

are independent of this system and also independent of ADP. Studies with human mammmary carcinoma cell lines suggest, however, that some tumours (MCF-7, T47-D) aggregate platelets via an ADP-dependent mechanism (Camez *et al.*, 1986). ADP-mediated platelet aggregation occurs through the binding of plasma fibrinogen to platelet membrane glycoprotein gpIIb/IIIa. Not surprisingly, tumour cells which aggregate platelets via ADP-dependent mechanisms fail to aggregate platelets from both thrombasthenic and afibrinogenemic patients lacking gpIIb/IIIa and fibrinogen respectively (Camez *et al.*, 1986). The possibility that some tumours may not aggregate platelets at all should not be overlooked, and indeed the human mammary carcinoma cell line MDA-MB 231 may actually inhibit ADP-induced aggregation, possibly through the release of adenosine (Camez *et al.*, 1986).

Attempts to characterize the platelet-aggregating activity of tumour cells began with studies by Pearlstein, Cooper and Karpatkin (1979) and Hara, Steiner and Baldini (1980). According to Pearlstein and his colleagues (1979), the platelet-aggregating material extractable from SV40-transformed 3T3 cells by urea/EDTA treatment is a sedimental sialo-lipoprotein complex. Much smaller quantities of this poorly characterized material are extractable from non-transformed 3T3 cells, but it is not a general feature of malignant cells since different tumours appear to utilize different avenues for inducing platelet aggregation. Interestingly, Pearlstein *et al.* (1980) speculated that the increased

---

particularly important in normal blood coagulation since patients deficient in factor XII lack bleeding problems.

The extrinsic pathway is initiated following blood vessel damage which leads to the expression of tissue factor (TF). Tissue factor is a phospholipoprotein which activates factor VII, and the two molecules together then activate the sequences of the common coagulation pathway.

The activities of factor VIII and factor V are inhibited by protein C, which is activated by thrombomodulin in conjunction with thrombin. Activated protein C, bound to the cell membrane through protein S, functions as an anticoagulant by inactivating factor VIIIa and factor Va.

Tumour products seem to be able to initiate the clotting cascade in a number of ways. Some, for example, act early in the cascade as tissue factors while others seem to be able to activate factor X directly. Other tumour products may act even later in the cascade, perhaps functioning as thrombin-like enzymes. Tumours may also produce products with factor XIIIa-like activity stabilizing fibrin deposits. On the other hand, tumours are known to produce plasminogen activator which, through generation of plasmin, will lead to fibrinolysis. The coagulation and fibrinolytic systems are best envisaged as being in dynamic equilibrium. This equilibrium may shift from fibrin deposition to lysis and *vice versa* , the outcome of which may influence numerous steps in the metastatic cascade.

sialylation of the cell surface components of some tumours may enhance their abilities to aggregate platelets. A sialic acid–nucleoside conjugate (KI-8110) has recently been shown to reduce the platelet-aggregating activity of NL-17 murine colon adenocarcinoma 26 cells, known to have high lung-colonizing ability (Kijima-Suda *et al.*, 1988). Pretreatment of NL-17 cells with KI-8110 inhibits lung colonization apparently through reducing the retention of tumour cells in the lungs of experimental animals. Since KI-8110 is a sialyltransferase inhibitor, these studies suport a possible role for sialic acid in colonization, presumably through involvement in tumour-induced platelet aggregation. Interestingly, KI-8110 appears to inhibit PDGF receptor phosphorylation, thereby having the bonus effect of inhibiting PDGF induced proliferation of NL-17 cells.

Since thromboxane synthetase inhibitors do not block tumour cell-induced platelet aggregation, it would seem that TXA_2 (despite its known platelet aggregatory activities) is not involved in this process. Interestingly, Sloane, Dunn and Honn (1981) suggested that the release of cathepsin B by malignant cells could promote platelet aggregation, which might have a bearing on the correlation of this enzyme with metastatic behaviour. How cathepsin B might exert its effects on platelet aggregation is not clear, but as a cysteine protease it might directly activate factor X.

The *in vitro* aggregation of platelets induced by tumour cells may be inhibited by PGI_2, and Honn *et al.* (1981) have shown that a single i.v. injection of this prostaglandin can reduce B16a melanoma lung colony formation by 70%. Furthermore, the inhibition of PGI_2 synthesis in recipient animals by treatment with 15-HPETE was associated with a relative increase in lung colonies following the subsequent injection of B16a cells. In other experiments, the effects of increasing the endogenous production of PGI_2 on the spontaneous metastasis of B16a and 3LL tumours was explored using nafazatrom. As anticipated, this stimulator of PGI_2 production reduced the extent of metastasis shown in these two model systems. Since PGI_2 appeared to have no significant effect on the initial lodgement of the tumour cells, however, it would seem that its antimetastatic effect operates at some stage after metastatic cell arrest.

Several lines of evidence suggest that Ca^{2+} serves as an intercellular messenger for platelet aggregation. Consequently, Honn *et al.* (1983) have tested the effects of the calcium blocker nifedipine on platelet aggregation and lung colonization. They found that nifedipine inhibited platelet aggregation induced by B16a melanoma cells, and mice pretreated with the drug also had reduced numbers of B16a lung colonies relative to controls. The mobilization of Ca^{2+} from intracellular stores is apparently inhibited by cAMP, and indeed PGI_2 is thought to primarily exert its effects via stimulation of adenylate cyclase. The dipyrimidole derivative RA233 suppresses phosphodiesterase activity which presumably accounts for its inhibitory effects on platelet aggregation. Its effects on metastasis, however, as judged from both experimental systems and preliminary clinical trials are somewhat contradictory. In an experimental study performed by Stackpole, Fornabaio and Alterman (1987), it was concluded that orally administered RA233 was cleared from experimental mice too rapidly to be effective in influencing tumour cell–platelet interactions.

Various studies, apart from that of Tohgo *et al.* (1984) reported above, support a role for the clotting cascade in metastasis, but conflicting results abound. Procoagulant activity does not always correlate with metastatic behaviour. Colucci *et al.* (1981 a and b), for example, failed to find any correlation between procoagulant activity and metastatic ability. Indeed, some non-metastatic sublines of the mFS6 murine sarcoma that they studied had more than six times the procoagulant activity of the parent line, while other sublines with metastatic ability had procoagulant activity which did not differ significantly from that displayed by the parent cell line. Defibrination or anticoagulant therapy with a variety of agents has been shown in some cases to reduce metastasis (Clifton and Agostino, 1965), but in other cases to promote it (Retik *et al.*, 1962). These results are perhaps not surprising given the contrasting effects of fibrin deposition referred to above. Pretreatment of animals with coumarin-type (Brown, 1973; Ryan *et al.*, 1968 a and b) or heparin-like (Lione and Bosmann, 1978; Kolenich *et al.*, 1972) anticoagulants can result in decreased experimental and spontaneous metastasis, although in some circumstances this may be due to toxic effects. Hilgard (1984 a and b) has shown that some coumarin derivatives could consistently inhibit lung colonization even when administered 24 h after tumour inoculation. The latter studies suggest that the effects of coumarin may be related to events which occur some time after entrapment, perhaps during extravasation. In a limited clinical trial of patients with small cell carcinoma of lung, Zacharski *et al.* (1981) reported that supplementary treatment with coumarin derivatives could prolong survival. Generally speaking, however, anticoagulants do not always inhibit metastasis (Maat and Hilgaard, 1981) and thus generalizations cannot be substantiated.

Other studies have shown that pretreatment of host mice with fibrinolysin may lead to decreased lung colonization by injected tumour cells (Chew and Wallace, 1976). Conversely, inhibitors of fibrinolysis such as ε-aminocaproic acid have been reported to have opposite effects (Boeryd *et al.*, 1974). We should note here that the conversion of fibrinogen to fibrin is dependent on thrombin, and that this molecule has diverse effects which add to our difficulties in interpreting many of the effects of perturbation of the clotting system on metastasis. For example, when bound to endothelial cells, thrombin loses its procoagulant activity and it may actually exert an anticoagulant effect through subsequent interactions involving protein C. Furthermore, thrombin appears to be able to stimulate the synthesis and release of a collagenase inhibitor from platelets and, as has been argued elsewhere, collagenase may be important in tumour cell extravasastion. Thrombin is also mitogenic for certain cells and some tumour types may be stimulated to divide in its presence. In an interesting experiment, Ostrowski *et al.* (1986) perfused mice with a thrombin inhibitor released via a minipump implant in order to sustain levels 100 × greater than its K_i. These authors found that B16F10 lung colonies were significantly increased under these conditions, and they concluded that thrombin may have a role in preventing metastasis, at least in this model system. On balance, however, it is difficult to interpret what the local effects of thrombin generation may be, and it is these local effects (rather than the more general phenomena which are usually measured experimentally) that are probably significant to the metastasizing cell.

Many of the studies referred to have been based on models employing lung colonization, and such a system is hardly representative of the natural history of a tumour. In fact, with respect to platelet aggregation phenomena, it is difficult to conceive of a more inadequate model than one involving the direct injection of tumour cells into the blood. The use of this technique represents a major limitation to many of the studies which form the experimental basis for our observations on the role of platelets and fibrin in metastasis, and we are thus cautioned against reaching unjustifiable conclusions.

So far in this discussion on the contribution of platelets and fibrin to metastasis we have been referring mainly to intravascular phenomena. When tumours growing *in situ* are examined histologically, most are found to be associated with extravascular fibrin (reviewed in Dvorak, Senger, and Dvorak, 1983). The tumour-associated fibrin network probably arises from the coagulation of plasma fibrinogen, which may diffuse from nearby vessels made leaky by a specific tumour permeability factor. A candidate permeability factor (M_r 38 kD) has been isolated from a guinea-pig bile duct carcinoma cell line, but further studies using a spectrum of tumours and involving more detailed analysis are required. When deposited about a growing tumour in significant quantity, extravascular fibrin is likely to have a protective effect against host defence cells in a manner similar to that decribed above.

There is no doubting the complexity of platelet–fibrin–tumour cell interactions! Coupled with the unique behaviour of many tumours, what we already know of their heterogeneity, and the limitations of model systems, it is perhaps not surprising that confusion pervades much of the literature. Although a small ray of hope is offered by the fact that some tumours do respond to antiplatelet and anticoagulant treatments, we still have a long journey ahead of us before we fully understand how these systems modulate the behaviour of individual metastasizing cells.

5.7.2.3 *The molecular mechanisms of platelet adhesion and aggregation*

This is a notoriously difficult area of current research, but as we shall see it has many links with processes involved in tumour cell adhesion. There are four principal adhesive proteins of platelets (reviewed in Plow, Ginsberg and Marguerie, 1986):

(a) Fibrinogen. This dimer of M_r 3.4×10^5 D is an important cofactor in platelet aggregation. Since it is bivalent, fibrinogen may simply act as a bridge linking gpIIb/IIIa receptor sites on adjacent platelets, although its role in aggregation is not yet fully resolved.

(b) Fibronectin. This dimer of M_r 4.5×10^5 D influences platelet spreading and antibodies directed against its RGD sequence are able to inhibit platelet aggregation. As will become clear below, a number of molecules implicated in platelet aggregation have at least one RGD sequence and thus the full significance of the antibody-mediated aggregation inhibition is not clear. The structure and adhesive functions of fibronectin are discussed in more detail in section 5.4.2.

(c) Von Willebrand factor (vWF). This multimeric molecule consists of a linear array of 4–50 subunits (270 kD) linked by disulphide bonds. In the plasma, vWF is non-covalently associated with factor VIII (antihaemophilic factor) to form complexes of M_r 1×10^3–12×10^6 D. Since vWF binds to collagen and to platelets (see below), it is likely to be involved in the binding of platelets to the basement membrane. It is synthesized by platelets, megakaryocytes (the platelet precursors) and endothelial cells. In the lattermost cell type vWF is either secreted constitutively (possibly as dimers, but certainly of relatively small molecular mass) or it is stored in intracellular vesicles known as Weibel-Palade bodies. These bodies are rod shaped vesicles about 2 μm in length which

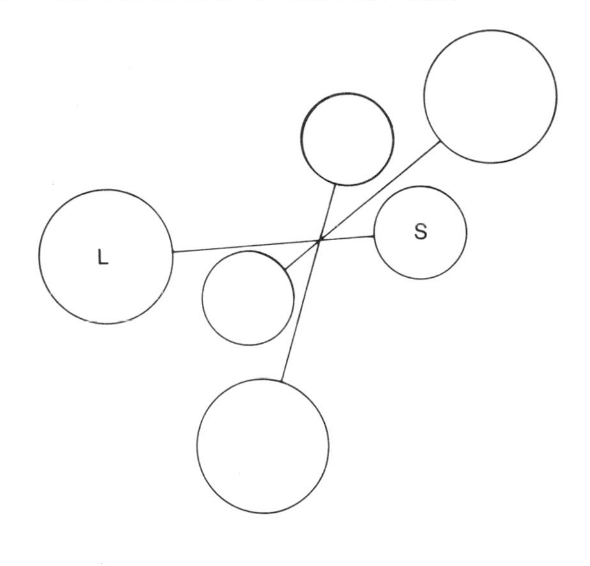

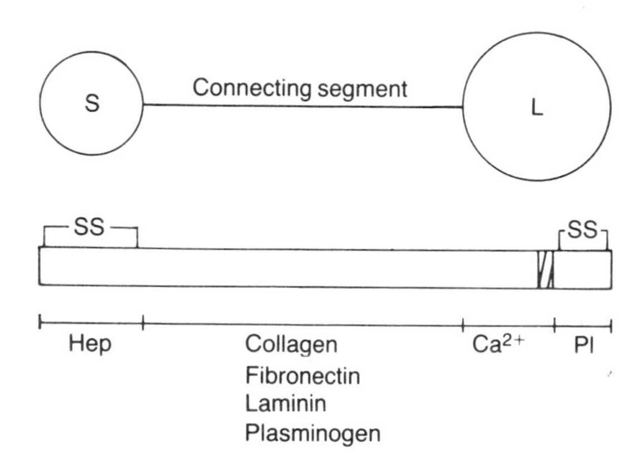

Figure 5.11 The structure of thrombospondin (after Frazier, 1987). (a) Thrombospondin (TSP) is composed of three asymmetric dumbbell-like subunits in which the weights represent small (S) and large (L) globular domains, and each bar represents a connecting segment. The subunits are hinged together near the smaller globular domains giving the molecule a high degree of flexibility which may be advantageous in promoting adhesion and aggregation. An analogy may be drawn with the bolas of South American cowboys, with one difference being extra balls in

appear to originate from the Golgi apparatus. Ultrastructurally, they are seen to contain oriented tubular material which is thought to represent tightly packed molecules of vWF. This material is released on stimulation (by thrombin or injury, for example) in a multimeric form which appears to be more active in binding to platelets than smaller forms (Sporn *et al.*, 1986). The cationic antibiotic ristocetin appears to promote platelet aggregation artificially by inducing the binding of multimeric vWF to gpIb (see below).

(d) Thrombospondin (TSP). This trimer of M_r 4.8×10^5 D appears to be able to bind to heparin, laminin, plasminogen, and fibronectin. It also binds to platelets and to both fibrinogen and fibrin, being incorporated into blood clots (reviewed in Silverstein, Leung and Nachman, 1986; Frazier, 1987). Like other adhesive molecules such as fibronectin and laminin, TSP is organized into discrete protease resistant domains (Fig. 5.11). Its main role in platelet aggregation appears to be stabilization of fibrinogen binding to the gpIIb/IIIa complex. Apart from platelets, TSP is produced by endothelial cells, monocytes and fibroblasts and other cell types, including those from tumours such as the HT1080 fibrosarcoma and the C32 and G361 melanomas. The production of TSP by fibroblasts is probably responsible for its relatively wide distribution in the ECM. Ultrastructurally, TSP appears as a molecular-sized bola but with large and small balls (domains) on each rope (connecting segment). Bolas are used by gauchos of the South American pampas for catching cattle. They are thrown at the beasts and bring them to the ground by wrapping around their legs. In some ways, this analogy may point to how TSP functions. The three subunits of TSP show considerable flexibility about their common point of association, and the molecule itself displays relatively promiscuous binding to a number of others. Although TSP contains an RGD sequence (near the platelet binding domain which is at the C-terminal end of each subunit), this sequence may be cryptic since it appears not to be involved in TSP functions.

---

the form of the smaller globular domains.
(b) The conformation of thrombospondin is Ca^{2+}-dependent, with multiple Ca^{2+} binding sites being present in the large globular domain. The platelet (Pl) binding site and an RGD sequence (hatched) are also present in this domain. The binding sites for collagen, fibronectin, laminin and plasminogen are found along the connecting segment, whereas that for heparin (Hep) is located within the small globular domain. Both globular domains contain internal disulphide bonds (SS).

All four of the molecules listed above are more or less absent from the surfaces of resting platelets, during which time they appear to be confined to the α-granules. Following stimulation, they take on a surface distribution. Plasma is a major source of all but TSP (present at only 20 ng ml^{-1}), and endothelial cells synthesize all but fibrinogen.

Considerable progress has been made recently in characterizing the receptors for the four platelet adhesive proteins, but the current position is still a little confusing. Platelet glycoprotein gpIb (M_r 170 kD) appears to be a receptor for vWF, while gpIIb/IIIa functions as a receptor for fibrinogen, fibronectin, thrombospondin and probably vWF (but with low affinity unless under high shear conditions). GPIIb is a dimer (M_r 132 kD and 23 kD) held together by disulphide bonds: it forms a Ca^{2+}-dependent complex with the gpIIIa monomer (M_r 90–105 kD) in detergent-containing solutions. Molecular sequencing has allowed classification of gpIIb/IIIa into the superfamily of integrins which includes receptors for fibronectin, vitronectin, laminin and the leukocyte adherence-related glycoproteins (see section 5.7.2.4). In fact gpIIIa appears to be identical to the β chain of the mammalian vitronectin receptor. The gpIIb/IIIa heterodimer is present on unstimulated platelets, but it is presumably functionally cryptic until platelet activation takes place. Immunological studies indicate that gpIIb/IIIa is expressed on endothelial cells following their activation, which suggests that in certain situations any of the four adhesive molecules may bind to the endothelial cell surface directly. Although this situation still requires further study, it should be evident that this possibility could have considerable consequences with respect to tumour cell binding. Interestingly, Santoro (1986) has recently reported that platelets can adhere to types I, III, and IV collagen by a Ca^{2+}-independent (but otherwise divalent cation-dependent) mechanism involving a molecule now thought to be gpIa/IIa which is believed to be identical with VLA2. This mechanism is apparently independent of platelet activation and may represent the initial attachment process to exposed basement membrane containing type IV collagen. Non-activated platelets can also bind to fibronectin in a Ca^{2+}-independent manner via gpIc/IIa which appears to be identical to VLA5 (Table 5.2).

We are now in a position to re-examine certain aspects of tumour cell behaviour which may be related to platelet adhesive proteins. First, we should note that platelet aggregation stimulated by ADP, thrombin or collagen appears to be inhibitable by RGD(S) (Gartner and Bennett, 1985). Furthermore, antibodies to fibrinogen, fibronectin, TSP and gpIIb/IIIa all inhibit platelet aggregation. We have seen that the RGD(S) sequence is found in several adhesive molecules including fibrinogen, fibronectin, TSP and vWF. It appears that RGD containing peptides will inhibit the binding to platelets of fibronectin, fibrinogen and vWF, but not TSP, indicating that this sequence may be cryptic in TSP as suggested above. Although TSP can bind to platelets in the presence of RGD peptides, the fact that aggregation does not occur under these circumstances suggests that TSP may not be directly involved in aggregation as a linking molecule. This interpretation is in accord with the postulated role of TSP in stabilizing fibrinogen-mediated interactions. Alternatively, TSP may aggregate platelets indirectly via another RGD-containing molecule serving as a bridge. This possibility remains to be tested, and further work is obviously required in this area before the situation is fully clear.

Now Humphries, Olden and Yamada (1986) have shown that co-injection of B16F10 melanoma cells with 3 mg per mouse of gly-arg-gly-asp-ser (GRGDS) inhibited lung colony formation. This inhibition appeared to be related to reduced tumour cell retention in the lungs and was not due to any obvious toxic effects (even though the dose was rather extreme). It is difficult to interpret these results reliably in the light of the multiple interactions described above which are likely to involve RGD related peptides, but they are at least consistent with a role for platelet associated events in metastasis. It is entirely conceivable that many of the adhesive molecules which are implicated in metastasis may exert their effects at an early stage in the process (platelet–tumour cell interaction or possibly tumour cell–endothelial cell interaction) rather than at a later stage (tumour cell–basement membrane interaction) as has been presupposed to be the case.

In other studies, the i.p. or i.v. injection of TSP 5 min before the i.v. injection of T241 sarcoma cells has been found to enhance lung colony formation, whereas vWF was without effect (reviewed in Gasic, Tuszynski and Gorelik, 1986). It should be noted here that vWF multimers of high molecular mass are more effective in promoting cell adhesion than are smaller ones, and that the nature of vWF in many experimental prepartions is not clear. Since the enhancing effects of TSP were not observed in thrombocytopenic mice or in mice anticoagulated with a coumarin derivative, it would seem that TSP enhances lung colony formation via complex interactions associated with platelet aggregation and blood clotting (Tuszynski *et al.*, 1987a). At the very least, the results suggest that TSP does not exert its effects simply by aggregating tumour cells alone and nor are they indicative of role for TSP in binding metastatic cells directy to the basement membrane, as has been suggested for laminin. This latter role for TSP might seem a possibility since tumour cells (including B16F10 and human melanoma cells) can bind directly to TSP (Tuszynski *et al.*, 1987b), but the *in vivo* experiments clearly point to a major involvement of platelets and fibrin in lung colonization. This conclusion is favoured intuitively, since the possibility of metastatic cells coming into contact with denuded basement membrane before platelets and fibrin is probably rather remote. Since there is ample evidence from a variety of sources that most tumour cells will adhere to the other platelet adhesive proteins fibronectin and fibrinogen (or fibrin), which can interact with TSP, it seems that there may be a role for a multimolecular complex involving at least three of the platelet adhesive proteins in tumour cell metastasis. According to Marcum *et al.* (1980), who employed a three way xenogenic model, vWF is important in the adhesion of tumour cells to platelets, but other studies with a human melanoma cell line (G361) have shown that these cells do not adhere directly to vWF (Roberts, Sherwood and Ginsburg, 1987). Similarly, in unpublished observations from my laboratory, we have shown that B16F10 cells do not adhere to purified human vWF. Taken together, the various lines of evidence suggest that the effects of RGD containing peptides in inhibiting metastasis may be due to the perturbation of tumour cell-platelet aggregation involving fibrinogen, TSP, fibronectin and possibly vWF.

5.7.2.4 *Immune cell–tumour cell interaction*

The interaction between cells of the immune system and tumour cells has two possible outcomes with respect to metastasis. On the one hand binding may lead to clump formation thereby promoting metastasis (Fidler, 1974a), while on the other adhesion may be a prelude to subsequent tumour cell destruction (Martz, 1975). Much of our understanding of how cytotoxic lymphocytes work has been gained from the use of the ^{51}chromium release assay. In this procedure, host cells labelled with radioactive chromium are mixed with cytotoxic cells and the amount of destruction over a defined period of time is assessed by counting radiolabelled proteins released from dead cells into the culture medium. The mechanism of destruction is independent of antibody and complement. Cytolysis requires adhesive contact between the target and effector cells since it does not occur if the two are separated by a filter. Many features of the cytolytic process have been determined with respect to cytotoxic T cells, and certain aspects of their behaviour may be used to model the cytolytic process in general. The process may be divided functionally into four stages:

1. *Recognition and adhesion.* Effector cells bind more or less specifically to particular target cells. For cytotoxic T cells, this specific recognition is almost certain to involve the T cell antigen

receptor. The T cell antigen receptor, referred to as Ti (for T cell idiotype), is a 90 kD dimer which associates with three small proteins (about 20 kD each) known as CD3 (formerly T3). The abbreviation CD (as in CD3) refers to numbered 'cluster of differentiation' designations which have been assigned to human leukocyte differentiation antigens by international agreement. Some designations are temporary, and these are recognized by the workshop designation CDw. The interaction between the T cell receptor and its appropriate antigen on the target cell appears to be relatively weak, and a stronger intercellular bond is required for lysis to occur. This strong adhesive interaction is temperature and energy dependent, requires physiological Mg^{2+} but not Ca^{2+}, and can be inhibited by cytochalasin B and colchicine indicating cytoskeletal involvement. Antibodies against CD3 inhibit killing but do not block adhesion, indicating that the latter process is not antigen dependent. Killing appears to be more specific than adhesion. Thus cytotoxic T cells may bind to a variety of potential targets, but killing requires specific target recognition.

We have seen already that one of the more important candidates for mediating cell adhesion during the cytotoxic process is the lymphocyte adherence-related protein LFA-1 (section 2.12.2). Three principal leukocyte adherence-related proteins have been identified:

(a) **LFA-1.** This was originally called lymphocyte function-associated antigen-1 because it was thought to be restricted to lymphocytes, but in fact it is found on virtually all peripheral blood leukocytes including B cells, T cells, PMNs, monocytes and NK cells. It appears to play a crucial role in the adhesive interaction of cytotoxic T cells and NK cells with their targets, since McAbs against LFA-1 can inhibit cytotoxicity. LFA-1 is also involved in the homotypic adhesions of haematopoietic cells, and the binding of lymphocytes to the endothelium (Mentzer, Burakoff and Faller, 1986). Since LFA-1 is expressed on many haematopoietic neoplasms, it could play a major role in the dissemination of these tumours. The ligand for LFA-1 binding has recently been identified as ICAM-1 (intercellular adhesion molecule-1), which shares some homology with the neural cell adhesion molecule N-CAM (Simmons, Makgoba and Seed, 1988). ICAM-1, like N-CAM, does not contain the RGD sequence, but it does contain a single RGE.

(b) **Mo1 antigen.** Also known as **Mac-1**, this antigen is found on neutrophils and monocytes. It is identical to the receptor (CR3) for a fragment of complement known as C3bi. Mo1 antigen is not common on B cells or T cells.

(c) **p150,95.** As with the Mo1 antigen, this antigen is generally lacking on B cells and T cells. Although both the Mo1 antigen and p150,95 are found on the surfaces of monocytes and neutrophils, they are actually present in higher amounts inside them. It seems that the binding of inflammatory mediators to the surfaces of these leukocytes induces mobilization of the two antigens from their cytoplasmic pools, and this correlates with increased adhesivnes to the endothelium. Thus experimental stimulation of PMNs and monocytes with phorbol esters or chemoattractants such as FMLP and C5a, for example, has shown an increase in the surface expression of the Mo1 antigen and p150,95, but not LFA-1 (Springer, Miller and Anderson, 1986). The up-regulation of these adhesive molecules is likely to be of considerable importance during the inflammatory response, since they seem to contribute to both PMN extravasation and phagocytosis.

The distinctive feature of these three leukocyte adherence-related molecules is that they all exist as heterodimers, with a common 95 kD β chain (CD18, mapped to human chromosome 21) non-covalently linked to different α chains (CD11) known as αL, αM and αX respectively. Although not identical, the three α chains nevertheless do show considerable homology with each other.

Anderson and colleagues (1986) have explored the role of leukocyte adherence-related proteins in various aspects of PMN behaviour. They reported that PMN aggregation in response to various chemoattractants was inhibited by antibodies against the common β chain>anti-αM>anti-αX>anti-αL (in order of inhibitory strength). Inhibition of chemotactic behaviour was noted using the under agarose technique (but not the Boyden chamber method) for both anti-β and anti-αM, whereas phagocytosis of C3-opsonized particles was inhibited by anti-αM. Te Velde, Keizer and Figdor (1987) studied the adhesion of human monocytes to melanoma cells and umbilical vein endothelial cells. These authors found that McAbs against αL, αM, αX and the common β chain inhibited adhesion of monocytes to both target cell types when carried out in suspension, but only McAbs against αX and the common β chain were effective in inhibiting the adhesion of monocytes to monolayers of the same cells. The most likely reason for these observations is that when suspended cells adhere and spread on the substrate, their adhesive molecules become polarized with many moving to the lower membrane surface. The adhesion of monocytes to cells in suspension or to cells which are spread may subserve different phenomena. Thus cytolytic interactions within the blood may involve aggregation, whereas extravasation (via adherence to endothelial cells) may be akin to binding to spread cells. On balance, the evidence suggests that the leukocyte adherence-related glycoproteins are able to subserve a number of functions in which adhesion is thought to play an important role.

With respect to the immune system, antibodies directed against LFA-1 block the adhesions between most effector and target cells, and successfully inhibit lysis mediated by cytotoxic T cells, NK cells and LAK cells. Furthermore, Strassman *et al.* (1986) have shown that macrophage–tumour cell adhesions can be blocked by antibodies against the LFA-1 α chain, although not by antibodies directed against the Mo1 α chain. It should be noted here that Oda and colleagues (1988) have recently provided evidence that the binding of tumour cells to activated mouse macrophages is mediated at least in part by a cell surface lectin of M_r 45–60 kD. This lectin is specific for Gal and GalNac residues, is not found on resident peritoneal macrophages, and seems to be involved in distinguishing transformed cells from normal ones. Similarly, although LFA-1 appears to play an important role in the adhesion between cytotoxic lymphoid cells and their targets, it is not the only molecule which may be involved. In an excellent review of T lymphocyte adhesions, Martz (1987) recognizes five possible adhesive mechanisms:

(i) The T cell receptor, which recognizes antigen in association with the MHC.
(ii) T4 (CD4) or T8 (CD8) on T cells only, which recognize class II or class I MHC antigens respectively.
(iii) LFA-2 (CD2), a 50 kD glycoprotein expressed on T cells, formerly referred to as the sheep red blood cell receptor. McAbs against LFA-2 bind to cytotoxic T cells and inhibit the killing of target cells, whereas McAbs against LFA-3 achieve the same effect but they bind to receptors on the target cells. On the basis of these results it has been suggested that LFA-3 (70 kD) which is present on various cell types (including epithelial cells) may be the cell surface ligand for LFA-2 on T cells (Dustin *et al.*, 1987).
(iv) LFA-1 on T cells and other haematopoietic cells, which probably recognizes ICAM-1 on target cells (Rothlein *et al.*, 1986).
(v) A gp90MEL^{-14}-like molecule on both B cells and T cells which is involved in their binding to high endothelial venules (section 2.12.2).

This list is probably not yet complete, and it is likely that other adhesive molecules remain to be identified. Of the mechanisms listed, it would seem that all but those whose activity is restricted to lymphocyte adhesion to HEVs probably play a role in cytolysis. Thus antibodies to T4, T8 and LFA-3 can inhibit the adhesion of cytotoxic T cells to target cells, although not all cytotoxic T cells are necessarily sensitive to anti-T4 and anti-T8. Since binding of LFA-2 and LFA-3 is not temperature sensitive and nor is it dependent on divalent cations, it is clear that another mechanism (involving LFA-1) probably plays the predominant role.

2. *Programming for lysis*. The molecular events which follow adhesive interaction prior to lysis are not clear, but an inverse relationship between effector cell cAMP and the extent of cytotoxic activity has been recorded. Thus agents such as PGE_2 inhibit cytotoxicity. Evidence is available indicating that various cytoskeletal elements including the Golgi apparatus and cytoplasmic granules polarize in the direction of the bound target cell.
3. *The lethal hit*. What is the agent or process that leads directly to target cell destruction? This answer to this question is made intriguing by the fact that lysis is one way. When an effector cell delivers a toxin, its effects are distinctly directional and confined to cells which have been in contact with it. Although cytotoxic cells can themselves be destroyed, when two cytotoxic cells interact only one is lysed. What protects the killer cell from cytolysis is not known, but it has been postulated that cell surface proteoglycans may have a protective role. Note that cytotoxic cell–effector cell contact is no longer necessary after delivery of the lethal hit, and the effector cell may move off to bind to other target cells. Several cytolytic molecules have now been described:
   (a) The pore-forming molecules.
   Perforin probably represents a group of related proteins which lyse target cells by inducing the formation of 15 nm diameter pores in their membranes. The large granular lymphocyte (LGL) and cytotoxic T cell lytic factors fall into this category. Mouse perforin (c. 534 amino acids long, reduced M_r 70 kD) is about 27% homologous in its primary structure with the complement component C9. Although perforin and C9 are also antigenically cross reactive, they are nevertheless distinct from each other. The complement component C9 interacts with the C5b–8 complex to bring about insertion into the target cell membrane. In the presence of Ca^{2+}, perforin secreted from cytoplasmic granules polymerizes within the membrane of the target cell to form extremely large tubules ($M_r > 10^6$ D) which associate to form the pores. Although the

presence of pores will lead to lysis through their effects on cell water, they may also be used as channels for the transmission of other molecules released from the effector cell granules, some of which may also be lytic in nature. Recent evidence indicates that phosphorylcholine head groups on plasma membrane lipids such as sphingomyelin and phosphatidylcholine serve as the receptor for perforin (Tschopp *et al.*, 1989).

(b) The non-pore-forming molecules.

(i) **Lymphotoxin.** This also represents a heterogeneous group of molecules. Cytotoxic T cells and NK cells produce a lymphotoxin which appears to be identical to tumour necrosis factor beta (see below). Other lymphotoxins are currently poorly defined.

(ii) **Tumour necrosis factor (TNF).** This exists in two forms (α and β) which are related but distinct molecular entities. They appear to bind to cells via the same receptor, but whereas the α form is produced primarily by macrophages, the β form is produced by lymphoid cells. According to Wang and colleagues (1985), TNFα has cytotoxic activity against some but not all (e.g. HeLa) tumour cell lines *in vitro*, and it is apparently inactive against many but not all normal cell lines (e.g. primary breast epithelial cells). There is some evience that TNFα may play a role in both cell-mediated and fluid-phase cytotoxicity.

(iii) **Leukoregulin** (M_r 135 kD). This appears to be able to kill tumour cells directly, or indirectly by enhancing NK cell cytolytic activity.

Other molecules which are associated with target cell destruction include interferon, interleukin-1, NK cell cytotoxic factor, various enzymes, and reactive oxygen intermediates.

4. *Target cell destruction.* Following delivery of the lethal hit, the membrane of the target cell becomes increasingly permeable and ultimately ruptures, presumably as a result of the excessive influx of water. Non-pore-forming toxins such as TNF appear able to induce fragmentation of target cell DNA.

Many of the studies which have formed the basis of the summary presented above have employed allogeneic combinations using selected cell types, since in this situation relatively high kill ratios may be obtained. However, the destruction of tumour cells *in vivo* actually involves an autochthonous interaction, which is modelled more closely by syngeneic cytotoxicity. In section 1.8.4, evidence is provided for the possible existence of autoreactive T cells which can lyse tumour cells that express class II MHC antigens. Evidence is also provided indicating that such effector cells may be stimulated by class II MHC positive cells, such as macrophages, to lyse bystander tumour cells through the release of cytokines including lymphotoxin and IFN-γ. The non-targeted destruction of cells by the general release of toxic molecules is in stark contrast to the more or less specific cytolytic effects described above, which is presumably more effective *in vivo*.

The relationship between substrate adhesivness, NK cell sensitivity and metastasis has been examined by Young, Duffie and Newby (1987), who worked with metastatic (C3) and non-metastatic (C8) clones of the Lewis lung carcinoma. The difference in metastatic ability to the lungs between these two clones was seen after subcutaneous injection, and it was also reflected in their different lung colonizing abilities, with the C3 clone forming more colonies than the C8 clone after i.v. injection. The C3 cells were found to be non-adherent to tissue culture surfaces and were relatively insensitive to NK cell activity, whereas the non-metastasizing C8 cells, on the other hand, both adhered and spread well *in vitro* and were more sensitive to NK cell mediated cytolysis. When treated for 72 h with 2% dimethyl-sulphoxide (DMSO), an agent which induces the differentiation of a number of cell types, the C3 cells became adherent and spread, sensitive to NK cell mediated cytolysis, and less metastatic. Conversely, treatment of C8 cells for 1 h with 10 $\mu g\ ml^{-1}$ of the microfilament-disrupting agent cytochalasin B caused them to become non-adherent, increased their resistance to NK cells, and improved their metastatic ability. The alterations in metastatic ability seen after the different treatments were not detectable in three-week-old, NK-cell-deficient mice. This result points towards the NK cell as the host defence cell most likely to be involved in affecting metastatic outcome in this particular model system. Despite the clear correlation between substrate adhesiveness, metastatic capability and NK cell activity, Young, Duffie and Newby (1987) reported that there were no differences between the two clones in their abilities to adhere to effector NK cells. Certainly, a number of reports have indicated that tumour cell–NK cell conjugates can form without necessarily resulting in lysis. The results of this study with Lewis lung clones suggest that a later stage in the lytic process was affected when using treated cells, perhaps one relating to the transfer of cytolytic molecules. It seems highly unlikely that DMSO and cytochalasin B would effect NK cell mediated lysis by directly opposing effects on the same key cellular system, and thus the molecular mechanisms involved remain unclear.

5.7.3 Detachment and metastasis

Since a metastasis is defined in relation to its loss of contiguity with its primary, it would seem clear that detachment represents an essential step in metastasis. An inverse relationship between detachment and metastasis was suggested, however, by Briles and Kornfeld (1978) who selected B16F10 cells using an EDTA-based detachment procedure. These authors found that the EDTA-resistant subpopulation formed more lung colonies than the EDTA-sensitive subpopulation, although both subpopulations adhered to Petri dishes at similar rates. A similar relationship was reported by Varani *et al.* (1980a) who explored the detachment characteristics of 3-methylcholanthrene-induced mouse fibrosarcoma variants of different spontaneous metastatic potential. Using an assay based on the percentage of tumour cells released from plastic flasks after a standard *in vitro* treatment with trypsin, these authors found that detachment correlated inversely with metastatic potential, although there was no difference in the rates of substrate attachment of the different metastatic variants. In another study, Varani *et al.* (1980b) reported that there was no difference between low and high metastasizing variants in their abilities to adhere to bovine endothelial cells, although once attached the low malignant cells were released more easily by trypsin treatment. One problem with this type of assay is that the high malignant cells might have penetrated through the endothelial monolayer and thus been sheltered from detachment by the overlying endothelium which is known to be more resistant to trypsin treatment during the short term (10 min) of the assay.

In other studies from this group, highly metastatic fibrosarcoma cells have been shown to bind more strongly to the I-B_4 isolectin of *Griffonia simplicifolia* (GSI-B_4) than do poorly metastatic ones (Grimstad, Varani and McCoy, 1984). The GSI-B_4 isolectin is specific for terminal, non-reducing α-D-galactopyranosyl groups, and the enzymatic removal of these groups with α-galactosidase inhibits the adhesion of both fibrosarcoma variants to types I and IV collagen. Although these results implicate a role for α-D-galactopyranosyl end groups in adhesion to collagen, their precise mechanism of involvment in metastasis remains to be clarified. Since GSI-B_4 binds to laminin, the possibility that the cellular interaction with collagen reported above was mediated via residual laminin contamination of the collagen preparations cannot be excluded.

5.7.4 Adhesion and invasion of the basement membrane

The principal molecular components of the basement membrane are type IV collagen, the proteins laminin, fibronectin and nidogen, and proteoglycans such as heparan sulphate and both the chondroitin sulphates (section 2.10.1). Whereas the polyanionic proteoglycans of the basement membrane are thought to influence its filtration properties, type IV collagen probably accounts for most of the mechanical stability of the basement membrane. The non-collagenous proteins also probably serve a mechanical role since they can be cross-linked in various sorts of combinations. Nidogen seems likely to act primarily as a cross-linker, while the other proteins are also able to function as cell adhesive molecules.

Basement membranes may either inhibit or facilitate metastatic spread:

(a) The presence of a basement membrane around a primary tumour might restrict metastatic spread by minimizing cell escape. Furthermore, there is ample evidence that the basement membrane can act as a barrier to invasion across structures such as blood vessels. Type IV collagen seems to be the principal component of the basement membrane which restricts tumour cell invasion. We have seen that the presence of a specific type IV collagenase correlates with metastatic potential (Liotta *et al.*, 1980), and although only a proportion of malignant cells (10–75% as judged by immunolabelling) may express type IV collagenase activity, Barsky and colleagues (1984) in a restricted study could find no such activity in benign cells. It is possible that basement membrane proteoglycans may also serve to restrict metastasis since the presence of heparan sulphatase correlates reasonably well with metastatic ability (Nakajima *et al.*, 1983, 1984).

(b) On the other hand, since it is composed of molecules which are adhesive for a wide range of cell types, the basement membrane may actually facilitate emigration from blood vessels. There is also considerable evidence that glycoproteins such as laminin and fibronectin are chemotactic or haptotactic for numerous tumour cells, and their presence in basement membranes could thus facilitate extravasation to form secondaries.

Murray *et al.* (1980) compared the adhesiveness of metastatic PMT cells (derived from the murine T241 fibrosarcoma) to a range of collagens with that of normal and spontaneously transformed adult murine connective tissue fibroblasts. These authors found that in the absence of serum the metastatic cells adhered preferentially to type IV collagen relative to other collagens, and they suggested that metastatic cells might possess a unique system for binding to basement membrane collagen. Although displaying a different adhesive profile to a range of collagens when compared to that of normal and spontaneously transformed fibroblasts, the metastatic PMT cells actually adhered to type IV collagen at a level more or less similar to that for the spontaneously transformed cells. Since this level was in fact markedly less than that shown by normal connective tissue fibroblasts, the significance of the binding reported to the metastatic process is open to question.

Metastatic potential is influenced both by the presence of free laminin receptors at high density (Hand *et al.*, 1985) and also by the presence of cell surface bound laminin which may be either of endogenous (McCoy *et al.*, 1984) or exogenous origin (Barsky *et al.*, 1984). These results suggest that metastatic cells may take either direct advantage of the presence of laminin in the basement membrane or, more indirectly, they may utilize laminin-binding sites on other molecules within the basement membrane. The enhancing influence of exogenous laminin on metastatic potential can be inhibited by antibodies against laminin as well as by various laminin fragments. Small laminin fragments which bind to the cell surface presumably inhibit metastatic potential because the cell receptor is occupied by a fragment which lacks the ability to bind to other molecules within the basement membrane (Barsky *et al.*, 1984).

Although the presence of cell surface laminin may correlate with increased metastatic potential, there is some evidence that the presence of this material may promote macrophage anti-tumour activity (Perri *et al.*, 1985). Interestingly, NK cells apparently have laminin-like material on their surfaces (Hiserodt *et al.*, 1985). If this proves to be indicative of the presence of laminin receptors then it is conceivable that NK cell–tumour cell interaction could be mediated through this specific form of receptor–ligand binding, but as yet this remains speculative (section 5.7.2.4).

It is clear from a general survey of the components of the basement membrane that many of these are adhesive for a wide range of cell types. Yet we have seen that certain tumour cells can metastasize selectively to particular organs. This would seem to suggest that adhesion to the basement membrane is probably of little significance in organ selective metastasis. Nevertheless, some tumours have been shown to adhere selectively to subendothelial ECMs with a pattern which correlates reasonably well with their metastatic or colonizing preferences. Thus Nicolson and his colleagues (reviewed in Nicolson,

1988) have shown that variants of the 13762NF rat mammary adenocarcinoma cell line bind poorly to lung endothelial cells irrespective of whether they can metastasize significantly to this organ or not. However, when tested on lung subendothelial ECM, a correlation with their metastatic behaviour was more evident. These results suggest that the subendothelial ECM contains adhesive molecules other than those immediately obvious (such as laminin and perhaps fibronectin), but their molecular nature remains to be elucidated.

The adhesion of tumour cells to the basement membrane does not necessarily always correlate with metastatic or colonizing potential. Korach *et al.* (1986), for example, found that whereas there was a correlation between the adhesion of nickel-induced rat rhabdomyosarcoma cells to porcine aortic endothelial cells and the colonizing ability of the former, this was not obvious when adhesion to the underlying endothelial matrix was examined. Adhesion to the matrix components was found to proceed at a rapid rate for cells with both low and high lung-colonizing potential. It could be argued that these results reflect the use of a xenogeneic system, but Nicolson and colleagues have reported that the adhesiveness of RAW117 murine lymphoma cell variants to syngeneic subendothelial ECM correlates only poorly with organ specificity of metastasis (reviewed in Nicolson, 1988). As explained previously (section 5.7.2.1) a better correlation is found when the adhesiveness of these variants to specific endothelial cells is examined. The obvious conclusion from the foregoing is that adhesion to the basement membrane does not always correlate with or predict metastatic ability, let alone organ selectivity.

Of interest here are the observations of Gehlsen *et al.* (1988) who reported that a range of RGD containing synthetic peptides showed only minimal inhibition of attachment of A375P and A375M human melanoma cells and RuGli human glioblastoma cells to the amnion basement membrane. All but one of the peptides they tested nevertheless inhibited invasion of the amnion. This particular peptide was a cyclic variant which inhibits attachment to vitronectin but not to fibronectin. A peptide with opposite specificity (inhibiting adhesion to fibronectin but not vitronectin) successfully blocked invasion, as did peptides which inhibit adhesion to both of these molecules or to both of these as well as type I collagen. The results of these workers and of other groups discussed elsewhere in this chapter allow us to make the following comments:

(a) Adhesion and invasion are clearly separable phenomena.
(b) Multiple processes are probably involved in the adhesion of cells

to the basement membrane, and thus it is difficult to make any valid generalizations when considering only one particular adhesive system.

(c) Vitronectin may not play a significant role in invasion, presumbly reflecting its minimal presence in the basement membrane.

(d) The inhibitory effects of RGD containing peptides on lung colonization (Humphries, Olden and Yamada, 1986) do not result from blocking adhesion of the circulating cells to the basement membrane (section 5.4.2).

5.8 HOMOLOGIES WITHIN THE DIFFERENT ADHESIVE SYSTEMS

Research in progress continues to identify homologies between the various ECM adhesive receptors originally identifed in different animals. In humans, the receptor for fibronectin appears to be equivalent to the very late antigen, VLA-5 (Takada, Huang and Hemler, 1987). Both subunits of human VLA-3 appear to be antigenically similar to parts of the integrin complex found in avian cells and recognized by JG22E and CSAT McAbs. Integrin runs on gels as three bands rather than two, and at first sight this would seem rather difficult to relate to the structure of the VLAs. However, it is possible that the three bands represent a heterodimer in which one subunit is common while the other may be in two forms. Using this heterodimer model of Buck and Horwitz (1987), the simplest interpretation of these observations is that the avian fibronectin receptor (integrin) consists of alpha and beta subunits which share some homology with the subunits of the human VLA-3 receptor, although the possibility of a heterotrimer has not yet been rigorously excluded. As will be seen below, it has turned out that one of the subunits of the integrin heterodimer is in fact a disulphide-linked dimer itself.

Since fibronectin contains RGD sequences which are important in binding to its receptor, it should prove worthwhile to examine the receptors of other molecules containing RGD such as vitronectin, collagen (particularly type I), fibrinogen, von Willebrand factor and thrombospondin. We have seen that the adhesions of cells and platelets to most of these may be inhibited by RGD, and evidence points towards a general heterodimer structure for their receptors. For the receptors studied so far, the α subunit seems to be composed of either a single polypeptide or two disulphide-linked polypeptides, whereas the β subunits are single polypeptides (90–140 kD). The light chain and heavy chain components of the α subunit are believed to arise from cleavage of a precursor molecule, indicating that the receptors may have evolved via a heterodimer of two single but different polypeptides. It was this possibility which originally drew attention to a possible relationship with the leukocyte adherence-related proteins discussed earlier. Significantly, the Mo1 antigen, which binds to the complement fragment C3bi, has been shown to recognize an RGD sequence on the latter (Wright *et al.*, 1987). Since this family has been implicated in cell–cell adhesion, then a relationship between receptor molecules implicated in cell–substrate adhesion and other molecules involved directly in cell–cell adhesion becomes apparent. In fact, the β subunit gene of the leukocyte adherence-related proteins has been shown to be 45% homologous with the β subunit of integrin (Tamkun *et al.*, 1986). Furthermore, N-terminal sequence analysis has shown considerable homology between the α subunits of the leukocyte adherence-related proteins, the VLAs, gpIIb/IIIa (i.e. gpIIb) and the vitronectin receptor (Takada *et al.*, 1987). A study by Springer, Teplow and Dreyer (1985) has also indicated considerable homology between the leukocyte adherence-related proteins and both human and mouse interferon (particularly IFN-α). Interestingly, although the cell adhesion molecule N-CAM belongs to the immunoglobulin superfamily it shares some homology with the α chains of the adherence-related molecules, giving rise to the speculation that many cellular interactions probably had their basis in a common, evolutionary distant adhesive process. Indeed, homology has not only been noted between these adhesion related molecules in mammals and birds, but also between them and the positional specific antigens of *Drosophila* implicated in compartmentalization during development. It should be noted here that although there are similarities between the leukocyte adherence-related proteins and the receptors for various adhesion related molecules which bind via RGD receptors, there is no evidence that the binding of cytotoxic lymphocytes to their targets is inhibited by RGD (section 5.7.2.4).

The lines of evidence described above point towards the existence of a superfamily of receptors for adhesion molecules (Table 5.2). The basic structure of this superfamily is that of a heterodimer, with a β subunit that is relatively constant in structure, and a more variable α subunit which largely determines their binding specificities. Despite similarities in the β subunits it is clear that they are not identical for all of the adhesion related molecules. Nevertheless, apparent similarities suggest grouping of the molecules into at least three families within the superfamily. One family (identified by the presence of the β1 subunit) is characterized by the VLAs and integrin, another (characterized by the presence of the β2 subunit) includes the leukocyte adherence-related proteins, while the third (recognized by the presence of the β3 subunit) includes gpIIb/IIIa and the mammalian vitronectin receptor (Hynes, 1987).

Since a common feature of many of the integrins is their ability to bind RGD-containing peptides, the possibility exists that these receptors share similar binding sequences. An attempt has been made recently by D'Souza and colleagues (1988) to identify the RGD recognition site in the platelet gpIIb/IIIa receptor. These authors used radiolabelled lys-tyr-gly-arg-gly-asp-ser (KYGRGDS) as a probe for locating the receptor site. Briefly, activated platelets were treated with radiolabelled peptide and bound peptide was then covalently linked to neighbouring residues (presumably in the recognition site of the receptor) by a bifunctional cross-linking reagent. Following membrane extraction, proteolytic cleavage and sequence analysis of the products, the radiolabelled peptide was found to be linked to a region corresponding to residues 109–171 in gpIIIa. When compared with the β subunits of four other members of the integrin family, the amino acid sequence encompassed by this region was found to be highly conserved indicating a functionally important role. Of course, the α subunit gpIIb is likely to be involved in RGD recognition as well as the β subunit gpIIIa, and in any case the technical limitations of these experiments do not allow the 109–171 sequence to be conclusively identified as the RGD binding site, only that it probably lies in close proximity to it.

5.9 GENERAL ASPECTS OF CELL LOCOMOTION

The motile or locomotory responses of cells are of undeniable importance in biology, beginning from the very first steps of embryogenesis (Trinkaus, 1984). The motility of cells within the body is obviously under some form of control, since if it were not it would be difficult to envisage how body form and organization could be maintained. It has been argued that these controlling systems could break down in certain developmental anomalies and neoplasia, giving rise to changes in cell positioning which might be reflected respectively in perturbed morphology and invasion. A good review of cell locomotion is provided in Lackie (1986), and papers on several aspects of cell behaviour can be found in Bellairs, Curtis and Dunn (1982).

5.9.1 The machinery for locomotion

Considerable similarities exist between the contractile proteins of muscle cells and non-muscle cells, and thus it is not unreasonable to conclude that the actomyosin system represents the locomotory machinery or motor of most motile cells. The two principal components of this system are actin and myosin. Actin is a highly conserved globular protein (G-actin, 43 kD) which can polymerize into a filamentous form (F-actin) that is 5–7 nm in diameter but of indeterminate length. Interconversion between G-actin and F-actin is a reversible phenomenon dependent on monomer supply, ion concentration and ATP, and the process of polymerization usually begins at specialized nucleation sites. Myosin is composed of two heavy chains (200 kD each) and two pairs of light chains (16–22 kD each), and several myosin molecules can assemble into polarized filaments. Each heavy chain has a rod like segment at the C-terminal end and a globular head region at the other. Treatment with trypsin splits the molecule into rods and heads referred to as light meromysin (LMM) and heavy meromysin (HMM) respectively. ATPase activity is localized in the head region as is the ability to bind F-actin. Since the ability to bind F-actin is retained by HMM, this molecule can be used to label or decorate actin filaments. Under the EM, decorated actin has a characteristic arrowhead pattern and this can be used to indicate the polarity of the filaments. In addition to actin and myosin, non-muscle cells contain other proteins such as tropomyosin which are likely to be involved in the motor although their precise role is not yet clear.

In essence, the motor of motile cells acts like the actomyosin system of striated muscle. Although actin and myosin in non-muscle cells lack the highly organized structure seen for the same components in skeletal muscle, they are not totally randomly arranged and spontaneous polarization should ensure that they operate appropriately, if only at the molecular level. Of course, the molecular contractions of a cell must be co-ordinated in some way to ensure movement, but exactly how this takes place is still the subject of considerable debate.

F-actin within non-muscle cells is often visualized as stress fibres, which are bundles of microfilaments that are usually associated with other proteins of the motor including myosin and tropomyosin. In many cases of spread cells *in vitro*, stress fibres can be seen to extend through the cytoplasm from points of adhesion either to a general region around the nucleus or to other points of adhesion. At one time it was thought that these particular fibres were probably involved in cell locomotion, but in fact they are found predominantly in non-motile cells and they are more likely to arise as a result of cell spreading.

The motility of most non-muscle cells is inexorably linked to changes in cell shape. Exceptions arise in cases where movement is by flagella, for example, but these need not concern us here. However, although changes in shape are associated with locomotion, the form of a moving cell is remarkably persistent. This suggests that the motile form is preserved in some way, presumably as a result of cytoskeletal involvement. Apart from actin, other components of the cytoskeleton, particularly the microtubular and intermediate filament networks, are also likely to contribute to cell locomotion and shape changes.

5.9.2 Models for cell locomotion

Because their behaviour is relatively easy to observe under the microscope, fibroblasts serve as a paradigm in the study of cell locomotion. Most tumours, however, are epithelial in origin and we may well have to question (at least in part) the relevance of models for tumour cell locomotion based on fibroblasts. Two principal models exist to describe the basic mechanisms by which cells locomote:

(a) *The cytoskeletal model*

This general model for locomotion assumes that a cell is adherent to a substrate and that it extends a lammella-like protrusion in the direction of locomotion (Huxley, 1973; Hitchcock, 1977). The cellular extension is associated with changes in actin polymerization and cross-linking which contributes to the gelated state of the cytoplasm, and it polarizes the cell giving it a front and a tail. Since several such protrusions may be extended at once, it is not immediately apparent in what direction the cell will eventually proceed (if at all). At some stage, one protrusion usually dominates over the others, but this needs to be anchored to the substrate before full cellular translocation can occur. After anchoring, contraction of actin filaments within the protrusion will lead to translocation if three events are satisfied: (1) the actin network is anchored somehow within the protrusion; (2) the force of contraction can be transmitted to parts of the cell posterior to the protrusion; and (3) posterior adhesion sites are disrupted in preference to the new adhesion sites formed in association with the cell extension.

The general features of the model described above suggest that the moving cell is pulled forward in a manner akin to the frontal contraction hypothesis proposed for the movement of amoeboid cells (Allen, 1961). In this model, conversion of endoplasmic sol to ectoplasmic gel is thought to pull the endoplasmic core forward. In modern parlance we can envisage the polymerization ('gelation') of G-actin to F-actin which then associates with myosin to generate a contractile force. This force is transmitted from an appropriate anterior adhesive site through the posterior cytoplasm and the cell moves forward detaching its posterior end as it does so. Formation of the initial cell protrusion may involve an alternative but related process based on the ectoplasmic tube contraction hypothesis for amoeboid locomotion (Mast, 1925; Kamiya, 1959). According to this hypothesis contraction of posterior actin filaments would generate the protrusion by exerting pressure on the endoplasm. Although there is no definitive evidence for either of these models, it seems likely that components of both contribute to cell locomotion. Regardless of the mechanism, cytoplasmic flow similar to that seen in amoebae and probably involving reversible sol–gel conversion is associated with cell protrusion. As mentioned above, there is some evidence that this flow is focused on several points (possibly by the microtubular system) rather than over the entire lamella. Indeed, the microtubular system has been thought to represent part of a recycling system by which contractile and membrane elements may be passed forward into the protrusion zone (see below).

The idea of a cell reaching out, adhering and pulling itself forward from its very tip should not be taken too literally, since studies on flexible substrata actually suggest that traction is exerted by a broad area of the lower cell surface some distance behind its leading edge (section 5.9.3). Chen (1981) has provided evidence that tail retraction of cells locomoting *in vitro* can involve scission of the trailing edge so that portions of the cell (including adherent focal contacts) are left behind. Of course, this observation may be due entirely to *in vitro* culture conditions.

(b) *The membrane flow model*

In this model cells are thought to locomote by a membrane cycling process somewhat akin to a modified caterpillar track. Early observations by Abercrombie *et al.* (1970) and Harris and Dunn (1972) had shown that a migrating cell transports small particles backwards over its surface, primarily on the dorsal aspect although rearward movement of the ventral cell surface occurs. It was proposed that the moving cell added new membrane at its leading edge to replace that which was lost rearwards, where it was presumed to be endocytosed into the cell, processed, and passed forward internally. Evidence in support of this possibility was supplied by Kupfer, Louvard and Singer (1982) and by Bergmann, Kupfer and Singer (1983). These authors were able to show that, within minutes of taking on a polarized locomotory profile (induced by monolayer wounding), both the Golgi apparatus and the prominent microtubular organizing centre (MTOC) of NRK cells were positioned forward of the nucleus in the direction of subsequent locomotion. One of the consequences of the repositioning of the Golgi apparatus was that newly synthesized integral membrane proteins were inserted preferentially into the leading edge of motile cells. In poorly motile cells within the bulk of the monolayer there was no evidence of polarized insertion of new membrane components and the Golgi and MTOC were more randomly dispersed. Disruption of the cytoplasmic microtubules leads to dispersion of Golgi elements within the cytoplasm, from which it was hypothesized that the microtubules might provide a cytoskeletal guiding system for directing the traffic of Golgi-derived membrane vesicles towards the leading edge.

Because of the likely requirement for new membrane in the anterior end of a moving cell, membrane cycling may well contribute to all models which attempt to describe the process of cell locomotion. In the membrane flow model, however, the cycling process itself (acting through appropriate substrate adhesion sites) is thought to generate the force leading to movement (Bretscher, 1988). Experiments to date fall short of resolving which if either of the two models is more applicable to describing the processes by which cells locomote and further work is obviously required.

5.9.3 Traction and locomotion

Harris and his colleagues have developed an interesting experimental system to visualize the forces generated by adherent and locomoting cells. Briefly, small quantities of silicone fluid are dispensed into dishes and a thin film of silicone rubber is formed on the top by cross-linking the outer surface of the fluid through exposure to a flame. When cells are seeded rather sparsely onto such surfaces they distort it in two ways:

(a) Tension wrinkles extend through the silicone rubber beyond the margins of the cells.
(b) Compression wrinkles form directly beneath the cells perpendicular to the direction of maximum traction. These wrinkles are thought to relate to sites on the cell surface where the traction forces are exerted. Interestingly, compression wrinkles form behind the front of the cell suggesting that traction is not exerted at the leading edge, but some distance behind it.

When different cell types were seeded onto a silicone rubber surface, some startling results were seen. Fibroblasts and platelets distorted the rubber the most, whereas macrophages distorted it poorly and PMNs hardly at all. These and similar results suggest that there may an inverse relationship between traction and motility. The possibility that traction might be inversely related to the invasive capacity of tumour cells was studied using transformed CHO cells. 'Reverse' transformation of these cells was seen to be associated with increased traction as evidenced by more wrinkling of the silicone rubber substrate (Leader, Stopak and Harris, 1983). The precise relation between traction and invasion, however, remains to be elucidated.

Cell traction can also be viewed on collagen gels, which may have more relevance than silicone rubber. When tissue explants (containing fibroblasts) are placed on such gels, collagen fibres are arranged to form tracts reaching out over 4 cm in length, and they also align and compact circumferentially around the explants. Such patterns are permanent, presumably as a result of cross-links formed between neighbouring fibres, and they are strikingly reminiscent of tendons and organ capsules. In fact, Stopak and Harris (1982) have suggested that these anatomical structures may be formed by the mechanical alignment of collagen fibres by the traction forces exerted by fibroblasts. Their hypothesis proposes that the traction forces of fibroblasts, which are far greater than that needed for their locomotion alone, are predominantly involved in structural morphogenetic events. Thus the locomotion shown by fibroblasts in culture may actually reflect frustrated attempts by the cells to rearrange their substrate. In essence, if the substrate cannot be moved by traction, then the cell is moved instead. These novel ideas relating to the locomotion of fibroblasts lead us to query the use of this cell type as a paradigm for locomotion. Interestingly, few malignant tumours are associated with a fibrous capsule and it would be intriguing if its deposition by fibroblasts was influenced by the malignancy. Certainly, Schor and his colleagues have suggested that apparently normal fibroblasts from individuals with malignant disease behave aberrantly (section 4.7).

5.10 MEASURING LOCOMOTION

Neutrophils typically move *in vitro* at speeds of about 10–15 μm min^{-1}. These are probably amongst the fastest moving cells in the body and they move roughly an order of magnitude faster than do fibroblasts. Quantifying the various aspects of cell locomotion, however, is not a simple case of measuring the speed of moving cells. Many cells move *in vitro* on a solid 2D surface, but this does not necessarily describe how they move *in vivo*. Cell adhesion to the substrate is an integral aspect of most forms of 2D locomotion, and thus the nature of the substrate will clearly influence the degree of locomotion. However, we have seen that cells need not necessarily adhere *per se* in order to move through a 3D environment as would be found *in vivo*. Cells moving in such an environment could gain purchase from hooking extensions around collagen fibrils and using this anchorage to draw themselves forward, somewhat analogous to the way a monkey swings through the trees. Cell locomotion is also influenced by the presence of other cells and by chemicals in the environment, which can serve to promote or inhibit locomotion, or guide it in a particular direction (section 5.12). Furthermore, not all cells within a population move to the same extent. Some may move poorly if at all, while others may be rapid movers. At least two populations of human monocytes are thought to exist, for example, one of which lacks receptors for the synthetic chemotactic substance FMLP and does not respond chemotactically to this agent (Fach, Harvath and Leonard, 1982). Human PMNs also have a non-migratory population, but in this case the defect is not due to a deficiency in cell surface receptors (Harvath and Leonard, 1982).

Cell movement is often assessed by measuring the rate of progression of the leading front of a cell population: in filter assays (section 4.5) the term 'leading front' usually refers to the two fastest moving cells which occur in the same plane as visualized by light microscopy. Varani, Orr and Ward (1978b) developed a simple system for quantifying locomotion by measuring the speed at which cells emigrated from an agar drop, but this too only relates to the migration of the cells in the leading edge. These cells are not representative of the population as a whole, and thus detailed cell-by-cell analysis is required in order to obtain a better estimate of the average movement of a cell population. If the analysis proceeds over some time, then cell division will influence the rate of locomotion since two cells cannot occupy the same space. It is not fully appreciated as to how the cell cycle affects locomotion, but since cells round up during mitosis locomotion is likely to be minimal during this phase.

The tracks taken by individually moving cells can show considerable variation. Under uniform (**isotropic**) conditions *in vitro* (usually on a 2D substrate and in serum-containing culture medium), cells do not show true random locomotion since they have a tendency to persist in a particular direction. This 'persistence' is probably a consequence of energy conservation and it is likely to be governed by cytoplasmic and membrane-located events within the cell. Interestingly, neutrophils seldom reverse direction by changing polarity through 180°: they prefer to turn in small steps (Zigmond, Levitsky and Kreel, 1981). Because cells persist in a given direction, at least in the short term, the path they take is described as a **biased random walk**. A description of cell locomotion in a uniform environment thus has two components: the speed at which the cells move, and the extent of their persistence (Gail and Boone, 1972; Dunn, 1983). A change in the speed or frequency of movement is referred to as **orthokinesis**, whereas a change in the frequency or magnitude of turning (persistence) is known as **klinokinesis**.

An interesting model system for studying cell locomotion has been developed by Albrecht-Buehler (1977). In this model, cells are allowed to settle and move on a glass substrate covered with gold particles. As the cells move, they remove the particles from the substrate leaving behind gold-free tracks which are easily seen under dark field optics. Since some of the gold particles appear to be phagocytosed by the moving cells, Albrecht-Buehler has coined the term **phagokinesis** to describe the process. The phagocytosed gold particles are probably not without an effect on cell behaviour, since they may accumulate in cells to a reasonable quantity before being released by exocytosis. For some cells at least, the rate of phagocytosis appears to exceed the rate of exocytosis and thus the particles inside the cells slowly increase with time, possibly bringing the cells to a standstill.

More recently, Partin and colleagues (1989) have used Fourier analysis to quantify cell motility. Working with the Dunning R3327 rat prostatic adenocarcinoma cell system these authors have shown a correlation between changes in motility and increased metastatic ability. The Dunning model is based on a number of histologically indistinguishable tumour variants which nevertheless display diverse metastatic capabilities. Although Partin *et al.* (1989) argue that cancer cell motility *in vitro* may be sufficient to characterize metastatic capability *in vivo*, at least for the Dunning model, we shall see later that the universality of this conclusion is questionable.

5.11 CHEMOTAXIS AND CHEMOKINESIS

Although the locomotion of cells *in vitro* can be studied under relatively uniform conditions, there is every likelihood that the environment *in vivo* is non-uniform (**anisotropic**). There are several possible sources of environmental variability *in vivo*. These include:

(a) *Biochemical gradients*

Gradients of biochemicals may influence locomotion directly, or indirectly by altering other aspects of cell behaviour which themselves impinge upon motility. Such gradients may serve two important functions with respect to the spread of tumours. First, they contribute to the process of angiogenesis, and second they may stimulate the directional invasion or extravasation of metastatic cells.

(b) *Structural discontinuities*

It is well known that cells tend to be guided along grooves and structures such as nerves and blood vessels (**contact guidance**). With a certain elegance lacking in most of modern science, Harrison (1914) grew cells on spider webs and fish scales, and showed that the cells became aligned along the fibres or in the grooves respectively. Cells also follow oriented collagen tracts and they often move preferentially in the plane or gap between two tissues. There is abundant histopathological evidence that many malignancies may invade along structural discontinuities.

Interestingly, sarcoma 180 cells will move along neural crest migration pathways when they are experimentally grafted into them, but normal fibroblasts will not. These results suggest that the presence of a pathway does not necessarily result in guided motility, and thus factors intrinsic to the cell are also likely to play an important role (Trinkaus, 1985).

(c) *Adhesive discontinuities*

Cells need to adhere to the substratum in order to locomote, and thus differences in substratum adhesiveness could guide cells to more adhesive areas (**haptotaxis**).

The directed locomotion of a cell in response to a gradient of some substance in the environment is known as **chemotaxis**. The term **chemokinesis**, on the other hand, refers to a change in the rate or frequency of movement and/or a change in the frequency or magnitude of turning in response to a chemical signal. This signal may or not be in a gradient form. Kineses thus differ from taxes by the lack of a directional component in the way they respond to an external signal. Cells moving kinetically can move relatively straight and they can assume a localized distribution if their movement ceases as they near the source of the chemical signal. Many aspects of chemotactic and chemokinetic behaviour, particularly for leukocytes, bacteria and slime moulds, are discussed in books by Gallin and Quie (1978), Lackie and Wilkinson (1981), and Wilkinson (1982).

What is the mechanism which enables a cell to respond chemotactically? By definition, we have seen that a cell moves chemotactically in response to a gradient of a chemical substance (the **chemoattractant**) and thus it must possess some mechanism for detecting the chemical gradient. Detection itself is not a major problem, since we need only envisage the presence of a receptor for the chemoattractant. However, how does a cell assess differences in the concentration of the chemoattractant so that it can move appropriately in response to the gradient? Calculations suggest that a cell can respond directionally when there is a difference of as little as 10–20 bound receptors between the front and tail of a cell. Two models have been proposed to account for gradient detection (Zigmond, 1974):

(a) *The spatial sensing model*

In this model, the concentration of a chemoattractant is assessed over all or part of the cell surface, and the concentrations are compared. The minimum requirement is two receptors spatially separated, but the presence of more receptors would improve resolution.

(b) *The temporal sensing model*

Here, the concentration of a chemoattractant is assessed (possibly by the same receptors) in different localities as a consequence of cell movement. The minimum requirements in this case are the need for discrete sampling and some form of memory for comparing the samples.

Which, if either, of these models more accurately explains the mechanisms of sensing is not known. Indeed, it is possible that some mechanism comprising aspects of both of these models may operate.

The final problems associated with responding to a chemotactic signal concern transducer and effector processes. Somehow the message must cross the cell membrane and activate the locomotory machinery so that the cells respond accordingly. These are clearly in the domain of intracellular signalling systems (section 1.7.22), but precisely how these systems operate to effect directed locomotion in response to a chemoattractant is not yet fully elucidated (Snyderman and Goetzl, 1981).

5.11.1 The chemotactic behaviour of tumour cells

We have seen that certain malignant cells have a tendency to metastasize to particular organs (section 2.14.6). This organ-selective metastatic behaviour led to the proposal that chemotactic gradients emanating from particular cells or organs might contribute to the observed metastatic pattern (Hayashi *et al.*, 1970; Hayashi and Ishimaru, 1981; Lam *et al.*, 1981). Indeed, tumour cells can be shown to respond chemotactically to normal tissue extracts (Orr *et al.*, 1979; Mundy, De Martino and Rowe, 1981; Mundy and Poser, 1983;

Magro *et al.*, 1985; Hujanen and Terranova, 1985), to extracts of tumour cells (Romualdez and Ward, 1975; Wass, Varani and Ward, 1980; Hayashi *et al.*, 1970), to synthetic chemoattractants such as FMLP (Wass *et al.*, 1981; Elvin, Wong and Evans, 1985), and to fragments of complement (Lam *et al.*, 1981).

Walker 256 carcinosarcoma cells, derived originally from a rat breast tumour, have a propensity to metastasize to bone. Experimentally, it has been shown that W256 cells respond chemotactically to products released from resorbing bone (Orr *et al.*, 1979) and more recently Magro *et al.* (1985) have shown that such products promote the adhesion of W256 cells. These authors speculated that as W256 cells pass through the microcirculation of the bone they may aggregate in response to products of local bone resorption. Such aggregates could lodge in the capillaries, and tumour cells would then emigrate from them in response to a gradient of the products of bone resorption. The precise molecular nature of these products is uncertain although that responsible for aggregation has been separated from that responsible for chemotaxis by gel filtration. Despite its appealing nature, this hypothesis requires further experimental support and nothing is known of what gradients, if any, exist in the vicinity of resorbing bone *in vivo*.

Hayashi's group have isolated a tumour-derived chemotactic factor which can promote site-selective colonization by injected tumour cells. When Ozaki *et al.* (1971) injected this chemotactic factor intradermally into host rats which had previously been inoculated with tumour cells directly into the carotid artery, samples of injected skin sites at various time intervals thereafter showed tumour cells adherent to the endothelium of venules within 24 h and extravascular tumour cells by 72 h. At later times, tumour invasion of underlying connective tissue and muscle in the injected areas became increasingly evident. Rats not injected with tumour cells showed only mild oedema and some localization of PMNs after injection of the chemotactic factor. Furthermore, the intradermal injection of permeability factors or PMN chemotactic agents into tumour-bearing animals promoted oedema and PMN localization, but did not promote tumour cell colonization.

In one of the more detailed studies of the organ-selective chemotactic behaviour of tumour cells to date, Hujanen and Terranova (1985) confirmed that some tumour cells display organ-selective chemotactic behaviour *in vitro* that correlates reasonably well with their metastatic or colonizing behaviour *in vivo*. Thus, as determined using a modified Boyden chamber assay system coated with type IV collagen, brain-colonizing B16-Br2 melanoma cells, liver- and ovary-colonizing M-5076 reticulum cell sarcoma cells, and lung-colonizing T341-PM2 fibrosarcoma cells responded best to brain, liver and lung derived factors respectively. Non-metastasizing mouse C3H fibrosarcoma cells failed to respond chemotactically to any of the organ extracts tested by these authors. The molecular nature of the putative chemotactic factors involved is far from clear, but preliminary studies using Sepharose CL-6B chromatography implicated a component of high molecular mass (c. 1–1.5 $\times$ 10^6 kD) in the chemotaxis of B16-Br2 cells to the brain extract, and one of lower mass (c. 1–1.5 $\times$ 10^5 kD) in the chemotaxis of M5076 cells to the liver extract.

Despite these interesting and informative studies on tumour cell chemotaxis, it should be emphasized that organ-directed motility is not an absolute pre-requisite for all tumours which display selective metastatic behaviour.

5.11.2 Autocrine motility factor

A new class of motility factors has recently been reported to be secreted by some tumour cells and embryonic cells (Guirguis *et al.*, 1987; Stoker *et al.*, 1987). The material isolated from human melanoma cells and rat breast carcinoma cells has been shown to be a protein of M_r 54 kD and it appears to induce pseudopodial extension and to increase locomotion in homologous cells. Although referred to as autocrine motility factor (AMF), this protein in fact does induce motile responses in heterologous cells, providing they express the appropriate receptors. The pseudopodia induced by AMF (<1 nM) from the human breast carcinoma cell line MDA 435 appear about 30 min before cell locomotion is induced. When pretreated with an appropriate antibody, the inducing effect on both pseudopodial extension and locomotion is inhibited suggesting some relationship between pseudopodial extension and locomotion. The mechanism for detecting pseudopodial formation involves a frustrated form of locomotion in which the cells are seeded on to 3 μm diameter pore size filters into which they may extend cellular extensions, but through which they cannot translocate. Interestingly, the pseudopodia which form have prominent stress fibres, and there is evidence from other systems that such fibres may be related more to adhesion than to locomotion. Certainly, there is roughly a 20-fold increase in the number of laminin and fibronectin (arg-gly-asp) recognition sites in pseudpodial membranes relative to whole cell plasma membrane. We shall see later that many tumour cells respond haptotactically to laminin and fibronectin, and thus if AMF increases the concentration of laminin and fibronectin receptors at the leading edge of the cell then it could promote a haptotactic response.

5.12 ADHESION [illegible]PTOTAXIS

The term **hap**[illegible]rected locomotion of cells in response to a [illegible]ess. Carter (1965, 1967) first provided evide[illegible]ation by coating a non-adherent cellulose acet[illegible] a glass coverslip) with a gradient of palladium [illegible]dhesive surface for cells. When mouse L cel[illegible]urfaces they tended to move from regions of [illegible]ions of higher adhesiveness (more palladium)[illegible] adhesiveness can also serve to trap cells in pa[illegible] because their rate of locomotion is slowed do[illegible]d more time on the adhesive surface (Wilkinso[illegible]use adhesiveness is so high that the cells are [illegible]ld be noted here, that cells which do move o[illegible] do so more or less randomly and do not bec[illegible]4a). This pattern of behaviour suggests kinetic [illegible]omena, a possibility which needs to be explor[illegible]

Ha[illegible]sponse to a gradient of adhesion is not neces[illegible]ration in response to a substrate-bound gradient [illegible] Consider the following example: Dierich, Wilhelmi and Till [illegible]ere able to demonstrate that PMNs could migrate in response to substrate-bound casein (which is chemotactic for these cells), and Wilkinson and Allan (1978) subsequently showed that such cells could move up a gradient of substrate bound chemoattractant. Two possibilities exist to explain this behaviour:

(a) The response of the cells to the gradient of surface-bound chemoattractant is essentially similar to their response to a fluid phase gradient i.e. they display chemotaxis to a bound signal.
(b) Cell binding to the substrate-located molecule is an adhesive event, and the cells move haptotactically in response to increasing adhesion up the gradient.

Which of these two possibilities might more reliably explain directional locomotion in response to a substrate bound chemical is not easily resolved experimentally. According to Wilkinson and Allan (1978) chemotactic peptides such as f-met-phe bind poorly to albumin-coated substrates, and they suggest that directed locomotion in this case reflects true chemotaxis to a fluid phase attractant rather than haptotaxis. Since most cells cannot locomote on a clean tissue culture surface, it seems that the albumin provides a suitable surface for PMN locomotion in response to an appropriate chemotactic signal. Gradients of protein molecules such as casein, on the other hand, not only provide an adequate surface for adhesion but they also stimulate the directional locomotory response. For such molecules it is still not clear whether it is the gradient of adhesiveness which generates the directional locomotory response or the gradient of chemotactic inducing activity. Nevertheless, the ability of cells to respond to substrate bound attractants whether chemotactic or haptotactic in nature may be of considerable significance *in vivo* since the gradient generated may be more stable than would be the case in the fluid phase.

Cell behaviour *in vivo* is believed to be considerably influenced by the extracellular matrix (ECM), particularly during development and in other processes involving tissue remodelling. Since many cells adhere to the ECM, its components could conceivably direct the movement of cells in a haptotactic manner. The treatment of fibroblastic cells with the ECM component fibronectin induces cytoskeletal organization, spreading and a polarized morphology which are reflected in enhanced cell locomotion (Ali and Hynes, 1978). Furthermore, the inclusion of fibronectin in collagen gels promotes invasion by Syrian hamster melanoma cells (Schor, Schor and Bazill, 1984). Perhaps not surprisingly in the light of these observations, Boyden chamber filters precoated with gradients of fibronectin promote the migration of a variety of tumour cells including B16F10 melanoma cells (McCarthy and Furcht, 1984). Similarly, laminin and vitronectin (which, like fibronectin, also promote cell adhesion and spreading) are able to induce directed locomotion in tumour cells (McCarthy and Furcht, 1984; Basara *et al.*, 1985). In these examples it is possible to precoat the substrate with a poorly adhesive substance such as albumin which blocks the binding of subsequenty applied proteins. In such cases the cells then fail to move up soluble gradients of the adhesive molecules. Clearly, the molecules need to be surface-bound and in gradient form in order to elicit directed locomotion, but there is still no definitive proof that the cells are responding haptotactically i.e. that they are actually driven directionally by adhesiveness. Be that as it may, a general hypothesis has emerged that tumour cells which lodge on the endothelium might extravasate by exchanging low-energy adhesion to the endothelium for high-energy adhesion to components in the basement membrane and subsequently in the interstitial stroma. There appears to be only a relatively small amount of fibronectin (and virtually no laminin) on the surface of the endothelium and this gives some credence to this hypothesis, but there is no firm evidence that an adhesive gradient could be perceived by emigrating tumour cells. Lacovara, Cramer and Quigley (1984) have suggested that invading tumour cells might enzymatically remove fibronectin thereby creating their own gradient but even if degradation is localized, it is hard to envisage how directionality might result.

It is worth noting that not all cells migrate equally well to vitronectin, fibronectin or laminin. MCF-7 breast carcinoma cells, for example, migrate in response to vitronectin but not to fibronectin (Basara *et al.*, 1985). For B16 melanoma cells the peak migratory response is seen around 25 μg ml^{-1} for both fibronectin and laminin, whereas RN22F rat Schwannoma cells respond to laminin (peak around 15 μg ml^{-1}) but not to fibronectin, even though these cells synthesize both molecules (McCarthy, Palm and Furcht, 1983). Interestingly, antibodies against laminin can inhibit the migration of RN22F cells on a gradient of this molecule, but they do not inhibit RN22F cell adhesion to the same substrate. These results suggest that there may be differences in the ways cells adhere and migrate on laminin, and indeed the motile behaviour observed on a gradient of this molecule may not be haptotaxis at all.

In a rather startling series of experiments, Newman *et al.* (1985) found that non-uniform matrices of collagen and fibronectin not only promoted the directional translocation of cells, but they also promoted the translocation of inert polystyrene-latex beads (provided they were of a certain size and could interact with fibronectin). The basic experiment involved preparing two contiguous gels of type I collagen, one containing the cells or particles under study and the other containing fibronectin. The cells or particles were seen to translocate at speeds of approximately 5 μm sec^{-1} about 10 sec after pouring the gels and continued moving for 20 min. The authors interpreted this locomotion-independent behaviour to events associated with fibrillogenesis (which occurred under physiological conditions), and speculated that some cell movements *in vivo* might reflect similar mechanisms, if only in part. Whether or not malignant cells might be able to express this behaviour is not known, although they could conceivably alter local patterns of fibrillogenesis.

Various cell types have been shown to adhere, spread and translocate on fibronectin. McCarthy, Hagen and Furcht (1986) have shown that B16F10 melanoma cells can display all three of these types of behaviour on a 75 kD trypsin-derived fragment of fibronectin, but they could not demonstrate motility on a smaller 11.5 kD pepsin-derived fragment, even though both contained the sequence RGDS. These results suggest that different sequences may be involved in generating adhesive and motile behaviour in B16F10 cells, and that RGDS is not involved in their motility. Furthermore, a 33 kD heparin-binding fragment of fibronectin stimulated melanoma cell adhesion and spreading but was without effect on motility. Cell binding by this sequence could conceivably be mediated by cell surface heparan sulphate proteoglycan, but Humphries *et al.* (1986) were unable to show any influence of heparin on B16F10 binding to fibronectin. The conclusion reached by McCarthy and colleagues (1986) is that the haptotactic response of B16F10 melanoma cells to fibronectin may not be due to a simple adhesion gradient of this protein, which rather intriguingly negates the original definition of haptotaxis as an adhesion based behavioural phenomenon. The studies of Humphries *et al.* (1987) decribed in section 5.4.2 indicate that B16 melanoma cells adhere preferentially to sequences in the type IIICS region of the longer fibronectin polypeptide. The most active of these sequences appears not to contain RGDS, and its activity also seems to be independent of heparan sulphate. Whether or not this sequence exists in the 33 kD fragment used by McCarthy *et al.* (1986) is unclear, and further work is required before these conflicting results can be resolved.

5.13 THE EFFECTS OF TUMOUR CELLS ON HOST LOCOMOTION-RELATED FUNCTIONS

Dizon and Southam (1963) first noted that macrophages represented a class of cells which were selectively deficient in the cellular response of cancer patients to dermal abrasion. Since this report, a considerable amount of evidence has accumulated confirming that the presence of a malignant tumour may affect monocyte/macrophage behaviour, possibly as a consequence of altered chemotactic responsiveness. D'Arrigo and her colleagues (1985), for example, explored the *in vivo* exudative responses of both macrophages and granulocytes in mice bearing the B16 melanoma. They reported that the numbers of cells in both resident and induced populations of peritoneal macrophages were more consistently depressed in tumour-bearing mice relative to that in normal control mice. These results seemed not to be a simple function of tumour burden, and whilst macrophage exudation in response to peritoneal inflammatory excitation was depressed in tumour bearing animals, there was little if any effect on PMNs.

It is not precisely clear how a tumour such as the B16 might influence leukocyte behaviour *in vivo*, but it has been speculated that retroviral products might be responsible. The retroviral envelope protein p15E, for example, has been shown to have immunosuppressive and anti-inflammatory activity. Thus purified p15E administered to mice inhibits macrophage accumulation at inflammatory sites (Cianciolo *et al.*, 1980), and the anti-inflammatory effects of a methylacholanthrene-induced fibrosarcoma were specifically negated by treatment with monoclonal anti-p15E antibody (Cianciolo *et al.*, 1983). Since p15E is produced by B16 melanoma cells (Cianciolo *et al.*,

1983), it could conceivably play a significant role in suppressing macrophage exudation in mice bearing this tumour (Snyderman and Cianciolo, 1984; D'Arrigo *et al.*, 1985). Although a possibility, why macrophages should be affected in this way and not PMNs is not clear.

The abnormal chemotactic behaviour of monocytes from certain cancer patients may be detectable *in vitro* before lymph node involvement is discernible, thus highlighting the possible clinical relevance of simple cell behavioural assays (Boetcher and Leonard, 1974). Furthermore, patients with malignant tumours whose monocytes show more normal chemotactic behaviour generally have a better prognosis than those whose monocytes display abnormal behaviour (Snyderman, Siegler and Meadows, 1977). Interestingly, surgical removal of malignant tumours was found to be associated with a return of monocyte chemotactic behaviour to normal levels on retesting about 3 weeks after surgery (Snyderman and Stahl, 1975). This observation clearly implies that abnormal monocyte chemotactic behaviour is related to the presence of the malignancy. Snyderman *et al.* (1978) studied three groups of patients with breast disease: the first group had benign lesions, the second group were clinically free of disease having had malignant tumours removed, and the third had active breast cancer. Monocyte chemotactic behaviour was essentially normal in the first two groups and abnormal only in the group with active breast cancer (regardless of stage). In many instances, the presence of a malignant tumour is without effect on granulocyte locomotory behaviour (Fauve *et al.*, 1974; Snyderman *et al.*, 1976; D'Arrigo *et al.*, 1985) although this appears not always to be the case (Hamby and Barrett, 1977; Maderazo, Anton and Ward, 1978). Since both granulocytes and macrophages can destroy tumour cells, at least *in vitro*, it would be of interest to know the prognosis for those cancer patients who showed abnormal chemotactic behaviour for both of these inflammatory cell types.

It should be noted that data from both animals and humans with malignant tumours suggest that individuals may vary considerably in the chemotactic responses of their inflammatory cells. Some tumour-bearing individuals may actually show a heightened monocyte chemotactic response. According to Bottazzi *et al.* (1983), culture supernatants of both murine and human tumour cells promote monocyte/macrophage chemotaxis *in vitro*, but results from clinical studies suggest that in the majority of patients (about 85%) this response is depressed. It has been proposed that the extent of chemotactic responsiveness may be a function of the immune status of individuals, but this remains speculative at the moment. Animal studies have suggested that the presence of a tumour may have a biphasic effect on macrophage activities, their locomotion and bactericidal activity being depressed early in tumour growth and enhanced in later growth (Snyderman *et al.*, 1978). This idea is akin to the concept of immune stimulation-inhibition formulated by Prehn (1972) and discussed in section 1.8.1. There is also some evidence to suggest that infection may increase the chemotactic ability of monocytes in patients with cancer (Boetcher and Leonard, 1974). Despite these sources of variation, since monocytes are potentially cytotoxic to tumour cells, it is not hard to envisage how inhibiting their chemotactic behaviour might confer some advantage on tumour growth.

5.14 LOCOMOTION, INVASION AND METASTASIS

It is clear from the above that cancer cells can move on various substrata and that they can respond to chemotactic and possibly haptotactic signals. However, how important is malignant cell locomotion in invasion and metastasis? One body of evidence equates increased cell locomotory capacity with enhanced metastatic ability. Volk *et al.* (1984), for example, have reported this correlation following their studies with the K1735 melanoma system. Similarly, Varani *et al.* (1980b) have reported a correlation between locomotory ability and metastatic potential using a range of clones of a chemically induced murine fibrosarcoma. These authors employed the agarose droplet assay technique and concluded (as judged by leading front analysis) that metastatic clones of this tumour migrated on the average about 66% as far as poorly metastatic clones. It is worth noting here that the parent population used in this study was moderately malignant and that the poorly metastatic clones were thus selected for 'negative' characteristics. Taniguchi and colleagues (1989) have explored the relationship between increased motility and experimental metastasis using v-*fos*-transfected, *src* virally-transformed 3Y1 rat fibroblastic cells. They found that the increased lung colonizing ability of the oncogene-transfected cells relative to that of control cells correlated with enhanced invasive ability through Matrigel (an artifical substrate primarily of type IV collagen and laminin) and also with increased motility. Significantly, many other behavioural parameters shown by other researchers to correlate with experimental metastasis failed to display any such relationship. These parameters included tumour cell arrest in the lungs, cell growth in the presence of a lung extract, sensitivity to NK cells, attachment to basement membrane components, and type IV collagenase production. A point worth noting here is that because of the short period of observation the 'motility' measurements of Taniguchi *et al.* (1989) actually reflect pseudopodial extension rather than significant cell movement *per se*.

Another body of evidence has suggested that, on the contrary, metastasis correlates with decreased locomotory ability (Werling *et al.*, 1986). Of course, these studies are based on too small a range of cell types to allow generalizations to be made, but since variant results have been reported it would seem that there is no universal locomotory mode which correlates with metastasis. This is in accord with the studies of Haemmerli *et al.* (1983, 1984) who reported both motile and non-motile cells with metastatic capabilities. Furthermore, other studies of Volk *et al.* (1984) in which no correlation could be found between locomotion and metastatic ability for the B16 melanoma and UV2237 fibrosarcoma model systems also support this conclusion. Nevertheless, when Netland and Zetter (1985) selected B16 cells for preferential adhesion to lung tissue they obtained cells with higher lung-colonizing capacity which displayed increased chemokinetic activity. If nothing else, these results confirm the nuances of *in vitro* study since in many ways the results are a function of the system employed. The studies of Schor (1980), in which he examined cell infiltration into collagen gels, have shown that both the ability to move on collagen and the ability to grow within a gel do not necessarily result in invasion. HeLa cells (from a human cervical carcinoma), for example, do not invade 3D collagen gels even though they move faster on collagen than on a plastic tissue culture surface.

It thus seems that while motility may be important for the spread of some tumours, it may not be so for others. Wood (1958) has provided unambiguous visual evidence for the importance of locomotion in the extravasation of V_2 carcinoma cells. The metastasis of single leukaemic cells can hardly be imagined to occur without recourse to locomotion, and this has been shown convincingly by Haemmerli and Strauli (1978) who filmed infiltration of the mesentery by L5222 cells. Because carcinomas often invade *en masse*, it has been implied that they can invade and metastasize without recourse to individual cell locomotion. Cells may be moving within the tumour mass, however, and individual cell locomotion could thus contribute to the invasion of a tongue or sheet of carcinoma cells. Furthermore, single cells or small clumps can detach and move away from the mass and these may be very important in establishing metastases. Interestingly, Werling *et al.* (1986) showed that non-motile but metastatic BSp73ASML rat tumour cells could be carried piggy-back by fibroblasts, but whether this form of passive locomotion has any relevance *in vivo* is not known.

Goodman, Vollmers and Birchmeier (1985) have reported selection of a murine McAb (SLOW-1) which can inhibit the the locomotion of chick embryo fibroblasts in tissue culture assays. It has also proven effective against mouse NIH 3T3 cells, human MRC5 fibroblasts and human TR126 epithelial cells, but it was inactive (or poorly active) against MDCK canine epithelial cells, the B16 mouse melanoma and CSG120/7 mouse carcinoma-like cells. The full molecular details of how SLOW-1 exerts its effects are not clear, but it appears to recognize two glycoproteins of 44 kD and 57 kD on responsive cells. Although these antigens are not enriched on the ventral surface, it still remains a distinct possibility that the McAb acts by perturbing cell adhesion. McAbs analogous to SLOW-1 could have applications in studying the role of motility in metastasis, but as yet no appropriate McAbs have been developed.

In an earlier study, Cohen and colleagues claimed to have prepared a non-toxic factor from lymphoid cells (including con-A-stimulated murine lymphocytes) which reversibly inhibited the *in vitro* migration of various mouse and rat tumour cells (Cohen *et al*., 1978). Although this is indicative of a process by which tumour cell motility may be modulated, the *in vivo* significance of these observations remains to be determined.

6 Reflections and new horizons

6.1 GENES, THEIR REGULATION AND MALIGNANCY

A detailed study of the association between genes and cancer is clearly beyond the scope of this book, although we have touched upon the importance of oncogenes in tumorigenesis. There is abundant evidence that genetic changes are associated with many cancers. Consider, for example, the association between chromosomal abnormalities and cancer, the effects of DNA changes induced by chemicals and irradiation, the increased incidence of cancer in some patients with defects in their DNA repair mechanisms, and the effects of oncogenic viruses.

We should remind ourselves here of two points: (a) whatever causes cancer is heritable, in the sense that it is passed on from the originally transformed cell to its daughters, and (b) although not all carcinogens are mutagenic, around 90% appear to be so from *in vitro* studies. That not all mutagens are carcinogens and not all carcinogens are mutagens leads us to suspect that epigenetic events (changes in gene activity) may be important in the development of some cancers. Thus the expression of the transformed state is likely to be affected by epigenetic processes and herein may lie the importance of such events in cancer, not so much in causing transformation but in influencing its outcome. This point illustrates again the complexity of the cancer cell, since although we may look towards the genome as the root of transformation, for some cancers we must also look towards epigenetic processes to fully understand the outcome of potentially transforming changes. Yet even this view may be somewhat simplistic, given the possible inheritability of epigenetic defects (Holliday, 1987) and the observation (described below) that some epigenetic changes may also lead to mutagenic change. The inheritability of epigenetic changes may have important consequences in the tansformation process. Holliday (1987) has suggested that the low level of spontaneous transformation of human cells in culture relative to that for rodent cells (even though both have similar mutation frequencies) may be due to the presence of an effective buffering system against epigenetic change in the former.

Hypomethylation of DNA is an epigenetic event that can result in the activation of specific genes. Existing evidence suggests that hypomethylation may persist for at least several cell generations before reversion occurs (Razin and Riggs, 1980; Riggs and Jones, 1985). Briefly, DNA methylation involves covalent modification of the genome, almost exclusively at 5′-cytosine-guanine-3′ sequences. In fact, about 70% of cytosines in normal, differentiated mammalian cells are methylated. Studies with a range of experimental tumours have revealed considerable variation in cytosine methylation levels. Nevertheless, the extent of DNA methylation has been correlated with the lung colonizing potential of B16 murine melanoma cells treated with the nucleoside analogue 5-azacytidine (which induces DNA hypomethylation). There is a possibility, however, that this drug may have mutagenic potential (see below) and thus epigenetic events alone may not be responsible for its reported effects on B16 cells. In a study of human tissue, Goelz *et al.* (1985) reported a correlation between the extent of hypomethylation and tumorigenicity. These authors studied normal colon tissue, benign adenomatous colonic polyps (which may become malignant), and fully malignant colon tissue, and they found that less DNA methylation occurred in both the benign and malignant tumours relative to normal controls. Thus hypomethylation was apparent before progression to malignancy. An earlier report involving one patient suggested that metastases may have even less methylated DNA than their primaries, but further studies are required to substantiate this possibility (Feinberg and Vogelstein, 1983).

How might DNA hypomethylation influence tumorigenicity? The relationship between hypomethylation and gene expression suggests that resting genes (perhaps normally involved in development) might be switched on and these could then lead to expression of the tumorigenic phenotype. Alternatively, DNA hypomethylation might interfere with chromosome condensation which could then result in chromosomal abnormalities that lead to transformation. Indeed, treatment of some cells with 5-azacytidine leads to their transformation which is associated with hypomethylation and a specific chromosome translocation (Harrison *et al.*, 1983).

Olsson and Forchhammer (1984) have examined the effects of treatment with 5-azacytidine on the expression of the metastatic phenotype in Lewis lung carcinoma cells. These authors first selected clonal variants which were tumorigenic but either expressed (T^+M^+) or failed to express (T^+M^-) metastatic potential. Treatment of the T^+M^- variant with the drug induced the metastatic phenotype, whereas treatment of the T^+M^+ variant resulted in the loss of metastatic potential. Both of these effects occurred without obvious changes in immunogenicity. Nevertheless, when McAbs were prepared against the T^+M^+ variant, one was found which recognized a 45 kD antigen (pI 6.7) that was not present on the T^+M^- cells. This specific distribution suggests that the 45 kD antigen may be directly associated with metastatic activity, but further work is required to substantiate this possibility.

Other studies (reviewed in Takenaga, 1986) have focussed on the effects of agents such as DMSO and butyric acid on enhancing the metastatic potential of malignant cells. DMSO is a polar agent which has been shown to induce differentiation in a range of malignant cell types, including murine erythroleukaemic and embryonal carcinoma cells, and human promyelocytic leukaemia cells and colon carcinoma cells. According to Takenaga (1986), DMSO alters the expression of normally silent genes, and when used on a poorly colonizing clonal variant (P-29) of the Lewis lung carcinoma it enhanced lung-colonizing ability, increased the activity of enzymes such as cathepsin B, promoted degradation of the subendothelial matrix, increased the rates of tumour cell adherence to plastic, the subendothelial matrix and to endothelial cells, and slightly increased homotypic aggregation ability. DMSO treatment of P-29 cells also enhanced trapping in the lung after i.v. injection, and the rate of clearance of the localized cells was slower than that of untreated control cells. Although the precise mechanism of action of DMSO is uncertain, its enhancing effect on lung colonizing ability was reversible, suggesting that the changes might be due to epigenetic alterations. Broadly similar effects were seen after treatment of P-29 cells with butyric acid, although this agent tended to enhance homotypic aggregation rather than heterotypic adhesion. Butyric acid probably acts in part through the hyperacetylation of histones, thereby affecting gene expression (Sealy and Charkley, 1978). Like DMSO, its effects are reversible.

Studies of the possible involvement of the spectrum of epigenetic processes on malignancy are still in their infancy. Apart from the examples mentioned above, other epigenetic events influencing regulation at the transcriptional and translational levels are likely to be important in malignancy, probably through their effects in generating phenotypic diversity. It seems likely that epigenetic events will prove to be involved in the expression of a number of aspects of the tumour phenotype, but we are not yet in a position to fully understand the contribution of such processes. Experimentally, it seems that epigenetic events have the potential to induce metastatic behaviour or to reduce it, presumably as a consequence of which particular silent genes are activated.

We have seen that tumorigenesis is a complex process in which cells progress to full malignancy. It is possible to identify four stages in this progression:

(a) *Immortality*

Experimentally, immortality is separable from tumorigenicity. Thus cells can become established in culture (rescued from senescence) without showing other traits of the transformed cell, such as anchorage independence and the ability to form tumours in experimental animals. The onset of immortality has been considered to be the result of an 'intitiating step' involving mutational events (Knudson, 1983).

(b) *Tumorigenicity*

The ability to form tumours in experimental animals represents a second stage (the 'completing step') in tumorigenicity, but for full expression of the malignant phenotype we may postulate two further stages.

(c) *Invasiveness*

Invasive behaviour characterizes malignancy, since although not all malignant tumours necessarily have metastastic ability it is difficult to conceive of metastasis not involving an invasive step.

(d) *Metastasis*

We have seen that most (but not all) malignant tumours are metastatic, and the expression of this phenotype seems to represent the culmination of tumour cell progression.

Can we identify any specific genes which might be responsible for the expression of these stages in progression towards full malignancy? Although this question cannot be completely answered as yet, evidence is available implicating possible roles for some viral genes in this progression. Within the papovaviruses, for example, polyoma large T induces immortality, but it requires the middle T product for completion of transformation (Rassoulzadegan *et al.*, 1982, 1983).

Although these results suggest that immortality is indeed separable from tumorigenicity, it does not follow that the genes referred to are the only ones involved in conferring immortality and tumorigenicity. The polyoma middle T product forms complexes with $pp60^{c\text{-}src}$, supporting the notion of a possible higher order of complexity in the transforming process (see below).

With respect to the retroviruses, we have seen that the availability of specific probes has enabled a number of genes to be identified whose expression is closely correlated with the transformed state. The protein products for many of these oncogenes have been characterized, but how they relate to transformation is still not particularly clear. The B chain of human PDGF, for example, is about 96% homologous with $p28^{sis}$ of the simian sarcoma virus and this product is thought to underly the tansforming ability of SSV. However, why doesn't exogenous PDGF transform cells? Even though PDGF can elicit many aspects of the transformed phenotype, such as stimulation of cell division and growth in soft agar (usually with other factors) and the induction of morphological transformation, treated cells are not tumorigenic. Recent evidence indicates that autocrine activation by PDGF (as occurs in *sis* transformed cells) takes place in intracellular compartments before the receptors are fully mature (Keating and Williams, 1988). While indicating a difference between autocrine and paracrine stimulation, this observation does not explain precisely why exogenous PDGF does not transform cells. Presumably other events, as yet undetermined, are necessary within the cells before full tumorigenic transformation ensues. It is clear that single oncogenes can transform particular cells under experimental conditions. This is of undoubted relevance to viral carcinogenesis, but how much does it tell us about the causes of most clinically relevant tumours? Relatively few human tumours are thought to express significant changes in c-*onc* genes, although specific examples exist where there is evidence for c-*onc* gene amplification (Little *et al.*, 1983) or activation (Duesberg, 1985). Furthermore, where single oncogenes have been shown experimentally to transform cells it must be borne in mind that a vast number of genes actually operate within a cell, and progressive effects following oncogene introduction must surely play a part in the tumorigenic process.

The functional diversity of both DNA and RNA viral gene products suggests that few, if any, are likely to act alone in tumorigenesis. Accordingly, the tumorigenic process is almost certainly multifactorial. Continuously stimulating a cell to divide through normal pathways, for example, does not directly induce tumour formation, but it may predispose the cell in this direction. Our knowledge is too fragmentary to understand the full sequence of events necessary for transformation, although progress continues to be made as may be illustrated by recourse to the DNA viruses. The adenoviruses, for example, have two domains (E1a and E1b) associated with transformation. E1a itself encodes two products, either of which will induce immortality. The two together, however, along with the products of E1b, appear to be required to ensure the completing step in tumorigenesis. E1b encodes about five proteins, none of which alone has any effects on the transformed phenotype (reviewed in Levine, 1984). Thus the adenovirus products E1a and E1b can together result in transformation, but what else is necessary within the cell? One of the E1a products stimulates transcription: are other genes affected by adenoviruses, and are these essential for transformation? It seems that the E1a coded proteins form complexes with about 10 other proteins ranging in M_r from 28 to 300 kD. One of these, a 105 kD molecule, has since been shown to be more or less identical to the retinoblastoma (*RB1*) gene product p105RB (Whyte *et al.*, 1988). Since this phosphoprotein is the product of an anti-oncogene, it may be that transformation involves a shift in control between gene products which influence cell growth both positively (E1a) and negatively (*RB1*). One of the products of the E1b domain binds and stabilizes p53, a protein which we have seen is implicated in cell division (section 1.7.1). The adenoviruses may overcome normal proliferative restraints by perturbing p53 stability, but this alone does not result in transformation. In fact, we have seen that E1b (presumably in association with p53) has no effect on the cellular phenotype, and thus stabilizing p53 cannot itself result in increased proliferation. As indicated elsewhere, evidence is rapidly accumulating that our current interpretation of the function of p53 may be completely wrong in that it may actually operate as an anti-oncogene. If this turns out to be the case, many functional interpretations involving p53 will have to be re-evaluated.

The current consensus is that E1a products promote immortalization by binding with $p105^{RB}$ thereby blocking the growth inhibitory effects of the latter, and that the immortalized cells are then influenced by the products of other genes (such as E1b or *ras*) to become fully transformed. More recent studies have shown that the product of the *E7* gene of human papilloma virus-16 forms similar complexes with the *RB1* gene product (Dyson *et al.*, 1989).

If genes act in concert during the tumorigenic process then by concentrating on only one or two oncogenes we may overlook other important events. Yokota *et al.* (1986) examined alterations in the expression of a battery of c-*onc* genes in over 100 fresh human tumours. Although they found changes in c-*myc*, c-*myb* and c-*ras*, their frequency of occurrence leaves some doubt as to their relevance in progression. Leibovitch and colleagues (1987) have performed an essentially similar study using normal rat myoblasts and their transformed counterparts of low and high malignant potential. Of the 15 c-*onc* genes they studied, the expression of c-K-*ras* was increased in the highly malignant cells, but all three lines studied failed to express one or more different c-*onc* genes. These results indicate that particular c-*onc* gene expression may be either suppressed or elevated in malignancy and it is therefore unwise to attribute malignancy to any one or even a group of such genes in the absence of more thorough studies.

Although *onc* genes act in a **dominant** manner as seen in transfection experiments (i.e. their presence is required for transformation), other experiments based on hybridization have suggested tumorigenicity to be under the control of **recessive** genes, the so-called **tumour suppressor genes**, whose absence is associated with transformation (Stanbridge, 1985; Sager, 1986). Thus when most normal cells are fused with malignant cells the resulting hybrid tends to be non-tumorigenic, although tumorigenic segregants can soon arise through the loss of specific chromosomes. How might tumour suppressor genes work? The answer to this question is not yet known, but they could conceivably act at any one of a number of levels including the suppression of *onc* gene expression. Indeed, Comings (1973) has suggested that oncogene expression may be prevented by trans-acting factors coded for by at least two suppressor alleles. There is clearly a need to further our understanding of how oncogenes and tumour suppressor genes function, and this should help clarify many of the confusing aspects of tumorigenesis that currently exist.

Cell fusion experiments have also been performed using metastatic tumour cells and normal cells. In such experimental situations, the resulting hybrids may display a suppressed metastatic phenotype (Ramshaw *et al.*, 1983; Sidebottom and Clark, 1983). Fusion of metastatic PG19 mouse melanoma cells with normal CBA T6 lymphocytes, for example, led to suppression of the metastatic phenotype. In some cases, however, the fusion of normal cells such as macrophages or T cells with non- or poorly-metastatic cells may result in increased metastatic potential (De Baetselier *et al.*, 1984). Thus fusion of normal T cells with non-invasive, non-metastatic BW5147 T lymphoma cells can generate highly invasive and metastatic variants. Larizza and Schirrmacher (1984) have suggested that the highly metastatic Esb lymphoma cells may have arisen from the fusion of a poorly metastatic Eb cell with a macrophage, coupled with the segregation and loss of particular chromosomes. This idea is in keeping with the notion that fusion of a malignant cell with a normal cell generates a phenotypically normal hybrid unless certain chromosomes are lost from the normal parent. The possibility that fusion of an Eb cell with a macrophage may have generated the Esb variant is supported by the observation that Esb cells express certain macrophage markers (such as Mac-1) and by the generation *in vitro* of highly malignant, Esb-like cells by fusion of Eb cells with macrophages. Interestingly, since Eb and Esb cells contain similar chromosome numbers, quite extensive segregation must have occurred if this explanation is correct. It is also interesting to note here that fusion of tumour cells with cells of haematopoietic origin may often generate invasive and metastatic variants. Both lymphoid and myeloid cells are often thought of as being 'naturally invasive', and they may express appropriate genes which can generate metastatic variants on fusion with tumour cells, whereas other cells (such as epithelial cells, which are normally stabilized *in vivo*) may be less able to do so. Cowell and Franks (1984) have reported that the tumorigenic potential of mouse bladder epithelial cells is suppressed by fusion with epithelial cells, but enhanced by fusion with mesenchymal cells. These results suggest that the outcome of fusion is influenced by the cells involved, although the universality of this conclusion remains to be tested.

The studies of Larizza and Schirrmacher (1984) suggesting that a metastatic variant may have been derived from fusion of a less metastatic tumour cell with a macrophage are of considerable interest, since they imply that cell fusion, perhaps by acting as a source of genetic variability, could promote progression (Warner, 1975). Some time ago, Goldenberg, Pavia and Tsao (1974) claimed that the metastatic cells which arose after the injection of human tumour cells into the cheek pouch of hamsters were actually human-hamster hybrids. Since this study, several reports have provided evidence for spontaneous cell fusion *in vivo* between normal host cells and tumour cells (reviewed in Lagarde and Kerbel, 1985). Kerbel and his colleagues (1984), for example, made use of drug-resistance markers to suggest that poorly metastatic WGA-resistant mutants of the highly metastatic cell line MDAY-D2 could become metastatic *in vivo* following fusion with normal mouse cells. Unfortunately, the relevance of cell fusion to the natural progression of tumours is not clear. Isoenzyme studies might be expected to turn up heterozygous tumours if fusion was relevant to malignancy, but in fact the vast majority of tumours and their metastases appear to be clonal in origin.

Nevertheless, as described earlier, both cell fusion and transfection experiments have suggested that there may be genes which can control the expression of the invasive and metastatic phenotypes. Proof that specific genes are involved in invasion and metastasis requires a particularly rigorous experimental approach. In order to investigate this problem properly we might refer to a modified version of the postulates of the 19th century microbiologist Robert Koch. Koch proposed that the following criteria should be satisfied in order to establish a causal relationship between a microorganism and a disease:

(a) A microorganism suspected of causing a disease should be associated with all cases of that disease.
(b) The microorganism should be able to be recovered from disease lesions.
(c) The microorganism should produce the disease when introduced into susceptible hosts.

With reference to genes and metastasis, we might propose that the demonstration of a causal relationship would require that:

(a) The product of the gene for metastasis should be found in metastatic cells.
(b) The product should be isolatable from metastatic lesions.
(c) The product (or treatments leading to its expression) should induce metastatic behaviour when introduced into non-metastatic cells.

Although they provide a framework for an experimental approach, Koch's postulates may be too restrictive for application to the study of metastasis. The reason for this is two-fold. First, the metastatic phenotype is probably based on normal cell behavioural patterns and second, metastasis probably involves multiple phenotypic changes. Most aspects of malignant cell behaviour are important at some stage in development or adult life, and thus the characteristic features of tumour cells may represent quantitative rather than qualitative changes (section 1.7).

Although it might seem unlikely that a single gene could be responsible for the expression of the entire malignant phenotype, it is possible nevertheless that a single genetic change could induce a cascade of events which leads ultimately to full malignancy. As will be seen in due course, the answer to the question as to whether particular genes are involved in metastasis is a qualified 'yes': qualified because although genes can be introduced into cells rendering them metastatic, this is only true for certain oncogenes and certain cells. Furthermore, we do not know how such oncogenes work, and nor do we fully appreciate their relevance in the clinical situation.

6.1.1 The *ras* Figure gene family and metastasis

Considerable attention has been directed towards the possible involvement of the *ras* gene product in metastasis. We have already seen that transfection of cells from the established (i.e. immortalized) cell line NIH 3T3 with oncogenes such as *myc* can induce tumorigenic conversion. Similar conversion of primary embryo fibroblasts, however, was thought to require the collaboration of another oncogene such as *ras* (Land, Parada and Weinberg, 1983 a and b). In general, it was found that the transfected cells formed non-metastasizing tumours in immunocompetent hosts.

Since these early studies, it has been shown that a single oncogene (Ha-*ras*-1 from the T24 bladder carcinoma cell line) can transform primary rodent cells providing it is linked to an enhancer sequence to boost the level of expression (Spandidos and Wilkie, 1984). In this latter study the transfected cells were shown to be tumorigenic in nude mice (and occasionally metastatic, but details were not given). Spandidos and Wilkie (1984) also showed that normal c-*ras* can induce immortality (but not tumorigenicity) in primary cultured cells, providing its high expression was ensured. More recently it has been shown that the increased expression of this proto-oncogene in NIH 3T3 cells is associated with tumour formation (Muschel *et al.*, 1985).

Several other groups have now shown that NIH 3T3 cells or diploid fibroblasts transfected with the activated *ras* oncogene (or with DNA sequences containing this activated oncogene) are metastatic when injected subcutaneously into nude mice, but not into immunocompetent, histocompatible mice (Bernstein and Weinberg, 1985; Pozzatti *et al.*, 1986; reviewed in Liotta, 1986). Taken together, these results provide striking support of the relevance of the immune system in metastasis, but they also illustrate how the use of specific experimental procedures can determine the ability to demonstrate particular events. That this is so provides a cautionary note for those who might rush too quickly into interpreting the clinical relevance of such experiments. In the studies of Pozzatti *et al.* (1986), transfection of primary rat embryo cells with *ras* induced the metastatic phenotype when tested in nude mice, but double transfection with *ras* and the adenovirus E1a oncogene actually reduced metastatic ability. The reasons for this apparently inhibitory effect of the E1a gene are not clear, but it was proposed that the decreased metastatic potential of the doubly transfected cells might have been due to an increased susceptibility to lysis by host defence cells.

Transfection of NIH 3T3 cells with the activated *ras* oncogene not only induces the capability of metastasizing in nude mice, but it also increases the production of type IV collagenase (Thorgeirsson *et al.*,

1985) and stimulates the motile responsiveness of the cells to fibronectin and laminin (Varani, Fligiel and Wilson, 1986). It should be noted here, however, that there was no evidence that the cells in the latter study were actually metastatic since only local invasive behaviour was assessed. Nevertheless, both the production of type IV collagenase and increased motility are likely to be advantageous to the metastasizing cell, and it is therefore of considerable relevance that these two behavioural responses are enhanced on transfection with *ras*. In their study of four pairs of transformed and non-transformed cells (see section 4.8.4), Bolscher and colleagues (1986) noted that invasiveness correlated with the introduction or activation of *ras* genes or of functionally analagous DNA sequences. Since invasiveness also correlated with altered carbohydrate content, the implication was made that *ras* activation may somehow affect glycosylation although exactly how this may be brought about is unknown.

Studies of clinical material using immunohistochemical techniques have shown that the p21 product of c-Ha-*ras* is present in larger quantities in metastases of human gastric carcinomas than in the corresponding primary tumours (Tahara *et al.*, 1986). Furthermore, patients with $p21^{ras}$ positive carcinomas were found to have a significantly worse prognosis than patients whose tumours were negative for $p21^{ras}$. Using similar immunological techniques, Thor *et al.* (1984) had already shown that deeply invading colon carcinoma displayed high levels of $p21^{ras}$, whereas normal colonic mucosa and benign tumours of the colon had low or undetectable levels of the protein. Superficially invading colon tumours expressed more or less intermediate levels of $p21^{ras}$, suggesting some relationship between $p21^{ras}$ expression and malignancy. It should be noted, however, that metastases from primary lesions which expressed high levels of $p21^{ras}$ showed considerable heterogeneity. Furthermore, a recent study by Chesa *et al.* (1987) almost completely confuses the picture. According to the detailed immunohistochemical studies of these authors, the binding of a McAb raised against a highly conserved region of $p21^{ras}$ suggests that this protein is widely expressed in normal tissues. Furthermore, expression of $p21^{ras}$ correlates more with terminal differentiation than with proliferation, and thus the authors proposed that its expression in tumours reflects differentiation rather than transformation or malignancy.

Despite this recent report, one might be led into concluding that expression of the *ras* oncogene is correlated in some way with invasive and/or metastatic behaviour. The majority of studies, however, have used cell lines which are subject to spontaneous changes in phenotype and thus some of the evidence may be questionable. Reference has already been made to the study of Bernstein and Weinberg (1985) in which they showed that Ha-*ras* transfected NIH 3T3 cells can form metastases in nude mice but almost not at all in syngeneic, immunocompetent NFS/NCr mice. In a rather tortuous experiment, these authors also examined the effects of the further transfection of these cells with DNA from the human metastatic tumour cell line ME-180. The doubly transfected cells formed only one metastasis in syngeneic, immunocompetent NFS/NCr mice, but the DNA from this metastasis enhanced metastatic ability when it was transfected into other Ha-*ras* transformed recipients. Examination of transfected DNA segments which appeared to correlate with metastatic ability in immunocompetent mice showed no relationship to any of the genes of the *ras* family, even though transfection with *ras* results in metastasis in nude mice.

A number of other reports suggest that the *ras* oncogene may not be involved directly in metastasis. Gallick *et al.* (1985), for example, actually found reduced levels of $p21^{ras}$ in four out of five metastases when compared to their primary colorectal tumours, and Albino *et al.* (1984) showed considerable heterogeneity in $p21^{ras}$ expression in cell lines derived from separate metastatic deposits of a single melanoma patient. Furthermore, elevated *ras* expression has been noted in benign colorectal lesions in humans (Spandidos and Kerr, 1984). Thorgeirsson *et al.* (1986) found that primary rat adenocarcinomas induced by *N*-methyl-*N* -nitrosourea expressed more c-Ha-*ras* DNA than did their metastases, and they also noted considerable variability between metastatic lesions in the same experimental animal (some actually having lower c-Ha-*ras* DNA content and less expression). Finally, it is worth noting that the various members of the *ras* gene family do not have similar effects, even when the same cell type is transfected. Thorgeirsson *et al.* (1985, 1986), for example, have reported that activated c-Ha-*ras* but not c-N-*ras* can induce the metastatic phenotype in NIH 3T3 cells. We are left with the conclusion that although *ras* expression may occasionally correlate with metastatic ability other factors (not necessarily related to oncogenes) are almost certainly involved.

6.1.2 Other genes affecting metastatic behaviour

Although a good deal of recent research has concentrated on the role of the *ras* oncogene in metastasis, a number of groups have examined the possible involvement of other oncogenes (reviewed in Mareel and Van Roy, 1986). Based on the studies reported above with the *ras* oncogene, Egan and colleagues (1987) have recently explored the role of kinase-encoding oncogenes in metastasis. These authors found that a clone of NIH 3T3 cells transformed with any of the kinase encoding oncogenes *ras, mos, raf, src, fes,* or *fms* could give rise to both experimental and spontaneous metastases in nude mice. Cells transformed with either the nuclear oncogene *myc* or with *p53*, however, were tumorigenic but only poorly metastatic (in fact, they only displayed lung-colonizing ability and no spontaneous metastatic ability after s.c. injection). The 3T3 clone used in these studies was chosen because it failed to show any signs of metastatic ability, although the parent NIH 3T3 line actually gave rise to a small number of metastases (roughly equivalent to that shown by cells transformed with *p53* or *myc*). The full significance of these observations remains to be evaluated, and it is complicated by the fact that the functional consequences of oncogene activation are likely to be extensive and varied. Note that there is current concern over whether *p53* is an oncogene, since some evidence suggests it may actually be a tumour suppressor gene.

In other studies, Yuhki *et al.* (1986), have found that the amount of c-*fos* mRNA correlates with metastatic potential in rat mammary carcinoma cell lines. As described in section 1.5.4, the *fos* gene product ($pp62^{fos}$) is a nuclear protein which complexes with $p39^{jun}$ to regulate transcription (Curran *et al.*, 1985). The expression of c-*fos* is rapidly induced when starved NIH 3T3 cells are stimulated with serum, showing a 50-fold increase within 30 min which drops to basal levels about 120 min after stimulation. This response precedes that shown by c-*myc* and it is also of a larger magnitude, but it does not last as long. The precise function of the *fos* gene product is not clear, since its appearance does not necessarily lead to increased proliferation. This could suggest that $p\text{-}p62^{fos}$ is involved in progression, which certainly is in agreement with its early production following the stimulation of cells to proliferate. How this might be related to metastatic potential is uncertain, but Taniguchi *et al.* (1986) have argued that the *fos* gene product might be involved in triggering the expression of a range of genes. These authors transfected an RSV transformed rat fibroblast cell line with v-*fos* and noted an increase in metastatic potential following intramuscular injection into syngeneic rats. The metastatic potential of the transfected lines appeared to depend on the manner of integration and the extent of transcription of the *fos* oncogene, with the most highly metastatic transfectant displaying the greatest expression of *fos* related mRNA.

From these and similar studies, it would seem that although the expression of some oncogenes may be related to metastasis, the situation is such that no simple, universal change is likely to be involved. It should be obvious, however, that further studies are required before the role played by genes in metastasis is fully understood.

Despite current limitations, experiments based on DNA transfection are likely to contribute significantly in the future to our understanding of malignancy at the molecular level. Attention has already been drawn to some of the current problems with transfection technology. Many experiments, for example, have been based on cell lines such as NIH 3T3 which can spontaneously transform *in vitro*, particularly if cultures are allowed to become overconfluent. Other experiments have used cells of embryonic origin, but these cells may possess many characteristics of transformed cells and it is uncertain as to whether the manner in which they respond to transfection is predictive of the behaviour expected when normal cells are treated similarly. Available evidence suggests that different cell types may respond differently to the same oncogene. Thus Muschel *et al.* (1985), for example, found that transfection with activated c-Ha-*ras* could induce the metastatic phenotype (tested in nude mice) in diploid fibroblasts and NIH 3T3 cells, but not in C127 murine mammary tumour cells even though they expressed high levels of $p21^{ras}$ after transfection. Regardless of what cell type is ultimately chosen for transfection, it is clear that the recipients must express a degree of heterogeneity and it is not known how this contributes to the eventual outcome of transfection. Subpopulations of cells may exist within the recipient population that are remarkably susceptible to transfection, perhaps because they already have other attributes which contribute towards malignancy. Furthermore, in some cases transfection with control vectors alone can result in the acquisition of metastatic behaviour by NIH 3T3 cells and the consequences of this have not yet been fully appreciated. We must also ask how the transfected gene inserts itself into the host DNA, since this is likely to affect the eventual outcome of transfection. Most transfection events result in multiple copies of the genes being inserted in tandem into several sites of the host DNA, and the consequences of this on normal host processes are not properly understood. Finally, we should recall that destabilization of clonal interactions within heterogeneic populations may generate diversification which could lead to the acquisition of increased malignant potential (section 1.7.26). If the process of destabilization is induced by transfection, then resulting changes in malignancy may arise only indirectly from the transfection process itself. This possibilty is, of course, sheer speculation but it serves to emphasize the point that as yet we know remarkably little about the process of transfection and how it alters cell behaviour. Of interest here are the experiments of Korczak *et al.* (1988) who infected non-metastatic and metastatic Sp1 tumour variants with retroviruses.

As expected, this generated an enormous number of uniquely labelled cells, clones of which could be identified by restriction enzyme mapping. Interestingly, when injected s.c. into syngeneic hosts the tumours which grew were composed of a relatively small number of clones. Furthermore, examination of metastases from these primaries showed that only a few of these dominant clones in the primary were able to form metastases.

Other genes may influence tumour progression and metastatic ability without necessarily being involved in transformation. Such genes, termed 'modulator genes' by Klein and Klein (1985), might include the MHC which may influence invasion, metastasis and immune sensitivity (Taniguchi, Karre and Klein, 1985; Fenyo and Klein, 1976).

6.1.3 Oncogenes as prognostic factors

Multiple copies of the oncogene N-*myc* have been found in some human neuroblastomas, and the level of amplification has been found to correlate with the stage of the disease (Brodeur *et al.*, 1984). Indeed, the more copies of N-*myc* that are present in the tumour cells, the worse the outcome is likely to be. Similarly, amplification of the *neu* oncogene in patients with breast cancer is associated with increased likelihood of relapse and shorter survival time (Slamon *et al.*, 1984). Furthermore, these authors have shown that *neu* amplification actually has greater predictive value than most other current prognostic factors, being equivalent to prognosis based on assessment of the extent of lymph node involvement. The point was made in Chapter 1 that treatment regimens for cancer patients are often based on parameters which reflect the clinical prognosis, and clearly data which both broaden and enhance these parameters are of considerable clinical relevance. The association of *neu* oncogene amplification with breast cancer progression would seem to indicate a significant role for it in this disease, but how it contributes to the development of this tumour is not yet clear.

6.2 CANCER IN TRANSGENIC MICE

We have seen that although we know a good deal about oncogenes and their products we are still remarkably ignorant of the processes involved in oncogenesis. Are there any other ways in which these processes may be studied? One possible avenue of research involves the introduction of specific DNA sequences into the germ lines of experimental animals, such as laboratory mice, with the subsequent analysis of the processes that follow. Mice which have inherited an acquired gene (**transgene**) in this way are known as **transgenic** mice, and they usually express the appropriate phenotype (Palmiter and Brinster, 1985). Gene transfer may accomplished by treatment of embryos or embryonal stem cells (reviewed in Jaenisch, 1988):

(a) *Embryos*

(1) Microinjection. This is accomplished by injecting a solution of DNA into one of the pronuclei of a fertilized egg. The embryo is then allowed to develop after insertion into the oviduct of a pseudopregnant female host (Gordon and Ruddle, 1983). One or more copies (commonly in a head to tail array) of the foreign DNA generally integrate randomly into a single chromosome.
(2) Retroviral transfer. Alternatively, genes may be introduced into embryos via retroviruses. With retroviral infection, usually only a single provirus is inserted at any given chromosomal site.

(b) *Embryonal stem cells*

Genes may be introduced into embryonal stem cells established in culture by DNA transfection (section 1.5.6), microinjection or retroviruses. The use of embryonal stem cells has advantages in that cells expressing certain phenotypes can be selected *in vitro* before injection into host blastocysts, from where they subsequently colonize the embryo giving rise to chimaeric animals.

There have been a number of developments based on the production of transgenic mice which look useful in furthering our understanding of oncogenesis:

(a) *The use of genes which predispose transgenic animals towards cancer*

This is the straightforward approach in which oncogenes are used in the production of transgenic mice. The large T antigen of the SV40 early region, for example, has been introduced as a transgene under the control of its own regulatory region (reviewed in Brinster and Palmiter, 1986). Under these circumstances, most of the transgenic mice display a characteristic phenotype, namely the development of tumours of the choroid plexuses (which are little tufts, rich in capillaries and lined with cuboidal epithelial cells, that project into four of the brain ventricles). These tumours express considerable amounts of large T antigen and yet adjacent normal tissue does not. It appears that the large T region is initially inactive in transgenic mice, but at some stage it undergoes activation with subsequent expression of the tumour phenotype. Why do only specific tissues express the tumour phenotype in such transgenic mice? The answer to this question is not yet clear, but one possibility relates to the ability of certain cell types to undergo activation.

(b) *Specific cell expression of the oncogenic transgene*

A transgene may be either under the control of its own regulatory region (as described above) or it may be under the control of an unrelated region. This is a useful ploy, since it can be used to target gene expression to particular tissues. Thus the gene for the large T antigen has also been inserted into mouse embryos coupled to particular regulatory sequences and this has allowed the detailed study of the expression of the oncogene in any one of several specific cell types. Hanahan (1985), for example, has shown that the insertion of SV40 (large T region) coupled to the rat insulin II regulatory region results in the production of transgenic mice with insulin-producing, pancreatic β cell tumours (insulinomas). β Cells are found in endocrine 'islands', known as the islets of Langerhans, within the pancreas. Interestingly, only a few percent of islets actually progress into full encapsulated and vascularized tumours, and even though large T expression is seen in all β cells during embryogenesis, the first tumours are not seen until the mice are about 9 weeks of age. Generally speaking, oncogene expression seems to result in hyperplasia of all affected cells, with tumours occurring only as rare clonal growths. Thus oncogene expression itself is not entirely sufficent for oncogenesis, and secondary events are probably involved.

Angiogenic activity seems to be crucial for tumour development in this model system since every tumour nodule is highly vascularized and capable of inducing blood vessel growth (Folkman *et al.*, 1989a). In fact, according to these authors angiogenic activity preceeds tumour formation since it first becomes apparent in a small number of hyperplastic islets. Since the frequency of angiogenic islets correlates with subsequent tumour nodule incidence, Folkman *et al.* (1989a) argue that the induction of angiogenesis may be a significant step in malignant change. The generality of this conclusion, however, remains to be substantiated.

(c) *Transmission of oncogenic transgenes*

Because transgenes are transmitted in a normal Mendelian manner, it is possible to produce strains of mice which mimic dominant hereditary cancer. Ornitz and colleagues (1987) have produced transgenic lines of mice using a rat elastase I-SV40 T antigen fusion gene. Expression of the normal elastase I gene begins in acinar cells of the exocrine pancreas around day 14 of fetal development, and transgenic mice bearing the fusion construct develop acinar cell pancreatic tumours. This development may be classified as a two stage process culminating in the death of all transgenic offspring at a predictable time. The first stage of development is thought to represent a preneoplastic state and it is characterized by progression of the exocrine pancreas from hyperplasia to dysplasia, by the increasing development of tetraploidy, and by the lack of acinar cell differentiation. In the second stage, hundreds of apparently monoclonal, aneuploid tumour nodules develop in the pancreas. The existence of aneuploidy in the nodules suggests that transformation by the fusion transgene may be coupled with chromosome loss. Ornitz *et al.* (1987) postulate that rapid division of tetraploid acinar cells may predispose them to the loss of chromosomes, some of which could contain tumour suppressor genes. Clearly, a model system in which the genetic lesion resulting in cancer is known is useful in studying oncogenesis. However, although the changes involved in progression to malignancy within this model system occur in a predictable fashion, the molecular mechanisms involved in transformation remain enigmatic. Furthermore, the expression of an activated *ras* gene in pancreatic acinar cells has been shown to lead to the development of fetal tumours, whereas expression of the T antigen leads to the formation of tumours in the adult. Presumably these differences reflect secondary genetic or epigenetic processes, but currently the exact mechanisms involved are not clear. Finally, there is little evidence of metastatic spread in this tumour, although this may be because the experimental animals die at too early an age for metastasis to have occurred.

(d) *Interbreeding of transgenic mice*

It is possible to interbreed mice with different transgenes in order to look at the consequences of combination in the hybrids. This has already been achieved with transgenic mice which carried either v-Ha-*ras* or c-*myc* genes linked to the mouse mammary tumour virus (MMTV) regulatory segment (Sinn *et al.*, 1987). Mice carrying either of these transgenes can develop tumours (the c-*myc* being deregulated), but hybrids formed by crossbreeding develop tumours at a rate greater than the additive effect of each oncogene alone. As in other studies with transgenic mice, normal tissue can be found adjacent to malignant tissue, with both tissues presumably expressing in this case the two oncogenes. It should be noted here, however, that technical limitations make it difficult as yet to prove that the same cell expresses both oncogenes.

We are not yet able to fully comprehend the significance of most of these results from experiments with transgenic mice. Nevertheless, they seem to parallel our understanding of oncogenesis based on experimental and epidemiological studies from which it has been proposed that the development of cancer takes place in a number of steps, possibly as few as three to four (Peto, 1977). The answer to the cancer problem undoubtedly lies in unravelling the nature and consequences of these steps at the molecular level, and transgenic mice may provide the experimental means for doing so. In this respect it is interesting to note that Fox and colleagues (1989) have recently reported the development of transgenic mice following the introduction of a hybrid gene composed of the SV40 T antigen fused with an α-amylase gene promoter. These mice developed malignant tumours of brown fat which displayed a high frequency of metastasis to a wide range of organs. The availablity of such transgenic mice offers a unique opportunity for studying many aspects of the metastatic process.

6.3 THE REVERSIBILITY OF CANCER

The spontaneous regression of benign tumours is not at all uncommon. Even pre-malignant tumours such as cervical carcinoma *in situ* show occasional indications of regression. Over the years there have been reports of the spontaneous regression of malignant human tumours, but the numbers are not that remarkable. Neuroblastomas, certain renal tumours such as Wilms' tumour, malignant melanomas and choriocarcinomas are the most commonly cited, whereas regression of tumours of the uterus, liver, lung, and pancreas, for example, are seldom reported. It is important to note that the term 'regression' is not necessarily synonymous with 'reversion', since in the former situation occasional dormant cells may still remain. Furthermore, the term 'reversion' is used rather loosely in this context, since tumour cells may only be blocked from continuing along their normal developmental pathway and a true reversion step may not be necessary to yield normalization.

The phenotypic instability of tumour cells suggests that certain treatments may return some phenotypic features of transformed cells more or less to normal. The term **reverse transformation** was coined by Hsie and Puck (1971) to describe the normalizing effects seen when transformed CHO cells were treated with cAMP. Many transformed cells (although often not those of epithelial origin) may regain aspects of the non-transformed cell phenotype following treatment with cAMP (Pastan and Willingham, 1978). Typically, such cells show an increase in cell–substrate adhesion and spreading, and a decrease in their lectin agglutinability and rate of cell division. The studies of Nielson and Puck (1980) indicate that some of these changes correlate with an increase in cell surface fibronectin.

However, does treatment with cAMP or its derivatives really lead to the *reversal* of cancer? If the changes induced are not permanent, as seems to be the case, then the answer is firmly in the negative. There is no evidence that treatment with agents such as cAMP directly affects the genome, and indeed Puck (1977) subsequently attributed its effects largely to interaction with the cytoskeletal system. Succinylated concanavalin A (succ-con A) was found by Mannino, Ballmer and Burger (1978) to increase the intercellular adhesiveness and to reversibly inhibit the growth of SV3T3 cells. When treated with this agent, transformed cells were found to accumulate in G_1 and to stop growing at lower culture densities. Similar results were found, however, with non-transformed 3T3 cells (Mannino and Burger, 1975) and thus the phenomenon is not unique to tumour cells. At low (2%) serum concentrations, however, a degree of specificity of growth inhibition was purported to have been seen. It may be that succ-con A interferes in some manner with the interaction between serum growth factors and the cell membrane, but its precise mechanism of action remains speculative. In any case, there is no evidence of any permanent reversal effect of treatment with succ-con A, and its *in vivo* relevance is suspect.

Other groups have explored the use of antibodies to influence cell transformation. Feramisco *et al.* (1985), for example, microinjected antibodies specific for the v-Ki-*ras* protein into Ki-NRK cells. They found that this treatment resulted in transient reversion (15–48 h) of the transformed cells to the normal phenotype, as assessed by return to flattened morphology and reduction in growth rate. Drebin *et al.* (1986) exposed *neu* oncogene transformed cells to McAbs reactive against p185, the product of the *neu* gene. The *neu* oncogene bears some homology to the *erbB* oncogene, and it was isolated originally from ethylnitrosourea-induced rat neuroblastomas. When exposed to anti-p185 McAbs, the *neu* transformed cells showed rapid and reversible loss of both cell surface and total cellular p185 in association with a return to anchorage dependency. Changes in p185 levels appeared to involve down modulation of cell surface p185 in conjunction with enhanced intracellular degradation. Interestingly, the loss of anchorage independency by the *neu* transformed cells could be restored by the presence of an unrelated oncogene product such as that of activated *ras*.

Such changes as those outlined above may very well alter the expression of tumorigenicity, but we have seen that unless this change is irreversible and occurs in all the cells then it is difficult to envisage complete eradication of the tumour. An alternative approach may be possible if tumours contain a proliferating stem cell compartment (section 1.2.2). The stem cell concept of malignancy suggests that transformed stem cells fail to respond to appropriate signals regulating division. It is these stem cells which give rise to the rapidly expanding population of more or less differentiated cells that make up the bulk of a tumour. Now cancer might be expected to be reversible if the malignant stem cells could be induced to shift from the self renewal to the differentiated compartments since the supply of stem cells would then be exhausted. An alternative possibility relates to conversion of the malignant stem cells to normal stem cells, but either process amounts to what may be called 'differentiation therapy'. When a stem cell divides it has three choices: it either yields two stem cells or two daughter cells (which enter the transitional compartment and proceed down the differentiation pathway), or one of each type. What governs the choice of developmental pathway following stem cell division remains unknown. With respect to malignant tumours, it would seem that their stem cells produce predominantly other malignant stem cells, which presumably accounts for the fact that most malignancies appear relatively undifferentiated. Some malignant tumours nevertheless express a degree of differentiation and this can be explained in the light of the options referred to above.

The idea of controlling tumour cell differentiation is clearly an attractive possibility for treating, if not curing, malignancy. Formulation of this idea followed the discovery that malignant cells could indeed differentiate into benign cells and that the process could be modulated experimentally (Pierce and Dixon, 1959; Pierce and Verney, 1961). Somewhat later Friend *et al.* (1971) reported that substances such as dimethyl sulphoxide (DMSO) could stimulate differentiation of murine erythroleukaemic cells *in vitro*, thereby confirming that some tumour cells (given an appropriate, although here artificial, signal) may be able to respond along normal regulatory pathways. With respect to the possible clinical use of DMSO in differentiation therapy, the studies of Takenaga (1986) in which this agent actually promoted lung colonization should not be overlooked (section 6.1). Some of the haematopoietic growth factors may also be useful in this direction with respect to the leukaemias. Granulocyte-colony stimulating factor (G-CSF), for example, not only induces differentiation of myeloid leukaemic cells, but it also appears to be able to amplify normal progenitor cells at the same time.

Jimenez and Yunis (1987) have studied the potential of using enhanced differentiation in the control of chloroleukaemia in the rat. This is a myelogenous leukaemia, showing some similarities to that in humans, which was originally induced by gastric instillation of 20-methylcholanthrene. These authors have found that rat lung conditioned medium contains a differentiation factor which induces chloroleukaemic cells to differentiate into macrophages. Now the successful transfer of chloroleukaemic cells in rats depends both on the age of the recipient and the number of leukaemic cells injected. Inoculates which are lethal to 7-day-old rats do not take in 21-day rats, and it has been tacitly assumed that this is associated with maturation of the immune system. Jimenez and Yunis (1987), however, provide evidence to support the notion that tumour rejection might be mediated by the production of higher levels of differentiation factor in the older mice, with consequent differentiation of the chloroleukaemic cells. Indeed, the development of leukaemia in 7-day-old rats can be inhibited by treatment with differentiation factor. Although the full significance of these results remains to be explored, there is considerable potential in using factors with this type of activity in clinical cases.

Another possible mechanism for inducing differentiation centres on modulating oncogene products. Symonds *et al.* (1984), for example, have shown that treatment of v-*myb* transformed haematopoietic cells with a phorbol ester can induce expression of the normal phenotype. This is associated with relocation of the *myb* product from the cell nucleus to the cytoplasm where its effects may be significantly altered. It is not yet known, however, whether such treatment directly affects the transformed haematopoietic stem cell compartment. In a later study, Symonds and colleagues (1986) reported that differentiation of haematopoietic cells of the myelomonocytic lineage could be coordinately regulated by changes in the expression of the oncogenes v-*myb* and v-*myc*. Thus a high level of v-*myb* expression correlated with the presence of an immature phenotype whereas that of v-*myc* was associated with a more mature phenotype. When the two oncogenes were expressed at similar levels an intermediate phenotype was observed.

Teratocarcinomas present an interesting model system for the study of many aspects of tumour differentiation. They are highly malignant tumours, thought to be derived from primordial germ cells, which can give rise to a wide range of differentiated tissues. When embryonal carcinoma (EC) cells (believed to be teratocarcinoma stem cells) were injected into normal mouse blastocysts which were then reimplanted into host mice, normal mice (expressing appropriate genetic markers) were produced (Brinster, 1974; Papaioannou *et al.*, 1975; Mintz and Illmensee, 1975; Illmensee and Mintz, 1976). The implication from these studies is that a malignant stem cell may produce either malignant or normal cell variants depending on its environment. Apparently, contact with the trophectoderm and the presence of blastocoele fluid are both essential for malignant reversal in this model system (Pierce *et al.*, 1984). The normalization of embryonal carcinoma cells following transplantation into blastocysts may unfortunately represent a special case, since not all malignant cells are similarly influenced by the embryonic environment. Although the implantation of melanoma cells into blastocysts does not stop the growth of melanoma lesions, it is of considerable interest that injection of melanoma cells into mouse embryos (at the time when premelanocytes migrate into the skin) can lead to a significant reduction in the number of melanomas which form compared to injection at later stages (Gerschenson *et al.*, 1986). Embryonic skin cultures apparently release soluble factors which are cytotoxic for melanoma cells. Additionally, there is some evidence that carcinomas which arise following skin implantation of retrovirus-infected keratinocytes may be suppressed by mixing with normal dermal fibroblasts, but we are far from understanding the processes involved (section 3.4.3.2).

Experimentally, EC cells can be induced to differentiate by a variety of agents including retinoic acid, dibutyryl cAMP, and DMSO. *In vitro* differentiation seems to be affected by various processes including growth factor availability, interaction with components of the ECM such as fibronectin, laminin, and type IV collagen, and cell–cell contact. Spontaneous differentiation in fact is not uncommon when EC cells are cultured at a density which encourages aggregation. The precise role that cell adhesion might play in differentiation is not clear, although it is known that two intercellular adhesion systems exist in EC cells. One system is Ca^{2+}-dependent and includes E-cadherin (section 5.3.1.2), while the other is Ca^{2+}-independent and is probably focussed on an endogenous lectin with fucose and mannose specificity (Grabel *et al.*, 1983).

One possible caveat concerning the work with teratocarcinomas relates to whether EC cells are actually transformed or not. Although the tendency is to think of them as transformed cells, it has been argued that they may actually represent a category of normal embryonic cells which continually proliferate until they are induced to differentiate (Martin, 1980). If this latter possibility gains credence, then the idea of using teratocarcinomas as models for the reversal of malignancy may be questionable. With most of these model systems for differentiation therapy little is known of the long-term effects, particularly with respect to the stability of the induced differentiation.

Currently, there is considerable interest in the influence of homeobox genes on differentiation. The **homeobox**, first discovered in the fruit fly *Drosophila*, is an evolutionary conserved 183 base pair DNA sequence. It is thought to play a key role in many aspects of development, probably by regulating transcription since it encodes a 61 amino acid protein domain which seems able to bind to DNA (Gehring, 1987). In vertebrates, the homeobox-containing genes fall into two major groups. One group (known as *Hox*) is related to the Antennapaedia complex (ANT-C) of *Drosophila* homeotic genes. The members of each vertebrate gene group are arranged into clusters and they are identified by numbers. Thus *Hox 1.2* refers to the second gene identified in the first *Hox* cluster. **Homeotic genes** in *Drosophila* determine the identity of the body segments generated in the embryo by **segmentation genes**. These latter genes act following the establishment of egg polarity by **maternal effect genes**. The roles of each of these gene classes in vertebrates is less certain. One problem underlying the difficulty of ascribing functional roles to vertebrate homeobox-containing genes is that they are also expressed in adult tissues, and thus a purely developmental role for them seems unlikely. Nevertheless, in the mouse embryo most homeobox-containing genes appear to be spatially expressed during growth, and this is consistent with a possible role in establishing pattern formation.

Is there a relationship between homeobox gene expression and cancer? The basic idea here is that homeobox expression might influence the developmental pathway of a cell, and that aberrant expression could be associated with tumorigenesis. Teratocarcinomas represent an obvious model system to utilize in exploring this possibility. The human embryonal carcinoma cell line NT2/D1 fails to express a number of genes from various *Hox* gene clusters, but does so when stimulated to differentiate by exposure to retinoic acid. *Hox* gene activation, however, is not simply related to tumour differentiation since some murine embryonal carcinomas express these genes and furthermore, differentiation of NT2/D1 cells by agents other than retinoic acid does not necessarily lead to *Hox* gene expression (Mavilio *et al.*, 1988).

Perhaps more relevant to tumorigenesis is the recent demonstration that nuclear accumulation of the *Hox 1.3* protein is rapidly and dramatically reduced in confluent cultures of mouse embryonic cells compared to sparse cultures of the same cell type (Odenwald *et al.*, 1987). This observation relates regulation of homeobox expression to density-dependent inhibition of growth, a phenomenon which we have seen is usually lacking in transformed cells (section 1.7.3). Expression of the *Hox 1.3* protein does not simply correlate with mitotic activity in all cell types, however, and thus a proper understanding of the significance of its reduced expression in density-inhibited growth must await the results of further research.

The discovery of homeobox genes extending from *Drosophila* to humans, and the general acknowledgement of a relationship between aspects of development and tumorigenesis may kindle enthusiasm for the use of the fruit fly as a model system for studying the biology of tumours. *Drosophila* is known to develop melanomas and haematopoietic tumours, and various oncogene products have recently been identified in this insect. Several of the genes involved in the hierarchical development of *Drosophila* have in fact shown to be homologous with vertebrate oncogenes. These include the segment polarity gene *wingless*, which is the homolog of the *int*-1 oncogene activated in mice following integration of the MMTV provirus (Rijsewijk *et al.*, 1987), and the maternal effect gene *dorsal* which shows homology with v-*rel*, the oncogene of the avian reticuloendotheliosis virus strain T (Steward, 1987). The homeotic gene *sevenless*, specific for determination of a particular photoreceptor cell (R7) in each ommatidium of *Drosophila*, has a tyrosine kinase domain homologous to *src* and *ros* oncogenes (Hafen *et al.*, 1987). Other homeotic genes of *Drosophila* have been shown to be homologous with specific growth factors: the product of the *Notch* gene, for example, displays homology with EGF (reviewed in Bender, 1985). Interestingly, homology has been noted between the adhesion related molecules of mammals and birds (section 5.8) and the positional specific antigens of *Drosophila* implicated in compartmentalization during development. An intriguing possibility here is that breakdown in compartmentalization after metamorphosis, possibly involving mutations in adhesive molecules, might lead to the insect equivalent of metastasis.

The idea that cancer might be reversible has proven attractive to clinicians because of the limitations in the current treatment regimens based primarily on radiotherapy and cytotoxic chemotherapy. Brugarolas and Gosalvez (1980) initially reported encouraging results in the treatment of squamous cell carcinoma of head and neck with thioproline, a drug originally selected because of its 'reverse transforming ability' *in vitro*. Thioproline was later shown to have considerable toxicity, however, and its beneficial clinical affects were not supported in other studies (Alberto, 1981).

6.4 THE EFFECTIVE TREATMENT OF CANCER: OBSERVATIONS AND SPECULATIONS

It should be clear that the extraordinary variability of the cancer cell presents a major problem for researchers and clinicians involved in cancer treatment. Cancer cells are both genetically and phenotypically unstable, and progression usually leads to the appearance of malignant cells that are markedly different from their cells of origin. One of the consequences of this variability is that there is a good chance that resistant cells will emerge during any course of treatment. Since chemotherapy represents a major focus for cancer treatment, it is important to understand the development of drug resistance (reviewed in Riordan and Ling, 1985).

How then do some cancer cells resist drugs?. Experimentally, it has been known for some time that the selection of cells for resistance against a specific drug, such as colchicine, is often associated with the emergence of cells which are cross-resistant to other unrelated drugs. This multidrug resistance is believed to be due to reduced cellular accumulation of the drugs involved, and experiments have shown that resistant cells can rapidly pump out the drugs, presumably before they can do any serious damage. Studies with a colchicine-resistant CHO transformed cell system have shown that multidrug resistance correlates invariably with the expression of a 170 kD plasma membrane glycoprotein referred to as P-glycoprotein (for permeability glycoprotein). P-glycoprotein has been found on multidrug-resistant cell lines of hamster, mouse or human origin and recent studies have also shown P-glycoprotein expression in human patients with drug-resistant ovarian cancer (Bell *et al.*, 1985). Transfection of hamster or human DNA containing the gene for P-glycoprotein into mouse cells renders them drug-resistant, but at this stage it is still uncertain whether other genes might also be involved in multidrug resistance. One exciting prospect arising from the discovery of cell surface location of P-glycoprotein is that antibodies directed against it could either block its activity or be used to target other drugs or toxic agents to which the cells are still sensitive. The presence of P-glycoprotein might also be useful as a prognostic indicator, but this possibility remains to be explored. Expression of P-glycoprotein is not the only mechanism by which cells may become resistant to drugs, but it represents an avenue of research with potentially important therapeutic benefits.

The prospects for favourable treatment for all cancers lies in early diagnosis. The treatment of localized disease might seem to be relatively straight forward since it can be effected by surgical removal of the offending tumour. However, it is usually not that simple since, for many cancers, metastases have already been seeded by the time of presentation. About 70% of patients presenting with small cell carcinoma of the lung, for example, are believed to already have metastatic disease. Many patients may show no signs of metastatic disease at presentation, only to suffer widespread disease some time after removal of the primary. The possible existence of clinically undetectable metastases has encouraged combined therapeutic regimens (adjuvant therapy) utilizing surgery, irradiation and chemotherapy.

Many surgically-based therapeutic regimens achieve nothing more than a substantial debulking of the solid tumour mass, and in many cases they fail to extend the life of the patient. Furthermore, in destroying the majority of tumour cells such treatment may lead to increased diversification and the emergence of more resistant or more malignant cells. Unlike localized solid tumours, the effective treatment of disseminated disease requires a modality that is systemic in action. Chemotherapy would seem to offer the best hope for treatment of metastases and non-solid tumours such as the leukaemias. However, many chemotherapeutic agents induce genetic damage and it has been suggested that their use could conceivably facilitate tumour progression. Although systemic radiotherapy requires whole body irradiation, this is regaining some favour now that it is possible to remove a sample of bone marrow and regraft it after irradiation of the patient. Prior to regrafting of the bone marrow, attempts may be made to purge it of any tumour cells by using various *in vitro* manipulative techniques, many based on McAb targetting of one form or another.

Immunotherapy as a systemic form of treatment has not lived up to early expectations. The association of viruses with some malignancies, such as hepatitis B virus with primary liver cancer and Epstein-Barr virus with Burkitt's lymphoma and nasopharyngeal carcinoma, for example, suggests nevertheless that immunological approaches to treating these types of cancers may prove beneficial. We have seen already, however, that infection with oncogenic DNA viruses need not necessarily lead to increased susceptibility to host immune effector cells (section 1.5.5), and infection with retroviruses or activation of c-*onc* genes may not alter the immunogenicity of the transformed cells at all. The inevitable conclusion is that whereas some tumours may respond to immunotherapy others probably will not.

Experimental studies suggest that the cancers most likely to be responsive to immunotherapy (on grounds of their immunogenicity) are in fact the cancers least likely to metastasize. Such tumours may be treated more effectively using one or more methods of the classical therapeutic triad: surgery, irradiation and chemotherapy. Another problem facing the use of immunotherapy is that although cancer cells can be killed *in vitro* through immunologically-based procedures, such procedures appear to be less effective in the *in vivo* situation. In some systems, as shown by Olsson and Ebbesen (1979) with polyclonal murine lymphomas, for example, immunotherapy may fail because it is directed against a dominant subpopulation in the tumour mass. This selectivity can leave minor subpopulations with the opportunity to proliferate at the expense of the host. Several hypotheses have in fact been put forward to explain the inadequacy of active immunotherapy *in vivo* but it must be admitted that for most tumours (particularly in the clinical situation) the precise reasons are not yet clear.

There are two principal types of immunotherapy (Mathe, 1970):

(a) *Adoptive or passive immunotherapy*

In these forms of immunotherapy the patient is treated either with immune cells or their products respectively.

(b) *Active immunotherapy*

Here, the patient's own immune system is geared into action.

One of the current main thrusts in passive immunotherapy is centred around the use of monoclonal antibodies (McAbs). However, there appears to be little hope of using McAbs to destroy tumour cells through relatively direct processes such as complement-mediated lysis. Instead, McAbs may be used as 'magic bullets' in the delivery of toxins and other drugs as described below. Monoclonal antibodies may also be coupled to various tracers in order to improve detection techniques.

Now one of the main problems with chemotherapy concerns the delivery of an effective dose that is not toxic to the host. All sorts of tricks can be played with the chemistry of drugs in efforts to see that they reach their appropriate destinations, but often dilution within the body coupled with progressive inactivation requires their delivery at very high levels. This is one area where immunological techniques may combine with classical tumour therapy. Thus McAbs coupled to appropriate drugs could be targetted towards tumour cells via antigens on their surfaces. This area is fraught with problems of its own, however, including the need to engineer McAbs which do not elicit immune reactions in the recipient indirectly. Furthermore, appropriate delivery requires restricted expression of the tumour antigens which serve as the targets, and we have seen that this may not always be the case.

Ehrlich (1913) first proposed that antibodies (or 'haptophores' as he called them) could be conjugated with toxic molecules ('toxophores') to destroy cells 'in the manner of magic bullets'. Of course, toxic molecules need not only be coupled to antibodies, and it is possible that selective targetting may be obtained using other molecules such as hormones and lectins, whose receptors may have restricted cell surface expression. Coupling to McAbs, however, has caught the imagination of both scientists and the general public. Scientifically at least, the reason for this lies predominantly in the exquisite specificity offered by McAbs.

Although *in vitro* studies show antibody–toxin conjugates to have impressive killing capacity, this has not been borne out to the same extent *in vivo*. The first therapeutic use of toxins conjugated to antibodies was reported bu Moolten *et al.* (1972), although the situation was rather artificial. These authors coupled dinitrophenyl groups to hamster sarcoma cells and then used anti-dinitrophenyl antibodies conjugated to diptheria toxin to treat hamsters in which the modified tumour cells were implanted subcutaneously. Treatment with the conjugate was found to prolong the lifespan of the experimental animals. Many subsequent studies have reported at least some beneficial effects of treatment with antibody–toxin conjugates, but the most encouraging results have been reported with very small tumour burdens (a few hundred cells). This probably reflects a delivery problem, suggesting that this form of therapy might ultimately be most successful in treating small metastases (perhaps those arising after surgical removal of the primary).

Several problems may be identified with the general concept of the magic bullet:

(a) *Potency*

The toxic drug must be remarkably potent because of potential problems of delivery. It has been claimed that a single molecule of diphtheria toxin, abrin (from seeds of *Abrus precatorius*) or ricin (from seeds of *Ricin communis*) may be enough to kill a single cell following penetration. Not surprisingly, these molecules are the favourites of most scientists working in this area. One problem with the use of diphtheria toxin is that it appears to lose its potency on conjugation. *In vivo*, the LD_{50} (the lethal dose required to kill half a test group) of these three toxins after i.p. injection into guinea pigs (diphtheria toxin) or mice is about 0.1, 0.5 and 2.6 μg kg^{-1} body weight respectively. Note that all animals are not equally susceptible to the same toxins. Mouse cells, for example, are not sensitive to diphtheria toxin, which is why its LD_{50} given above relates to guinea pigs.

Another problem in manifesting potency concerns the possible presence of inhibitors *in vivo*. These may include various inactivating enzymes, shed tumour antigens, as well as the neutralizing antibodies which would probably develop against foreign determinants on the conjugate.

(b) *Specificity*

The toxic drug must be directed relatively specifically to the target cell. Conjugate formation may lead to a reduction of the antigen binding capacity of the McAb thereby reducing specificity, but several chemical linking processes are now available to overcome this problem. Absolute specificity need not be required in those circumstances where doses could be selected to kill the principal target cell without otherwise being severely detrimental to the host. Destruction of aberrant blood cells would be possible, for example, if a McAb-drug conjugate (perhaps cross-reacting with tissue cells such as fibroblasts) showed only limited diffusion from the circulation. Progress is being made in treating some leukaemias by a process involving removal of the bone marrow, purging of neoplastic cells with targetted toxins, and then re-introducing the cleansed marrow back into the patient, who meanwhile has undergone ablative radiotherapy. A similar process (using antibodies directed against lymphocytes) can be used to treat bone marrow prior to grafting.

(c) *Delivery*

Delivery of conjugates to large tumour masses often with avascular centres may limit the application of magic bullets. Delivery would be enhanced if the conjugate itself was of low molecular mass. Cytotoxic molecules such as abrin, ricin and the diphtheria toxin actually exist as dimers (A and B chains) of native M_r about 62–65 kD. It seems that the B chain recognizes the receptor and then inserts itself in the membrane to allow the intracellular translocation of the A chain which exerts the cytotoxic effect. If the toxic molecules work in this manner, it may be possible to use only the A chain subunit in conjugate formation, providing association with a McAb allows it to enter the cell. The A chains of these toxins vary from 21 to 31 kD which would considerably reduce the size of the conjugate.

(d) *Side reactions*

The Fc portion of McAbs in the conjugate may be able to interact with phagocytic cells bearing Fc receptors engendering side reactions including inflammation which may be either advantageous or disadvantageous to the host. If necessary, reactions such as these may be overcome by the use of McAb fragments which lack the Fc region. When delivered *in vivo*, the chemical linkage between the McAb and the toxin may be disrupted allowing the toxin to exert non-specific effects. Apart from forming relatively resistant conjugates, the use of A chains rather than the complete toxin molecule could ensure specificity since when free, this subunit alone cannot penetrate a cell. Interestingly, the seeds of some plants contain toxins similar to the A chains of abrin and ricin. Since these molecules lack a B chain they cannot bind onto cells to exert their toxic effects. When conjugated to antibodies, however, some of these molecules (such as gelonin from the seeds of *Gelonium multiflorum*) acquire cytotoxic activity (Thorpe *et al.*, 1981).

(e) *Variant generation*

Both HeLa and CHO variant cell lines have been selected on the basis of resistance to diphtheria toxin, and the possibility exists that such variants may arise *in vivo*. This is not too much of a problem if the variants are based on a toxin recognition defect, since this may be overcome by using an appropriate antibody in conjugate formation. If the variants arise on the basis of intracellular inactivation of the toxin, however, then the problem may be more difficult to surmount. Of course, variants which fail to bind the antibody used in the conjugate may also arise, thereby diminishing the effectiveness of this form of treatment.

Recently, attention has been drawn towards the possibility of harnessing the lytic potential of cytoxic T cells by using McAbs to couple them to target cells (Liu, Nussbaum and Eisen, 1988). Human cytotoxic T cells express CD3 (part of the T cell receptor complex) on their surfaces. Monoclonal antibodies against this moiety can be coupled to specific ligands which bind to cell surface receptors on particular cell types. Coupling of an anti-CD3 McAb to an αMSH analogue, for example, was shown by Liu and colleagues (1988) to promote cytotoxic T cell binding to and lysis of human melanoma cells *in vitro*. In this situation, lysis would be generated by the binding of the αMSH analogue to cell surface receptors on the melanoma cells followed by binding of cytotoxic T cells to the attached anti-CD3. As outlined above, there are obvious problems in trying to adapt this to clinical use. The ligand chosen, for example, would have to be specific to a particular cell type, and problems with delivery will need to be overcome. However, this approach is not devoid of potential.

Another novel approach towards tumour eradication involves

infection with particular parvoviruses which can inhibit both spontaneous and virus-induced tumours (Tattersall and Ward, 1978). In most cases, these parvoviruses apparently destroy tumour cells almost exclusively, having no observable effect on normal cells. The mechanism underlying this antitumour effect is unknown, but an intriguing possibility has been raised by the studies of Mousset and Rommelaere (1982) using the mouse minute parvovirus and SV40-transformed Balb/c derived cells. These authors showed that the antitumour effect of this parvovirus occurred *in vitro* (as determined by a reduction in soft agar colony formation), and thus it was unlikely to be attributable to host defence activities. Instead, Mousset and Rommelaere (1982) suggested that changes in gene expression occurring on transformation might remove inhibitory signals for parvovirus division. Thus transformed cells, but not normal cells in which the inhibitory signals remain, would become susceptible to destruction by the parvovirus. This is an interesting hypothesis, but further investigation is required to substantiate the proposed mechanism of action.

The existence of tumour suppressor genes, as seen in retinoblastoma and Wilms' tumour (section 1.7.2), raises the possibility of controlling malignancy through suppressor products. As yet this possibility remains speculative, since we are still uncertain as to how suppressor genes operate at the molecular level.

There are two other related aspects of research based on immunotherapy worth considering here, both of which involve lymphokines (the hormone-like products of lymphocytes). High hope was once held for the interferons (IFNs) as effective anti-cancer drugs, but this now appears to have been overly optimistic. Nevertheless, there are some relatively rare forms of cancer (such as hairy cell leukaemia) in which IFN-α may prove to be an effective form of treatment (reviewed in Smyth *et al.*, 1987). The optimum method of administration, however, is not yet resolved and side-effects such as fever, general malaise and some neurotoxicity have been reported.

Recent studies support the idea that the interferons act as transcription regulators stimulating various quiescent genes. The effects of the interferons are often short-lived, however, because they also stimulate the production of controlling factors which bring about a decrease in transcriptional activity. The precise mode of action of the interferons at the cellular level is unclear. There is evidence that they can augment NK cell and macrophage activity, but they may also be able to operate via a non-immune mechanism. As transcription regulators, interferons can also have direct effects upon tumour cells, either enhancing or inhibiting the production of cellular products including those of some oncogenes. More recently, it has become clear that interferons can interact synergistically with a number of cytotoxic drugs, and this may have potential in future clinical use.

Another lymphokine, interleukin 2 (IL-2), may be useful in treating cancer since it may stimulate the host's immune response against malignant cells. IL-2 stimulates phenotypically diverse lymphoid cells to differentiate into cytotoxic effector cells known as lymphokine-activated killer (LAK) cells (section 1.9.2). *In vitro*, LAK cells destroy tumour cells (including NK-resistant cells) but not normal tissue cells. *In vivo*, the administration of LAK cells in association with IL-2 (**combined adoptive immunotherapy**) has been shown to result in the regression of both primary tumours and metastases in mice (Mule *et al.*, 1984). Systemic administration of LAK cells and IL-2 in human patients with metastatic cancer can result in tumour regression, but high doses of IL-2 result in the development of vascular leakage. Recent studies by Damle *et al.* (1987) suggest that LAK cells bind to and lyse endothelial cells, thereby generating the vascular leakage which characterizes this form of adoptive immunotherapy. Lymphocytes primed by soluble antigens adhere to endothelial cells but do not cause their lysis. The binding of both LAK cells and activated lymphocytes to endothelial cells is probably mediated via the leukocyte adherence-related proteins (section 5.7.2.4), but the precise molecular details are yet to be elucidated. Whether methods will be developed to overcome LAK cell toxicity for endothelial cells remains to be seen, but it does present a major problem in IL-2 based therapy. Interestingly, LAK cells adhere better to endothelial cells than to tumour cells, even though the latter represent a target for lysis, and it may be of some interest to explore ways of enhancing this interaction. Finally, patients treated with IL-2 do not display significant circulating LAK cells, possibly because of their avid binding to the endothelium. In this respect, it would be advantageous to develop ways of activating LAK cells after they had emigrated from the blood, but exactly how this might achieved is uncertain.

The side-effects of IL-2 treatment (fluid accumulation in the lungs, anaemia, fever, rigors) are severe enough to have brought about a call for halting clinical treatment based on this kind of lymphokine therapy. Yet despite the side effects involved, the evidence suggests that IL-2 therapy in one form or another can result in improvement for a proportion of patients with advanced tumours. Thus Rosenberg et al. (1985) have described objective tumour regression (more than 50% reduction in volume) for 11 out of 25 patients treated with autologous LAK cells and recombinant IL-2. More recently, it has been shown that the side effects may be reduced and some patient improvement retained if IL-2 is used alone, albeit at extremely high doses (Lotze *et al.*, 1986). In the latter trial, 3 out of 10 patients showed objective responses following i.p. or i.v. treatment wih 30 000 U kg^{-1} or more of IL-2. Interestingly, all three patients had malignant melanoma and the course of treatment was ineffective in patients with colorectal and ovarian cancer. It must be borne in mind, however, that these studies involve small numbers of patients and that the long term effects of the treatment regimens remain to be evaluated.

An alternative approach using chronic indomethacin therapy in conjunction with IL-2 treatment is being developed by Lala and colleagues in Canada (Lala and Parhar, 1988). These authors have reported that although either treatment alone is effective in reducing B16F10 metastasis by up to 66%, combined treatment is much more effective with nearly total eradication of metastases. The therapeutic potential of this treatment is brought home by the fact that it is active when commenced after allowing for the establishment of lung colonies from as many as 10^6 i.v. injected melanoma cells. The rationale behind this treatment schedule is that cytotoxic defence cell activity *in vivo* may be inactivated by prostaglandin (primarily PGE_2) release from host macrophages. The chronic indomethacin therapy (administered in the drinking water at 14 μg ml^{-1}) is designed to block prostaglandin synthesis *in vivo*, whereas the IL-2 treatment (25 000 U every 8 h i.p. for 5 days in one or two rounds) is presumed to enhance the activity of the cytotoxic cells. The future of this therapeutic regimen, like that of all others, rests on the results of clinical trials and will presumably also have to overcome some of the side effects mentioned above.

Considerable effort is currently being directed towards the possible therapeutic use of natural cytotoxins such as tumour necrosis factor alpha (TNFα) in the treatment of cancer (reviewed in Old, 1985). TNFα, also known as cachectin, is composed of unglycosylated 17 kD subunits which associate in different ways to form molecules as large as pentamers in some species, although the trimer probably predominates. TNFα is produced by activated macrophages and by some con A activated T cell clones (although whether this occurs *in vivo* is questionable). The related cytokine TNFβ is thought to be produced primarily by lymphocytes of B and T origin. TNF*a* was originally found in the serum of mice stimulated with bacille Calmette-Guerin (BCG) and bacterial lipopolysaccharide (LPS), and it was identified by its ability to destroy the meth A sarcoma growing in host mice (Carswell *et al.*, 1975). It turns out that these sarcoma cells are not killed *in vitro* by TNFα, and thus the suggestion has been made that it may exert its effects *in vivo* by inducing an inflammatory exudate which then destroys the tumour. Indeed, TNFα appears to have chemotactic activity for monocytes and PMNs and it can also act on endothelial cells to enhance the expression of molecules adhesive for both lymphocytes and PMNs thereby promoting their extravasation. However, some tumours appear to be directly destroyed by TNFα and thus the molecule may have two modes of action. In fact, the effects of TNFα are widespread (particularly through the induction of cytokine release), and it may utilize various pathways to ultimately bring about cell destruction. A rather bizarre twist in this saga is that some tumours actually produce TNFα autonomously, and this may be the prime basis for the wasting (**cachexia**) seen in individuals bearing these tumours. Wasting asociated with other tumours may be due to macrophage-derived TNFα, but it should be made clear that a number of cytokines including IL-1 and IFN-γ are also likely to contribute to this phenomenon.

As with most other anti-cancer drugs, hope for TNFα lies in the possibility that it may destroy a tumour before this kills the host. Its clinical effectiveness, however, is still under study. It is worth noting that the majority of human tumour cell lines are not destroyed by TNFα and that much of its reputation is actually based on the use of a small number of selected animal models. Furthermore, TNFα is a potent pyrogen (causing fever), it induces anorexia and weight loss and it is the principal mediator of endotoxin induced shock. These properties hardly render it suitable for *in vivo* use, but it may be possible to chemically alter the molecule to obtain an agent capable of lysing tumours without these potentially severe side effects. Interestingly, although TNFα is chemotactic for phagocytes it is not active against LGL cells (Ming, Bersani and Mantovani, 1987) and it has actually been shown to inhibit tumour-induced migration of bovine capillary endothelial cells as well as their proliferation *in vitro* (Mano-Hirano *et al.*, 1987). We have seen that many events involved in the biology of tumours have two sides to their nature: TNFα may exert a cytotoxic effect through stimulating phagocyte exudation, but it may also abrogate this effect to some degree by inhibiting the very process (i.e. angiogenesis) which is essential for phagocyte cell delivery. On the other hand, anti-angiogenic activity may also serve to limit metastatic spread. It is appropriate to consider the alternative view of Liebovich at al (1987) here again (see section 1.7.20) since this represents a classical scientific dilemma, albeit of mini-proportions. Thus two different groups have obtained diametrically opposed results on the angiogenic effects of TNFα, leading to the formulation of two mutually exclusive hypotheses. We have seen that this is not particularly unusual in cancer research, possibly as a consequence of the complexity of this area of study combined with the capricious nature of cancer cells, and resolution awaits the outcome of further independent studies.

McIntosh *et al.* (1988) have recently reported on the effects of TNFα and IL-2 combination therapy on inducing regression of a chemically induced mouse sarcoma. Synergistic activity was seen only when tested against an immunogenic form (MCA-106) of the tumour, and not when tested against a non-immunogenic form (MCA-102). The sequence of timing and dosage appeared crucial for maximal effects. According to the authors, TNFα may induce necrosis and an inflammatory response which would involve lymphoid cells with anti-tumour activity that might be activated by IL-2.

Despite its possible side-effects and uncertainties about how it manifests its toxicity, clinical trials are under way with TNFα. Since its half-life in the blood is about 6 min, it would appear that any possible beneficial effects will require continuous infusion.

Another possible treatment regimen might be one based on the clinical application of factors which modulate tumour cell division independently of normal cell division. Various sorts of growth factors have been shown to stimulate normal cell division while inhibiting that of transformed cells, although it must be admitted that only a limited range of cell types have been studied so far (section 1.7.1). Two interesting factors in this respect have been isolated from culture medium of the human rhabdomyosarcoma cell line A637 by Todaro's group (Fryling *et al.*, 1985). These factors, known as tumour inhibitory factors (TIFs) 1 and 2, inhibit the growth of a number of tumour cell lines while stimulating the growth of normal fibroblasts. One of the factors (TIF-2) has recently been shown by Fryling and colleagues (1989) to be more or less identical with IL-1 which has potent immunoregulatory effects (reviewed in Dinarello *et al.*, 1986). The clinical relevance of these factors as potential anti-tumour agents remains to be tested, and given the wide ranging activities of IL-1 it may be inappropriate to do so for TIF-2. Some of the other factors may be more resticted in their activities, however, and thus more appropriate for testing.

We have seen that for many solid types of tumours which spread predominantly via the blood, metastasis is correlated with angiogenesis (section 1.7.20). If new blood vessels could be inhibited from growing into the primary tumour, then the possibilty of metastases occurring may be less likely. One approach likely to be of some use here involves the development of anti-angiogenic drugs. Such drugs may act at several levels:

(a) They may block the secretion of angiogenic factors by tumour cells.

(b) They may interfere with the factors themselves.
(c) They may block the interaction of the factors with the endothelial cells.

An interesting development in the possible use of anti-angiogenic therapy for cancer treatment occurred when Taylor and Folkman (1982) showed protamine to be an inhibitor of angiogensis. Protamine (M_r 4300 D) is a positively charged, arginine-rich protein (found only in sperm) which binds to heparin. Protamine was tested as a potential angiogenic inhibitor because heparin had been found to potentiate the activity of certain angiogenic factors and it was thus reasoned that an agent which bound to heparin might block its potentiating effects. That heparin itself might be involved in the activity of some (but not all) angiogenic factors was first mooted because mast cells (which release heparin) had been shown to accumulate at tumour sites before angiogenesis began. Although it potentiates angiogenesis, heparin is not angiogenic itself. Instead, it appears to work by increasing the migration of capillary endothelial cells which is an important step in the outgrowth of new vessels. Although it was initially reasoned that protamine might reduce the level of angiogenesis to that seen in the absence of heparin, in fact at increased levels it was capable of completely blocking angiogenesis. When mice with lung metastases were treated with a reasonably well tolerated dose of protamine (60 mg kg^{-1}, every 12 h) there was a 77–92% decrease in the mean tumour volume of the lung metastases compared to controls. Protamine was shown not to be cytostatic nor cytotoxic for tumour cells in tissue culture studies (except at relatively high levels), and when given to mice with solid and ascites tumours only the solid form regressed. These and other results suggest that protamine does not act directly on tumour cells and that it may exert its effects on tumour growth through interfering with angiogenesis. There are, however, limitations to the systemic use of protamine as an anti-angiogenic drug, particularly on account of its toxic effects which include lethargy, weakness and occasionally sudden death. Furthermore, not all solid tumours (particularly sub-cutaneous forms) responded to treatment with protamine. The reason for this remains unclear. Obviously, protamine has limited clinical value, but the study by Taylor and Folkman (1982) does suggest that some potential lies in pursuing the anti-angiogenic approach to treating tumour growth and spread.

More recently, Folkman *et al.* (1983) have shown that a combination of cortisone and heparin can markedly inhibit angiogenesis. Serendipity played its part well in this study, since cortisone was originally used as a means of suppressing background inflammation in

CAM preparations so that the potentiation of angiogenesis by heparin might be more conspicuous. By using different fragments or variants of the two principal molecules, it has been possible to show that the anti-angiogenic properties of cortisone and heparin are not due to the glucocorticoid or mineralocorticoid properties of cortisone, nor to the anticoagulant activity of heparin. Exactly how the combination of cortisone and heparin manifests its activities thus remains enigmatic, but the effects can be quite marked. The subcutaneous growth of a variety of tumours (including the B16 melanoma and the Lewis lung carcinoma) can be eradicated in more than 50% of mice when they are given heparin in their drinking water along with injections of cortisone acetate. The treatment is not effective for all tumours, however, and four 3-methylcholanthrene-induced tumour lines (two sarcomas and two gliomas) growing *in vivo* failed to respond (reviewed in Folkman, 1985). In more recent studies Folkman and his colleagues (1989b) have shown that certain angiostatic steroids are particularly effective when they are delivered to endothelial cells complexed with cyclic oligosaccharides such as β-cyclodextrin tetradecasulphate. It is thought that each steroid sits within the centre of the doughnut-shaped oligosaccharide and that this complex is then adsorbed via hydrophobic interactions to the endothelial cell surface.

Several research groups have attempted to extract angiogenesis inhibitors from tissues such as hyaline cartilage, the underlying assumption being that its avascular nature might result from the production of anti-angiogenic material. Interestingly, tumours also often fail to invade cartilage (osteosarcoma, for example, rarely invades local cartilage) and chondrosarcoma is one of the least vascularized of all tumours. When human bone fragments are cultured *in vitro* with tumour cells, the cells invade and erode the bone but not the cartilage, unless it had been previously occupied by capillary loops from the growth plate (Kuettner and Pauli, 1983). When cartilage is extracted by guanidine hydrochloride it loses its resistance to vascular invasion (Eisenstein *et al.*, 1973). Material extracted from cartilage in this way can be incorporated into sustained-release polymers which, when implanted into the rabbit cornea, can inhibit tumour induced angiogenesis (Langer *et al.*, 1976). The material extracted from cartilage by several groups seems to have trypsin and collagenase inhibitory activity (Sorgente *et al.*, 1976; Kuettner *et al.*, 1977) as discussed in section 2.2.

A similar approach has been applied to epithelial tissues which are also avascular. Waxler, Kuettner and Pauli (1982) isolated crude material from bovine urinary bladder by extraction with 1 M NaCl. This extract was found to be cytotoxic for cultured endothelial cells, but not for various other normal and neoplastic cell types. An active, low molecular mass fraction contained trypsin inhibitor activity, but this was apparently not responsible alone for the specific cytotoxic effects. Although further research is required to elucidate the nature of all of the active components from epithelial tissues, with respect to cartilage it seems that its enzyme inhibitory activities are the key anti-angiogenic factors. The process of angiogenesis begins with the local degradation of the basement membrane of the parent vessel, followed by the outward migration of endothelial cells to form a sprout. These early events bear all the hallmarks of invasion, and thus we can suspect a crucial role for various proteolytic enzymes in angiogenesis. Indeed, capillary endothelial cells stimulated *in vitro* with angiogenic substances have been shown to secrete high levels of various enzymes including collagenase and plasminogen activator. Similarly, the invasion of some tumours is clearly destructive in nature and involves the secretion of proteolytic enzymes. When considered together, it is not too difficult to envisage that the anti-invasive behaviour of cartilage with respect to both bood vessel and tumour invasion probably resides in its enzyme inhibitory activity.

The question we now must ask concerns the possible therapeutic relevance of enzyme inhibitors in the treatment of cancer. When cartilage extract was infused into the right carotid artery of experimental animals, it could inhibit both tumour-induced angiogenesis and the growth of a tumour in the cornea of the right eye relative to the left (Langer *et al.*, 1980; reviewed in Maught, 1981). By the time blood from the infused right carotid artery reached the left eye, the cartilage extract would have been considerably diluted and thus in principle the events occuring in the left eye can be used as controls for those occurring in the right. At present, it is not clear whether such treatment will be similarly effective when more indirect infusion methods are used and, furthermore, any adverse systemic effects remain to be fully evaluated.

In more recent studies, Rastinejad, Polverini and Bouck (1989) have identified a 140 kD protein in the conditioned medium of BHK21 cl13 hamster cells which acts as an angiogenic inhibitor. Transformation of these cells by chemical carcinogens is associated with inactivation of a tumour suppressor gene and the induction of angiogenic ability. The association between anti-angiogenic activity and expression of the suppressor gene was confirmed in studies with transformants, revertants and segregating hybrids. This linkage of anti-angiogenic activity with a tumour suppressor gene is particularly intriguing given the apparently essential role for angiogenesis in neoplastic transformation in some model systems such as the transgenic insulinoma described

Table 6.1 Possible strategies for improving the treatment of cancer

1. Development of techniques for early detection e.g. monoclonal antibodies; imaging systems
2. Improvement in techniques associated with the classic triad of surgery, chemotherapy and irradiation
3. Selective inhibition of tumour cell proliferation e.g. new cytostatic drugs
4. Development of methods for controlling progression in tumours
5. Development of anti-angiogenesis regimens thereby limiting tumour growth and reducing the possibility of metastasis
6. Development of anti-invasive techniques, perhaps involving inhibition of beneficial enzymes such as type IV collagenase
7. Inhibition of tumour cell motility, which might inhibit invasion and metastasis
8. Inhibition of tumour cell adhesion to endothelial cells which might restrict the possibility of extravasation during haematogenic and lymphatic spread
9. Potentiation of the host's defence systems:
   (a) Activation of defence cells by interferons, interleukins, macrophage activating factor and other agents
   (b) Enhancement of tumour cell immunogenicity through cell surface modification by chemical or biological means
   (c) Passive immunity with appropriate antibodies
   (d) Adoptive transfer of anti-tumour cells such as NK cells or specifically reactive immune effector cells
   (e) Utilization of natural cytotoxins
10. Improving our understanding of dormancy; inhibition of the activation of dormant cells, for example, may prove useful in preventing 'recurrence'
11. Gaining further insight into the molecular basis and potential of differentiation therapy

earlier (section 6.2).

A summary of possible strategies which might prove useful in the treatment of cancer is provided in Table 6.1.

6.5 AN ULTIMATE UNDERSTANDING?

Although we have a reasonable understanding of certain types of malignant disease, many aspects of cancer, particularly metastasis, remain enigmatic. We can identify certain agents that cause cancer, we can recognize certain individuals or groups that may have a high risk of developing cancer either for familial or environmental reasons, and with some cancers we have a reasonably high chance of treating them successfully. Epidemiological studies have contributed much to our current understanding of what causes cancer, and research into therapeutic techniques based largely on drugs and irradiation underlies most successful treatments. Such studies should continue to yield advances in our understanding and treatment of cancer, but the future for cancer research now seems to lie more in how we view processes within the cancer cell itself.

Recent research into what biochemical changes may cause cancer has already taken us from the cancer cell itself to its genes. This is characterized by the current burst of activity (likely to continue for some time) in the molecular biology of cancer. However, research into cancer should not stop at genes, their products and their actions, for it should also reach into the world of sub-molecular and indeed sub-atomic particles. Scientists of extraordinary vision, like the late Albert Szent-Gyorgi for example, have long championed the need to grapple with a truly fundamental understanding of cancer, for only then will we be able to say with some confidence that we fully understand the nature of the disease.

References

AA Abdel-Latif (1986) *Pharm Rev* **38**: 227–272

GL Abelev, SD Perova, MI Khramkova, ZA Postnikova and IS Irlin (1963) *Transplant* **1**: 174–180

M Abercrombie (1960) *Exp Cell Res Suppl* **8**: 188–198

M Abercrombie (1965) In *Cells and tissues in culture* Ed EN Willmer Academic Press New York Vol **1** pp 177–202

M Abercrombie (1967) *Nat Cancer Inst Monogr* **26**: 249–277

M Abercrombie (1970a) *In vitro* **6**: 128–142

M Abercrombie (1970b) *Eur J Cancer* **6**: 7–13

M Abercrombie (1979) *Nature* **281**: 259–262

M Abercrombie and EJ Ambrose (1958) *Exp Cell Res* **15**: 332–345

M Abercrombie and EJ Ambrose (1962) *Cancer Res* **22**: 525–545

M Abercrombie and JEM Heaysman (1953) *Exp Cell Res* **5**: 11–131

M Abercrombie and JEM Heaysman (1954) *Exp Cell Res* **6**: 293–306

M Abercrombie and JEM Heaysman (1976) *J Nat Cancer Inst* **56**: 561–570

M Abercrombie, JEM Heaysman and HM Karthauser (1957) *Exp Cell Res* **13**: 276–291

M Abercrombie, JEM Heaysman and SM Pegrum (1970) *Exp Cell Res* **62**: 389–398

DC Adelman, RA Miller and HS Kaplan (1980) *Int J Cancer* **25**: 467–473

KC Agarwal and RE Parks (1983) *Int J Cancer* **32**: 801–804

SK Akiyama and KM Yamada (1985) *J Biol Chem* **260**: 4492–4500

P Alberto (1981) *Eur J Cancer* **17**: 1061–1062

DS Alberts, HSG Chen, B Saehnlen, SE Salmon, EA Surwit, L Young and TE Moon (1980) *Lancet* ii 340–342

AP Albino, R Le Strange, AI Oliff, ME Furth and LJ Old (1984) *Nature* **308**: 69–72

G Albrecht-Buehler (1977) *Cell* **11**: 395–404

L Alby and R Auerbach (1984) *Proc Nat Acad Sci* **81**: 5739–5743

P Alexander (1974) *Cancer Res* **34**: 2077–2082

P Alexander (1977) *Cancer* **40**: 467–470

GH Algire, JM Weaver and RT Prehn (1954) *J Nat Cancer Inst* **15**: 493–507

IU Ali and RO Hynes (1978) *Biochim Biophys Acta* **510**: 140–150

IU Ali, VM Mantner, RP Lanza and RO Hynes (1977) *Cell* **11**: 115–126

K Alitalo (1984) *Med Biol* **62**: 304–317
K Alitalo, M Kurkinen, A Vaheri, T Krieb and R Timpl (1980) *Cell* **19**: 1053–1062
RD Allen (1961) *Exp Cell Res Suppl* **8**: 17–31
LE Allred and KR Porter (1979) In *Morphology of normal and transformed cells* Ed RO Hynes John Wiley and Sons Chichester pp 21–61
J Alroy, BU Pauli and RS Weinstein (1981) *Cancer* **47**: 104–112
AL Alterman, DM Fornabaio and CW Stackpole (1985) *J Nat Cancer Inst* **75**: 691–702
P Altevogt, M Fogel, R Cheingsong-Popov, J Dennis, P Robinson and V Schirrmacher (1983) *Cancer Res* **43**: 5138–5144
EJ Ambrose (1967) In *Mechanisms of invasion in cancer* Ed P Denoix Springer-Verlag Berlin pp 130–139
EJ Ambrose and DM Easty (1976) In *Human tumours in short term culture* Ed PP Dendy Academic Press London pp 45–55
JL Ambrus, LM Ambrus, JW Byron, ME Goldberg and JWE Harrison (1956) *Ann NY Acad Sci* **63**: 938–961
AO Anderson and ND Anderson (1976) *Immunol* **31**: 731–748
DC Anderson, LJ Miller, FC Schmalstieg, R Rothlen and TA Springer (1986) *J Immunol* **137**: 15–27
JM Anderson, BR Stevenson, LA Jesaitis, DA Goodenough and MS Mooseker (1988) *J Cell Biol* **106**: 1141–1149
MA Anzano, AB Roberts and MB Sporn (1986) *J Cell Physiol* **126**: 312–318
JD Aplin, RC Hughes, CL Jaffe and N Sharon (1981) *Exp Cell Res* **134**: 488–494
WS Argraves, S Suzuki, H Arai, K Thompson, MD Pierschbacher and E Ruoslahti (1987) *J Cell Biol* **105**: 1183–1190
HA Armelin (1973) *Proc Nat Acad Sci* **70**: 2702–2706
HA Armelin, MCS Armelin, K Kelly, T Stewart, P Leder, BH Cochran and CD Stiles (1984) *Nature* **310**: 655–660
MT Armstrong and PB Armstrong (1978) *J Cell Sci* **33**: 37–52
PB Armstrong (1977) *Bioscience* **27**: 803–808
PB Armstrong and MT Armstrong (1973) *Dev Biol* **35**: 187–209
PB Armstrong, JP Quigley and E Sidebottom (1982) *Cancer Res* **42**: 1826–1837
C Artigas, DMP Thomson, M Durko, M Sutherland, R Scanzano, G Shenouda and AEJ Dubois (1986) *Cancer Res* **46**: 1874–1881
F Ashall, ME Bramwell and H Harris (1982) *Lancet* ii: 1–6
SM Astrin and C Costanzi (1989) *Sem Oncol* **16**: 138–147
T Atsumi and M Takeichi (1980) *Dev Growth Diff* **22**: 133–142
JC Aub, BH Sanford and MN Cote (1965) *Proc Nat Acad Sci* **54**: 396–399
R Auerbach, L Alby, LW Morrissey, M Tu and J Joseph (1985) *Microvasc Res* **29**: 401–411
R Auerbach and W Auerbach (1982) *Science* **215**: 127–134
R Auerbach, HC Lu, E Pardon, F Gumkowski, G Kaminska and M Kaminski (1987) *Cancer Res* **47**: 1492–1496
R Auerbach, LW Morrissey and YA Sidky (1978) *Nature* **274**: 698–699
N Auersperg and A Worth (1966) *Int J Cancer* **1**: 219–238
P Aulenbacher, H-O Weiling, N Paweletz and E Spiess (1984) *Anticancer Res* **4**: 75–82
D Axelrod (1981) *J Cell Biol* **89**: 141–145
R Ayesh, JR Idle, JC Ritchie, MJ Cruthers and R Hetzel (1984) *Nature* **312**: 169–170

R Azarnia and WR Loewenstein (1973) *Nature* **241**: 455–457
R Azarnia and WR Loewenstein (1984) *J Membr Biol* **82**: 191–205
R Azarnia and WR Loewenstein (1987) *Mol Cell Biol* **7**: 946–950

U Bagge, A Blixt and K-G Strid (1983) *Int J Microcirc* **2**: 215–222
CM Balch, S Soong, R Murad, AC Ingalls and WA Maddox (1979) *Surgery* **86**: 343–356
RW Baldwin and MJ Embleton (1974) *Int J Cancer* **13**: 433–443
D Baltimore (1970) *Nature* **226**: 1209–1211
AD Bangham and BA Pethica (1960) *Proc Roy Soc Edin* **28**: 43–52
M Baniyash, T Netanel and IP Witz (1981) *Cancer Res* **41**: 433–437
JB Bard and ED Hay (1975) *J Cell Biol* **67**: 400–418
BE Barker and KK Sanford (1970) *J Nat Cancer Inst* **44**: 39–51
D Barnes (1984) In *The use of serum-free and hormone supplemented media* Ed JP Mather Plenum New York pp 195–237
DW Barnes and J Reing (1985) *J Cell Physiol* **125**: 207–214
F Barre-Sinoussi and 11 others (1983) *Science* **220**: 868–871
JC Barrell, BD Crawford, LO Mixter, LM Schechtman, POP Ts'o and R Pollack (1979) *Cancer Res* **39**: 1504–1510
G Barski (1967) In *Mechanisms of invasion in cancer* Ed P Denoix Springer-Verlag Berlin pp 40–46
G Barski and J Belehradek (1965) *Exp Cell Res* **37**: 464–480
SH Barsky, CN Rao, JE Williams and LA Liotta (1984) *J Clin Invest* **74**: 843–848
ML Basara, JB McCarthy, DW Barnes and LT Furcht (1985) *Cancer Res* **45**: 2487–2494
RC Bast, B Zbar, GB Mackanen and HJ Raff (1975) *J Nat Cancer Inst* **54**: 749–756
OH Beahrs and MH Meyers (1983) Eds *Manual for staging of cancer* 2nd ed Lippincott Philadelphia
RP Beasley, CC Lin and LU Hwang (1981) *The Lancet* ii: 1129–1133
J Behrens, W Birchmeier, S Goodman and BA Imhof (1985) *J Cell Biol* **101**: 1307–1315
DR Bell, JH Garland, N Kartner, RN Buick and V Ling (1985) *J Clin Oncol* **3**: 311–315
PB Bell (1977) *J Cell Biol* **74**: 963–982
PB Bell (1978) In *The molecular basis of cell–cell interaction* Eds RA Lerner and D Bergsma Alan R Liss Inc New York pp 177–194
R Bellairs, A Curtis and G Dunn (1982) Eds *Cell behaviour– a tribute to Michael Abercrombie* CUP Cambridge
W Bender (1985) *Cell* **43**: 559–560
WF Benedict, BE Weissman, C Mark and EJ Stanbridge (1984) *Cancer Res* **44**: 3471–3479
DC Bennett, TJ Dexter, EJ Ormerod and IR Hart (1986) *Cancer Res* **46**: 3239–3244
A Ben-Ze'ev and A Raz (1981) *Cell* **26**: 107–115
A Ben-Ze'ev and A Raz (1985) *Cancer Res* **45**: 2632–2641
WE Berdel, WRE Bausert, U Fink, J Rasteller and PG Munder (1981) *Anticancer Res* **1**: 345–352
JW Berg, AG Huvos, LM Axtell and GF Robbins (1973) *Ann Surg* **177**: 8–16
JE Bergmann, A Kupfer and SJ Singer (1983) *Proc Nat Acad Sci* **80**: 1367–1371
S Bernstein and R Weinberg (1985) *Proc Nat Acad Sci* **82**: 1726–1730

MJ Berridge (1985) *Sci Amer* **253**: 124–134
MJ Berridge (1987) *Ann Rev Biochem* **56**: 159–193
MP Bevilacqua, JS Pober, DL Mendrick, RS Cotran and MA Gimbrone (1987) *Proc Nat Acad Sci* **84**: 9238–9242
MP Bevilacqua, S Stengelin, MA Gimbrone and B Seed (1989) *Science* **243**: 1160–1165.
EC Beyer, DA Goodenough and DE Paul (1988) In *Gap junctions* Eds EL Hertzberg and RG Johnson AR Liss Inc New York pp 167–175
RJ Beyth and LA Culp (1984) *Exp Cell Res* **155**: 537–548
RJ Beyth and LA Culp (1985) *Med Ageing Dev* **29**: 151–169
VP Bhavanandan, JG Kemper and J-C Bystryn (1980) *J Biol Chem* **255**: 5145–5153
C Birchmeier, TE Kreis, HM Eppenberger, KH Winterhalter and W Birchmeier (1980) *Proc Nat Acad Sci* **77**: 4108–4112
CR Birdwell, D Gospodarowicz and GL Nicolson (1978) *Proc Nat Acad Sci* **75**: 3273–3277
JM Bishop (1982) *Sci Amer* **246**: 68–79
JM Bishop (1983) *Ann Rev Biochem* **2**: 301–354
JM Bishop (1985) *Cell* **42**: 23–38
JM Bishop (1987) *Science* **235**: 305–311
C Biswas (1982) *Bichem Biophys Res Comm* **109**: 1026–1034
C Biswas (1984) *Cancer Letters* **24**: 201–207
JJ Bittner (1942) *Science* **95**: 462–463
JL Bixby, RS Pratt, J Lilien and LF Reichardt (1987) *Proc Nat Acad Sci* **84**: 2555–2559
R Bjerkvig, OD Laerum and O Mella (1986) *Cancer Res* **46**: 4071–4079
PH Black (1980) *Adv Cancer Res* **32**: 75–199
WF Bodmer, CJ Bailey, J Bodmer, HJR Bussey, A Ellis, P Gorman, FC Lucibello, VA Murday, SH Rider, P Scambler, D Scheer, E Solomon and NK Spurr (1987) *Nature* **328**: 614–616
B Boeryd, B Hagmar, G Johnsson and W Ryd (1974) *Pathol Eur* **9**: 119–123
DA Boetcher and EJ Leonard (1974) *J Nat Cancer Inst* **52**: 1091–1099
E Bogenmann, C Mark, H Isaacs, HB Neustein, YA De Clerck, WE Laug and RA Jones (1983) *Cancer Res* **43**: 1176–1186
E Bogenmann and B Sordat (1979) *J Cell Biol* **83**: c293
JGM Bolscher, DCC Schallier, LA Smets, H van Rooy, JG Collard, EA Bruyneel and MMK Mareel (1986) *Cancer Res* **46**: 4080–4086
CW Boone (1975) *Science* **188**: 68–78
CJ Boreiko, DJ Abernethy and DB Stedman (1987) *Carcinogenesis* **8**: 321–324
JL Bos (1989) *Cancer Res* **49**: 4682–4689
HB Bosmann, GF Bieber, AE Brown, KR Case, DM Gersten, TW Kimmerer and A Lione (1973) *Nature* **246**: 487–489
HB Bosmann and TC Hall (1974) *Proc Nat Acad Sci* **71** 1833–1837
HB Bosmann and A Lione (1974) *Biochem Biophys Res Comm* **61** 564–567
K Bosslet and V Schirrmacher (1981) *J Exp Med* **154** 557–562
B Bottazzi, N Polentarutti, A Balsari, D Boraschi, P Ghezzi, M Salmona and A Mantovani (1983) *Int J Cancer* **31** 55–63
PD Bowman and CW Daniel (1975) *Med Ageing Dev* **4** 147–158
S Boyden (1962) *J Exp Med* **115** 453–466
EA Boyse and LJ Old (1969) *Ann Rev Genet* **3** 269–290
EA Boyse, E Stockert and LJ Old (1967) *Proc Nat Acad Sci* **58** 954–957

ME Bracke, RM-L van Cauwenberge and MM Mareel (1984) *Clin Exp Metast* **2**: 161–170
R Brackenbury (1985) *Cancer Metast Rev* **4** 41–58
R Brackenbury, ME Greenberg and EM Edelman (1984) *J Cell Biol* **99**: 1944–1954
RO Brady and PH Fishman (1974) *Biochim Biophys Acta* **355**: 121–148
ME Bramwell (1985) In *Receptors in tumour biology* Ed CM Chadwick CUP Cambridge pp 189–220
D Branton, S Bullivant, NB Gilula, MJ Karnovsky, MJ Moor, K Muhlethaler, DH Northcote, L Packer, P Satir, V Speth, LA Staehelin and RS Weinstein (1975) *Science* **190**: 54–56
RS Bresalier, ES Hujanen, SE Raper, FJ Roll, SH Itzkowitz, GR Martin and YS Kim (1987) *Cancer Res* **47**: 1398–1400
MS Bretscher (1988) *J Cell Biol* **106**: 235–237
EB Briles and S Kornfeld (1978) *J Nat Cancer Inst* **60**: 1217–1222
RL Brinster (1974) *J Exp Med* **140**: 1049–1056
RL Brinster and RD Palmiter (1986) *Harvey Lect* 1–38
GM Brodeur, RC Seeger, M Schwab, HE Varmus and JM Bishop (1984) *Science* **224** 1121–1124
IDJ Bross and LE Blumenson (1976) In *Fundamental aspects of metastasis* Ed L Weiss North Holland Amsterdam pp 359–375
AF Brown (1982) *J Cell Sci* **58** 455–467
JM Brown (1973) *Cancer Res* **33** 1217–1224
JP Brown, RM Hewick, I Hellstrom, KE Hellstrom, RF Doolittle and WJ Dreyer (1982) *Nature* **296** 171–173
A Brugarolas and M Gosalvez (1980) *Lancet* i: 68–70
KW Brunson, G Beattie and GL Nicolson (1978) *Nature* **272**: 543–545
KW Brunson and GL Nicolson (1979) *J Supramol Struct* **11**: 517–528
KW Brunson and GL Nicolson (1980) In *Brain metastasis* Eds L Weiss, HA Gilbert and JB Posner Hall Boston pp 50–65
J Bubenik, P Perlmann, EM Fenyo, T Jandlova, E Suhajova and M Malkovsky (1979) *Int J Cancer* **23**: 392–396
J Bubenik, P Perlmann, K Helmstein and G Moberger (1970) *Int J Cancer* **5**: 310–319
CA Buck, MC Glick and L Warren (1971a) *Science* **172**: 169–171
CA Buck, MC Glick and L Warren (1971b) *Biochem* **10**: 2176–2180
CA Buck and AF Horwitz (1987) *J Cell Sci Suppl* **8**: 231–250
I Buckley (1988) *Adv Cancer Res* **50**: 71–94
SA Buhrow, S Cohen and JV Stavros (1982) *J Biol Chem* **257**: 4019–4022
RN Buick and MN Pollak (1984) *Cancer Res* **44**: 4909–4918
S Bullivant (1981) In *Epithelial ion and water transport* Eds AC Macknight and JP Leader Raven Press New York pp 265–275
WS Bullough and EB Laurence (1960) *Proc Roy Soc Lond B* **151**: 517–536
M Bundgaard (1984) *J Ultrastruct Res* **88**: 1–17
MM Burger (1969) *Proc Nat Acad Sci* **62**: 994–1001
MM Burger (1970) *Nature* **227**: 170–171
MM Burger (1973) *Fed Proc* **32**: 91–101
MM Burger and AR Goldberg (1967) *Proc Nat Acad Sci* **57**: 359–366
D Burk, M Woods and J Hunter (1967) *J Nat Cancer Inst* **38**: 839–863
RR Burk (1968) *Nature* **219**: 1272–1275
RR Burk, JD Pitts and JH Subak-Sharpe (1968) *Exp Cell Res* **53**: 297–306

FM Burnet (1970) *Prog Exp Tumour Res* **13**: 1–27
K Burridge and L Connell (1983a) *J Cell Biol* **97**: 359–367
K Burridge and L Connell (1983b) *Cell Motility* **3**: 405–417
K Burridge and L McCullough (1980) *J Supramolec Struct* **13**: 53–65
K Burridge, L Molony and Kelly (1987) *J Cell Sci Suppl* **8**: 211–229
SJ Burwen and AL Jones (1987) *Trends Biochem Sci* **12**: 159–162
E Butcher, R Scollay and I Weissman (1979) *Nature* **280**: 496–498
EC Butcher, RG Scollay and IL Weissman (1980) *Eur J Immunol* **10**: 556–561
TP Butler and PM Gullino (1975) *Cancer Res* **35**: 512–516
JC Bystryn (1976) *J Immunol* **116**: 1302–1305
JC Bystryn, RS Bart, P Livingston and AW Kopf (1974) *J Invest Dermatol* **63**: 369–373
JC Bystryn, I Schenkein, S Baur and JW Uhr (1974) *J Nat Cancer Inst* **52**: 1263–1269

J Cairns (1981) *Nature* **289**: 353–357
D Camerini, SP James, I Slamenkovic and B Seed (1989) *Nature* **342**: 78–82
A Camez, E Dupuy, J Bellucci, F Calvo, MC Bryckaert and G Tobelem (1986) *Invasion Metastasis* **6**: 321–334
IL Cameron and TB Pool (1981) *The transformed cell* Academic Press, New York
ID Capel, M Jenner, MH Pinnock, HM Dorrell, DC Payne and DC Williams (1979) *Oncology* **36**: 242–244
G Carbone and G Parmiani (1975) *J Nat Cancer Inst* **55**: 1195–1197
B Carlin, R Jaffe, B Bender and AE Chung (1981) *J Biol Chem* **256**: 5209–5214
G Carpenter and S Cohen (1979) *Ann Rev Biochem* **48**: 193–216
I Carr and J Carr (1982) In *Tumour invasion and metastasis* Eds LA Liotta and IR Hart Martinus Nijhoff The Hague pp 189–205
EA Carswell, LJ Old, RL Kassel, S Green, N Fiore and B Williamson (1975) *Proc Nat Acad Sci* **72**: 3666–3670
R Carter (1982) *J Clin Pathol* **35**: 1041–1049
SB Carter (1965) *Nature* **208**: 1183–1187
SB Carter (1967) *Nature* **213**: 256–260
WG Carter and EA Wayner (1988) *J Biol Chem* **263**: 4193–4201
JJ Cassiman and MR Bernfield (1975) *Exp Cell Res* **91**: 31–35
WK Cavanee and 8 others (1983) *Nature* **305**: 779–784
JG Chafouleas, RL Pardue, JR Brinkley, JR Dedman and AR Means (1980) In *Calcium-binding proteins: structure and function* Eds FL Sieger, E Carafoli, RH Kretsinger, DH MacLennan and RH Wasserman Elsevier/Nth Holland Amsterdam pp 189–196
JG Chafouleas, RL Pardue, BR Brinkley, JR Dedman and AR Means (1981) *Proc Nat Acad Sci* **78**: 996–1000
JP Chalcroft and S Bullivant (1970) *J Cell Biol* **47**: 49–60
M Chambard, J Gabrion and J Mauchamp (1981) *J Cell Biol* **91**: 157–166
AF Chambers, R Shafir and V Ling (1982) *Cancer Res* **42**: 4018–4025
R Chambers and GS Renyi (1925) *Amer J Anat* **35**: 385–402
C-C Chang, JR Trosko, H-J Kung, D Bombick and F Matsumura (1985) *Proc Nat Acad Sci* **82**: 5360–5364
EH Chang, ME Furth, EM Scolnick and DR Lowy (1982) *Nature* **297**: 479–483
IF Charo, C Yuen, HD Perez and IM Goldstein (1986) *J Immunol* **136**: 3412–3419
AS Charonis and 7 others (1988) *J Cell Biol* **107**: 253–264

SK Chattergee (1979) *Eur J Cancer* **15**: 1351–1356
SK Chattergee and U Kim (1977) *J Nat Cancer Inst* **58**: 273–280
SK Chattergee and U Kim (1978) *J Nat Cancer Inst* **61**: 151–162
R Cheingsong—Popov, P Robinson, P Altevogt and V Schirrmacher (1983) *Int J Cancer* **32**: 359–366
J-M Chen and W-T Chen (1987) *Cell* **48**: 193–203
LP Chen and JM Buchanan (1975) *Proc Nat Acad Sci* **73**: 4457–4461
W-T Chen (1981) *J Cell Biol* **90**: 187–200
W-T Chen, J-M Chen, SJ Parsons and JT Parsons (1985) *Nature* **316**: 156–158
W-T Chen and SJ Singer (1980) *Proc Nat Acad Sci* **72**: 1132–1136
DA Cheresh, MD Pierschbacher, MA Herzig and K Mujoo (1986) *J Cell Biol* **102**: 688–696
DA Cheresh, R Pytela, MD Pierschbacher, FG Klier, E Ruoslahti and RA Reisfeld (1987) *J Cell Biol* **105**: 1163–1173
PG Chesa, WJ Rettig, MR Melamed, LJ Old and HL Niman (1987) *Proc Nat Acad Sci* **84**: 3234–3238
EC Chew, RL Josephson and AC Wallace (1976) In *Fundamental aspects of metastasis* Ed L Weiss N Holland Amsterdam pp 121–150
EC Chew and AC Wallace (1976) *Cancer Res* **36**: 1904–1909
Y-H Chin, RA Rasmussen, JA Woodruff and TG Easton (1986) *J Immunol* **136**: 2556–2561
S Chipowsky, YC Lee and S Roseman (1973) *Proc Nat Acad Sci* **70**: 2309–2312
R Chiquet-Ehrismann, P Kalla, CA Pearson, K Beck and M Chiquet (1988) *Cell* **53**: 783–790
R Chiquet-Ehrismann, EJ Mackie, CA Pearson and T Sakakura (1986) *Cell* **47**: 131–139
YS Cho-Chung (1982) In *Prolonged arrest of cancer* Ed BA Stoll John Wiley and Sons Chichester pp 199–220
AS-F Chong, CR Parish and DR Coombe (1987) *Immunol Cell Biol* **65**: 85–95
H Chopra and 9 others (1988) *Cancer Res* **48**: 3782–3800
IN Chou, PH Black and RO Roblin (1974) *Proc Nat Acad Sci* **71**: 1748–1752
GJ Cianciolo, ME Lostrom, M Tarn and R Snyderman (1983) *J Exp Med* **158**: 885–900
GJ Cianciolo, TJ Matthews, DP Bolognesi and R Snyderman (1980) *J Immunol* **124**: 2900–2905
MA Cifone and IJ Fidler (1981) *Proc Nat Acad Sci* **78**: 6949–6952
S Citi, H Sabanay, R Jakes, B Geiger and J Kendrick—Jones (1988) *Nature* **333**: 272–276
ER Clark and EL Clark (1935) *Amer J Anat* **57**: 385–438
ER Clark and EL Clark (1936) *Amer J Anat* **59**: 123–173
RA Clark and SJ Klebanoff (1979) *J Immunol* **122**: 2605–2610
EE Clifton and D Agostino (1965) *Vasc Dis* **2**: 43–52
MJ Cline and DC Livingston (1971) *Nature (New Biol)* **232**: 155–156
BH Cochran (1985) *Adv Cancer Res* **45**: 183–216
BH Cochran, J Zullo, IM Verma and CD Stiles (1984) *Science* **226**: 1080–1082
JF Codington, AG Cooper, MC Brown and RW Jeanloz (1975) *Biochem* **14**: 855–859
JH Coggin and NG Anderson (1974) *Adv Cancer Res* **19**: 105–165
HJ Cohen and BB Gilbertsen (1975) *J Clin Invest* **55**: 84–93
MC Cohen, A Gross, T Yoshida and S Cohen (1978) *J Immunol* **121**: 840–843

S Cohen (1959) *J Biol Chem* **234**: 1129–1137
S Cohen (1962) *J Biol Chem* **237**: 1555–1562
CALS Colaco and WH Evans (1983) In *Electronmicroscopy of proteins* Ed JR Harris Academic Press New York Vol 4 pp 332–336
GJ Cole, D Schubert and L Glaser (1985) *J Cell Biol* **100**: 1192–1199
JG Collard, JF Schijven, A Bikker, G La Riviere, JGM Bolscher and E Roos (1986) *Cancer Res* **46**: 3521–3527
JG Collard and JHM Temmink (1975) *J Cell Sci* **19**: 21–32
JG Collard and JHM Temmink (1976) *J Cell Biol* **68**: 101–112
JG Collard, WP van Beek, JWG Janssen and JF Schijven (1985) *Int J Cancer* **35**: 207–214
M Colucci, R Giavazzi, G Alessandri, N Semeraro, A Mantovani and MB Donati (1981a) *Blood* **57**: 733–735
M Colucci, R Giavazzi, G Alessandri, N Semeraro, A Mantovani and MB Donati (1981b) *Br J Cancer* **43**: 100–104
DR Coman (1944) *Cancer Res* **4**: 625–629
DR Coman (1953) *Cancer Res* **13**: 397–404
DR Coman (1961) *Cancer Res* **21**: 1436–1438
DE Comings (1973) *Proc Nat Acad Sci* **70**: 3324–3328
KG Conn and KA Knudsen (1989) *Cancer Res* **49**: 7098–7105
CG Connor, PB Moore, RC Brady, JP Hurn, RB Arlinghaus and JR Dedman (1983) *Biochem Biophys Res Comm* **112**: 647–654
DR Coombe, CR Parish, IA Ramshaw and JM Snowden (1987) *Int J Cancer* **39**: 82–88
GM Cooper (1982) *Science* **218**: 801–806
SF Cotmore and RL Carter (1973) *Int J Cancer* **11**: 725–738
M Cottler-Fox, W Ryd, B Hagmar and CH Fox (1980) *Int J Cancer* **26**: 689–694
JR Couchman, M Hook, DA Rees and R Timpl (1982) *J Cell Biol* **96**: 177–183
EV Cowdry (1940) *Arch Pathol* **30**: 1245–1274
JK Cowell and LM Franks (1984) *Int J Cancer* **33**: 657–667
P Cowin, H-P Kapprell, WW Franks, J Tamkun and RO Hynes (1986) *Cell* **46**: 1063–1073
EB Cramer, LC Milks and GK Ojakian (1980) *Proc Nat Acad Sci* **77**: 4069–4073
CO Criborn, S Franzen, G Unsgaard and J Zajieck (1974) *Science J Haematol* **1**: 272–279
JD Crissman, JS Hatfield, DG Menter, B Sloane and KV Honn (1988) *Cancer Res* **48**: 4065–4072
JD Crissman, J Hatfield, M Schaldenbrand, BF Sloane and KV Honn (1985) *Lab Invest* **53**: 470–478
DR Critchley (1979) In *Surfaces of normal and malignant cells* Ed RO Hynes Wiley Chichester pp 63–101
C Crone and O Christensen (1981) *J Gen Physiol* **77**: 349–371
SJ Cross and I ap Gwynn (1987) *Cytobios* **50**: 41–62
R Crum, S Szabo and J Folkman (1985) *Science* **230**: 1375–1378
LA Culp (1976) *J Supramolec Struct* **5**: 239–255
LA Culp, BJ Rollins, J Buncet and S Hitri (1978) *J Cell Biol* **79**: 788–796
BA Cunningham, JJ Hemperly, BA Murray, EA Prediger, R Brackenbury and GM Edelman (1987) *Science* **236**: 799–806
DD Cunningham and AB Pardee (1969) *Proc Nat Acad Sci* **64**: 1049–1056
LJ Cuprak and WF Lever (1974) *Proc Soc Exp Biol Med* **146**: 309–315

T Curran and JI Morgan (1987) *BioEssays* **7**: 255–258
T Curran, C Van Beveren, N Ling and IM Verma (1985) *Ann Rev Cell Biol* **5**: 167–169
GA Currie and P Alexander (1974) *Br J Cancer* **29**: 72–75
GA Currie and I Basham (1972) *Br J Cancer* **26**: 427–438
ASG Curtis (1964) *J Cell Biol* **20**: 199–215
ASG Curtis (1967) *The cell surface its molecular role in morphogenesis* Academic Press New York
ASG Curtis (1969) *J Embryol Exp Morph* **22**: 305–325
ASG Curtis (1973) *Prog Biophys Molec Biol* **27**: 315–386
ASG Curtis (1987) *J Cell Sci* **87**: 609–611
ASG Curtis, JV Forrester, C McInnes and F Lawrie (1983) *J Cell Biol* **97**: 1500–1506
ASG Curtis and P Rooney (1979) *Nature* **281**: 222–223
ASG Curtis and GM Seehar (1978) *Nature* **274**: 52–53
F Cuttitta, DN Carney, J Mulshine, TW Moody, J Fedorko, A Fischler and JD Minna (1985) *Nature* **316**: 823–826

G Dahl, T Miller, D Paul, R Voellmy and R Werner (1987) *Science* **236**: 1290–1293
NK Damle, LV Doyle, JR Bender and EC Bradley (1987) *J Immunol* **138**: 1779–1785
NK Damle, LV Doyle and EC Bradley (1986) *J Immunol* **137**: 2814–2822
CH Damsky, KA Knudsen, D Bradley, CA Buck and AF Horwitz (1985) *J Cell Biol* **100**: 1528–1539
CH Damsky, J Richa, D Solter, K Knudsen and CA Buck (1983) *Cell* **34**: 455–466
CV Dang, WR Bell, O Kaiser and A Wong (1985) *Science* **227**: 1487–1490
K Dano, PA Andreasen, J Grondahl-Hansen, P Kristensen, LS Nielsen and L Skriver (1985) *Adv Cancer Res* **44**: 139–266
WL D'Arrigo, P Elvin and CW Evans (1985) *Cancer Letters* **28**: 47–54
P Datta and CV Natraj (1980) *Exp Cell Res* **125**: 431–439
T David-Pfeuty and SJ Singer (1980) *Proc Nat Acad Sci* **77**: 6687–6691
M Dean, RA Levine, W Ran, MS Kindy, GE Sonenshein and J Campisi (1986) *J Biol Chem* **261**: 9161–9166
P De Baetselier, E Roos, L Brys, L Remels, M Gioberti, D Dekegel, S Segal and M Feldman (1984) *Cancer Metast Rev* **3**: 5–24
PPH De Bruyn and Y Cho (1979) *J Nat Cancer Inst* **62**: 1221–1227
S Dedhar, E Ruoslahti and MD Pierschbacher (1987) *J Cell Biol* **104**: 585–593
JF DeLarco and GJ Todaro (1978) *Proc Nat Acad Sci* **75**: 4001–4005
DT Denhardt, DR Edwards and CLJ Parfett (1986) *Biochim Biophys Acta* **865**: 83–125
JW Dennis, S Laferte, C Waghorne, ML Breitman and RS Kerbel (1987) *Science* **236**: 582–585
S De Petris (1974) *Nature* **250**: 54–56
CJ Der, TG Krontiris and GM Cooper (1982) *Proc Nat Acad Sci* **79**: 3637–3640
L De Ridder, OD Laerum (1981) *J Nat Cancer Inst* **66**: 723–728
L De Ridder, M Mareel and L Vakaet (1975) *Cancer Res* **35**: 3164–3171
L De Ridder, M Mareel and L Vakaet (1977) *Arch Geschw* **47**: 7–27
BV Derjaguin and LD Landau (1941) *Acta Physicochemica URSS* **14**: 633–662
TF Deuel, JS Huang, SS Huang, P Stroobant and MD Waterfield (1983) *Science* **221**: 148–150

TF Deuel, JS Huang, RT Proffitt, D Chang and BB Kennedy (1981) *J Supramolec Struct Suppl* **5**: 128–137
DL Dexter, ES Lee, DJ De Fusco, NP Libbey, EN Spermulli and P Calabresi (1983) *Cancer Res* **43**: 1733–1740
TM Dexter and NG Testa (1976) *Methods Cell Biol* **14**: 387–405
WD De Wys (1972) *Cancer Res* **32**: 374–379
MP Dierich, D Wilhelmi and GO Till (1977) *Nature* **270**: 351–352
CA Dinarello and 12 others (1986) *J Clin Invest* **77**: 1734–1739
KP Dingemans, E Roos, MA Van den Bergh Weerman, IV van de Pavert (1978) *J Nat Cancer Inst* **60**: 583–598
KP Dingemans and MA Van den Bergh Weerman (1980) In *Metastasis clinical and experimental aspects* Eds K Hellmann, P Hilgard and S Eccles Martinus Nijhoff The Hague pp 194–198
KP Dingemans, R van Spronsen and E Thunnissen (1985) *Invas Metast* **5**: 50–60
LD Dion, JE Blalock and GE Gifford (1977) *J Nat Cancer Inst* **58**: 795–801
LD Dion, JE Blalock and GE Gifford (1978) *Exp Cell Res* **117**: 15–22
JA DiPaolo (1983) *J Nat Cancer Inst* **70**: 3–8
A DiPasquale and PB Bell (1974) *J Cell Biol* **62**: 198–214
QS Dizon and CM Southam (1963) *Cancer* **16**: 1288–1292
L Dobrossy, Z Pavelic and R Bernacki (1981) *Cancer Res* **41**: 2262–2266
AJP Docherty, A Lyons, BJ Smith, EM Wright, PE Stephens, TJR Harris, G Murphy and JJ Reynolds (1985) *Nature* **318**: 66–69
RJ Docherty, JV Forrester and JM Lackie (1987) *J Cell Sci* **87**: 399–409
RF Doolittle, MW Hunkapiller, LE Hood, SG DeVare, KC Robbins, SA Aaronson and HN Antoniades (1983) *Science* **221**: 275–276
JK Dorsey and S Roth (1973) *Dev Biol* **33**: 249–256
J Downward, Y Yarden, E Mayes, GT Scrace, N Totly, P Stockwell, A Ullrich, J Schlessinger and MD Waterfield (1984) *Nature* **307**: 521–527
C Doyle and JL Strominger (1987) *Nature* **330**: 256–259
PR Dragsten, R Blumethal and JS Handler (1981) *Nature* **294** 718–721
PR Dragsten, JS Handler, R Blumenthal (1982) *Fed Proc* **41**: 48–53
JA Drebin, VC Link, RA Weinberg and MI Greene (1986) *Proc Nat Acad Sci* **83**: 9129–9133
MH Dresden, SA Heilman and JD Schmidt (1972) *Cancer Res* **32**: 993–996
RW Drummond and LM Silverman (1982) *Clin Lab Med* **2**: 469–477
SE D'Souza, MH Ginsberg, TA Burke, SC-T Lam and EF Plow (1988) *Science* **242**: 91–93
R Dudler, C Schmidhauser, RW Parish, REH Wettenhall and T Schmidt (1988) *EMBO J* **7**: 3963–3970
PH Duesberg (1983) *Nature* **304**: 219–226
PH Duesberg (1985) *Science* **228**: 669–677
PH Duesberg and PK Vogt (1970) *Proc Nat Acad Sci* **67**: 1673–1680
R Dulbecco and J Elkington (1973) *Nature* **246**: 197–199
R Dulbecco and MGP Stoker (1970) *Proc Nat Acad Sci* **66**: 204–210
GA Dunn (1983) In *Leukocyte locomotion and chemotaxis* Eds H-U Keller and GO Till Birkhauser Basel pp 14–33
GA Dunn and GW Ireland (1984) *Nature* **312**: 63–65
ML Dustin, P Selvaraj, RJ Mattaliano and TA Springer (1987) *Nature* **329**: 846–848
HF Dvorak, SC Quay, NS Orenstein, AM Dvorak, P Hahn, AM Bitzer and AC Carvalho (1981) *Science* **212**: 923–924

HF Dvorak, DR Senger and AM Dvorak (1983) *Cancer Metast Rev* **2**: 41–73
N Dyson, PM Howley, K Munger and E Harlow (1989) *Science* **243**: 934–937
M Dziadek and R Timpl (1985) *Dev Biol* **111**: 372–382

DM Easty and GC Easty (1974) *Br J Cancer* **29**: 36–49
DM Easty and GC Easty (1976) In *Organ culture in biomedical research* Eds M Balls and M Monnichendamm CUP Cambridge pp 379–392
DM Easty, GC Easty, A Baici, RL Carter, SA Cederholm-Williams, H Felix, B Gusterton, G Haemmerli, I Hauser-Urfer, CW Heizmann, M Mareel, B Stehrenberger and P Strauli (1986) *Eur J Cancer Clin Oncol* **22**: 617–634
GC Easty (1975) In *Biology of cancer* Eds EJ Ambrose and FJC Roe Ellis Horwood Chichester pp 58–73
GC Easty and DM Easty (1963) *Nature* **199**: 1104–1105
GC Easty and DM Easty (1973) In *Chemotherapy of cancer dissemination and metastasis* Eds S Garattini and G Franchi Raven Press New York pp 45–50
G Eaves (1973) *J Pathol* **109**: 233–237
H Ebendal and JP Heath (1977) *Exp Cell Res* **110**: 469–473
SA Eccles and P Alexander (1974) *Nature* **250**: 667–668
SA Eccles and P Alexander (1975) *Nature* **257**: 52–53
GM Edelman (1983) *Science* **219**: 450–457
GM Edelman (1984) *Sci Amer* **250**: 118–129
D Edgar, R Timpl and H Thoenen (1984) *EMBO J* **3**: 1463–1468
M Edidin and A Weiss (1974) In *Control of proliferation in animal cells* Eds B Clarkson and R Baserga Cold Spring Harbor Labs New York pp 213–220
M Edward, JA Gold and RM MacKie (1989) *J Cell Sci* **93**: 155–161
M Edward and RM MacKie (1989) *J Cell Sci* **94**: 537–543
JG Edwards, JA Campbell, RT Robson and MG Vicker (1975) *J Cell Sci* **19**: 653–657
JG Edwards, J McK Dysart, DH Edgar and RT Robson (1979) *J Cell Sci* **35**: 307–320
SE Egan, GA McClarty, L Jarolim, JA Wright, I Spiro, G Hager and AH Greenberg (1987) *Mol Cell Biol* **7**: 830–837
P Ehrlich (1908) *Ned Tijdschr Geneesk* **1**: 273–290
P Ehrlich (1913) In *The collected papers of Paul Ehrlich* Ed F Himmelweit Pergamon Press Oxford Vol 3 p 510
RL Ehrmann and GO Gey (1956) *J Nat Cancer Inst* **16**: 1375–1390
M Einat, D Resnitzky and A Kimchi (1985) *Nature* **313**: 597–600
L Eisenbach, N Hollander, L Greenfield, H Yakor, S Segal and M Feldman (1984) *Int J Cancer* **34**: 567–572
L Eisenbach, S Segal and M Feldman (1983) *Int J Cancer* **32**: 113–117
L Eisenbach, S Segal and M Feldman (1985) *J Nat Cancer Inst* **74**: 87–93
R Eisenstein, N Sorgente, LW Soble, A Miller and KE Kuettner (1973) *Am J Pathol* **73**: 765–774
MJ Elices and ME Hemler (1989) *Proc Nat Acad Sci* **86**: 9906–9910.
R Eliyahu, D Michalovitz and M Oren (1985) *Nature* **316**: 158–160
J Ellison and DR Garrod (1984) *J Cell Sci* **72**: 163–172
E Elmore, T Kakunaga and JC Barrett (1983) *Cancer Res* **43**: 1650–1655
T Elsdale and J Bard (1972) *J Cell Biol* **54**: 626–637
P Elvin, BL Drake and CW Evans (1983) *J Immunol Methods* **64**: 295–301
P Elvin and CW Evans (1982) *Eur J Cancer Clin Oncol* **18**: 669–675
P Elvin and CW Evans (1983) *Biol Cell* **48**: 1–10

P Elvin and CW Evans (1984) *Eur J Cancer Clin Oncol* **20**: 107–114
P Elvin and CW Evans (1985) In *Molecular basis of cancer, Part A macromolecular structure, carcinogens and oncogenes* Ed R Rein Alan R Liss Inc pp 537–547
P Elvin, V Wong and CW Evans (1985) *Exp Cell Biol* **53**: 9–18
J Enenstein and L Furcht (1984) *J Cell Biol* **99**: 464–470
CA Erickson (1978a) *Exp Cell Res* **115**: 303–315
CA Erickson (1978b) *J Cell Sci* **33**: 53–84
HP Erickson and JL Inglesias (1984) *Nature* **311**: 267–269
WB Ershler, E Berman and AL Moore (1986) *J Nat Cancer Inst* **76**: 81–85
WB Ershler, JA Stewart, MP Hacker, AL Moore and BH Tindle (1984) *J Nat Cancer Inst* **72**: 161–164
J Estrada and GL Nicolson (1984) *Int J Cancer* **34**: 101–105
CW Evans and MDJ Davies (1977) *Cell Immunol* **33**: 211–218
CW Evans and J Proctor (1978) *J Cell Sci* **33**: 17–36
CW Evans, JE Taylor, JD Walker and NL Simmons (1983) *Br J Exp Path* **64**: 644–654
EP Evans, MD Burtenshaw, BB Brown, R Hennion and H Harris (1982) *J Cell Sci* 56 113–130
J Ewing (1928) *Neoplastic diseases: a treatise on tumours* Saunders Philadelphia

W Fach, L Harvath and EJ Leonard (1982) *Infect Immunity* **36**: 450–454
S Fairbairn, R Gilbert, G Ojakian, R Schwimmer and JP Quigley (1985) *J Cell Biol* **101**: 1790–1798
A Falanga and SG Gordon (1985) *Biochem* **24**: 5558–5567
R Falcioni, SJ Kennel, P Giacomini, G Zupi and A Sacchi (1986) *Cancer Res* **46**: 5772–5778
E Farber (1984) *Cancer Res* **44**: 4217–4223
MG Farquhar and GE Palade (1963) *J Cell Biol* **17**: 375–412
PM Faustmann and R Dermietzel (1985) *Cell Tissue Res* **242**: 399–407
RM Fauve, B Hevin, H Jacob, JA Gaillard and F Jacob (1974) *Proc Nat Acad Sci* **71**: 4052–4056
ER Fearon, SR Hamilton and B Vogelstein (1987) *Science* **238**: 193–195
J Feder, JC Marasa and JV Olander (1983) *J Cell Physiol* **116**: 1–6
AP Feinberg and B Vogelstein (1983) *Nature* **301**: 89–92
EM Fenyo and G Klein (1976) *Nature* **260**: 355–357
JR Feramisco, M Gross, T Kamata, M Rosenberg and RW Sweet (1984) *Cell* **38**: 109–117
V Ferro, C D'Arrigo, D Ogden and CW Evans (1988) *Biochem Soc Trans* **16**: 144–146
JW Fett, DJ Strydom, RR Lobb, EM Alderman, JL Bethune, JF Riordan and BL Vallee (1985) *Biochem* **24**: 5480–5486
PJ Fialkow (1974) *New Eng J Med* **291**: 26–35
IJ Fidler (1970) *J Nat Cancer Inst* **45**: 775–782
IJ Fidler (1973a) *Nature (New Biol)* **242**: 148–149
IJ Fidler (1973b) *Eur J Cancer* **9**: 223–227
IJ Fidler (1974a) *Cancer Res* **34**: 491–498
IJ Fidler (1974b) *Cancer Res* **34**: 1074–1078
IJ Fidler (1975a) *Cancer Res* **35**: 218–224
IJ Fidler (1975b) *J Nat Cancer Inst* **55**: 1159–1164
IJ Fidler (1978a) *Meth Cancer Res* **15**: 399–439

IJ Fidler (1978b) *Isr J Med Sci* **14**: 177–191
IJ Fidler (1980) *Science* **208**: 1469–1471
IJ Fidler and C Bucana (1977) *Cancer Res* **37**: 3945–3956
IJ Fidler, DM Gersten and MB Budmen (1976) *Cancer Res* **36** 3160–3165
IJ Fidler, DM Gersten and IR Hart (1978) *Adv Cancer Res* **28**: 149–250
IJ Fidler, DM Gersten and CW Riggs (1977) In *Cancer invasion and metastasis biologic mechanisms and therapy* Eds SB Day, WPL Myers, P Stansly, S Garattini and MG Lewis Raven Press New York pp 277–30
IJ Fidler and IR Hart (1981) *Cancer Res* **41**: 3266–3267
IJ Fidler and ML Kripke (1977) *Science* **197**: 893–895
IJ Fidler and ML Kripke (1980) *Cancer Immunol Immunother* 7: 201–213
IJ Fidler and GL Nicolson (1976) *J Nat Cancer Inst* **57**: 1199–1202
IJ Fidler and GL Nicolson (1977) *J Nat Cancer Inst* **58**: 1867–1872
IJ Fidler and GL Nicolson (1978) *Isr J Med Sci* **14**: 38–50
IJ Fidler and AJ Schroit (1988) *Biochim Biophys Acta* **948**: 151–173
IJ Fidler and I Zeidman (1972) *J Med* 3: 172–177
ME Finbow, TEJ Buultjens, S John, E Kam, L Meagher and JD Pitts (1987) *Ciba Fdn Symp* **125**: 92–107
J Finne, MM Burger and J-P Prieels (1982) *J Cell Biol* **92**: 277–282
J Finne, T-W Tao and MM Burger (1980) *Cancer Res* **40**: 2580–2587
B Fisher and ER Fisher (1959) *Science* **130**: 918–919
B Fisher and ER Fisher (1966) *Science* **152**: 1397–1398
B Fisher, ER Fisher and N Fednska (1967) *Cancer* **20**: 23–30
B Fisher and EA Saffer (1978) *J Nat Cancer Inst* **60**: 687–691
MS Fisher and ML Kripke (1978) *J Immunol* **121**: 1139–1144
FA Fitzpatrick and DA Stringfellow (1979) *Proc Nat Acad Sci* **76**: 1765–1769
J Flagg-Newton, I Simpson and WR Loewenstein (1979) *Science* **205**: 404–407
SEG Fligiel, KA Laybourn, BP Peters, RW Ruddon, JC Hiserodt and J Varani (1986) *Clin Exp Metast* **4**: 259–272
O Fodstad, CT Hansen, GB Cannon, CN Statham, GR Lichtenstein and MR Boyd (1984) *Cancer Res* **44**: 4403–4408
M Fogel, P Altevogt and V Schirrmacher (1983) *J Exp Med* **157**: 371–376
M Fogel, E Gorelik, S Segal and M Feldman (1979) *J Nat Cancer* Inst **62**: 585–588
EJ Foley (1953) *Cancer Res* **13**: 835–837
J Folkman (1985) *Adv Cancer Res* **43**: 175–203
J Folkman (1987) *Eur J Cancer Clin Oncol* **23**: 361–363
J Folkman and C Haudenschild (1980) *Nature* **288**: 551–556
J Folkman and M Klagsbrun (1987a) *Science* **235**: 442–447
J Folkman and M Klagsbrun (1987b) *Nature* **329**: 671–672
J Folkman, R Langer, R Lindhart, S Haudenschild and S Taylor (1983) *Science* **221**: 719–725
J Folkman and A Moscona (1978) *Nature* **273**: 345–349
J Folkman, K Watson, D Ingber, and D Hanahan (1989a) *Nature* **339**: 58–61
J Folkman, PB Weisz, MM Joullie, WW Li and WR Ewing (1989b) *Science* **243**: 1490–1493
KA Foon and RF Todd (1986) *Blood* **68**: 1–31
DD Foster and AB Pardee (1969) *J Biol Chem* **244**: 2675–2681
L Foulds (1975) *Neoplastic development* Acdemic Press New York
N Fox, R Crooke, L–HS Hwang, U Schibler, BB Knowles and D Solter (1989) *Science* **244**: 460–463
WW Franke, P Cowin, M Schmelz and H–P Kapprell (1987) *Ciba Fdn Symp* **125**: 26–44

WA Frazier (1987) *J Cell Biol* **105**: 625–632
M Fridkin and VA Najjar (1989) *Crit Rev Biochem Molec Biol* **24**: 1–40
M Fried (1965) *Proc Nat Acad Sci* **53**: 486–491
HS Friedman, J Kurtzberg, B Crokar, JM Fallet and T Kinney (1984) *Clin Pediatr* **23**: 184–187
JM Friedman and PJ Fialkow (1976) *Transpl Rev* **28**: 17–33
C Friend, W Scher, JG Holland and T Sato (1971) *Proc Nat Acad Sci* **68**: 378–382
LMS Fritze, CF Reilly and RD Rosenberg (1985) *J Cell Biol* **100**: 1041–1049
P Frost, J Smith and H Frost (1978) *Proc Soc Exp Biol Med* **157**: 61–67
C Fryling, M Dombalagian, W Burgess, N Hollander, AB Schreiber and J Haimovich (1989) *Cancer Res* **48**: 3333–3337
C Fryling, KK Iwata, PA Johnson, WB Knott and GJ Todaro (1985) *Cancer Res* **45**: 2695–2699
J Fujita, SK Srivastava, MH Kraus, JS Rhim, SR Tronick, and SA Aaronson (1985) *Proc Nat Acad Sci* **82**: 3849–3853
K Furukawa and VP Bhavanandan (1983) *Biochim Biophys Acta* **740**: 466–475

H Gabbert, R Wagner, R Moll and C-D Gerharz (1985) *Clin Exp Metast* 3: 257–279
BJ Gaffney (1975) *Proc Nat Acad Sci* **72**: 664–668
MH Gail and CW Boone (1971) *Exp Cell Res* **64**: 156–162
MH Gail and CW Boone (1972) *Exp Cell Res* **70**: 33–40
JT Gallagher, M Lyon and WP Steward (1986) *Biochem J* **236**: 313–325
WM Gallatin, IL Weissman and EC Butcher (1983) *Nature* **304**: 30–34
GE Gallick, R Kurzrock, WS Kloetzer, RB Arlinghaus and JG Gutterman (1985) *Proc Nat Acad Sci* **82**: 1795–1799
JI Gallin and PG Quie (1978) Eds *Leukocyte chemotaxis* Raven Press New York
WJ Gallin, GM Edelman and BA Cunningham (1983) *Proc Nat Acad Sci* **80**: 1038–1042
RC Gallo, PS Sarin, WA Blattner, F Wong-Stahl and M Popovic (1984) *Semin Oncol* **11**: 12–17
JR Gamble and MA Vadas (1988) *Science* **242**: 97–99
MB Ganz, RB Sterzel and WF Boron (1989) *Nature* **337**: 648–651
J Gao, X Kexue, L Baogui, D Huayi, W Lisheng, Z Zhaotie, T Shen and X Zhengue (1984) *Clin Exp Metast* **2**: 205–212
C Garbi and SH Wollman (1982) *J Cell Biol* **94**: 489–492
B Garlin, R Jaffe, B Bender and AE Chung (1981) *J Biol Chem* **256**: 5209–5214
DR Garrod and MS Steinberg (1973) *Nature* **244**: 568–569
DR Garrod and MS Steinberg (1975) *J Cell Sci* **18**: 405–425
TK Gartner and JS Bennett (1985) *J Biol Chem* **260**: 11891–11894
GJ Gasic, D Boettiger, JL Catalfamo, TB Gasic and GJ Stewart (1978) *Cancer Res* **38**: 2950–2955
GJ Gasic, TB Gasic, N Galanti, T Johnson and S Murphy (1973) *Int J Cancer* **11**: 704–718
GJ Gasic, TB Gasic and CC Stewart (1968) *Proc Nat Acad Sci* **61**: 46-52
GJ Gasic, GP Tuszynski and E Gorelik (1986) *Int Rev Exp Pathol* **29**: 173–212
KR Gehlsen, WS Argraves, MD Pierschbacher and E Ruoslahti (1988) *J Cell Biol* **106**: 925–930
WJ Gehring (1987) *Science* **236**: 1245–1252
B Geiger (1979) *Cell* **18**: 193–205
B Geiger, KY Tokuyasu, AH Dutton and SJ Singer (1980) *Proc Nat Acad Sci* **77**: 4127–4137

B Geiger, T Volk and T Volberg (1985) *J Cell Biol* **101**: 1523–1531
B Geiger, T Volk, T Volberg and R Bendori (1987) *J Cell Sci Suppl* **8**: 251–272
AG Geiser, CJ Der, CJ Marshall and EJ Stanbridge (1986) *Proc Nat Acad Sci* **83**: 5209–5213
M Gerschenson, K Graves, SD Carson, RS Wells and GB Pierce (1986) *Proc Nat Acad Sci* **83**: 7307–7310
H Gershman (1982) In *Tumour invasion and metastasis* Eds LA Liotta and IR Hart Martinus Nijhoff, The Hague pp 231–250
DM Gersten, VJ Hearing and JJ Marchalonis (1981) *Proc Nat Acad Sci* **78**: 5109–5112
DM Gersten and JJ Marchalonis (1978) *Biochem Biophys Res Comm* **90**: 1015–1029
BM Gesner and V Ginsburg (1964) *Proc Nat Acad Sci* **52**: 750–755
R Giavazzi, DE Campell, JM Jessup, K Cleary and IJ Fidler (1986) *Cancer Res* **46**: 1928–1933
LC Gilbert and SG Gordon (1983) *Cancer Res* **43**: 536–540
PX Gilbert and H Harris (1988) *J Cell Sci* **90**: 433–446
NB Gilula (1987) *Ciba Fdn Symp* **125**: 128–139
NB Gilula, OR Reeves and A Steinbach (1972) *Nature* **235**: 262–265
AG Gilman (1984) *Cell* **36**: 577–579
AG Gilman (1987) *Ann Rev Biochem* **56**: 615–649
MA Gimbrone and PM Gullino (1976) *Cancer Res* **36**: 2611–2620
D Gingell (1981) *J Cell Sci* **49**: 237–247
D Gingell, I Todd and J Bailey (1985) *J Cell Biol* **100**: 1334–1338
G Girmann, H Pees, G Schwarze and PG Scheurlen (1976) *Nature* **259**: 399–401
D Glaves (1983) *Invas Metast* **3**: 160–173
K Glenn, DF Bowen-Pope and R Ross (1982) *J Biol Chem* **257**: 5172-5176
B Glimelius, K Nilsson and J Ponten (1975) *Int J Cancer* **15**: 888–896
B Glimelius, B Westermark and J Ponten (1974) *Int J Cancer* **14**: 314–325
SE Goelz, B Vogelstein, SR Hamiton and AP Feinberg (1985) *Science* **228**: 187–190
P Gold and SO Freedman (1965) *J Exp Med* **122**: 439–462
M Goldberg and F Escaig-Haye (1986) *Eur J Cell Biol* **42**: 365–368
N Goldberg, MK Haddox, E Durham, C Lopez and JW Hadden (1974) In *Control of proliferation in animal cells* Eds B Clarkson and R Baserga Cold Spring Harbour Labs New York pp 609-626
DM Goldenberg, RA Pavia and MC Tsao (1974) *Nature* **250**: 649–651
RD Goldman, C Chang and JF Williams (1974) *Cold Spring Harbor Symp Quant Biol* **39**: 601–614
RD Goldman, AJ Yerna and JA Schloss (1976) *J Supramolec Struct* **5**: 155–183
HL Goldsmith (1971) *Fed Proc* **30**: 1578–1588
PN Goodfellow and PW Andrews (1982) *Nature* **300**: 107–108
SL Goodman, R Deutzmann and K von der Mark (1987) *J Cell Biol* **105**: 589–598
SL Goodman, HP Vollmers and W Birchmeier (1985) *Cell* **41**: 1029–1038
G Gorbsky and MS Steinberg (1981) *J Cell Biol* **90**: 243–248
JW Gordon and FH Ruddle (1983) *Meth Enzymol* **101c**: 411–433
E Gorelik, M Feldman and S Segal (1982) *Cancer Immunol Immunother* **12**: 105–109
E Gorelik, M Fogel, M Feldman and S Segal (1979) *J Nat Cancer Inst* **63**: 1397–1404
RR Gorman (1979) Fed Proc **38**: 83–88

D Gospodarowicz (1985) In *Receptors in tumour biology* Ed CM Chadwick CUP Cambridge pp 72–94
D Gospodarowicz, AL Mescher and CR Birdwell (1978) *NCI Monogr* **48**: 109–130
MM Gottesman and RD Fleischmann (1986) *Cancer Surveys* 5: 291–308
MM Gottesman, C Roth, G Vlahakis and I Pastan (1984) *Mol Cell Biol* **4**: 2639–2642
AS Goustin, EB Leof, GD Shipley and HL Moses (1986) *Cancer Res* **46**: 1015–1029
LB Grabel, MS Singer, SD Rosen and GR Martin (1983) In *Teratocarcinoma stem cells* Eds LM Silver, GR Martin and S Strickland Cold Spring Harbor Laboratory Press New York pp 145–161
J Graf, Y Iwamoto, M Sasaki, GR Martin, HK Kleinman, FA Robey and Y Yamada (1987) *Cell* **48**: 989–996
DI Grdina, WM Hittelman, RA White and ML Meistrich (1977) *Br J Cancer* **36**: 659–669
MF Greaves (1986) *Science* **234**: 697–704
ME Greenberg, R Brackenbury and GM Edelman (1984) *Proc Nat Acad Sci* **81**: 969–973
EL Green (1968) *Handbook of genetically standardized Jax mice* Jackson Lab Bar Harbor
H Gregory (1975) *Nature* **257**: 325–327
RG Greig, TP Koestler, DL Trainer, SP Corwin, L Miles, T Kline, R Sweet, S Yokoyama and G Poste (1985) *Proc Nat Acad Sci* **82**: 3698–3701
EB Griepp, WJ Dolan, ES Robbins and DD Sabatini (1983) *J Cell Biol* **96**: 693–702
JD Griffiths and AJ Salsbury (1963) *Br J Cancer* **17**: 546–557
JD Griffiths and AJ Salsbury (1965) *Circulating cancer cells*. Thomas, Springfield, Illinois
G Grignani, L Pacchiarini, P Almasio, M Pagliarino, G Gamba, SC Rizzo and F Ascari (1986) *Int J Cancer* **38**: 237–244
EA Grimm, LM Muul and DJ Wilson (1985) *Transpl* **39**: 537–540
IA Grimstad, J Varani and JP McCoy (1984) *Exp Cell Res* **155**: 345–358
F Grinnell (1978) *Int Rev Cytol* **53**: 65–144
F Grinnell (1984) *J Cell Sci* **65**: 61–72
F Grinnell (1986) *J Cell Biol* **103**: 2697–2706
F Grinnell and MK Feld (1980) *Cell* **17**: 117–129
F Grinnell, DG Hays and D Minter (1977) *Exp Cell Res* **110**: 175–190
F Grinnell, MQ Toblemen and CR Hackenbrock (1975) *J Cell Biol* **66**: 470–479
L Gross (1943) *Cancer Res* **3**: 326–333
L Gross (1945) *J Immunol* **50**: 91–99
N Grosser and DM Thomson (1975) *Cancer Res* **35**: 2571–2579
M Groudine and H Weintraub (1980) *Cell* **24**: 393–401
M Grumet, S Hoffman, KL Crossin and GM Edelman (1985) *Proc Nat Acad Sci* **82**: 8075–8079
VI Guelstein, OY Ivanova, LB Margolis, JM Vasilier and IM Gelfand (1973) *Proc Nat Acad Sci* **70**: 2011–2044
R Guirguis, I Margulies, G Taraboletti, E Schiffmann and L Liotta (1987) *Nature* **329**: 261–263
B Gumbiner (1988) *Trends Biochem Sci* **13**: 75–76
B Gumbiner and K Simons (1986) *J Cell Biol* **102**: 457–468

B Gumbiner and K Simons (1987) *Ciba Fdn Symp* **125**: 168–186
T Gustafson and L Wolpert (1967) *Biol Rev* **42**: 442–498
D Guy-Grand, C Griscelli and P Vassalli (1974) *Eur J Immunol* **4**: 435–443

G Haemmerli, B Arnold and P Strauli (1982) *Int J Canc* **29**: 223–227
G Haemmerli, B Arnold and P Strauli (1983) *Cell Biol Int Rep* **7**: 708–725
G Haemmerli, B Arnold and P Strauli (1984) *Cell Biol Int Rep* **8**: 689–702
G Haemmerli and P Strauli (1978) *Virchows Arch B Cell Path* **29**: 167–177
G Haemmerli and P Strauli (1981) *Int J Cancer* **27**: 603–610
E Hafen, K Basler, JE Edstroem and GM Rubin (1987) *Science* **236**: 55–63
B Hagmer and W Ryd (1977) *Int J Cancer* **19**: 576–580
HT Haigler, JA McKanna and S Cohen (1979) *J Cell Biol* **83**: 82–90
S Hakomori (1970) *Proc Nat Acad Sci* **67**: 1741–1747
S Hakomori (1973) *Adv Cancer Res* **18**: 265–315
S Hakomori, J Koscielak, H Bloch and R Jeanloz (1967) *J Immunol* **98**: 31–38
S Hakomori and WT Murakami (1968) *Proc Nat Acad Sci* **59**: 254–261
J Hall (1985) *Immunol Today* **6**: 149–152
WJ Halliday and S Miller (1972) *Int J Cancer* **9**: 477–483
B Halpern, B Pejsachowicz, HL Febvre and G Barski (1966) *Nature* **209**: 157–159
A Hamann, D Jablonski-Westrich, AM Duijvestijn, EC Butcher, H Baisch, R Harder and H-G Thiele (1988) *J Immunol* **140**: 693–699
CV Hamby and JT Barrett (1977) *Oncology* **34**: 13–15
F Hammersen and E Hammersen (1984) *J Cardiovasc Pharm* **6**: *Suppl* 2 S289–S303
H Hanahan (1985) *Nature* **315** 115–122
PH Hand, A Thor, J Schlom, CN Rao and L Liotta (1985) *Cancer Res* **45** 2713–2719
N Hanna (1980) *Int J Cancer* **26** 675–680
N Hanna (1982a) *Cancer Metast Rev* **1** 45–64
N Hanna (1982b) In *Tumour invasion and metastasis* Eds LA Liotta and IR Hart Martinus Nijhoff The Hague pp 29–41
N Hanna, TW Davis and IJ Fidler (1982) *Int J Cancer* **15**: 371–376
N Hanna and IJ Fidler (1980) *J Nat Canc Inst* **65**: 801–809
N Hanna and IJ Fidler (1981) *J Nat Canc Inst* **66**: 1183–1190
YA Hannun and RM Bell (1989) *Science* **243**: 50–507
M Hansson, K Karre, T Bakacs, R Kiessling and G Klein (1978) *J Immunol* **121**: 6–12
Y Hara, M Steiner and MG Baldini (1980) *Cancer Res* **40**: 1217–1221
IK Hariharan and JM Adams (1987) *EMBO J* **6**: 115–119
DG Harnden (1976) *Proc R Soc Med* **69**: 41–43
AK Harris (1973) *Dev Biol* **35**: 97–114
A Harris (1974a) *Exp Cell Res* **77**: 285–297
A Harris (1974b) In *Cell communication* Ed RP Cox John Wiley and Sons New York pp 147–185
A Harris and G Dunn (1972) *Exp Cell Res* **73**: 519–523
H Harris (1988) *Cancer Res* **48**: 3302–3306
JJ Harrison, A Anisowiez, IK Gadi, M Raffeld and R Sager (1983) *Proc Nat Acad Sci* **80**: 6606–6610
RG Harrison (1914) *J Exp Zool* **17**: 521–544
DA Hart and R Smith (1987) *Cancer Letts* **35**: 27–38
IR Hart (1979) *Am J Pathol* **97**: 587–600

IR Hart (1982a) *Cancer Metast Rev* **1**: 5–16
IR Hart (1982b) In *Tumour invasion and metastasis* Eds LA Liotta and IR Hart Martinus Nijhoff The Hague pp 1–14
IR Hart and IJ Fidler (1978) *Cancer Res* **38**: 3218–3224
IR Hart and IJ Fidler (1980a) *Cancer Res* **40**: 2281–2287
IR Hart and IJ Fidler (1980b) *Quart Rev Biol* **55**: 121–142
IR Hart and IJ Fidler (1981) *Biochim Biophys Acta* **651**: 37–50
IR Hart, NT Goode and EJ Ormerod (1989) *Biochim Biophys Acta* **989**: 65–84
IR Hart, A Raz and IJ Fidler (1980) *J Nat Cancer Inst* **64**: 891–900
IR Hart, JE Talmadge and IJ Fidler (1981) *Cancer Res* **41**: 1281–1287
IR Hart, JE Talmadge and IJ Fidler (1983) *Cancer Res* **43**: 400–402
L Harvath and EJ Leonard (1982) *Infect Immun* **36**: 443–449
T Hasegawa, E Hasegawa, W-T Chen and KM Yamada (1985) *J Cell Biochem* **28**: 307–318
D Haskard, D Cavender, P Beatty, T Springer and M Ziff (1986) *J Immunol* **137**: 2901–2906
JR Hassell, PG Robey, H–J Barrach, J Wilczek, SI Rennard and GR Martin (1980) *Proc Nat Acad Sci* **77**: 4494–4498
WS Haston, JM Shields, PC Wilkinson (1982) *J Cell Biol* **92**: 747–752
J Hata, Y Ueyama, N Tamaoki, T Furukawa and K Morita (1978) *Cancer* **42**: 468–473
M Hatakana (1974) *Biochim Biophys Acta* **355**: 77–104
VB Hatcher, N Fadl-Allah, MA Levitt, A Brown, SS Margossian and PB Gordon (1986) *J Cell Physiol* **128**: 353–361
K Hatta and M Takeichi (1986) *Nature* **320**: 447–449
CC Haudenschild, D Zahniser, J Folkman and M Klagsbrun (1976) *Exp Cell Res* **98**: 175–183
RE Hausman (1983) *Int J Cancer* **32**: 603–608
H Hayashi and Y Ishimaru (1981) *Int Rev Cytol* **70**: 139–215
H Hayashi, K Yoshida, T Ozaki and K Ushijima (1970) *Nature* **226**: 174–175
L Hayflick and PS Moorhead (1961) *Exp Cell Res* **25**: 585–621
EG Hayman, E Engvall, E A'Hearn, D Barnes, M Pierschbacher and E Ruoslahti (1982) *J Cell Biol* **95**: 20–23
EG Hayman and MD Pierschbacher (1985) *Exp Cell Res* **160**: 245–258
EG Hayman, MD Pierschbacher, Y Ohgren and E Ruoslahti (1983) *Proc Nat Acad Sci* **80**: 4003–4007
GR Haywood and LF McKhann (1971) *J Exp Med* **133**: 1171–1182
JP Heath and GA Dunn (1978) *J Cell Sci* **29**: 197–212
W Heath and W Boyle (1985) *Transpl Proc* **27**: 1658–1660
JEM Heaysman (1970) *Experientia* **26**: 1344–1345
JEM Heaysman (1978) *Int Rev Cytol* **55**: 49–66
JEM Heaysman and SM Pegrum (1973a) *Exp Cell Res* **78**: 71–78
JEM Heaysman and SM Pegrum (1973b) *Exp Cell Res* **78**: 479–481
JEM Heaysman and L Turin (1976) *Exp Cell Res* **101**: 419–422
MH Heggeness, J Ash and SJ Singer (1978) *Ann NY Acad Sci* **312**: 414–417
S Heim, N Mandahl and F Mitelman (1988) *Cancer Res* **48**: 5911–5916
RL Heimark and SM Schwartz (1985) *J Cell Biol* **100**: 1934–1940
MA Heisel, PA Jones and WE Lang (1981) *Proc Amer Assoc Cancer Res* **22**: 248–257
CH Heldin and B Westermark (1984) *Cell* **7**: 9–20
CG Hellerqvist (1979) *J Cell Biol* **82**: 682–687

I Hellstrom amd KE Hellstrom (1971) *J Reticuloendothel Soc* **10**: 131–136
KE Hellstrom and I Hellstrom (1969) *Adv Cancer Res* **12**: 167–223
KE Hellstrom and I Hellstrom (1974) *Adv Immunol* **18**: 209–277
JJ Hemperley, GM Edelman and BA Cunningham (1986) *Proc Nat Acad Sci* **83**: 9822–9826
GH Heppner (1984) *Cancer Res* **44**: 2259–2265
GH Heppner and BE Miller (1983) *Cancer Metast Rev* **2**: 5–23
RB Herberman, ME Nunn and DH Lavrin (1975) *Int J Cancer* **16**: 216–229
ZL Herd (1987) *Cancer Res* **47**: 2696–2703
B Herman and WJ Pledger (1985) *J Cell Biol* **100**: 1031–1040
C Heussen and EB Dowdle (1980) *Anal Biochem* **102**: 196–202
AT Hewitt, HH Varner, MH Silver, W Dessau, CM Wilkes and GR Martin (1982) *J Biol Chem* **257**: 2330–2334
HB Hewitt (1953) *Br J Cancer* **7**: 367–383
HB Hewitt (1978) *Adv Cancer Res* **27**: 149–200
HB Hewitt, ER Blake and AS Walder (1976) *Br J Cancer* **33**: 241–259
NJ Hicks, RV Ward and JJ Reynolds (1984) *Int J Cancer* **33**: 835–844
P Hilgard (1984a) In *Hemostatic mechanisms and metastasis* Eds KV Honn and BF Sloane Nijhoff Boston pp 259–265
P Hilgard (1984b) In *Cancer invasion and metastasis* Eds GL Nicolson and L Milas Raven Press New York pp 353–360
RP Hill, AF Chambers, V Ling and JF Harris (1984) *Science* **224**: 998–1001
JC Hiserodt, KA Laybourn and J Varani (1985) *J Immunol* **135**: 1484–1487
SE Hitchcock (1977) *J Cell Biol* **74**: 1–15
RW Holley (1974) In *Control of proliferation in animal cells* Eds B Clarkson and R Baserga Cold Spring Harbor Lab New York pp 13–18
RW Holley (1975) *Nature* **258**: 487–490
RW Holley, R Armour, JH Baldwin and S Greenfield (1983) *Cell Biol Int Rep* **7**: 141–147
RW Holley, P Bohlen, R Fava, JH Baldwin, G Kleeman and R Armour (1980) *Proc Nat Acad Sci* **77**: 5989–5992
RW Holley and JA Kiernan (1968) *Proc Nat Acad Sci* **60**: 300–304
R Holliday (1987) *Science* **238**: 163–170
R Holmes (1967) *J Cell Biol* **32**: 297–308
B Holzmann, BW McIntyre and IL Weissman (1989) *Cell* **56**: 37–46
Y Honma, T Kaskabe, J Okabe and M Hozumi (1979) *Cancer Res* **39**: 3167–3171
KV Honn, B Cicone and A Skoff (1981) *Science* **212**: 1270–1272
KV Honn, JM Onoda, CA Diglio and BF Sloane (1983) *Proc Soc Exp Biol Med* **174**: 16–19
HC Hoover and AS Ketcham (1975) *Am J Surg* **130**: 405–411
E Horak, DL Darling and D Tarin (1986) *J Nat Cancer Inst* **76**: 913–922
A Horwitz, K Duggan, C Buck, MC Beckerle and K Burridge (1986) *Nature* **320**: 531–533
A Houweling, P van der Elsen and A van der Eb (1980) *Virology* **105**: 537–550
A Hsie and TT Puck (1971) *Proc Nat Acad Sci* **68**: 358–361
YM Hsu and JL Wang (1986) *J Cell Biol* **102**: 362–369
F Hu and PF Lesney (1964) *Cancer Res* **24**: 1634–1643
RJ Huebner and GJ Todaro (1969) *Proc Nat Acad Sci* **64**: 1087–1094
RC Hughes, TD Butters and JD Aplin (1981) *Eur J Cell Biol* **26**: 198–207
ES Hujanen and VP Terranova (1985) *Cancer Res* **45**: 3517–3521
MJ Humphries, SK Akiyama, A Komoriya, K Olden and KM Yamada (1986) *J*

Cell Biol **103**: 2637–2647
MJ Humphries, A Komoriya, SK Akiyama, K Olden and KM Yamada (1987) *J Biol Chem* **262**: 6886–6892
MJ Humphries, K Matsumoto, SL White and K Olden (1986) *Proc Nat Acad Sci* **83**: 1752–1756
MJ Humphries, K Olden and KM Yamada (1986) *Science* **233**: 467–470
T Hunter (1984) *J Nat Cancer Inst* **73**: 773–786
T Hunter (1986) *Nature* **322**: 14–16
T Hunter (1987) *Cell* **49**: 1–4
T Hunter and JA Cooper (1985) *Ann Rev Biochem* **54**: 897–930
T Hunter and BM Sefton (1980) *Proc Nat Acad Sci* **77**: 1311–1315
HE Huxley (1973) *Nature* **243**: 445–449
RO Hynes (1973) *Proc Nat Acad Sci* **70**: 3170–3174
RO Hynes (1974) *Cell* **1**: 147–156
RO Hynes (1986) *Sci Amer* **254**: 32–41
RO Hynes (1987) *Cell* **48**: 549–554
RO Hynes and JM Bye (1974) *Cell* **3**: 113–120
RO Hynes and AT Destree (1978) *Cell* **15**: 895–886
RO Hynes, JA Wyke, JM Bye, KC Humphryes and ES Pearlstein (1975) In *Proteases and biological control* Eds E Reid, E Shaw and DB Rifkin Cold Spring Harbor New York pp 931–944

T Ichumura and PH Hashimoto (1984) *J Ultrastruct Res* **86**: 220–227
CR Ill and E Ruoslahti (1985) *J Biol Chem* **260**: 15610–15615
K Illmensee and B Mintz (1976) *Proc Nat Acad Sci* **73**: 549–553
M Inbar, H Ben-Bassat and L Sachs (1971) *Proc Nat Acad Sci* **68**: 2748–2751
T Irimura, R Gonzalez and GL Nicolson (1981) *Cancer Res* **41**: 3411–3418
T Irimura and GL Nicolson (1981) *J Supramol Struct Cell Biochem* **17**: 325–336
F Ishikawa and 9 others (1989) *Nature* **338**: 557–562
M Ishikawa, Y Koga, M Hosokawa and H Kobayashi (1986) *Int J Cancer* **37**: 919–924
KJ Isselbacher (1972) *Proc Nat Acd Sci* **69**: 585–589
Y Iwamoto, FA Robey, J Graf, M Sasaki, HK Kleinman, Y Yamada and GR Martin (1987) *Science* **238**: 1132–1134
CS Izzard and LR Lochner (1976) *J Cell Sci* **21**: 129–159
CS Izzard and LR Lochner (1980) *J Cell Sci* **42**: 81–116

R Jaenisch (1988) *Science* **240**: 1468–1474
S Jalkanen, RF Bargatze, J de los Toyos and EC Butcher (1987) *J Cell Biol* **105**: 983–990
S Jalkanen, RA Reichert, WM Gallatin, RF Bargatze, IL Weissman and EC Butcher (1986b) *Immunol Rev* **91**: 39–60
S Jalkanen, AC Steere, R Fox and EC Butcher (1986a) *Science* **233**: 556–558
J Janne, H Poso and A Raina (1978) *Biochim Biophys Acta* **473**: 241–293
JR Jenkins, K Rudge and GA Currie (1984) *Nature* **317**: 651–654
D Jenne and KK Stanley (1985) *EMBO J* **4**: 3153–3157
HM Jensen, I Chen, MR DeVanu and AE Lewis (1983) *Science* **218**: 293–295
JM Jessup, R Giavazzi, D Campbell, KR Cleary, K Morikawa, R Hostetter, EN Atlemson and IJ Fidler (1989) *Cancer Res* **49**: 8906–8910
JJ Jimenez and AA Yunis (1987) *Science* **238**: 1278–1280
GS Johnson, RM Friedman and I Pastan (1971) *Proc Nat Acad Sci* **68**: 425–429

GS Johnson and I Pastan (1972) *Nature (New Biol)* **236**: 247–249
A Johnsson, C-H Heldin, A Wasteson, B Westermark, TF Deuel, JS Huang, P-H Seeburg, A Gray, A Ullrich, G Scrace, P Stroobant and MD Waterfield (1984) *EMBO J* **3**: 921–928
GE Jones, RG Arumugham and ML Tanzer (1986) *J Cell Biol* **103**: 1663–1670
PA Jones (1982) In *Tumour invasion and metastasis* Eds LA Liotta and IR Hart Martinus Nijhoff The Hague pp 251–265
JCR Jones (1988) *J Cell Sci* **89**: 207–216
JCR Jones, KM Yokoo and RD Goldman (1986) *Proc Nat Acad Sci* **83**: 7282–7286
PA Jones and YA De Clerck (1980) *Cancer Res* **40**: 3222–3227
PA Jones, WE Laug, A Gardner, CA Nye, LM Fink and WF Benedict (1976) *Cancer Res* **36**: 2863–2867
PA Jones, HB Neustein, F Gonzales and E Bogenmann (1981) *Cancer Res* **41**: 4613–4620
DG Jose and RA Good (1973) *Cancer Res* **33**: 807–812
B Kachar and P Pinto da Silva (1981) *Science* **213**: 541–544

L Kaczmarek, JK Hyland, R Watt, M Rosenberg and R Baserga (1985) *Science* **228**: 1313–1315
JL Kadish, KM Wenc and HF Dvorak (1983) *J Nat Cancer Inst* **70**: 551–557
R Kalish and NI Brady (1980) *J Invest Dermatol* **75**: 275–278
N Kamata (1986) *Cancer Res* **46**: 1648–1653
B Kamenov and BM Longenecker (1985) *Leukemia Res* **9**: 1529–1537
N Kamiya (1959) *Plasmatologia* **8**: 1–199
Y Kanno, T Enomoto, Y Shiba and H Yamasaki (1984) *Exp Cell Res* **152**: 31–37
H-P Kapprell, K Owaribe and WW Franke (1988) *J Cell Biol* **106**: 1679–1691
J Kautz (1952) *Cancer Res* **12**: 180–187
T Kawaguchi, M Kawaguch, KM Dulski and GL Nicolson (1985) *Invas Metast* **5**: 16–30
T Kawaguchi and K Nakamura (1986) *Cancer Metast Rev* **5**: 77–94
T Kawaguchi, S Tobai and K Nakamura (1982) *Invas Metast* **2**: 40–50
H Kawakami and H Terayama (1981) *Biochim Biophys Acta* **646**: 161–166
Y-I Kawano, K Taniguchi, A Toshitani and K Nomoto (1986) *J Immunol* **136**: 4729–4734
N Kay, J Allen and JE Morley (1984) *Life Sci* **35**: 53–59
MT Keating and LT Williams (1988) *Science* **239**: 914–916
NA Kefalides, R Alper and CC Clark (1979) *Int Rev Cytol* **61**: 167–228
HU Keller, A Zimmerman, M Schmitt and H Cottier (1985) *Prog Appl Micros* **7**: 1–14
R Keller (1976) *J Nat Cancer Inst* **57**: 1355–1361
S Kellie and CW Evans (1981) *Br J Exp Path* **62**: 158–164
S Kellie, CW Evans and GD Kemp (1980) *Cell Biol Int Rep* **4**: 777
S Kellie, B Patel, NM Wigglesworth, DR Critchley and JA Wyke (1986) *Exp Cell Res* **165**: 216–228
K Kelly, BH Cochran, CD Stiles and P Leder (1983) *Cell* **35**: 603–610
WM Kent, FM Funderburg and LA Culp (1986) *Mech Ageing Dev* **3**: 115–137
RS Kerbel, A Lagarde, JW Dennis and TP Donaghue (1983) *Mol Cell Biol* **3**: 523–538
RS Kerbel, AE Lagarde, JW Dennis, FP Nestel, TP Donaghue, L Siminovitch and MR Fulchignoni–Latand (1984) In *Cancer invasion and metastasis* Eds GL

Nicolson and L Miles Raven Press New York pp 47–79
RS Kerbel, MS Man and D Dexter (1984) *J Nat Cancer Inst* **72**: 93–108
RS Kerbel, C Waghorne, MS Man, B Elliot and MC Breitman (1987) *Proc Nat Acad Sci* **84**: 1263–1267
Z Keren, F Leland, M Nakajima and SJ Le Grue (1989) *Cancer Res* **49**: 295–300
RE Key, JE Talmadge, WE Fogler, C Bucana and IJ Fidler (1982) *J Nat Cancer Inst* **69**: 1189–1198
R Khokha, P Waterhouse, S Yagel, PK Lala, CM Overall, G Norton and DT Denhardt (1989) *Science* **243**: 947–950
J Kieler, P Briand, MC Van Peteghem and M Mareel (1979) *In Vitro* **15**: 758–771
I Kijima-Suda, T Miyazawa, M Itoh, S Toyoshima and T Osawa (1988) *Cancer Res* **48**: 3728–3732
JJ Killion and IJ Fidler (1989) *Semin Oncol* **16**: 106–115
U Kim (1970) *Science* **164**: 72–74
U Kim, A Baumler, C Carruthers and K Bielat (1975) *Proc Nat Acad Sci* **72**: 1012–1016
AK Kimura, P Mehta, J Xiang, D Lawson, D Dugger, K-J Kao and L Lee-Ambrose (1987) *Clin Exp Metast* **5**: 125–133
AK Kimura and J Xiang (1986) *J Nat Cancer Inst* **76**: 1247–1254
DL Kinsey (1960) *Cancer* **13**: 674–676
DL Kinsey and RR Smith (1959) *Surg Gynecol Obstet* **109**: 539–543
J Kistler, D Christie and S Bullivant (1988) *Nature* **331**: 721–723
RJ Klebe, JR Hall, SL Naylor and WD Dickey (1978) *Exp Cell Res* **115**: 73–78
E Klein (1955) *Exp Cell Res* **8**: 188–212
G Klein and E Klein (1985) *Nature* **315**: 190–195
G Klein, HO Sjogren, E Klein and KE Hellstrom (1960) *Cancer Res* **20**: 1561–1572
PA Klein, J Xiang and AK Kimura (1984) *Clin Exp Metast* **2**: 287–295
HK Kleinman, GR Martin and PH Fishman (1979) *Proc Nat Acad Sci* **76**: 3367–3371
HK Kleinman, ML McGarvey, JR Hassell and GR Martin (1983) *Biochem* **22**: 4969–4974
TM Kloppel, TW Keenan, MJ Freeman and JD Morse (1977) *Proc Nat Acad Sci* **74**: 3011–3013
P Knox and S Griffiths (1979) *Exp Cell Res* **123**: 421–424
KA Knudsen, AF Horwitz and CA Buck (1985) *Exp Cell Res* **157**: 218–226
AG Knudson (1971) *Proc Nat Acad Sci* **68**: 820–823
AG Knudson (1983) *Prog Clin Biol Res* **132C**: 351–360
FE Koch (1939) *Z Krebsforsch* 48 495–507
JJ Kolenich, EG Mansour and A Flynn (1972) *Lancet* ii 714
T Komuro (1985) *Cell Tissue Res* **239**: 183–188
L Kopelovich (1982) *Cancer Genet Cytogenet* **5**: 333–351
S Korach, M-F Poupon, J-A Du Villard and M Becker (1986) *Cancer Res* **46**: 3624–3629
B Korczak, IB Robson, C Lamarche, A Bernstein and RS Kerbel (1988) *Mol Cell Biol* **41**: 720–726
AR Kornblihtt, K Vibe-Pedersen and FE Baraule (1985) *Proc Nat Acad Sci* **80**: 3218–3222
RE Kouri, CE McKinney, OJ Slomiany, DR Snodgrass, NP Wray and TL McLemore (1982) *Cancer Res* **42**: 5030–5037
JM Kozlowski, IR Hart, IJ Fidler and N Hanna (1984) *J Nat Cancer Inst* **72**: 913–917

PM Kraemer (1979) In *Surfaces of normal and malignant cells* Ed RO Hynes J Wiley and Sons Chichester pp 149–198
RH Kramer, R Gonzalez and GL Nicolson (1980) *Int J Cancer* **26**: 639–645
RH Kramer and GL Nicolson (1979) *Proc Nat Acad Sci* **76**: 5704–5708
RH Kramer and KG Vogel (1984) *J Nat Cancer Inst* **72**: 889–899
RH Kramer, KG Vogel and GL Nicolson (1982) *J Biol Chem* **257**: 2678–2686
JW Kreider, M Rosenthal and N Lengle (1973) *J Nat Cancer Inst* **50**: 555–558
TE Kreis, Z Avnur, J Schlessinger and B Geiger (1984) In *Molecular biology of the cytoskeleton* Eds G Borisy, D Cleveland and D Murphy Cold Spring Harbor Lab Press New York pp 45–57
TG Krontiris and GM Cooper (1981) *Proc Nat Acad Sci* **78**: 1181–1184
KE Kuettner and BU Pauli (1981) In *Bone metastasis* Eds L Weisss and HA Gilbert Hall Boston pp 131–165
KE Kuettner and BU Pauli (1983) *Ciba Fdn Symp* **100**: 163–173
KE Kuettner, L Soble, RL Croxen, B Marczynska, J Hiti and E Harper (1977) *Science* **196**: 653–654
FC Kull, DA Brent, I Parikh and P Cuatrecasas (1987) *Science* **236**: 843–845
A Kupfer, D Louvard and SJ Singer (1982) *Proc Nat Acad Sci* **79**: 2603–2607
K Kurachi, EW Davie, DJ Strydom, JF Riordan and BL Vallee (1985) *Biochem* **24**: 5494–5499
M Kurkinen, A Taylor, JI Garrels and BL Hogan (1984) *J Biol Chem* **259**: 5915–5922

N Labateya and DMP Thomson (1985) *J Nat Cancer Inst* **75**: 987–994
JM Lackie (1986) *Cell movement and cell behaviour* Allen and Unwin, London
JM Lackie and PB Armstrong (1975) *J Cell Sci* **19**: 645–652
JM Lackie and D DeBono (1977) *Microvasc Res* **13**: 107–112
JM Lackie, CM Urquhart, AF Brown and JU Forrester (1985) *Br J Haematol* **60**: 567–581
JM Lackie and PC Wilkinson (1981) Eds *Biology of the chemotactic response* Cambridge University Press, Cambridge
J Lacovara, EB Cramer and JP Quigley (1984) *Cancer Res* **44**: 1657–1663
OD Laerum, R Bjerkvig, SK Steinsvag and LI De Ridder *Cancer Metast Rev* **3**: 223–236
AE Lagarde and RS Kerbel (1985) *Biochim Biophys Acta* **823**: 81–110
M Laiho and J Keski-Oja (1989) *Cancer Res* **49**: 2533–2553
PK Lala and RS Parhar (1988) *Cancer Res* **48**: 1072–1079
WC Lam, EJ Delikatny, FW Orr, J Wass, J Varani and PA Ward (1981) *Am J Pathol* **104**: 69–76
H Land, LF Parada and RA Weinberg (1983a) *Nature* **304**: 596–602
H Land, LF Parada and RA Weinberg (1983b) *Science* **222**: 771–778
R Langer, H Brem, K Falterman, M Klein and J Folkman (1976) *Science* **193**: 70–72
R Langer, H Conn, J Vacanti, C Haudenschild and J Folkman (1980) *Proc Nat Acad Sci* **77**: 4331–4335
WM Landschulz, PF Johnson and SL McKnight (1988) *Science* **240**: 1759–1764
L Larizza and V Schirrmacher (1984) *Cancer Metast Rev* **3**: 193–222
MW Lark and LA Culp (1982) *J Biol Chem* **257**: 14073–14080
MW Lark and LA Culp (1983) *Biochem* **22**: 2289–2296
MW Lark and LA Culp (1984a) *J Biol Chem* **259**: 212–217
MW Lark and LA Culp (1984b) *J Biol Chem* **259**: 6773–6782
MW Lark, J Laterra and LA Culp (1985) *Fed Proc* **44**: 394–403

AL Latner, E Longstaff and JM Lunn (1971) *Br J Cancer* **25**: 568–573
J Laterra, EK Norton, CS Izzard and LA Culp (1983a) *Exp Cell Res* **146**: 15–27
J Laterra, JE Silbert and LA Culp (1983b) *J Cell Biol* **96**: 112–123
CC Lau, IK Gadi, S Kalvonjian, A Anisowicz and R Sager (1985) *Proc Nat Acad Sci* **82**: 2839–2843
EB Laurence (1979) In *Chemical messengers of the inflammatory process* Ed J Houck Elsevier/North-Holland Amsterdam pp 353–413
GW Laurie, CP Leblond, S Inoue, GR Martin and A Chung (1984) *Am J Anatomy* **169**: 463–481
GW Laurie, CP Leblond and GR Martin (1982) *J Cell Biol* **95**: 340–344
WM Leader, D Stopak and AK Harris (1983) *J Cell Sci* **64**: 1–11
MM Le Beau (1986) *Blood* **67**: 849–858
SJ LeGrue (1982) *Cancer Res* **42**: 2126–2134
SJ LeGrue (1985) *Cancer Metast Rev* **4**: 209–219
SJ LeGrue and DR Hearn (1983) *Cancer Res* **43**: 5106–5111
SJ Leibovich, PJ Polverini, HM Shepard, DM Wiseman, V Shively and N Nuseir (1987) *Nature* **329**: 630–632
J Leighton (1951) *J Nat Cancer Inst* **12**: 545–562
J Leighton (1959) *Science* **129**: 466–467
J Leighton (1968) *Methods Cancer Res* **4**: 86–104
J Leighton, G Just, M Esper and R Kronenthal (1967) *Science* **155**: 1259–1261
J Leighton, RL Kalla, I Klein and M Belkin (1959) *Cancer Res* **19**: 23–27
J Leighton, I Klise, M Belkin and Z Tetenbaum (1956) *J Nat Cancer Inst* **16**: 1353–1373
SK Lemmon, MC Riley, KA Thomas, GA Hoover, T Maciag and RA Bradshaw (1982) *J Cell Biol* **95**: 162–169
WA Lerner, E Pearlstein, C Ambrogio and S Karpatkin (1983) *Int J Cancer* **31**: 463–469
BR Lester, RG Greig, C Buscarino, JR Sheppard, SP Corwin and G Poste (1986) *Int J Cancer* **38**: 405–411
BR Lester, JB McCarthy, Z Sun, RS Smith, LT Furcht and AM Spiegel (1989) *Cancer Res* **49**: 5940–5948
JP Leung, CM Bordin, RM Nakamura, DH Delteer and TS Edington (1979) *Cancer Res* **39**: 2057–2061
JP Leung, EF Plow, RM Nakamura and TS Edington (1978) *J Immunol* **121**: 1287–1296
AJ Levine (1984) *Curr Top Microbiol Immunol* **110**: 143–161
AM Lewis and JL Cook (1985) *Science* **227**: 15–20
WH Lewis (1935) *Science* **81**: 545–553
MA Lieberman, D Raben and L Glaser (1981) *Exp Cell Res* **133**: 413–419
R Lindberg (1972) *Cancer* **29**: 1446–1449
E Linder, S Stenman, V–P Lehto and A Vaheri (1978) *Am NY Acad Sci* **312**: 151–159
D Linnemann and E Bock (1986) *Med Biol* **64**: 345–349
H Lione and HB Bosmann (1978) *Cell Biol Int Rep* **2**: 81–86
LA Liotta (1986) *Eur J Cancer Clin Oncol* **22**: 345–348
LA Liotta, S Abe, PG Robey and GR Martin (1979) *Proc Nat Acad Sci* **76**: 2268–2272
LA Liotta, J Kleinerman, P Catanzaro and D Rynbrandt (1977) *J Nat Cancer Inst* **58**: 1427–1431
LA Liotta, J Kleinerman and GM Saidel (1974) *Cancer Res* **34**: 997–1004

LA Liotta, J Kleinerman and GM Saidel (1976) *Cancer Res* **36**: 889–894
LA Liotta, CW Lee and DJ Morakis (1980) *Cancer Letts* **11**: 141–152
LA Liotta, CN Rao and UM Wewer (1986) *Ann Rev Biochem* **55**: 1037–1057
LA Liotta, G Saidel and J Kleinerman (1976) *Biometrics* **32**: 535–550
LA Liotta, VP Thorgeirrson and S Garbisa (1982) *Cancer Metast Rev* **1**: 277–288
LA Liotta, K Tryggvason, S Garbisa, IR Hart, CM Foltz and S Shafie (1980) *Nature* **284**: 67–68
M Lipinski, M-R Hirsch, H Deagostini-Bazin, O Yamada, T Tursz and C Goridis (1987) *Int J Cancer* **40**: 81–86
G Lipkin and M Knecht (1974) *Proc Nat Acad Sci* **71**: 849–853
G Lipkin, M Rosenberg and V Klaus–Kovtun (1986) *J Invest Dermatol* **87**: 305–308
CD Little, MM Nau, DN Carney, AF Gazdar and JD Minna (1983) *Nature* **306**: 194–196
MA Liu, SR Nussbaum and HN Eisen (1988) *Science* **239**: 395–398
WR Loewenstein, Y Kanno and S Socolar (1978) *Nature* **274**: 133–136
M-L Lohmann-Matthes, A Schleich, G Shantz and V Schirrmacher (1980) *J Nat Cancer Inst* **64**: 1413–1425
F London (1937) *Trans Faraday Soc* **33**: 8–26
DT Loo, JI Fuquay, CL Rawson and DW Barnes (1987) *Science* **236**: 200–202
R Lotan and GL Nicolson (1981) In *Molecular actions and targets for cancer chemotherapeutic agents* Eds AC Sartorelli, JS Lazo and JR Bertino Academic Press New York pp 527–539
R Lotan and A Raz (1983) *Cancer Res* **43**: 2088–2093
MT Lotze, AE Chang, CA Seipp, C Simpson, JT Vetto and SA Rosenberg (1986) *J Am Med Assoc* **256**: 3117–3124
RB Low, C Chapponier and G Gabbiani (1981) *Lab Invest* **44**: 359–367
MG Low (1987) *Biochem J* **244**: 1–13
FC Lowe and JT Issacs (1984) *Cancer Res* **44**: 744–752
RJ Ludford (1932) *Proc Roy Soc Lond B* **112**: 250–263
ES Luria and M Delbruck (1943) *Genetics* **28**: 491–511

B Maat and P Hilgard (1981) *J Cancer Res Clin Oncol* **101**: 275–283
DK MacCallum, JH Lillie, LJ Scaletta, JC Occhino, WG Frederich and SR Ledbetter (1982) *Exp Cell Res* **139**: 1–13
WJ Mackillop, A Ciampi, JE Tiel and RN Buick (1983) *J Nat Cancer Inst* **70**: 9–16
JP MacManus, JF Whitfield and DJ Stewart (1984) *Cancer Letts* **21**: 39–315
I Macpherson and L Montagnier (1964) *Virology* **23**: 291–294
JL Madara (1988) *Cell* **53**: 497–498
RE Madden and L Gyure (1968) *Oncology* **22**: 281–289
EG Maderazo, TF Anton and PA Ward (1978) *Clin Immunol Immunopathol* **7**: 166–176
C Magro, FW Orr, WJ Manishen, K Sivananthan and SS Mokashi (1985) *J Nat Cancer Inst* **74**: 829–838
PA Maher, EB Pasquale, JYJ Wang and SI Singer (1985) *Proc Nat Acad Sci* **82**: 6776–6580
MF Maignan (1979) *Biol Cell* **35**: 229–232
CL Mainardi, SN Dixit and AH Kang (1980) *J Biol Chem* **255**: 5435–5441
A Maiorana and PM Gullino (1978) *Cancer Res* **38**: 4409–4414
T Makinodan and MM Kay (1980) *Adv Immunol* **29**: 287–330

AB Malik (1983) *Physiol Rev* **63**: 1114–1207
HL Malinoff, JP McCoy, J Varani and MS Wicha (1984) *Int J Cancer* **33**: 651–655
CA Maniglia, JJ Gomez, SD Luikart and AC Sartorelli (1985) *J Nat Cancer Inst* **75**: 111–120
CA Maniglia and AC Sartorelli (1981) *Ann NY Acad Sci* **359**: 322–330
CA Maniglia, G Tudor, J Gomez and AC Sartorelli (1982) *Cancer Letts* **16**: 253–260
RJ Mannino, K Ballmer and MM Burger (1978) *Science* **201**: 824–826
RJ Mannino and MM Burger (1975) *Nature* **256**: 19–22
Y Mano-Hirano, N Sato, Y Sawasaki, K Haranaka, N Satomi, H Nariuchi and T Goto (1987) *J Nat Cancer Inst* **78**: 115–120
JJ Marchalonis, CB Schwabe, DM Gersten and J Hearing (1984) *Biochem Biophys Res Comm* **121**: 196–202
VT Marchesi and JL Gowans (1964) *Proc Roy Soc Lond B* **159**: 283–290
PC Marchisio, O Capasso, L Nitsch, R Cancedda and E Gionti (1984) *Exp Cell Res* **151**: 332–343
JM Marcum, M McGill, E Bastida, A Ordinas and GA Jamieson (1980) *Lab Clin Med* **96**: 1046–1053
MMK Mareel (1982) In *Tumour invasion and metastasis* Eds LA Liotta and IR Hart Nijhoff The Hague pp 207–230
M Mareel and M De Brabander (1978a) *Oncology* **35**: 5–7
M Mareel and M De Brabander (1978b) *J Nat Cancer Inst* **61**: 787–792
MMK Mareel, GK De Bruyne, F Vandesande and C Dragonetti (1981) *Invas Metast* **1**: 195–204
MM Mareel and M De Mets (1984) *Int Rev Cytol* **90**: 125–168
M Mareel, L De Ridder, M De Brabander and L Vakaet (1975) *J Nat Cancer Inst* **54**: 923–929
M Mareel, J Kint and C Meyvisch (1979) *Virchows Arch B Cell Pathol* **30**: 95–111
MM Mareel and FM Van Roy (1986) *Anticancer Res* **6**: 419–436
NG Maroudas (1973) *Nature* **244**: 353–354
H Marquardt, MW Hunkapiller, LE Hood, DR Twardszik, JE De Larco, JR Stephenson and GJ Todaro (1983) *Proc Nat Acad Sci* **80**: 4684–4688
H Marquardt, MW Hunkapiller, LE Hood and GJ Todaro (1984) *Science* **223**: 1079–1082
CJ Marshall, LM Franks and AW Carbonell (1977) *J Nat Cancer Inst* **58**: 1743–1751
JH Marti, N Grosser and DM Thomson (1976) *Int J Cancer* **18**: 48–57
GR Martin (1980) *Science* **209**: 768–776
GS Martin (1970) *Nature* **227**: 1021–1023
E Martz (1973) *J Cell Physiol* **81**: 39–48
E Martz (1975) *J Immunol* **115**: 261–267
E Martz (1987) *Human Immunol* **18**: 3–37
E Martz and MS Steinberg (1972) *J Cell Physiol* **81**: 25–38
JL Marx (1982) *Science* **215**: 1380–1383
DE Maslow and E Mayhew (1975) *J Nat Cancer Inst* **54**: 1097–1102
DE Maslow, E Mayhew and J Feldman (1980) *J Nat Cancer Inst* **64**: 635–637
DE Maslow, E Mayhew and J Minowada (1976) *Cancer Res* **36**: 2707–2709
DE Maslow and L Weiss (1978) *J Cell Sci* **29**: 271–275
SO Mast (1925) *J Morph* **41**: 347–425

H Masui, T Kawamoto, JD Sato, B Wolf, G Sato and J Mendelsohn (1984) *Cancer Res* **44**: 1002–1007
G Mathe (1970) *Brit Med J* **4**: 487–488
A Matsuzawa (1986) *Int Rev Cytol* **103**: 303–340
DL Mattey, A Suhrbier, E Parrish and DR Garrod (1987) *Ciba Fdn Symp* **125**: 49–65
C Mauch, K Von der Mark, O Helle, J Mollenhauer, M Pfaffle and T Krieg (1988) *J Cell Biol* **106**: 205–211
TH Maugh (1981) *Science* **212**: 1374–1375
P Maupin and TD Pollard (1983) *J Cell Biol* **96**: 51–62
F Mavilio, A Simeone, E Boncinelli and PW Andrews (1988) *Differentiation* **37**: 73–79
A Mazumder and SA Rosenberg (1984) *J Exp Med* **159**: 495–507
JB McCarthy and LT Furcht (1984) *J Cell Biol* **98**: 1474–1480
JB McCarthy, ST Hagen and LT Furcht (1986) *J Cell Biol* **102**: 179–188
JB McCarthy, SL Palm and LT Furcht (1983) *J Cell Biol* **97**: 772–777
JP McCoy, R Lloyd, MS Wicha and J Varani (1984) *J Cell Sci* **65**: 134–151
M McCutcheon, DR Coman and FB Moore (1948) *Cancer* **1**: 460–467
JK McIntosh, JJ Mule, MJ Merino and SA Rosenburg (1988) *Cancer Res* **48**: 4011–4017
NS McNutt, LA Culp and PH Black (1971) *J Cell Biol* **50**: 691–708
NS McNutt, LA Culp and PH Black (1973) *J Cell Biol* **56**: 412–428
NS McNutt, RA Hershberg and RS Weinstein (1971) *J Cell Biol* **51**: 805–825
NS McNutt and RS Weinstein (1967) *Science* **156**: 597–599
NS McNutt and RS Weinstein (1973) *Prog Biophys Mol Biol* **26**: 45–101
MJ Mead, SD Nathanson, M Lee and E Peterson (1985) *J Surg Res* **38**: 319–327
PP Mehta, JS Bertram and WR Loewenstein (1986) *Cell* **44**: 187–196
DG Menter, JS Hatfield, C Harkins, BF Sloane, JD Taylor, JD Crissman and KV Honn (1987) *Clin Exp Metast* **5**: 65–78
SJ Mentzer, SJ Burakoff and DV Faller (1986) *J Cell Physiol* **126**: 285–290
WE Mercer, D Nelson, AB DeLes, LJ Old and R Baserga (1982) *Proc Nat Acad Sci* **79**: 6309–6312
WD Merritt, DJ Morre and TW Keenan (1978a) *J Nat Cancer Inst* **60**: 1329–1337
WD Merritt, CL Richardson, TW Keenan and DJ Morre (1978b) *J Nat Cancer Inst* **60**: 1313–1327
A-M Mes-Masson, S Masson, D Banville and L Chalifour (1989) *J Cell Sci* **94**: 517–525
DJ Meyer, SB Yancey and J-P Revel (1981) *J Cell Biol* **91**: 505–523
C Meyvisch (1983) *Cancer Metast Rev* **2**: 295–306
C Meyvisch and M Mareel (1979) *Virchows Arch B Cell Path* **30**: 113–122
CA Middleton (1976) *Nature* **259**: 311–313
CA Middleton (1977) *Exp Cell Res* **109**: 349–359
CA Middleton (1982) In *Cell behaviour: a tribute to Michael Abercrombie* Eds R Bellairs, A Curtis and G Dunn CUP Cambridge pp 159–192
P Mignatti, E Robbins and DB Rifkin (1986) *Cell* **47**: 487–498
Y Mikuni-Takagaki and J Gross (1981) *In Glycoconjugates: proceedings of the 6th international symposium on glycoconjugates* Eds T Yamakaira, T Osawa and S Hanada Japan Scientific Societies Press Tokyo pp 491–492
BE Miller, FR Miller, J Leith and GH Heppner (1980) *Cancer Res* **40**: 3977–3981
EC Miller and JA Miller (1971) In *Chemical mutagens* Ed A Hollaender Plenum Press New York Volume 1 pp 83–119

FR Miller and GH Heppner (1979) *J Nat Cancer Inst* **63**: 1457–1463
RA Miller, DG Maloney, R Warnke and R Levy (1982) *N Eng J Med* **306**: 517–522
KM Miner, J Klostergaard, GA Granger and GL Nicolson (1983) *J Nat Cancer Inst* **70**: 717–724
KM Miner, T Kawaguchi, GW Uba and GL Nicolson (1982) *Cancer Res* **42** 4631–4638
WJ Ming, L Bersani and A Mantovani (1987) *J Immunol* **138**: 1469–1474
B Mintz and K Illmensee (1975) *Proc Nat Acad Sci* **72**: 3585–3589
F Mitelman (1983) *Cytogenet Cell Genet* **36**: 1–6
J Mollenhauer and K von der Mark (1983) *EMBO J* **2**: 45–50
S Mondal, DW Brankow and C Heidelberger (1976) *Cancer Res* **36**: 2254–2260
R Montesano, C Drevon, T Kuroki, L Saint Vincent, S Handleman, KK Sanford, D De Feo and IB Weinstein (1977) *J Nat Cancer Inst* **59**: 1651–1658
WH Moolenaar, LGJ Tertoolen and SW de Laat (1984) *Nature* **312**: 371–374
WH Moolenaar, RY Tsien, PT van der Saag and SW de Laat (1983) *Nature* **304**: 645–648
FL Moolten, NH Capparell and SR Cooperband (1972) *J Nat Cancer Inst* **49** 1057–1062
GE Moore, AA Sandberg and AL Watne (1960) *J Am Med Assoc* **172**: 1729–1733
J Morgan and D Garrod (1984) *J Cell Sci* **66**: 133–145
SJ Mork, LI De Ridder and OD Laerum (1982) *Anticancer Res* **2**: 1–10
HP Morris (1965) *Adv Cancer Res* **9**: 227–302
PR Morrison, JT Edsall and SG Miller (1948) *J Am Chem Soc* **70**: 3103–3108
A Moscona (1952) *Exp Cell Res* **3**: 535–539
A Moscona (1957) *Proc Nat Acad Sci* **43**: 184–194
A Moscona and H Moscona (1952) *J Anat* **86**: 287–301
MW Mosesson and RA Umfleet (1970) *J Biol Chem* **245**: 5728–5736
M Moskowitz (1963) *Nature* **200**: 854–856
DM Mott, PH Fabisch, BP Sani and S Sorof (1974) *Biochem Biophys Res Commun* **61**: 621–627
S Mousset and J Rommelaere (1982) *Nature* **300**: 537–539
W Mueller-Klieser (1987) *J Cancer Res Clin Oncol* **113**: 101–122
G Mugnai and LA Culp (1987) *Exp Cell Res* **169**: 328–344
LS Mulcahy, MR Smith and DW Stacey (1985) *Nature* **313**: 241–243
JJ Mule, S Shu, SL Schwartz and SA Rosenberg (1984) *Science* **225**: 1487–1489
R Muller and IM Verma (1984) *Curr Top Microbiol Immunol* **112**: 73–115
R Muller, IM Verma and ED Adamson (1983) *EMBO J* **2**: 679–684
DG Mullins and ST Rohrlich (1983) *Biochim Biophys Acta* **695**: 177–214
GR Mundy, S De Martino and DW Rowe (1981) *J Clin Invest* **68**: 1102–1105
GR Mundy and JW Poser (1983) *Calcif Tissue Int* **35**: 164–168
AJ Murgo (1985) *J Nat Cancer Inst* **75**: 341–344
G Murphy, TE Cawston and JJ Reynolds (1981) *Biochem J* **195**: 167–170
JC Murray, S Garbisa and L Liotta (1980) In *Metastasis. Clinical and experimental aspects* Eds K Hellmann, P Hilgard and S Eccles Martinus Nijhoff The Hague pp 169–173
JC Murray, L Liotta, SI Rennard and GR Martin (1980) *Cancer Res* **40**: 347–351
RJ Muschel, JE Williams, DR Lowry and LA Liotta (1985) *Am J Pathol* **121**: 1–8

A Nagafuchi, Y Shirayoshi, K Okazaki, K Yasuda and M Takeichi (1987) *Nature* **329**: 341–343

B Nagy, J Ban and B Brdar (1977) *Int J Cancer* **19**: 614–620
S Naito, R Giavazzi, SM Wacker, K Itoh, J Mayo and IJ Fidler (1987) *Clin Exp Metast* **5**: 135–146
S Naito, M Kinjo, S Nanno, S Kohja, K Oka and K Tanaka (1981) *Jpn J Cancer Res (Gann)* **72**: 1–7
S Naito, AC Von Eschenbech, R Giavazzi and IJ Fidler (1986) *Cancer Res* **46**: 4109–4115
M Nakache, EL Berg, PR Streeter and EC Butcher (1989) *Nature* **337**: 179–181
M Nakajima, T Irimura, D Di Ferrante, N Di Ferrante and GL Nicolson (1983) *Science* **220**: 611–613
M Nakajima, T Irimura, N Di Ferrante and GL Nicolson (1984) *J Biol Chem* **259**: 2283–2290
M Nakajima, T Irimura and GL Nicolson (1988) *J Cell Biochem* **36**: 157–167
M Nakajima, D Lotan, MM Baig, RM Carralero, RW Wood, JC Hendrix and R Lotan (1989) *Cancer Res* **49**: 1698–1706
S Nakano and PO Ts'o (1981) *Proc Nat Acad Sci* **78**: 4995–4999
P Nanni, C De Giovanni, P-L Lollini, G Nicoletti and G Prodi (1986) *J Nat Cancer Inst* **76**: 87–93
A Neri and GL Nicolson (1981) *Int J Cancer* **28**: 731–738
PA Netland and BR Zetter (1984) *Science* **224**: 1113–1114
PA Netland and BR Zetter (1985) *J Cell Biol* **101**: 720–724
SA Newman, DA Frenz, JJ Tomasek and DD Rabuzzi (1985) *Science* **228**: 885–888
J Neyton and A Trautmann (1985) *Nature* **317**: 331–335
C Nguyen, M-G Mattei, J-F Mattei, M-J Santoni, C Goridis and BR Jordan (1986) *J Cell Biol* **102**: 711–715
B Nicholson, R Dermietzel, D Teplow, O Traub, K Willecke and J-P Revel (1987) *Nature* **329**: 732–734
A Nicol and MV Nermut (1987) *Eur J Cell Biol* **43**: 348–357
GL Nicolson (1971) *Nature (New Biol)* **243**: 218–220
GL Nicolson (1973) *Nature (New Biol)* **243**: 218–220
GL Nicolson (1976) *Biochim Biophys Acta* **457**: 57–108
GL Nicolson (1978) *Biochim Biophys Acta* **458**: 1–72
GL Nicolson (1982a) *Biochim Biophys Acta* **695**: 113–176
GL Nicolson (1982b) *J Histochem Cytochem* **30**: 214–220
GL Nicolson (1984a) *Exp Cell Res* **150**: 3–22
GL Nicolson (1984b) *Cancer Metast Rev* **3**: 25–42
GL Nicolson (1987a) *Cancer Res* **47**: 1473–1487
GL Nicolson (1987b) *Exp Cell Res* **168**: 572–577
GL Nicolson (1988) *Biochim Biophys Acta* **948**: 175–224
GL Nicolson, CR Birdwell, KW Brunson, JC Robbins, G Beattie and IJ Fidler (1977) In *Cell and tissue interactions* Eds J Lash and MM Burger Raven Press New York pp 225–241
GL Nicolson and KW Brunson (1977) *Jpn J Cancer Res (Gann)* **20**: 15–24
GL Nicolson, KW Brunson and IJ Fidler (1977) *Acta Histochem Cytochem* **10**: 114–133
GL Nicolson, KW Brunson and IJ Fidler (1978) *Cancer Res* **38**: 4105–4111
GL Nicolson and SE Custead (1982) *Science* **215**: 176–178
GL Nicolson and KM Dulski (1986) *Int J Cancer* **38**: 289–294
GL Nicolson, K Dulski, C Basson and DR Welch (1985) *Invas Metast* **5**: 144–158
GL Nicolson, A Neri, CL Reading and KM Miner (1980) In *Metastasis. Clinical*

and experimental aspects Eds K Hellmann, P Hilgard and S Eccles Martinus Nijhoff The Hague pp 163–168
GL Nicolson and JL Winkelhake (1975) *Nature* **255**: 230–232
GL Nicolson, JL Winkelhake and AC Nussey (1976) In *Fundamental aspects of metastasis* Ed L Weiss North Holland Amsterdam pp 291–303
MJ Niedbala, K Crickard and RJ Bernacki (1985) *Exp Cell Res* **160**: 499–513
SE Nielson and TT Puck (1980) *Proc Nat Acad Sci* 77: 985–989
EA Nigg, BM Sefton, T Hunter, G Walter and SJ Singer (1982) *Proc Nat Acad Sci* **79**: 5322–5326
RM Niles and MP Logue (1979) *J Supramolec Struct* **11**: 251–258
RM Niles and JS Makarski (1978) *J Cell Physiol* **96**: 355–360
S Nir, J Bentz, J Wilschutz and N Duzgunes (1983) *Prog Surf Sci* **13**: 1–124
S Nishimura, JS Huang and TF Deuel (1982) *Proc Nat Acad Sci* **79**: 4303–4307
S Nishimura and T Sekiya (1987) *Biochem J* **243**: 313–327
C Nishio, U Ishii, K Jimbow and K Kikuchi (1981) *Jpn J Cancer Res (Gann)* **72**: 213–219
PD Noguchi, JB Johnson, R O'Donnel and JC Petricciani (1978) *Science* **199**: 980–983
EK Norton and CS Izzard (1982) *Exp Cell Res* **139**: 463–467
A Nose and M Takeichi (1986) *J Cell Biol* **103**: 2649–2658
PC Nowell (1976) *Science* **194**: 23–28
PC Nowell (1986) *Cancer Res* **46**: 2203–2207
WL Nyhan (1987) *BioEssays* **6**: 5–8

S Oda, M Sato, S Toyoshima, T Osawa (1988) *J Biochem* **104**: 600–605
WF Odenwald, CF Taylor, FJ Palmer-Hill, V Friedrich, M Tani and RA Lazzarini (1987) *Genes Dev* **1**: 482–496
S-I Ogou, C Yoshida-Noro and M Takeichi (1983) *J Cell Biol* **97**: 944–948
R Ohnishi (1981) *Biomed Res* **2**: 1–12
LJ Old (1962) *Ann NY Acad Sci* **101**: 80–87
LJ Old (1985) *Science* **230**: 630–632
LJ Old, DA Clarke, B Benacerraf and E Stockert (1962) *Experientia* **18**: 335-336
K Olden, RM Pratt and KM Yamada (1979) *Proc Nat Acad Sci* **76**: 3343–3347
K Olden and KM Yamada (1977) *Cell* **11**: 957–969
FE Oldfield (1963) *Exp Cell Res* **30**: 125–138
K Olsson and P Ebbesen (1979) *J Nat Cancer Inst* **62**: 623–627
L Olsson and J Forchhammer (1984) *Proc Nat Acad Sci* **81**: 3389–3393
DM Ornitz, RE Hammer, A Messing, RD Palmiter and RL Brinster (1987) *Science* **238**: 188–193
FW Orr and S Mokashi (1985) *Int J Cancer* **35**: 101–106
FW Orr, J Varani, MD Gondek, PA Ward and GR Mundy (1979) *Science* **203**: 176–179
L Ossowski and E Reich (1980) *Cancer Res* **40**: 2310–2315
L Ossowski and E Reich (1983a) *Cell* **33**: 323–333
L Ossowski and E Reich (1983b) *Cell* **35**: 611–619
LF Ostrowski, A Ahsan, BP Suthar, P Pagast, DL Bain, C Wong, A Patel and RM Schultz (1986) *Cancer Res* **46**: 4121–4128
J Overton (1977) *Dev Biol* **55**: 103–116
J Overton (1979) *J Exp Zool* **209**: 135–142
T Ozaki, K Yoshida, K Ushijima and H Hayashi (1971) *Int J Cancer* **7**: 93–100

SW Paddock and GA Dunn (1986) *J Cell Sci* **81**: 163–187
S Paget (1889) *Lancet* i 571–573
RD Palmiter and RL Brinster (1985) *Cell* **41**: 343–345
ST Pals, A den Otter, F Miedema, P Kabel, GD Keizer, RJ Scheper and CJLM Meijer (1988) *J Immunol* **140**: 1851–1853
VE Papaioannou, MW McBurney, RL Garder and RL Evans (1975) *Nature* **258**: 70–73
LF Parada, H Land, RA Weinberg, D Wolf and V Rotter (1984) *Nature* **312**: 649–651
L F Parada, CJ Tabin, C Shih and RA Weinberg (1982) *Nature* **297**: 474–478
RE Parchment and GB Pierce (1989) *Cancer Res* **49**: 6680–6686
AB Pardee (1975) *Biochim Biophys Acta* **417**: 153–172
AB Pardee (1987) *Cancer Res* **47**: 1488–1491
RW Parish, C Schmidhauser, T Schmidt and RK Dudler (1987) *J Cell Sci Suppl* **8**: 181–197
EK Parkinson and JG Edwards (1978) *J Cell Sci* **33**: 103–120
NP Parratto and AK Kimura (1988) *J Cell Biochem* **36**: 311–322
EP Parrish, DR Garrod, DL Maltey, L Hand, PV Steart and RO Weller (1986) *Proc Nat Acad Sci* **83**: 2657–2661
AW Partin, JS Schoeniger, JL Mohler and DS Coffey (1989) *Proc Nat Acad Sci* **86**: 1254–1258
I Pastan and GS Johnson (1974) *Adv Cancer Res* **19**: 303–329
I Pastan, GS Johnson and WB Anderson (1975) *Ann Rev Biochem* **44**: 491–522
I Pastan and M Willingham (1978) *Nature* **274**: 645–650
LM Patt and WJ Grimes (1974) *J Biol Chem* **249**: 4157–4165
DL Paul (1986) *J Cell Biol* **103**: 123–134
BU Pauli, SM Cohen, J Alroy and RS Weinstein (1978) *Cancer Res* **38**: 3276–3285
BU Pauli and KE Kuettner (1984) In *Invasion. Experimental and clinical implications* Eds MM Mareel and KC Calman OUP Oxford pp 205–227
BU Pauli, VH Memoli and KF Kuettner (1981) *Cancer Res* **41**: 2084–2091
BU Pauli, DE Schwartz, EJ Thomas and KE Kuettner (1983) *Cancer Metast Rev* **2**: 129–152
BU Pauli and RS Weinstein (1982) *Cancer Res* **42**: 2289–2297
M Paulsson, R Deutzmann, M Dziadek, H Nowack, R Tumps, S Weber and J Engel (1986) *Eur J Biochem* **156**: 467–478
RJ Paxton, G Mooser, H Pande, TD Lee and JE Shively (1987) *Proc Nat Acad Sci* **84**: 920–924
E Pearlstein (1976) *Nature* **262**: 497–500
E Pearlstein, C Ambrogio and S Karpatkin (1984) *Cancer Res* **44**: 3884–3887
E Pearlstein, LB Cooper and S Karpatkin (1979) *J Lab Clin Med* **93**: 332–344
E Pearlstein, PL Sack, G Yogeeswaran and S Karpatkin (1980) *Proc Nat Acad Sci* **77**: 4336–4339
ES Pearlstein, RO Hynes, LM Franks and VJ Hemmings (1976) *Cancer Res* **36**: 1475–1480
PL Pedersen (1978) *Prog Exp Tumour Res* **22**: 190–274
RM Perkins, S Kellie, B Patel and DR Critchley (1982) *Exp Cell Res* **141**: 231–243
R Perona and R Serrano (1988) *Nature* **334**: 438–440
RT Perri, G Vercellotti, J McCarthy, RL Vesella and LT Furcht (1985) *J Lab Clin Med* **105**: 30–35
B Persky, LE Ostrowski, P Pagast, A Ahsan and RM Schultz (1986) *Cancer Res* **46**: 4129–4134

H Persson and P Leder (1984) *Science* **225**: 718–721
J Peterson (1983) *J Theoret Biol* **102**: 41–53
R Peto (1977) *Cold Spring Harbor Symp Quant Biol* **4**: 1463–1428
N Peyrieras, F Hyafil, D Louvard, HL Ploegh and F Jacob (1983) *Proc Nat Acad Sci* **80**: 6274–6277
GP Phondke, KR Madyastha, PR Madyastha and RF Barth (1981) *J Nat Cancer Inst* **66**: 643–647
AH Pickaver, NA Ratcliffe, AE Williams and H Smith (1972) *Nature (New Biol)* **235**: 186–187
GB Pierce, D Aguilar, C Hood and RS Wells (1984) *Cancer Res* **44**: 3987–3996
GB Pierce and JF Dixon (1959) *Cancer* **12**: 573–583
GB Pierce and EL Verney (1961) *Cancer* **14**: 1017–1029
MD Pierschbacher and E Ruoslahti (1984a) *Nature* **309**: 30–33
MD Pierschbacher and E Ruoslahti (1984b) *Proc Nat Acad Sci* **81**: 5985–5988
HI Pilgrim (1971) *Proc Soc Exp Biol Med* **138**: 178–180
MV Pimm and RW Baldwin (1977) *Int J Cancer* **20**: 37–43
MV Pimm, MJ Embleton and RW Baldwin (1980) *Int J Cancer* **25**: 621–629
WJ Pledger, CA Hart, KL Locatell and CD Scher (1981) *Proc Nat Acad Sci* **78**: 4358–4362
WJ Pledger, CD Stiles, HN Antoniades and CD Scher (1978) *Proc Nat Acad Sci* **75**: 2839–2843
EF Plow, MH Ginsberg and GA Marguerie (1986) In *Biochemistry of platelets* Eds DR Phillips and MA Shuman Academic Press New York pp 225–256
RJ Poiesz, FW Ruscetti, AF Gazdar, PA Bunn, JD Minna, and RC Gallo (1980) *Proc Nat Acad Sci* **77**: 7415–7419
R Pollack (1977) *Jpn J Cancer Res (Gann)* **20**: 37–46
RE Pollack and MM Burger (1969) *Proc Nat Acad Sci* **62**: 1074–1076
J Pollanen, K Hedman, LS Nielsen, K Dano and A Vaheri (1988) *J Cell Biol* **106**: 87–95
PJ Polverini, RS Cotran, MA Gimbrone and ER Unanue (1977) *Nature* **269**: 804–806
AR Poole, KJ Tiltman, AD Recklier and TAM Stoker (1978) *Nature* **273**: 545–547
KR Porter and JB Tucker (1981) *Sci Amer* **244**: 40–51
G Poste, J Doll, AE Brown, J Tzeng and I Zeidman (1982a) *Cancer Res* **42**: 2770–2778
G Poste, J Doll and IJ Fidler (1981) *Proc Nat Acad Sci* **78**: 6226–6230
G Poste, J Doll, IR Hart and IJ Fidler (1980) *Cancer Res* **40** 1636–1644
G Poste and IJ Fidler (1980) *Nature* **283**: 139–146
G Poste and GL Nicolson (1980) *Proc Nat Acad Sci* **77**: 399–403
G Poste, J Tzeng, J Doll (1982b) *Proc Nat Acad Sci* **79**: 6574–6578
N Pourreau-Schneider, H Felix and G Haemmerli (1977) *Virchows Arch B Cell Pathol* **23**: 257–264
J Pouyssegur, RPC Shiu and I Pastan (1979) *Cell* **11**: 941–947
R Pozzatti, R Muschel, J Williams, R Padmanabhan, B Howard, L Liotta and G Khoury (1986) *Science* **232**: 223–227
RT Prehn (1970) *J Nat Cancer Inst* **45**: 1039–1044
RT Prehn (1972) *Science* **176**: 170–171
RT Prehn (1975) *J Nat Cancer Inst* **58**: 189–190
RT Prehn and JM Main (1957) *J Nat Cancer Inst* **18**: 769–778
M Prinzmetal, EM Ornitz, B Simkin and H-C Bergman (1948) *Am J Physiol* **152**: 48–52

JW Proctor (1976) *Br J Cancer* **34**: 651–654
JW Proctor, BG Auclair and CM Rudenstam (1976) *Int J Cancer* **18**: 255–262
JW Proctor, BG Auclair and L Stokowski (1971) *J Nat Cancer Inst* **57**: 1197–1198
TT Puck (1977) *Proc Nat Acad Sci* **74**: 4491–4495
R Pytela, MD Pierschbacher and E Ruoslahti (1985) *Proc Nat Acad Sci* **82**: 5766–5770

M Quintanilla, K Brown, M Ramsden and A Balmain (1986) *Nature* **322**: 78–80

TH Rabbitts, PH Hamlyn and R Baer (1983) *Nature* **309**: 760–764
D Rabeh, MA Lieberman and L Glaser (1981) *J Cell Physiol* **108**: 35–45
M Rabinovitch and MJ De Stefano (1973) *Exp Cell Res* **77**: 323–334
EW Raines and R Ross (1982) *J Biol Chem* **257**: 5154–5159
CK Ramachandran, K Sanders and G Melnykovych (1986) *Cancer Res* **46**: 2520–2525
IA Ramshaw, S Carlsen, HC Wang and P Badenoch–Jones (1983) *Int J Cancer* **32**: 471–478
CN Rao, IMK Margulies, TS Tracka, VP Terranova, JA Madri and LA Liotta (1982) *J Biol Chem* **257**: 9740–9744
RA Rasmussen, Y-H Chin, JJ Woodruff and TG Easton (1985) *J Immunol* **135**: 19–24
M Rassoulzadegan, A Cowie, A Carr, N Glaichenhous and R Kamen (1982) *Nature* **300**: 713–715
M Rassoulzadegan, Z Naghashfar, A Cowie, A Carr, M Grison, R Kamen and F Cuzin (1983) *Proc Nat Acad Sci* **80**: 4354–4358
F Rastinejad, PJ Polverini and NP Bouck (1989) *Cell* **56**: 345–355
A Raz and A Ben-Ze'ev (1982) *Int J Cancer* **29**: 711–715
A Raz and A Ben-Ze'ev (1983) *Science* **221**: 1307–1310
A Raz and A Ben-Ze'ev (1987) *Cancer Metast Rev* **6**: 3–21
A Raz, C Bucana, W McLellan and IJ Fidler (1980) *Nature* **284**: 363–364
A Raz and B Geiger (1982) *Cancer Res* **42**: 5183–5190
A Raz and R Lotan (1981) *Cancer Res* **41**: 3642–3647
A Raz, WL McLellan, IR Hart, CD Bucana, LC Hoyer, B-A Sela, P Dragsten and IJ Fidler (1980) *Cancer Res* **40**: 1645–1651
A Raz, L Meromsky, P Carmi, R Karakash, D Lotan and R Lotan (1984) *EMBO J* **3**: 2979–2983
A Raz, L Meromsky and R Lotan (1986) *Cancer Res* **46**: 3667–3672
A Razin and AD Riggs (1980) *Science* **210**: 604–610
CL Reading, PN Belloni and GL Nicolson (1980a) *J Nat Cancer Inst* 1241–1249
CL Reading, KW Brunson, M Torrianni and GL Nicolson (1980b) *Proc Nat Acad Sci* **77**: 5943–5947
DA Rees, RA Badley, CW Lloyd, D Thom and CG Smith (1978) *Symp Soc Exp Biol* **32**: 241–260
TA Reh, T Nagy and H Gretton (1987) *Nature* **330**: 68–70
HS Reinhold (1971) *Euro J Cancer* **7**: 273–280
AB Retik, MS Arons, AS Ketcham and N Mantel (1962) *J Surg Res* **2**: 49–53
JP Revel and MJ Karnovsky (1967) *J Cell Biol* **33**: C7–C12
JP Revel, SB Yancey, B Nicholson and J Hoh (1986) *Ciba Fdn Symp* **125**: 108–127
JA Rhodin (1967) *J Ultrastruct Res* **18**: 181–223
JA Rhodin (1968) *J Ultrastruct Res* **25**: 452–500

C Riccardi, P Puccetti, A Santoni and R Herberman (1979) *J Nat Cancer Inst* **63**: 1041–1045
GE Rice and MP Bevilacqua (1989) *Science* **246**: 1303–1306
M Rieber and MA Castillo (1984) *Int J Cancer* **33**: 765–770
M Rieber and MS Rieber (1981) *Nature* **293**: 74–76
AD Riggs and PA Jones (1985) *Adv Cancer Res* **40**: 1–30
F Rijsewijk, M Schuermann, E Wagenaar, P Parren, D Weigel and R Nusse (1987) *Cell* **50**: 649–657
V Riley (1981) *Science* **212**: 1100–1109
JR Riordan and V Ling (1985) *Pharmacol Ther* **28**: 51–75
H Ris (1985) *J Cell Biol* **100**: 1474–1487
A Rizzino, P Kazakoff, E Ruff, C Kuszynski and J Nebelsick (1988) *Cancer Res* **48**: 4266–4271
H Robenek, W Jung and R Gebhardt (1982) *J Ultrastruct Res* **78**: 95–106
AB Roberts, MA Anzano, LC Lamb, JM Smith, CA Frolik, H Marquardt, GJ Todaro and MB Sporn (1982) *Nature* **295**: 417–419
DD Roberts, JA Sherwood and V Ginsburg (1987) *J Cell Biol* **104**: 131–139
M Robertson (1983) *Nature* **306**: 733–736
S Rockwell (1981) *Cancer Res* **41**: 527–531
S Rogelj, RA Weinberg, P Fanning and M Klagsburn (1988) *Nature* **331**: 173–175
LR Rohrschneider (1980) *Proc Nat Acad Sci* **77**: 3514–3518
M Rola-Pleszczynski (1985) *Immunol Today* **6**: 302–307
AG Romualdez and PA Ward (1975) *Proc Nat Acad Sci* **72**: 4128–4132
E Roos (1980) In *Metastasis. Clinical and experimental aspects* Eds K Hellmann, P Hilgard and S Eccles Martinus Nijhoff The Hague pp 199–203
E Roos (1984) *Biochim Biophys Acta* **738**: 263–284
E Roos and KP Dingemans (1979) *Biochim Biophys Acta* **560**: 135–166
E Roos and FF Roossien (1987) *J Cell Biol* **105**: 553–559
E Roos and IV Van de Pavert (1982) *J Cell Sci* **55**: 233–245
E Roos, IV Van de Pavert and OP Middlekoop (1981) *J Cell Sci* **48**: 385–397
B Rose (1988) In *Gap junctions* Eds EL Hertzberg and RG Johnson AR Liss Inc New York pp 411–421
S Roseman (1970) *Chem Phys Lipids* **5**: 270–297
SD Rosen, MS Singer, TA Yednock and LM Stoolman (1985) *Science* **228**: 1005–1007
MD Rosenberg (1960) *Biophys J* 1 137–159
MD Rosenberg (1962) *Proc Nat Acad Sci* **48**: 1342–1349
SA Rosenberg and 12 others (1985) *New Eng J Med* **313**: 1485–1492
R Ross, EW Raines and DF Bowen-Pope (1986) *Cell* **46**: 155–169
R Rothlein, ML Dustin, SD Marlin and TA Springer (1986) *J Immunol* 137: 1270–1274
V Rotter, H Abutbul and A Ben-Ze'ev (1983) *EMBO J* **2**: 1041–1047
V Rotter, H Abutbul and D Wolf (1983) *Int J Cancer* **31**: 315–320
V Rotter and D Wolf (1985) *Adv Cancer Res* **43**: 113–141
PJ Rous (1911) *J Exp Med* **13**: 397–413
PJ Rous (1914) *J Exp Med* **20**: 433–451
E Rozengurt (1983) *Molec Biol Med* **1**: 169–181
E Rozengurt (1986) *Science* **234**: 161–166
E Rozengurt and SA Mendoza (1985) *J Cell Sci Suppl* 3: 229–242
H Rubin (1961) *Cancer Res* **21**: 1244–1253
H Rubin (1966) In *Major problems in developmental biology* Ed M Locke

Academic Press New York pp 315–337
H Rubin (1985) *Cancer Res* **45**: 2935–2942
P Rubin (1973) *Cancer* **31**: 963–982
P Rubin (1974) In *Clinical oncology* Ed P Rubin American Cancer Society Rochester pp 1–25
E Ruoslahti (1988) *Ann Rev Biochem* **57**: 375–413
E Ruoslahti and MD Pierschbacher (1986) *Cell* **44**: 517–518
RG Russo, LA Liotta, U Thorgeirsson, R Brundage and E Schiffmann (1981) *J Cell Biol* **91**: 459–467
RG Russo, H Thorgeirsson and LA Liotta (1982) In *Biology of cancer invasion and metastasis* Eds LA Liotta and IR Hart Martinus Nijhoff New York pp 173–187
U Rutishauser (1983) *Cold Spring Harbor Symp Quant Biol* **48**: 501–514
U Rutishauser, S Hoffman and GM Edelman (1982) *Proc Nat Acad Sci* **79**: 685–689
JJ Ryan, AS Ketcham and H Wexler (1968a) *Ann Surg* **168**: 163–168
JJ Ryan, AS Ketcham and H Wexler (1968b) *Science* **162**: 1493–1494
U Ryan (1986) *Fed Proc* **45**: 101–108
WL Ryan and ML Heidrick (1968) *Science* **162**: 1484–1485

S Saga, K Nagata, W-T Chen and KM Yamada (1987) *J Cell Biol* **105**: 517–527
R Sager (1986) *Cancer Res* **46**: 1573–1580
R Sager, IK Gadi, L Stephens and CT Grabwy (1985) *Proc Nat Acad Sci* **82**: 7015–7019
N Saijo, A Ozaki, H Nakano, M Sakurai, H Takahashi, Y Sasaki and A Hoshi (1986) *J Cancer Res Clin Oncol* **111**: 182–186
T Sakakura (1983) In *Understanding breast cancer* Eds MA Rich, JC Hager and P Furmaski Marcel Dekker New York pp 261–284
M Sakurai, M Zigo, A Ozaki, JR Jett, Y Sasaki, A Hoshi and N Saijo (1986) *Jpn J Cancer Res (Gann)* **77**: 774–781
T Salo, T Turpeenniemi-Hujanen and K Tryggvason (1985) *J Biol Chem* **260**: 8526–8531
AJ Salsbury (1975) *Cancer Treat Rev* **2**: 55–72
RHC San, MF Laspia, AI Soiefer, CJ Maslansky, JM Rice and GM Williams (1979) *Cancer Res* **39**: 1026–1034
MG Santoro, GW Philpott and BM Jaffe (1976) *Br J Cancer* **39**: 408–415
SA Santoro (1986) *Cell* **46**: 913–920
M Sasaki, S Kato, K Kohno, GR Martin and Y Yamada (1987) *Proc Nat Acad Sci* **84**: 935–939
H Sato, J Khato, T Sato and M Suzuki (1977) *Jpn J Cancer Res (Gann)* **20**: 3–13
H Sato and M Suzuki (1972) *Excpt Med Int Congr Ser* **269**: 168–176
H Sato and M Suzuki (1976) In *Fundamental aspects of metastasis* Ed L Weiss North Holland New York pp 311–317
JW Saunders (1966) *Science* **154**: 604–612
PJ Saxon, ES Srivatsan and EJ Stanbridge (1986) *EMBO J* **5**: 3461–3466
CD Scher, C Haudenschild and M Klagsbrun (1976) *Cell* **8**: 373–382
F Schapira, JC Dreyfus and G Schapira (1963) *Nature* **200**: 995–997
V Schirrmacher (1980) *Immunobiol* **157**: 89–98
V Schirrmacher (1985) *Adv Cancer Res* **43**: 1–73
V Schirrmacher, P Altevogt, M Fogel, J Dennis, CA Waller, D Barz, R Schwartz, R

Cheingsong-Popov, G Springer, PJ Robinson, T Nebe, W Brossmer, I Vlodavsky, N Paweletz, HP Zimmerman and G Uhlenbruck (1982) *Invas Metast* **2**: 313–360

V Schirrmacher and B Appelhans (1985) *Clin Exp Metast* **3**: 29–43

V Schirrmacher, R Cheingsong-Popov and H Arnheiter (1980) *J Exp Med* **151**: 984–989

V Schirrmacher, G Shantz, K Clauer, D Komitowski, HP Zimmerman and MC Lohmann-Matthes (1979) *Int J Cancer* **23**: 233–244

A Schleich (1973) In *Chemotherapy of cancer dissemination and metastasis* Eds S Garattini and G Franchi Raven Press New York pp 51–58

AB Schleich, M Frick and A Mayer (1976) *J Nat Cancer Inst* **56**: 221–225

P Schmialek, A Geyer, V Miosga, M Nundel and B Zapf (1977) *Cell Tissue Kinetics* **10**: 195–202

HP Schnebli and MM Burger (1972) *Proc Nat Acad Sci* **69**: 3825–3827

EE Schneeberger and RD Lynch (1984) *Circ Res* **55**: 723–733

SL Schor (1980) *J Cell Sci* **41**: 159–175

SL Schor, TD Allen and CJ Harrison (1980) *J Cell Sci* **46**: 171–186

SL Schor and AM Schor (1987) *J Cell Sci Supp* **8**: 165–180

SL Schor, AM Schor, TD Allen and B Winn (1985a) *Int J Cancer* **36**: 93–102

SL Schor, AM Schor and GW Bazill (1984) *J Cell Sci* **48**: 301–314

SL Schor, AM Schor, P Durning and G Rushton (1985b) *J Cell Sci* **73**: 235–244

SL Schor, AM Schor, G Rushton and L Smith (1985c) *J Cell Sci* **73**: 221–234

SL Schor, AM Schor, B Winn and G Rushton (1982) *Int J Cancer* **29**: 57–62

L Schour and RP Faraci (1974) *J Surg Oncol* **6**: 109–115

PI Schrier, R Bernards, RTMJ Vaessen, A Houweling and AJ van der Eb (1983) *Nature* **305**: 771–775

F Schroeder and JM Gardiner (1984) *Cancer Res* **44**: 3262–3269

D Schubert and M La Corbiere (1985) *J Cell Biol* **100**: 56–83

M Schwab (1985) *Trends Genet* **1**: 271–275

S Schwartz (1978) *In Vitro* **14**: 966–980

L Schweigerer, G Neufeld and D Gospodarowicz (1987) *J Clin Invest* **80**: 1516–1520

L Sealy and R Charkley (1978) *Cell* **14**: 115–121

HF Seigler and BF Setter (1977) *Ann Surg* **186**: 1–12

B Sela, H Lis, N Sharon and L Sachs (1971) *Biochim Biophys Acta* **249**: 564–568

P Selby, RN Buick and I Tannock (1983) *N Eng J Med* **308**: 129–134

GC Sephel, K-I Tashiro, M Sasaki, D Greatorex, GR Martin, Y Yamada and HK Kleinman (1989) *Biochem Biophys Res Comm* **162**: 821–829

FE Sharkey and J Fogh (1979) *Int J Cancer* **24**: 733–738

LM Shaw, N Yang, M Neat and W Croop (1982) *Ann NY Acad Sci* **390**: 73–88

N Sharon and H Lis (1989) *Science* **246**: 227–234

PJ Shearman, WM Gallatin and BM Longnecker (1980) *Nature* **286**: 267–269

G Shenouda, DM Thomson and JK MacFarlane (1984) *Cancer Res* **44**: 1238–1245

JR Sheppard, TP Koestler, SP Corwin, C Buscarino, J Doll, BR Lester, RG Greig and G Poste (1984) *Nature* **308**: 544–547

R Shields and K Pollock (1974) *Cell* **3**: 31–38

C Shih, LC Padhy, M Murray and RA Weinberg (1981) *Nature* **290**: 261–264

C Shih, BZ Shilo, MP Goldfarb, A Dannenberg and RA Weinberg (1979) *Proc Nat Acad Sci* **76**: 5714–5718

C Shih and RA Weinberg (1982) *Cell* **29**: 161–169

S Shih, VH Freedman, R Risser and R Pollack (1975) *Proc Nat Acad Sci* **72**: 4435–4439
Y Shing, J Folkman, C Haudenschild, D Lund, R Crum and M Klagsbrun (1985) *J Cell Biochem* **29**: 275–287
Y Shing, J Folkman, R Sullivan, C Butterfield, J Murray and M Klagsbrun (1984) *Science* **223**: 1296–1298
K Shinkai, M Mukai, K Komatsu and H Akido (1988) *Cancer Res* **48**: 3760–3764
T Shiohara, G Moellmann, K Jacobson, E Kuklinska, NH Ruddle and AB Lerner (1987b) *J Immunol* **138**: 1979–1986
T Shiohara, NH Ruddle, M Horowitz, GE Moellmann and AB Lerner (1987a) *J Immunol* **138**: 1971–1978
Y Shirayoshi, K Hatta, M Hosoda, S Tsunasawa, F Sakiyama and M Takeichi (1986) *EMBO J* **5**: 2485–2488
BD Shur (1983) *Dev Biol* **99**: 360–372
BD Shur and S Roth (1975) *Biochem Biophys Acta* **415**: 473–512
MS Shure, ND Young, J Kolega and W-T Chen (1979) *J Cell Biol* **83**: 491a
E Sidebottom and SR Clark (1983) *Br J Cancer* **47**: 399–406
BV Siegel (1985) *Int Rev Cytol* **96**: 89–120
M Siegelman, MW Bond, WM Gallatin, T St. John, HT Smith, VA Fried and IL Weissman (1986) *Science* **231**: 823–829
MH Siegelman, M van de Rijn and IL Weissman (1989) *Science* **243**: 1165–1172
MF Sigot-Luizard (1974) In *Neoplasia and cell differentiation* Ed GV Sherbet Karger Basel pp 350–379
J Silnutzer and DW Barnes (1984) In *Methods for preparation of media supplements and substrata for serum-free animal cell culture* Eds DW Barnes, DA Sirbasky and GH Sato Alan R Liss New York pp 245–268
J Silnutzer and DW Barnes (1985) *In Vitro* **21**: 73–78
RL Silverstein, LLK Leung and RL Nachman (1986) *Arteriosclerosis* **6**: 245–253
N Simionescu, M Simionescu and GE Palade (1978) *Microvasc Res* **15**: 17–36
D Simmons, MW Makgoba and B Seed (1988) *Nature* **331**: 624–627
SR Simon and WB Ershler (1985) *J Nat Cancer Inst* **74**: 1085–1088
DE Sims (1986) *Tissue Cell* **18**: 153–174
WF Sindelar, TS Tralka and AS Ketcham (1975) *J Surg Res* **18**: 137–161
II Singer (1979) *Cell* **16**: 675–685
II Singer (1982) *J Cell Biol* **92**: 398–408
BK Sinha and GJ Goldenberg (1974) *Cancer* **34**: 1956–1961
E Sinn, W Muller, P Pattenga, I Tepler, R Wallace and P Leder (1987) *Cell* **49**: 465–475
CJ Skerrow and AG Matoltsy (1974a) *J Cell Biol* **63**: 515–523
CJ Skerrow and AG Matoltsy (1974b) *J Cell Biol* **63**: 524–530
G Skolnik, U Bagge, A Dahlstrom and H Ahlman (1984) *Int J Cancer* **33**: 519–523
GR Skuse and PT Rowley (1989) *Sem Oncol* **16**: 128–137
DJ Slamon, JB Dekernion, IM Verna and MJ Kline (1984) *Science* **224**: 256–262
BF Sloane, JR Dunn and KV Honn (1981) *Science* **212**: 1151–1153
BF Sloane and KV Honn (1984) *Cancer Metast Rev* **3**: 249–263
BF Sloane, KV Honn, JG Sadler, WA Turner, JJ Kimpson and JD Taylor (1982) *Cancer Res* **42**: 980–986
BF Sloane, J Rozhin, K Johnson, H Taylor, JD Crissman and KV Honn (1986) *Proc Nat Acad Sci* **83**: 2483–2487
LA Smets (1973) *Nature (New Biol)* **245**: 113–115
LA Smets (1980) *Biochim Biophys Acta* **605**: 93–111

LA Smets and WP van Beek (1984) *Biochim Biophys Acta* **738**: 237–249
HS Smith, DC Scher and GJ Todaro (1971) *Virology* **44**: 359–370
JF Smyth, FR Balkwill, F Cavalli, A Kimchi, K Mattson, NE Nierderle and RJ Spiegel (1987) *Eur J Cancer Clin Oncol* **23**: 887–889
R Snyderman and GJ Cianciolo (1984) *Immunol Today* **5**: 240–244
R Snyderman and J Goetzl (1981) *Science* **213**: 830–837
R Snyderman, L Meadows, W Holder and S Wells (1978) *J Nat Cancer Inst* **60**: 737–740
R Snyderman, MC Pike, BL Blaylock and P Weinstein (1976) *J Immunol* **116**: 585–589
R Snyderman, H Siegler and L Meadows (1977) *J Nat Cancer Inst* **58**: 37–41
R Snyderman and C Stahl (1975) In *The phagocytic cell and host resistance* Eds JA Bellantini and DH Dayton Raven Press New York pp 267–281
OV Solesvik, EK Rofstad and T Brustad (1985) *Eur J Cancer Clin Oncol* **21**: 499–505
S Sone and IJ Fidler (1981) *Cell Immunol* **57**: 42–50
A Sonnenberg, PW Modderman and F Hogervorst (1988) *Nature* **336**: 487–489.
BC Sordat, Y Ileyama and J Fogh (1982) In *The nude mouse in experimental and clinical research* Academic Press New York Vol 2 pp 95–143
LK Sorge, BT Levy and PF Maness (1984) *Cell* **36**: 249–257
N Sorgente, DL Bullard, L Jakovljevic and K Dorey (1984) *Cell Tissue Kinetics* **17**: 573–582
N Sorgente, KE Kuettner and R Eisenstein (1976) *Protides Biol Fluids* **23**: 227–230
PD Sorlie and M Feinleib (1982) *J Nat Cancer Inst* **69**: 989–996
PJ Southern and P Berg (1982) *J Molec Appl Genet* **1**: 327–341
DA Spandidos and IB Kerr (1984) *Br J Cancer* **49**: 681–688
DA Spandidos and NM Wilkie (1984) *Nature* **310**: 469–475
S Spiegel, KM Yamada, BE Hom, J Moss and PH Fishman (1985) *J Cell Biol* **100**: 721–726
S Spiegel, KM Yamada, BE Hom, J Moss and PH Fishman (1986) *J Cell Biol* **102**: 1898–1906
LA Sporn, VJ Marder and DD Wagner (1986) *Cell* **46**: 185–190
MB Sporn, DL Newton, AB Roberts, JE De Larco and GJ Todaro (1981) In *Molecular actions and targets for cancer chemotherapeutic agents* Eds AC Sartorelli, JR Bertino and JS Lazo Academic Press New York pp 541–544
MB Sporn and AB Roberts (1985) *Nature* **313**: 745–747
MB Sporn and GJ Todaro (1980) *New Eng J Med* **303**: 878–880
DC Spray, RD Ginzberg, EA Morales, Z Gatmaitan and IM Arias (1986) *J Cell Biol* **103**: 135–144
DC Spray, AL Harris and MVL Bennett (1981) *Science* **211**: 712–715
TA Springer, LJ Miller and DC Anderson (1986) *J Immunol* **136**: 240–245
TA Springer, DB Teplow and WJ Dreyer (1985) *Nature* **314**: 540–542
A Srivastava, P Laidler, LE Hughes, J Woodcock and EJ Shedden (1986) *Eur J Cancer Clin Oncol* **22**: 1205–1209
CW Stackpole (1981) *Nature* **289**: 798–800
CW Stackpole (1983) *Cancer Res* **43**: 3057–3065
CW Stackpole, AL Alterman and DM Fornabaio (1985) *Invas Metast* **5**: 125–143
CW Stackpole, DM Fornabaio and AL Alterman (1987) *Clin Expl Metast* **5**: 165–180
CW Stackpole, JB Jacobson and MP Lardis (1974) *J Exp Med* **140**: 939–953

LA Staehelin (1974) *Int Rev Cytol* **39**: 191–278
HB Stamper and JJ Woodruff (1976) *J Exp Med* **144**: 828–833
HB Stamper and JJ Woodruff (1977) *J Immunol* **119**: 772–780
EJ Stanbridge (1985) *BioEssays* **3**: 252–256
EJ Stanbridge, RR Flandermeyer, DW Daniels and WA Nelson–Rees (1981) *Somatic Cell Genet* **7**: 699–712
EJ Stanbridge and J Wilkinson (1978) *Proc Nat Acad Sci* **75**: 1466–1469
EJ Stanbridge and J Wilkinson (1980) *Int J Cancer* **26**: 1–8
DR Stanford, JR Starkey and JA Magnuson (1986) *Int J Cancer* **37**: 435–444
JR Starkey, HL Hosick, DR Stanford and HD Liggitt (1984a) *Cancer Res* **44**: 1595–1594
JR Starkey, HD Liggitt, W Jones and HL Hosick (1984b) *Int J Cancer* **34**: 535–543
PA Steck and GL Nicolson (1983) *Exp Cell Res* **147**: 255–267
D Stehelin, HE Varmus, JM Bishop and PK Vogt (1976) *Nature* **260**: 170–173
MS Steinberg (1963) *Science* **141**: 401–408
MS Steinberg (1964) In *Cellular membranes in development* Ed M Locke Academic Press New York pp 321–366
MS Steinberg (1970) *J Exp Zool* **173**: 395–434
MS Steinberg and DR Garrod (1975) *J Cell Sci* **18**: 385–403
MS Steinberg, H Shida, GJ Gindice, M Shida, NH Patel and OW Blaschuk (1987) *Ciba Fdn Symp* **125**: 3–25
MS Steinberg and LL Wiseman (1972) *J Cell Biol* **55**: 606–615
C Steineman, M Fenner, H Binz and RW Parish (1984) *Proc Nat Acad Sci* **81**: 3747–3750
SK Steinsvag, OD Laerum and R Bjerkvig (1985) *J Nat Cancer Inst* **74**: 1095–1104
K Stellner, S Hakomori and G Warner (1973) *Biochem Biophys Res Comm* **55**: 439–445
S Stenman and A Vaheri (1981) *Int J Cancer* **27**: 427–435
S Stenman, J Wartiovaara and A Vaheri (1977) *J Cell Biol* **74**: 453–467
KS Stenn (1981a) *Proc Nat Acad Sci* **78**: 196–204
KS Stenn (1981b) *Proc Nat Acad Sci* **78**: 6907–6911
EM Stephenson and NG Stephenson (1978) *J Cell Sci* **32**: 389–418
SK Stevens, IL Weissman and EC Butcher (1982) *J Immunol* **128**: 844–851
BR Stevenson (1987) *Ciba Fdn Symp* **125**: 188–191
R Steward (1987) *Science* **238**: 692–694
T St John, WM Gallatin, M Siegelman, HT Smith, VA Fried and IL Weissman (1986) *Science* **231**: 845–850
M Stoker (1964) *Virology* **24**: 165–174
M Stoker (1967) In *Mechanisms of invasion in cancer* Ed P Denoix Springer Verlag Berlin pp 193–203
MGP Stoker (1973) *Nature* **246**: 200–203
M Stoker, E Gherardi, M Perryman and J Gray (1987) *Nature* **327**: 239–242
M Stoker and D Piggott (1974) *Cell* **3**: 207–215
MGP Stoker, D Piggott and P Riddle (1978) *Int J Cancer* **21**: 268–274
MGP Stoker and H Rubin (1967) *Nature* **215**: 171–172
LM Stoolman and SD Rosen (1983) *J Cell Biol* **96**: 722–729
LM Stoolman, TS Tenforde and SD Rosen (1984) *J Cell Biol* **99**: 1535–1540
D Stopak and AK Harris (1982) *Dev Biol* **90**: 383–398
MP Stoppelli, A Corti, A Soffientini, G Cassani, F Blasi and RK Assoian (1985)

Proc Nat Acad Sci **82**: 4939–4943
GA Storme, WE Berdel, W J van Blitterswijk, EA Bruyneel, GK De Bruyne and MM Mareel (1985) *Cancer Res* **45**: 351–357
G Storme and MM Mareel (1980) *Cancer Res* **40**: 943–948
G Storme and MM Mareel (1981) *Oncology* **38**: 182–186
G Strassman, TA Springer, SD Somers and DO Adams (1986) *J Immunol* **136**: 4328–4333
L Strauch (1972) In *Tissue interactions in carcinogenesis* Ed D Tarin Academic Press London pp 399–433
P Strauli (1980) In *Cell movement and neoplasia* Eds M De Brabander, M Mareel and L De Ridder Pergamon Press Oxford pp 187–191
P Strauli (1980) In *Proteinases and tumour invasion* Eds P Strauli, AJ Barrett and A Baici Raven Press New York pp 1–15
P Strauli and L Weiss (1977) *Eur J Cancer* **13**: 1–12
HB Streeter and DA Rees (1987) *J Cell Biol* **105**: 507–515
PR Streeter, EL Berg, BTN Rouse, RF Bargatze and EC Butcher (1988) *Nature* **331**: 41–46
PR Streeter, BTN Rouse and EC Butcher (1988) *J Cell Biol* **107**: 1853–1862
DA Stringfellow and FA Fitzpatrick (1979) *Nature* **282**: 76–78
DJ Strydom, JW Fett, RR Lobb, EM Alderman, JL Bethune, JF Riordan and BL Vallee (1985) *Biochem* **24**: 5486–5494
O Stutman (1978) In *The nude mouse in experimental and clinical research* Eds J Fogh and BC Giovanella Academic Press New York pp 411–435
H Subak-Sharpe, RR Burk and JD Pitts (1969) *J Cell Sci* **4**: 353–361
EV Sugarbaker (1981) In *Cancer biology reviews* Vol 2 Eds JJ Marchalonis, MG Hanna and IJ Fidler Marcel Dekker Inc New York and Basel pp 235–278
EV Sugarbaker and AM Cohen (1972) *Surgery* **72**: 155–164
EV Sugarbaker, AM Cohen and AS Ketcham (1971) *Ann Surg* **174**: 161–166
EV Sugarbaker and AS Ketcham (1977) *Semin Oncol* **4**: 19–32
PS Sunkara, CC Chanj, NJ Prakash and PJ Lachman (1985) *Cancer Res* **45**: 4067–4070
PS Sunkara, NJ Prakash, GD Mayer and A Sjoerdsma (1983) *Science* **219**: 851–853
PS Sunkara and AC Rosenberger (1987) *Cancer Res* **47**: 933–935
FR Susi, WD Belt and DE Kelly (1967) *J Cell Biol* **34**: 686–690
M Sussman and RR Sussman (1961) *Exp Cell Res* **8**: 91–106
RM Sutherland, J Carlsson, RE Durand and J Yuhas (1981) *Cancer Res* **41**: 2980–2994
N Suzuki, M Frapart, D Gidima, ML Meistrich and HR Withers (1977) *Cancer Res* **37**: 3690–3693
S Suzuki, A Oldberg, EG Hayman, MD Pierschbacher and E Ruoslahti (1985) *EMBO J* **4**: 2519–2524
S Suzuki, MD Pierschbacher, EG Hayman, K Nguyen, Y Ohgren and E Ruoslahti (1984) *J Biol Chem* **259**: 15307–15314
EA Swabb, J Wei and PM Gullino (1974) *Cancer Res* **34**: 2814–2822
B Sylven and I Bois-Svenssen (1965) *Cancer Res* **25**: 458–468
B Sylven, O Snellman and P Strauli (1974) *Virchows Arch B Cell Path* **17**: 97–112
G Symonds, K-H Klempnauer, G Evan and JM Bishop (1984) *Mol Cell Biol* **4**: 2587–2593
G Symonds, K-H Klempnauer, M Snyder, G Moscovici, C Moscovici and JM Bishop (1986) *Mol Cell Biol* **6**: 1796–1802

CJ Tabin, SM Bradley, CI Bargmann, RA Weinberg, AG Papageorge, EM Scolnick, R Dhar, DR Lowy and EH Chang (1982) *Nature* **300**: 143–149
E Tahara, W Yasui, K Taniyama, A Ochia, T Yamamoto, S Nakajo and M Yamamoto (1986) *Jpn J Cancer Res (Gann)* 77: 517–512
Y Takada, C Huang and ME Hemler (1987) *Nature* **326**: 607–609
Y Takada, JL Strominger and ME Hemler (1987) *Proc Nat Acad Sci* **84**: 8239–3243
Y Takada, EA Wayner, WG Carter and ME Hemler (1988) *J Cell Biol* **37**: 385–393
M Takeichi (1988) *Development* **102**: 639–655
D Takemoto and C Jilta (1983) *Leukemia Res* **7**: 97–100
K Takenaga (1986) *Cancer Metast Rev* **5**: 67–75
K Takenaga and K Takahashi (1986) *Cancer Res* **46**: 375–380
JE Talmadge and IJ Fidler (1982) *Nature* **297**: 593–594
JE Talmadge, KM Meyers, DJ Prieur and JR Starkey (1980) *J Nat Cancer Inst* **65**: 929–935
JE Talmadge, JR Starkey, WC Davis and AL Cohen (1979) *J Supramolec Struct* **12**: 227–243
JE Talmadge, JR Starkey and DR Stanford (1981) *J Supramolec Struct* **15**: 139–151
JE Talmadge, SR Wolman and IJ Fidler (1982) *Science* **217**: 361–363
JE Talmadge and B Zbar (1986) *J Nat Cancer Inst* **78**: 315–320
JW Tamkun, SW DeSimone, D Fonda, RS Patel, C Buck, AF Horwitz and RO Hynes (1986) *Cell* **46**: 271–282
K Tanaka, KJ Isselbacher, G Khoury and G Jay (1985) *Science* **228**: 26–30
N Tanaka, S-I Ashida, A Tohgo and H Ogawa (1982) *Invas Metast* **2**: 289–298
NG Tanaka, A Tohgo and H Ogawa (1986) *Invas Metast* **6**: 209–224
K Taniguchi, K Karre and G Klein (1985) *Int J Cancer* **36**: 503–510
S Taniguchi, T Kawano, T Mitsudomi, G Kimura and T Baba (1986) *Jpn J Cancer Res (Gann)* **77**: 1193–1197
S Taniguchi, M Tatsuka, K Nakamatsu, M Inoue, H Sadano, H Okazaki, H Iwamoto and T Baba (1989) *Cancer Res* **49**: 6738–6744
IF Tannock and D Rotin (1989) *Cancer Res* **49**: 4373–4384
T-W Tao and MM Burger (1977) *Nature* **270**: 437–438
T-W Tao and MM Burger (1982) *Int J Cancer* **29**: 425–430
T-W Tao and LK Johnson (1982) *Int J Cancer* **30**: 763–766
T-W Tao, A Matter, K Vogel and MM Burger (1979) *Int J Cancer* **23**: 854–857
N Taptiklis (1968) *Eur J Cancer* **4**: 59–66
N Taptiklis (1969) *Eur J Cancer* **5**: 445–457
D Tarin (1985) *Biochim Biophys Acta* **780**: 227–235
D Tarin and JE Price (1981) *Cancer Res* **41**: 3604–3609
YS Tatarinov (1964) *Vopr Med Khim* **10**: 90–91
P Tattersall and DC Ward (1978) In *Replication of mammalian parvoviruses* Eds DC Ward and P Tattersall Cold Spring Harbor Laboratory New York pp 3–12
AC Taylor (1961) *Exp Cell Res Suppl* **8**: 154–173
DD Taylor and PH Black (1986) In *The cell surface in development and cancer* Ed MS Steinberg Plenum Press New York pp 33–58
RB Taylor, WPH Duffus, MC Raff and S De Petris (1971) *Nature (New Biol)* **233**: 225–229
RF Taylor, TH Price, SM Schwartz and DC Dale (1980) *J Clin Invest* **67**: 584–587

S Taylor and J Folkman (1982) *Nature* **297**: 307–312
J Taylor-Papadimitriou, J Burchell and J Hurst (1981) *Cancer Res* **41**: 2491–2500
JM Teitel (1986) *J Cell Physiol* **128**: 329–336
HM Temin and S Mizutani (1970) *Nature* **226**: 1211–1213
HM Temin and H Rubin (1958) *Virology* **6**: 669–688
VP Terranova, CN Rao, T Kalebie, IM Margulies and LA Liotta (1983) *Proc Nat Acad Sci* **80**: 444–448
VP Terranova, JE Williams, LA Liotta and GR Martin (1984) *Science* **226**: 982–984
AH Terry and LA Culp (1974) *Biochem* **13**: 414–425
AA Te Velde, GD Keizer, CC Figdor (1987) *Immunol* **61**: 261–267
JP Thiery, JL Duband, U Rutishauser and GM Edelman (1982) *Proc Nat Acad Sci* **79**: 6737–6741
KA Thomas, MC Riley, SK Lemmon, NC Baglan and RA Bradshaw (1980) *J Biol Chem* **255**: 5517–5520
L Thomas (1959) In *Cellular and humoral aspects of the hypersensitivity states* Ed HS Lawrence Cassell London pp 529–531
DMP Thomson (1984) *J Nat Cancer Inst* **73**: 595–605
A Thor, PH Hand, D Wunderlich, A Caruso, R Muraro and J Schlom (1984) *Nature* **311**: 562–565
UP Thorgeirsson, LA Liotta, T Kalebic, IM Margulies, K Thomas, M Rios-Candelore and RG Russo (1982) *J Nat Cancer Inst* **69**: 1049–1054
UP Thorgeirsson, T Turpeenniemi-Hujanen, JE Talmadge and LA Liotta (1986) In *Cancer metastasis experimental and clinical strategies* Eds DR Welch, BK Bhuyan and LA Liotta Alan R Liss New York pp 77–93
UP Thorgeirsson, T Turpeenniemi-Hujanen, JE Williams, EH Westin, CA Heilman, JE Talmadge and LA Liotta (1985) *Mol Cell Biol* **5**: 259–262
RD Thornes, DW Edlow and S Wood (1968) *Johns Hopkins Med J* **123**: 305–316
PE Thorpe, ANF Brown, WCJ Ross, AJ Cumber, SI Detre, DC Edwards, AJS Davies and F Stirpe (1981) *Eur J Biochem* **116**: 447–454
C Tickle and A Crawley (1979) *J Cell Sci* **40**: 257–270
C Tickle, A Crawley and M Goodman (1978a) *J Cell Sci* **31**: 293–322
C Tickle, A Crawley and M Goodman (1978b) *J Cell Sci* **33**: 133–155
L Timpe, E Martz and MS Steinberg (1978) *J Cell Sci* **30**: 293–304
R Timpl and M Dziadek (1986) *Int Rev Exp Path* **29**: 1–112
R Timpl, J Engel and GR Martin (1983) *Trends Biochem Sci* **8**: 207–209
R Timpl and GR Martin (1982) In *Immunochemistry of the extracellular matrix* Ed H Furthmayr CRC Press Florida pp 119–150
GJ Todaro and H Green (1963) *J Cell Biol* **17**: 299–313
GJ Todaro, GK Lazar and H Green (1965) *J Cell Physiol* **66**: 325–333
GJ Todaro, SR Wolman and H Green (1963) *J Cell Comp Physiol* **62**: 257–265
A Tohgo, N Tanaka, S-I Ashida and H Ogawa (1984) *Invas Metast* **4**: 134–145
Y Tokada, C Huang and ME Hemler (1987) *Nature* **326**: 607–609
ZA Tokes, N Sorgente and T Okigaki (1977) *Proc Clin Biol Res* **7**: 615–624
BR Tomasini and DF Mosher (1986) *Blood* **68**: 737–742
LD Tomei, I Noyes, D Blocker, J Holliday and R Glaser (1987) *Nature* **329**: 73–75
Y Tomita, PM Montague and VJ Hearing (1985) *Int J Cancer* **35**: 543–547
A Toshitani, K Taniguchi, YI Kawano and K Nomoto (1987) *Cell Immunol* **108**: 188–202
PL Townes and J Holtfretter (1955) *J Exp Zool* **128**: 53–118
DL Trainer, T Kline, F Mallon, R Greig and G Poste (1985) *Cancer Res* **45**:

6124–6130
R Triesman, V Novak, J Favolora and R Kamen (1981) *Nature* **292**: 595–600
JP Trinkaus (1984) *Cells into organs. The forces that shape the embryo* Prentice Hall, New Jersey
JP Trinkaus (1985) *J Neurosci Res* **13**: 1–19
JP Trinkaus and JP Lentz (1964) *Dev Biol* **9**: 115–136
OA Trowell, B Chir and EN Willmer (1940) *J Exp Biol* **16**: 60–70
J Tschopp, S Schafer, D Masson, MC Peitsch and C Heusser (1989) *Nature* **397**: 272–274
K Tsukamoto and Y Sugino (1979) *Cancer Res* **39**: 1305–1309
S Tsukita and S Tsukita (1985) *J Cell Biol* **101**: 2070–2080
RF Tucker, GD Shipley, HL Moses and RW Holley (1984) *Science* **226**: 705–707
RW Tucker, KK Sanford, SL Handelman and GM Jones (1977) *Cancer Res* **37**: 1571–1579
KF Tullberg and MM Burger (1985) *Invas Metast* **5**: 1–15
M Turbitt and ASG Curtis (1974) *Nature* **249**: 453–454
L Turin and AE Warner (1977) *Nature* **270**: 56–57
T Turpeenniemi–Hujanen, UP Thorgeirsson, IR Hart, S Grant and LA Liotta (1985) *J Nat Cancer Inst* **75**: 99–103
GP Tuszynski, TB Gasic, VL Rothman, KA Knudsen and GJ Gasic (1987a) *Cancer Res* **47**: 4130–4133
GP Tuszynski, V Rothman, A Murphy, K Siegler, L Smith, S Smith, J Karczewski and KA Knudsen (1987b) *Science* **236**: 1570–1573
DR Twardzik, CA Sherwin, JE Ranchalis and GJ Todaro (1982) *J Nat Cancer Inst* **69**: 793–798

A Uchida and M Micksche (1983) *Int J Cancer* **31**: 1–5
JN Umbreit and RW Erbe (1979) *Cancer Res* **39**: 2001–2005
CB Underhill and JM Keller (1975) *Biochem Biophys Res Comm* **63**: 448–454
PA Underwood and FA Bennett (1989) *J Cell Sci* **93**: 641–649
JC Unkeless, A Tobia, L Ossowski, JP Quigley, DB Rifkin and E Reich (1973) *J Exp Med* **137**: 85–111
PN Unwin and PD Ennis (1984) *Nature* **307**: 609–613
PNT Unwin and G Zampighi (1980) *Nature* **283**: 545–549
TV Updyke and GL Nicolson (1986) *Clin Exp Metast* **4**: 273–284

J Vaage (1980) *Cancer Res* **40**: 3495–3501
BL Vallee, JF Riordan, RR Lobb, N Higachi, JW Felt, G Crossley, R Buhler, G Budzik, K Breddam, JL Bethune and EM Alderman (1985) *Experientia* **41**: 1–15
JM Van Alstine, P Sorenson, TJ Webber, R Greig, G Poste and DE Brooks (1986) *Exp Cell Res* **164**: 366–378
WP Van Beek, A Tulp, JGM Bolscher, GJ Blacker, KJ Roozendaal and M Egbes (1984) *Blood* **63**: 170–176
AJ van der Eb and 10 others (1980) *Cold Spring Harbor Symp Quant Biol* **44**: 383–399
J Varani (1982) *Cancer Metast Rev* **1**: 17–28
J Varani, SEG Fligiel and B Wilson (1986) *Invas Metast* **6**: 335–346
J Varani, EJ Lovett, S Elgebaly, J Lundy and PA Ward (1980b) *Am J Path* **101**: 345–352
J Varani, W Orr and PA Ward (1978a) *J Cell Sci* **34**: 133–144

J Varani, W Orr and PA Ward (1978b) *Am J Pathol* **90**: 159–171
J Varani, W Orr and PA Ward (1980a) *J Nat Cancer Inst* **64**: 1173–1178
J Varani and PA Ward (1983) *Agents Actions* **12**: 134–151
H Varmus (1988) *Science* **240**: 1427–1435
HE Varmus (1984) *Ann Rev Genet* **18**: 553–612
JM Vasiliev and IM Gelfand (1978) *Nature* **274**: 710–711
DE Vega-Salas, PJI Salas, D Gundersen and E Rodriguez-Boulan (1987) *J Cell Biol* **104**: 905–916
ML Veigl, TC Vanaman and WD Sedwick (1984) *Biochim Biophys Acta* **738**: 21–48
TJ Velu, L Beguinot, WC Vass, MC Willingham, GT Merlino, I Pastan and DR Lowy (1987) *Science* **238**: 1408–1410
H Verschueren (1985) *J Cell Sci* **75**: 279–301
FJW Verwey and JTG Overbeek (1948) *Stability of lyophobic colloids* Elsevier Amsterdam
P Vesely and RA Weiss (1973) *Int J Cancer* **11**: 64–76
E Viadana and KL Au (1975) *J Med* **6**: 1–14
E Viadana, IDJ Bross and JW Pickren (1978) In *Pulmonary metastasis* Eds L Weiss and HA Gilbert Hall Boston pp 143–167
I Virtanen, T Varho, RA Badley and V-P Lehto (1982) *Nature* **298**: 660–663
I Vlodavsky, LK Johnson and D Gospodarowicz (1979) *Proc Nat Acad Sci* **76**: 2306–2310
A Vogel, R Ross and E Raines (1980) *J Cell Biol* **85**: 377–385
B Vogelstein, ER Fearon, SR Hamilton and AP Feinberg (1985) *Science* **227**: 642–645
T Volk and B Geiger (1984) *EMBO J* **3**: 2249–2260
T Volk and B Geiger (1986a) *J Cell Biol* **103**: 1441–1450
T Volk and B Geiger (1986b) *J Cell Biol* **103**: 1451–1464
T Volk, B Geiger and A Raz (1984) *Cancer Res* **44**: 811–824
HP Vollmers and W Birchmeier (1983) *Proc Nat Acad Sci* **80**: 3729–3733
HP Vollmers, BA Imhof, I Wieland, A Hiesel and W Birchmeier (1985) *Cell* **40**: 547–557
DD Von Hoff, J Casper, E Bradley, D Jones and R Makuch (1981) *Am J Med* **70**: 1027–1032

BT Walther, R Ohman and S Roseman (1973) *Proc Nat Acad Sci* **70**: 1569–1573
AM Wang, AA Creasey, MB Ladner, LS Lin, J Strickler, JN Van Arsdell, R Yamamoto and DF Mark (1985) *Science* **228**: 149–154
BS Wang, GA McLoughlin, JP Richie and JA Mannick (1980) *Cancer Res* **40**: 288–292
TY Wang and GL Nicolson (1983) *Clin Exp Metast* **1**: 327–339
O Warburg (1930) *The metabolism of tumours* Constable London
HM Warenius, LS Freedman and NM Bleehen (1980) *Br J Cancer* **41**: 128–132
TFCS Warner (1975) *Med Hypoth* **1**: 51–56
JF Warner and G Dennert (1982) *Nature* **300**: 31–34
BA Warren (1976) *Z. Krebsforsch* **87**: 1–15
BA Warren (1979) In *Tumour blood circulation angiogenesis, vascular morphology and blood flow of experimental and human tumours* Ed HI Peterson CRC Press West Palm Beach Florida pp 1–47
BA Warren (1980) In *Brain metastasis* Eds L Weiss, HA Gilbert and JB Posner GK Hall Boston Mass pp 81–99

L Warren and CA Buck (1980) *Clin Biochem* **13**: 191–197
L Warren, I Zeidman and CA Buck (1975) *Cancer Res* **35**: 2186–2190
S Warren and O Grates (1936) *Am J Cancer* **27**: 485–492
JA Wass, J Varani, GE Piontek, PA Ward and FW Orr (1981) *Cell Diff* **10**: 329–332
JA Wass, J Varani and PA Ward (1980) *Cancer Letts* **9**: 313–318
MD Waterfield (1985) *Prog Cancer Res Therapy* **32**: 71–85
MD Waterfield, GT Scrace, N Whittle, P Stroobaut, A Johnsson, A Wasteson, B Westermark, CH Heldin, JS Huang and T Deuel (1983) *Nature* **304**: 35–39
B Waxler, KE Kuettner and BU Pauli (1982) *Tissue and Cell* **14**: 657–667
G Weber (1977a) *New Eng J Med* **296**: 486–492
G Weber (1977b) *New Eng J Med* **296**: 541–551
MJ Weber, AH Hale and L Losasso (1977) *Cell* **10**: 45–51
B Weimar and U Delvos (1986) *Arteriosclerosis* **6**: 139–145
G Weinbaum and MM Burger (1973) *Nature* **244**: 510–512
RA Weinberg (1985) *Science* **230**: 770–776
R Weindruch and RL Walford (1982) *Science* **215**: 1415–1418
S Weinhouse (1955) *Adv Cancer Res* **3**: 269–297
S Weinhouse (1983) *Adv Enz Reg* **21**: 369–386
RS Weinstein, FB Merk and J Alroy (1976) *Adv Cancer Res* **23**: 23–89
RS Weinstein and BU Pauli (1987) *Ciba Fdn Symp* **125**: 240–254
MM Weiser and JR Wilson (1981) *CRC Crit Rev Clin Lab Sci* **14**: 189–239
L Weiss (1976) In *Fundamental aspects of metastasis* Ed L Weiss North Holland New York pp 311–317
L Weiss (1977) *Int J Cancer* **20**: 87–92
L Weiss (1978) *Int J Cancer* **22**: 196–203
L Weiss (1979a) *Cell Biophys* **1**: 331–343
L Weiss (1979b) *Am J Path* **97**: 601–608
L Weiss (1983) *Invas Metast* **3**: 193–207
L Weiss and DS Dimitrov (1984) *Cell Biophys* **6**: 9–22
L Weiss, B Fisher and ER Fisher (1974) *Cancer* **34**: 680–683
L Weiss and JP Harlos (1972) *J Theoret Biol* **37**: 169–179
L Weiss and JC Holmes (1980) In *Proteinases and tumor invasion* Eds P Strauli, AJ Barrett and A Barchi Raven Press New York pp 181–200
L Weiss, JC Holmes and PM Ward (1983) *Br J Cancer* **47**: 81–89
L Weiss, E Mayhew, D Glaves Rapp and JC Holmes (1982) *Br J Cancer* **45**: 44–53
L Weiss and JR Subjeck (1974a) *J Cell Sci* **14**: 215–223
L Weiss and JR Subjeck (1974b) *Int J Cancer* **13**: 143–150
L Weiss, PM Ward, JP Harlos and JC Holmes (1984) *Int J Cancer* **33**: 825–830
BE Weissman, PJ Saxon, SR Pasquale, GR Jones, AG Geiser and EJ Stanbridge (1987) *Science* **236**: 175–180
DR Welch and SP Tomasovic (1985) *Clin Exp Metast* **3**: 151–188
H-O Werling, E Spiess, P Aulenbacher and N Paweletz (1985) *Invas Metast* **5**: 270–294
H-O Werling, E Spiess and N Paweletz (1986) *Invas Metast* **6**: 257–269
JA Weston and M Abercrombie (1967) *J Exp Zool* **164**: 317–323
JA Weston, KM Yamada and KL Hendricks (1979) *J Cell Physiol* **100**: 445–455
EF Wheelock, KJ Weinhold and J Levich (1981) *Adv Cancer Res* **34**: 107–140
B Whittenberger and L Glaser (1977) *Proc Nat Acad Sci* **74**: 2251–2255
B Whittenberger and L Glaser (1978) *Nature* **272**: 821–823
B Whittenberger, D Raben, MA Lieberman and L Glaser (1978) *Proc Nat Acad Sci*

75: 5457–5461
P Whyte, KJ Buchkovich, JM Horowitz, SH Friend, M Raybuck, RA Weinberg and E Harlow (1988) *Nature* **334**: 124–129
I Wieland, G Muller, S Braun and W Birchmeier (1986) *Cell* **47**: 675–685
EC Wiener and WR Loewenstein (1983) *Nature* **305**: 433–435
BC Wightman, EA Weltman and LA Culp (1986) *Biochem J* **235**: 469–479
M Wigler, R Sweet, GK Sim, B Weld, A Pellicier, E Lacy, T Maniatis, S Silverstein and R Axel (1979) *Cell* **16**: 777–785
CJ Wikstrand, SH Bigner and DD Bigner (1983) *Cancer Res* **43**: 3327–3334
GD Wilbanks and RM Richart (1966) *Cancer Res* **26**: 1641–1647
JA Wilkins and S Lin (1986) *J Cell Biol* **102**: 1085–1092
PC Wilkinson (1982) *Chemotaxis and inflammation* Churchill Livingstone Edinburgh
PC Wilkinson and RB Allan (1978) *Exp Cell Res* **117**: 403–412
PC Wilkinson, JM Lackie, JV Forrester and GA Dunn (1984) *J Cell Biol* **99**: 1761–1768
MC Willingham and I Pastan (1974) *J Cell Biol* **63**: 288–294
MC Willingham and I Pastan (1975) *Proc Nat Acad Sci* **72**: 1263–1267
MC Willingham, KM Yamada, SS Yamada, J Pouyssegur and I Pastan (1977) *Cell* **10**: 375–380
RA Willis (1973) *The spread of tumours in the human body* Butterworth, London
RH Wiltrout, RB Herberman, S-R Zhang, MA Chirigos, JR Ortaldo, KM Green and JE Talmadge (1985) *J Immunol* **134**: 4267–4275
JL Winkelhake and GL Nicolson (1976) *J Nat Cancer Inst* **56**: 285–291
DJ Winterbourne and PT Mora (1981) *J Biol Chem* **256**: 4310–4320
LL Wiseman (1977a) *Develop Biol* **58**: 204–211
LL Wiseman (1977b) *Experientia* **33**: 734–735
LL Wiseman and MS Steinberg (1973) *Exp Cell Res* **79**: 468–471
LL Wiseman and J Strickler (1981) *J Cell Sci* **49**: 217–223
WH Woglom (1929) *Cancer Rev* **4**: 129–214
E Wolff (1967) In *Le mecanisme de l'invasion du cancer en culture organotypique* Ed P Denoix pp 204–211
E Wolff and N Schneider (1956) *CRS de la soc de biol* **150**: 845–846
JJ Wolosewick and KR Porter (1979) *J Cell Biol* **82**: 114–139
GW Wood and GY Gillespie (1975) *Int J Cancer* **16**: 1022–1029
S Wood (1958) Arch Pathol **66**: 550–568
S Wood, RR Baker and B Marzocchi (1967a) In *Locomotion of cancer cells* in vivo *compared with normal cells* Ed P Denoix pp 26–30
S Wood, RR Baker and B Marzocchi (1967b) In *Endogenous factors influencing host–tumour balance* Eds RW Wissler, TL Dao and S Wood University of Chicago Press Chicago pp 223–237
S Wood, ED Holyoke and JH Yardley (1961) In *Canadian cancer conference* 4 Eds RW Begg, A Ham, CP Leblond, RC Noble and RJ Rossiter Academic Press New York pp 167–223
S Wood, R Lewis and JH Mulholland (1966) *Bull Hopkins Hosp* **119**: 1–5
MFA Woodruff (1980) *The interaction of cancer and host* Grune and Stratton New York
RI Woodruff and WH Telfer (1980) *Nature* **286**: 84–86
JJ Woodruff and LM Clarke (1987) *Ann Rev Immunol* **5**: 201–222
JJ Woodruff, IM Katz, LG Lucas and HB Stamper (1977) *J Immunol* **119**: 1603–1610

A Woods, JR Couchman, S Johansson and M Hook (1986) *EMBO J* **5**: 665–670
DE Wooley and CA Grafton (1980) *Brit J Cancer* **42**: 260–274
LC Wright, GL May, P Gregory, M Dyne, KT Holmes, PG Williams and CE Mountford (1988) *J Cell Biochem* **37**: 49–59
SD Wright, PA Reddy, MTC Jong and BW Erickson (1987) *Proc Nat Acad Sci* **84**: 1965–1968
TC Wright, TE Ukena, R Campbell and MJ Karnovsky (1977) *Proc Nat Acad Sci* **74**: 258–262

J Xiang and AK Kimura (1986) *Clin Exp Metast* **4**: 293–309

I Yahara and GM Edelman (1972) *Proc Nat Acad Sci* **69**: 608–612
KM Yamada (1983) *Ann Rev Biochem* **52**: 761–799
KM Yamada and DW Kennedy (1985) *J Cell Biochem* **28**: 99–104
KM Yamada, SH Ohanian and I Pastan (1976) *Cell* **9**: 241–245
KM Yamada and K Olden (1978) *Nature* **275**: 179–184
KM Yamada, K Olden and I Pastan (1978) *Ann NY Acad Sci* **312**: 256–277
KM Yamada, SS Yamada and I Pastan (1976) *Proc Nat Acad Sci* **73**: 1217–1221
H Yamasaki and F Katoh (1988) *Cancer Res* **48**: 3203–3207
K Yamashina and GH Heppner (1985) *Cancer Res* **45**: 4015–4019
MM Yarnell and EJ Ambrose (1969) *Eur J Cancer* **5**: 255–263
TM Yau, T Buckman, AH Hale and MJ Weber (1976) *Biochem* **15**: 3212–3219
G Yogeeswaran (1980) *Biochem Biophys Res Comm* **95**: 1452–1460
G Yogeeswaran (1983) *Adv Cancer Res* **38**: 289–350
G Yogeeswaran and P Salk (1981) *Science* **212**: 1514–1516
G Yogeeswaran, BS Stein and H Sebastian (1978) *Cancer Res* **38**: 1336–1343
J Yokota, Y Tsunetsugu-Yokota, H Battifora, C LeFevre and MJ Cline (1986) *Science* **231**: 261–265
K Yoshida, T Ozaki, K Ushijima and H Hayashi (1970) *Int J Cancer* **6**: 123–132
LP Yotti, CE Chang and JE Trasko (1979) *Science* **206**: 1089–1091
JS Young (1959) *J Path Bact* **77**: 321–339
MR Young, GP Duffie and M Newby (1987) *Invas Metast* **7**: 96–108
JD-E Young, ZA Cohn and NB Gilula (1987) *Cell* **48**: 733–743
SD Young and RP Hill (1986) *Clin Exp Metast* **4**: 153–176
N Yuhki, J Hamada, N Kuzumaki, N Takeuchi and H Kobayaski (1986) *Jpn J Cancer Res (Gann)* **77**: 9–12
JJ Yunis (1983) *Science* **221**: 227–236

LR Zacharski and 12 others (1981) *J Amer Med Assoc* **245**: 831–835
PO Zamora, KG Danielson and HL Hosick (1980) *Cancer Res* **40**: 4631–4639
L Zardi, C Cecconi, O Barberi, B Carnemolla, M Picca and L Santi (1979) *Cancer Res* **39**: 3774–3779
I Zeidman (1957) *Cancer Res* **17**: 157–162
I Zeidman (1961) *Cancer Res* **21**: 38–39
I Zeidman (1965) *Acta Cytologia* **9**: 136–140
I Zeidman and JM Buss (1952) *Cancer Res* **12**: 731–733
I Zeidman and JM Buss (1954) *Cancer Res* **14**: 403–405
SH Zigmond (1974) *Nature* **249**: 450–452
SH Zigmond, HI Levitsky and BJ Kreel (1981) *J Cell Biol* **89**: 585–592
DB Zimmer, CR Green, WH Evans and NB Gilula (1987) *J Biol Chem* **262**: 7751–7763

S Zucker, G Beck, JF Distefano and RM Lysik (1985) *Br J Cancer* **52**: 223–232
E Zwilling (1959) *Transpl Bull* **6**: 115–116

Index